Infanticide by males and its implications

CAREL P. VAN SCHAIK
Duke University, North Carolina

CHARLES H. JANSON
State University of New York, Stony Brook

CAMBRIDGE
UNIVERSITY PRESS

PUBLISHED BY THE PRESS SYNDICATE OF THE UNIVERSITY OF CAMBRIDGE
The Pitt Building, Trumpington Street, Cambridge, United Kingdom

CAMBRIDGE UNIVERSITY PRESS
The Edinburgh Building, Cambridge CB2 2RU, UK
40 West 20th Street, New York, NY 10011–4211, USA
10 Stamford Road, Oakleigh, VIC 3166, Australia
Ruiz de Alarcón 13, 28014 Madrid, Spain
Dock House, The Waterfront, Cape Town 8001, South Africa

http://www.cambridge.org

© Cambridge University Press 2000

First published 2000

Printed in the United Kingdom at the University Press, Cambridge

Typeface TEFFLexicon 9/13 pt   *System* QuarkXPress® [SE]

A catalogue record for this book is available from the British Library

Library of Congress Cataloguing in Publication data
Infanticide by males and its implications/edited by Carel P. van Schaik and Charles H. Janson
    p. cm.
  Includes bibliographical references.
  ISBN 0 521 77295 8 – ISBN 0 521 77498 5 (pbk.)
    1. Primates–Behavior.   2. Infanticide in animals.   I. Schaik, Carel van.   II. Janson,
Charles Helmar.
  QL762.5 I535 2000
  599.8'156–dc21   00-020380

ISBN 0 521 77295 8 hardback
ISBN 0 521 77498 5 paperback

Contents

Contributors

DANIEL T. BLUMSTEIN
Departments of Biology and Psychology, Macquarie University, Sydney,
NSW 2109, Australia

CAROLA BORRIES
Abt. Verhaltensforshung & Ökologie, Deutsches Primatenzentrum,
Kellnerweg 4, D-37077 Göttingen, Germany
Present address:
Department of Anthropology, State University of New York, Stony Brook,
Stony Brook, NY 11794-4364, USA

DOROTHY L. CHENEY
Department of Biology, University of Pennsylvania, Philadelphia, PA
19104, USA

CAROLINE M. CROCKETT
Regional Primate Research Center, Box 357330, University of
Washington, Seattle, WA 98195-7330, USA

LESLIE DIGBY
Department of Biological Anthropology and Anatomy, Duke University,
Box 90383, Durham, NC 27708-0383

J. KEITH HODGES
Department of Reproductive Biology, Deutsches Primatenzentum,
Kellnerweg 4, D-37077 Göttingen, Germany

SARAH B. HRDY
Department of Anthropology, University of California, Davis, Davis, CA
95616, USA

JULIA FISCHER

Department of Psychology, University of Pennsylvania, Philadelphia, PA 19104 - 6196, USA

CHARLES H. JANSON

Department of Ecology and Evolution, State University of New York, Stony Brook, Stony Brook, NY 11794-5245, USA

SARA JOHNSON

Department of Anthropology, University of New Mexico, Albuquerque, NM 87131, USA

ANDREAS KOENIG

Abt. Verhaltensforshung & Ökologie, Deutsches Primatenzentrum, Kellnerweg 4, D-37077 Göttingen, Germany

AMANDA H. KORSTJENS

Department of Ethology and Socioecology, Utrecht University, PO Box 80086, 3508 TB Utrecht, The Netherlands

CHARLES L. NUNN

Department of Biology, University of Virginia, Gilmer Hall, Charlottesville, VA 22903-2477, USA

RYNE A. PALOMBIT

Department of Anthropology, Center for Human Evolutionary Studies, Rutgers University, 131 George Street, Rutgers University, New Brunswick, NJ 08901-1414, USA

ANDREAS PAUL

Institut für Zoologie und Anthropologie, Universität Göttingen, Bürgerstrasse 50, D-37073 Göttingen, Germany

SIGNE PREUSCHOFT

Living Links, Emory University, Yerkes Regional Primate Research Center, Field Station, 2409 Taylor Lane, Lawrenceville, GA 30043, USA

DREW RENDALL

Department of Psychology and Neuroscience, University of Lethbridge, Lethbridge, Alberta, T1K 3M4, Canada

ROBERT M. SEYFARTH

Department of Psychology, University of Pennsylvania, Philadelphia, PA 19104-6196, USA

JOAN B. SILK
Department of Anthropology, University of California, Los Angeles, Los Angeles, CA 90095-1553, USA

VOLKER SOMMER
Department of Anthropology, University College London, Gower Street, London WC1E 6BT, UK

ROMY STEENBEEK
Department of Ethology and Socioecology, Utrecht University, PO Box 80086, 3508 TB Utrecht, The Netherlands

PETER STEPHAN
Bahnstrasse 22, D-06484 Ditfurt, Germany

ELISABETH H. M. STERCK
Department of Ethology and Socioecology, Utrecht University, PO Box 80086, 3508 TB Utrecht, The Netherlands

ADRIAN TREVES
Department of Zoology, University of Wisconsin, Madison, 250 North Mills Street, Madison, WI 53706, USA

MARIA A. VAN NOORDWIJK
3323 Ridge Road, Durham, NC 27705-5535, USA

CAREL P. VAN SCHAIK
Department of Biological Anthropology and Anatomy, Duke University, Box 90383, Durham, NC 27708-0383, USA

JOSÉ P. VEIGA
Museo Nacional de Ciencias Naturales, C.S.I.C. José Gutiérrez Abascal 2, 28006 Madrid, Spain

ECKART VOLAND
Zentrum für Philosophie und Grundlagen der Wissenschaft, Universität Giessen, Otto-Behaghel Strasse 10C, D-35394, Giessen, Germany

Foreword

Science does not deal in certainty, so "fact" can only mean a proposition affirmed to such high degree of certainty that it would be perverse to withhold one's provisional assent.

(S. J. GOULD, 1999)

"Quite possibly, readers ten years from now may take for granted the occurrence of infanticide in various animal species," Glenn Hausfater and I rashly conjectured back in 1984, in a preface to the first book on this subject, "and [they] may even be unaware of the controversies and occasionally heated debate that have marked the last decade of research on this topic. . .". For biologists, that projection turned out to be more or less accurate. For those with backgrounds in the social sciences, perhaps especially in my own field of anthropology, it was wildly optimistic.

Most animal behaviorists now take for granted that the killing of infants by conspecifics can be found throughout the natural world and that, for many primate species, the arrival in their group of unrelated males represents a threat to infant survival. Many anthropologists, however, remain skeptical of the proposition that a propensity to attack infants born to unfamiliar females evolved in non-human primate males because it increased their chances to breed. This would require accepting that a behavior obviously detrimental to the survival of the group or even the species could evolve in males through Darwinian sexual selection because it provided the killers with a reproductive edge in their competition with rival males. To them it reeks of "selfish genes" and would mean that this extraordinarily destructive behavior is adaptive rather than pathological or accidental behavior. Those wishing to familiarize themselves with the various positions in this debate will find a large, discursive, discouraging literature for the years 1974–1999 reviewed in Volker Sommer's chapter for this book (Chapter 1).

To introduce *this* book, then, I begin – with the wisdom of hindsight – as follows: for those familiar with the previous two volumes about the evolution of infanticide (Hausfater & Hrdy 1984; Parmigiani & vom Saal 1994) and who remain unconvinced by evidence compiled there, this book – even Carel van Schaik's summary (Chapter 2) – will do little to change minds. Indeed, it does not set out to do so. However, for those who are already convinced that infanticidal propensities did evolve in many animals, and that strange or newly dominant males represent a special liability for unweaned infants, it should logically follow that infanticide must have acted as an important selection pressure shaping the behavior and reproductive physiologies of mothers as well as the protective responses by fathers and other relatives. Over millions of years, infanticide would have shaped the behaviors of mothers and infants as well as of the infanticidal males coevolving in this dynamic system. Any recurring pattern of behavior that results in the death of one third or more of all infants born (as is the case for the populations of chacma baboons and langur monkeys studied long term at Moremi in Botswana and Jodhpur, India) is not likely to have been insignificant in evolutionary terms. Going further, even when observed mortality from infanticide is currently near zero, this does not necessarily mean that infanticide was not important, shaping present responses and selecting for the behaviors that keep it rare – an idea likely to elicit incredulous guffaws from skeptics who will regard this suggestion as akin to the White Knight's Rejoinder. Recall the White Knight in *Through the Looking Glass* who tells Alice that the anklets his horse wears are to keep sharks from biting. When Alice observes *that there are no sharks,* the White Knight replies that his prophylaxis must work.

Readers undeterred by venturing through the looking glass into worlds where observations are often sketchy will probably find themselves stimulated to ask novel questions. Why is the occurrence of infanticide so variable across species? Why have not infanticidal "precautions" gone to fixation in all species? This book provides a series of wide-ranging, well-reasoned, often state-of-the art chapters full of new observations and compilations, novel interpretations, exciting and often unabashedly speculative ideas. Some of these, I wager, will alter the way that even veteran researchers think about the evolution of the social behavior and reproductive physiologies of organisms they thought they knew well. For Carel van Schaik and Charlie Janson made a bold (no doubt some will say "rash") editorial decision. They focus on the larger implications of evolved infanticidal propensities, rather than continuing to be mired in what has become an increasingly sterile controversy over whether or not infanticidal behavior evolved. Van Schaik and Janson recognize that at some point, after a critical evaluation of available evidence, scientists make a decision to elevate an idea from speculative hypothesis to an accepted proposition in order to move on. The hypothesis just accepted can then be used to generate new hypotheses with a new series of predictions to test – a process that does not preclude re-examination or revision of established "theory" down the line. The result makes for a compelling read.

I know it sounds like an oxymoron to describe an edited volume as exciting. But consider the circumstances. The previous two volumes set out to document and describe a protean and disturbing phenomenon, to evaluate competing explanations for it at the ultimate level, to address arguments raised by those convinced that infanticide is most often a non-adaptive artifact of events that "just happen", and to explore proximate causes and cues. Because the strongest evidence for the sexual selection hypothesis has always come from rodents, not primates, primatologists were forced to start off from defensive postures, curtailing their speculations about the implications, because by disciplinary convention they were housed in (and depended for promotion on) anthropology departments in which many of their colleagues were not convinced that what they study even exists. (For biologists who assume I must be exaggerating, see Shea 1999.)

By taking for granted that infanticide is an evolved behavior, however, the editors of this volume liberate and even encourage contributors to explore heretofore neglected questions and to focus on the evolutionary consequences of infanticide for physiology and group structure. In what taxonomic groups are we most likely to find infanticide? And why? What particular demographic features (such as male migration rates or weaning age) affect the likelihood that infanticidal behavior will evolve or be maintained in a population? The volume's token rodent expert asks a totally new question: which came first, infanticide by strange males or "Bruce effects" (the capacity of female rodents to reabsorb fetuses when they sense the presence of a strange male)? Others inquire how infanticide has affected male–female friendships in species such as baboons. And what might it have to tell us about human capacities to form pair bonds?

The title notwithstanding, several chapters deal with the killing of infants by females, and in doing so highlight thorny issues raised when Darwin's theory of sexual selection – originally constructed with competition between males in mind – is applied to breeding competition among females. This is an excellent case in point illustrating how, long after a theory has proved itself time and again and been widely accepted, researchers may still be prompted to adjust or expand certain key elements to a theory, even one so venerable as sexual selection. This is an area where much research remains to be done. Just how female–female competition can be incorporated into the theory of sexual selection is still problematic, and new theoretical models are needed.

It is ironic that criticism of the sexual selection hypothesis for infanticide (Dagg 1999a: 947) focuses on the fact that the hypothesis was devised before all the information critical for testing it was in hand (e.g., before we knew for sure that unweaned infants were most likely to be attacked, or before DNA evidence for langurs confirmed that males were not killing their own infants). It is ironic because this is precisely what good scientific theory should do: make use of what we do know to make accurate predictions about what we do not yet know.

What makes the science in this volume exciting then is that field observations and other sources of information are combined with carefully reasoned inference to construct model worlds – sometimes partly imagined worlds – but models that generate predictions that are already being, or we hope soon can be, tested in the "real" world of fieldsites, laboratories and library. In this way we not only summarize the facts at hand, but rephrase what we already know. Books such as this position us to learn something new.

Sarah B. Hrdy
University of California – Davis

Infanticide by males: prospectus

Infanticide refers to the killing of dependent offspring, or more formally, to "any behavior that makes a direct and significant contribution to the immediate death of an embryo or newly hatched or born member of the perpetrator's own species" (Hrdy & Hausfater 1984). Although this seemingly bizarre behavior is often rare in the species in which it occurs, it is nonetheless remarkably widespread in the animal kingdom. Gathered under the unitary label "infanticide" is a diverse set of behaviors that differ in social context, the sex of the perpetrators, and the relationship between the perpetrators and the infant. This variability suggests similar variability in the nature of the selective factors that favored the evolution of infanticide. Hrdy (1979), therefore, distinguished five functionally distinct "classes" of infanticide. Infanticide could be an adaptation that (1) increases resource acquisition (cannibalism), (2) excludes competitors for the agent or its kin, (3) increases the inclusive fitness of one or both parents, or (4) increases the reproductive success of adult males unrelated to the infant. Alternatively (5), it could be a reflection of social pathology, thus not serving any adaptive function.

Among primates (and in this book "primates" refers to non-human primates unless otherwise specified), infanticide is relatively more homogeneous. Most infanticide is committed by males, and in most of these cases, adult males kill unweaned infants not related to them, usually without consuming them. After the first and often fragmentary observations of infanticide by adult males in primates in the 1960s (Sugiyama 1965, 1966), a protracted and heated debate ensued about the proper interpretation or interpretations of infanticide in primates, which focused on its possible selective advantage (Angst & Thommen 1977; Boggess 1979; Hrdy 1979). For most biologists, this debate was satisfactorily resolved in the mid 1980s, with the appearance of the first edited volume devoted largely to the problem (Hausfater & Hrdy 1984), in favor of the male reproductive strategy hypothesis. Better known as the sexual selection hypothesis, this idea proposes that a male increases his reproductive success by killing unrelated dependent infants if the infant's death makes the female return to receptivity sooner than would otherwise have been the case and if he has a higher probability of siring her next offspring (Hrdy 1979).

This idea is thoroughly embedded in contemporary evolutionary biology. Infanticide by males is only one expression of a more general conflict between the sexes that has its origins in fundamental reproductive asymmetries between the two sexes. Where females have internal fertilization and gestation, and in mammals also lactation, females are associated with the young much more intensely and much longer than males. As first noted by Darwin (1871), this sex difference is the basis for sexual selection, because the differential parental investment and thus differential latencies following fertile matings (Clutton-Brock & Parker 1992) produce operational sex ratios that are heavily skewed toward the sex with the shorter association with the offspring. For each female that is ready to engage in fertile mating, there will be many males. As a result, male reproductive success tends to be limited by access to mates, and selection will favor traits that favor success in competing with other males in acquiring mating access to females, resulting in the evolution of enlarged body size, weaponry, etc.

Female reproductive success is usually limited by access to resources that would enhance the number of offspring they can produce. With respect to mating, females usually do not suffer from a lack of access to mates. However, they can enhance their reproductive success by mating preferentially with mates of superior intrinsic viability, provided there is heritable variation in this trait and it can be assessed by females. Consistent female choice has therefore selected for male traits that indicate the intrinsic viability of the males (e.g., Petrie 1994).

Most of the time, female preferences and male mating interests diverge. The profound impact of this conflict of interest was not realized until relatively recently (Parker 1979; Smuts & Smuts 1993; Gowaty 1997a). Natural selection will favor traits that make males more likely to mate with females who do not prefer them, whereas selection will favor traits in females that will allow them to express their preferences. This conflict has set up an "arms race", where male traits are favored that allow them to force females into mating with them (e.g., grasping appendages or penetrating sperm delivery organs) and female traits are favored that allow them to overcome such male tactics. Where males can gain the upper hand, sexual coercion may ensue. Sexual coercion, as defined by Smuts & Smuts (1993) is "use by a male of force, or threat of force, that functions to increase the chances that a female will mate with him at a time when she is likely to be fertile, and to decrease the chances that she will mate with other males, at some cost to the female". The most obvious expressions of sexual coercion in mammals such as primates are harassment and forced matings of sexually attractive females, and infanticide. Thus infanticide by males can now be regarded as one component of nearly ubiquitous intersexual conflict.

Nonetheless, many still regard infanticide by males in primates as a non-adaptive phenomenon, if it exists at all. In the Foreword to this volume, Sarah Hrdy mentioned the long and acrimonious debate over infanticide, in particular among anthropologists. In Chapter 1, Sommer examines the various positions, from those who disagree with the interpretation of the data to those who disagree with the data or even with the fact that infanticide is a valid object for study in the first place.

The first, albeit subsidiary, aim of this book is to review the recent primate literature to reassess the evidence for sexually selected infanticide by males, to summarize and respond to the criticisms raised against it, and to identify remaining weaknesses. In Chapter 2, van Schaik expands the careful review of Struhsaker & Leland (1987), to evaluate the sexual selection hypothesis in detail, and to contrast it with other possibilities. The emphasis on primates in this book is not due to the taxonomic myopia of which primatologists so often stand accused. In Chapter 3, van Schaik proposes that this primate bias has a biological basis, by linking vulnerability to infanticide by males to features of life history and reproductive biology.

Subsequent chapters present detailed studies of red howlers (Crockett & Janson, Chapter 4), hanuman langurs (Borries & Koenig, Chapter 5), chacma baboons (Palombit *et al.*, Chapter 6) and Thomas's langurs (Steenbeek, Chapter 7). Each of them supports the sexual selection hypothesis, but also examines in some depth some puzzling aspects of the phenomenon of the sexual selection hypothesis, such as infanticide in highly seasonal breeders and the roles of diet, of group size and of group composition, or the effect of infanticide on the social organization.

Although the phenomenon is both best documented and most studied in primates, sexually selected infanticide is not limited to this taxon. Excellent work on lions has demonstrated that infanticide is an adaptive male reproductive strategy in that species, and has documented some of the effects on social organization (Pusey & Packer 1994). Outstanding work on the regulation of infanticidal behavior has been done in rodents (Perrigo & vom Saal 1994). In this volume, Blumstein (Chapter 8) takes an evolutionary approach, and reconstructs the phylogenetic history of infanticide by males in rodents in relation to the evolution of infanticide by females and of Bruce effects. Veiga (Chapter 9) reviews the evidence for sexually selected infanticide by male birds, and finds that it may be more common than is generally accepted. We hope that the obvious gaps in taxonomic coverage spark new attempts to document sexually selected infanticide in other taxa.

The continuing controversy has made many students of primate behavior conservative in their interpretation. While it is a scientifically proper attitude to avoid rushing prematurely to judgment over competing hypotheses, it has also prevented us from fully appreciating the pervasive importance of the threat of infanticide for the social lives of the animals vulnerable to it. Because critics insisted on the importance of directly observed cases, few realistic estimates of the rate of infanticide have been published. Now that realistic estimates show that infanticide can be a major source of infant mortality (see Chapters 2, 4–7), there is no reason to postpone any longer a thorough exploration of the consequences of infanticide: the various counterstrategies and infanticide's impact on other aspects of physiology and behavior, especially on primate social systems. Ironically, 20 years ago Hrdy (1979) had defined this as a pressing research agenda, but relatively little progress has been made since then in studies of infanticide by males. Similarly, Sherman (1981) proposed the importance of infanticide by females for territoriality in rodents, but serious studies of this suggestion have appeared only in the past few years (Wolff 1993; Hoogland 1995; Wolff & Peterson 1998). The second, and major, aim of this book, then, is to take up this challenge.

While the detailed case studies of Chapters 4 through 7 already hint at the possible consequences of infanticide by primate males on social organization, most of the following chapters have this possible impact as their major theme. We should stress that none of these explorations requires that infanticide by males be adaptive, only that its context be predictable enough to provide cues to favor counterstrategies. Both critics and proponents of the adaptive nature of infanticide by males agree that infanticide is far more frequent shortly after replacement of a top breeding male (Bartlett *et al.* 1993; van Schaik, Chapter 2). However, a more serious obstacle for such explorations is that, paradoxically, where counterstrategies are quite effective, the *rate* of infanticide will be so low that infanticide by males may never be observed in the average field study. In various chapters, authors escape from this by defining the *risk* of infanticide independently from direct observations of actual infanticide.

Infanticide by males may have affected the evolution of infant features and aspects of infant care (Treves, Chapter 10), of male–female relationships (Palombit, Chapter 11), of male–infant and male–male relationships (Paul *et al.*, Chapter 12), of female dispersal (Sterck & Korstjens, Chapter 13), of sexual behavior (van Noordwijk & van Schaik, Chapter 14) and features of female reproductive physiology (van Schaik *et al.*, Chapter 15), and finally of aspects of social organization as a whole (Nunn & van Schaik, Chapter 16). All these chapters take an explicitly comparative approach, although, with the exception of Chapters 11 and 14, their taxonomic scope is limited to primates.

The book's focus on infanticide by males allows us to maintain a reasonable functional homogeneity. However, infanticide by females is quite common in some mammalian taxa, and functionally more diverse. Digby (Chapter 17) examines the multiple contexts in which it is observed in mammals, and establishes a framework for functional interpretation. Voland & Stephan (Chapter 18) connect infanticide by females to this volume's main theme, infanticide as a sexually selected reproductive strategy, by arguing that in humans infanticide by mothers can reasonably be interpreted as a mating strategy. This new approach complements the more traditional approach of considering infanticide by mothers as an adaptive response to resource scarcity and the associated reduced probability of survival of the current offspring (Daly & Wilson 1984).

Most of these chapters are exploratory, in the sense that far more questions are raised than are answered, and we encouraged authors to speculate. The possible impacts of infanticide on other aspects of social behavior and individual physiology and life history need to be tested both on a broader comparative scale and against competing hypotheses within smaller clades. In the final chapter, we try to take stock and develop priorities for further research. The success of this book will be measured by the speed with which new studies will replace or refine the insights we have gained so far, and by the extent to which it inspires students of non-primate mammals and of birds to explore infanticide and its implications more than in the past. We hope that in the future, researchers exploring the evolutionary consequences of infanticide can publish their work without the controversy that has dogged the issue of the adaptive value of infanticide to males.

Acknowledgments

The first thought of a volume devoted to the consequences of infanticide arose during discussions at the German Primate Center (DPZ), where C.v.S. was a Humboldt Preisträger during 1995–6. C.v.S. especially thanks Carola Borries, Jörg Ganzhorn, Keith Hodges, Peter Kappeler, Andreas Paul, Signe Preuschoft, Maria van Noordwijk and Dietmar Zinner for profitable discussions. The actual genesis of this volume occurred during the very congenial visit to watch Japanese macaques in the hot springs of Nagano, arranged and hosted by Drs J. Yamagiwa and M. Huffman, following the conference "Frontiers in Primate Socioecology" in January 1998. We thank them for this opportunity to converse at length and engender the enthusiasm that carried us through the making of this book. We thank Maria van Noordwijk for considerable logistical support.

Part I

Introduction

1

The holy wars about infanticide. Which side are you on? And why?

Introduction

The topic of infanticide has been a staple theorem of sociobiology ever since this discipline – the study of social behavior from an evolutionary perspective – was born two and a half decades ago (Wilson 1975). The killing of conspecific young is still hotly debated. Does it occur at all, does it reflect an adaptation, a pathology or even a political agenda? Infanticide – observed among such varied taxa as birds, rodents, carnivores, pinnipeds and primates (Hausfater & Hrdy 1984; Parmigiani & vom Saal 1994) – therefore remains a litmus test upon which the validity of a sociobiological interpretation of behavior depends. I attempt to trace some intellectual roots of the controversy: those of defenders of adaptationist explanations, those of critics from within the paradigm of evolutionary biology, and those of critics who operate from other paradigms such as the social sciences. My ultimate aim is to defend the adaptationist interpretation as a valid and fruitful approach, while acknowledging that its narrative is anchored in a time-dependent framework of interpretation.

Cute and brute

People are fascinated by animals, not least because people are, in their own right, animals who can empathize with similar organisms. The average viewer of a natural history documentary will feel good if a monkey mother cuddles her newborn: "It's so cute." But different emotions flare up if, over television dinner, wild chimpanzees eat an infant of their own kind: "It's so brute."

These complementary sets of emotions are readily served by our brains and will often grow into thinly veiled judgments. The cute stuff animals do is "natural" because we like it, whereas the brute stuff is "animalistic"because we do not like it. Nevertheless, egg cannibalism in wasps will upset us less than seeing a little chimpanzee being torn apart. The repugnance is stronger if we are phylogenetically close to the victim.

Natural scientists are, of course, supposed to shrug their shoulders no matter what behavior is at stake and take refuge to the advice offered by David Hume in his 1740 *A Treatise of Human Nature* not to stroll from "Is to Ought", or else be in danger of committing the *naturalistic fallacy*. Still, scientists are governed by the same mechanisms of empathy that lay nature lovers possess. I will not forget 9 July 1981, the first time I witnessed a male monkey sinking his canines into an infant I had grown fond of during a study of hanuman langurs in India. Later in the fieldwork I shouted and threw stones at the aggressor. It did not prevent infant-killing. The first attack took me by surprise. My academic mentor, Christian Vogel of Germany's Göttingen University, had instilled in me disapproval (Vogel 1979) for the "out-of-America" hypothesis that infanticide occurs regularly amongst langurs and is caused by male competition over females (Hrdy 1974). Vogel's views still reverberated with the idea that animal behavior serves the good of the species. Accordingly, monkeys were expected to perform "group serving" and "group bonding" acts (Vogel 1976). As an evolutionary biologist, Vogel represented a *within-paradigm critic*. Data subsequently gathered by his students and Indian colleagues changed Vogel's *Weltanschauung* radically: he transformed into a vigorous defender of the theorem that infanticide amongst animals including humans reflects evolutionary adaptation (Vogel 1989), such as exploitation of the infant for cannibalistic purposes, or parental manipulation of progeny (cf. Hausfater & Hrdy 1984; Parmigiani & vom Saal 1994). With respect to langurs, the theory (Hrdy 1974) maintains that infanticidal males increase their relative genetic representation in future generations by eliminating unweaned offspring of other males, particularly those of their predecessors as harem residents in populations with one-male/multifemale group structures. Infanticide will shorten the waiting time of a new male until he can impregnate a female, because the loss of an infant terminates the period of temporary infertility associated with lactation. In addition, infanticide may be adaptive if it reduces resource competition for a male's kin.

I for my part learned to rationalize the gruesome events (Sommer 1987, 1994, 1996; Böer & Sommer 1992). I now publicly lecture and write about infant-killing in more or less the same way as about grooming, presenting both as functional behaviors. However, occasionally somebody from the audience or readership will call me a fascist (cf. Schües & Ostbomk-Fischer 1993: 17). I tend to reply that few people hold meteorologists responsible for the destruction and grief caused by tornadoes; by the same token, I should not be held morally accountable for the aggressive behavior of the monkeys that I study.

This excuse is an easy escape when dealing with benign minds who accept that they are committing the naturalistic fallacy. However, the route from "Is to Ought" is a two-way street and various apostles actually travel in the opposite direction: from "Ought to Is". They preach that our values construct the reality around us, and that it is imperative to possess the right values. Cute mother–infant interactions are OK, acceptable testimony to how the world should be, but brute male–infant interactions are not OK because reports about aggression are borne out of aggressive minds and breed more violence. This can be labeled as the *moralistic fallacy*: what should not be, cannot be.

Donna Haraway, American scholar of History of Consciousness, figures prominently as an *outside-paradigm critic* sympathetic to such conviction: "To center the debate on the biological meanings of infanticide among primates too easily plays into the culturally overdetermined lust for sexualized violence" (Haraway 1989: 311). There is some truth to this if we look at how the popular media disseminate findings about infanticide: as a story about sex and crime in which the theory is often not only trivialized but distorted. Headlines of, for example, German magazines were only at times acceptable ("Der neue Chef des Harems tötet seine Stiefkinder" [The harem's new boss kills his stepchildren]), but more often barely bearable ("Affen morden ihre Kinder" [Monkeys murder their young], "Mord im Harem" [Murder in the harem]) and at times blatantly sensational ("Anklage Mord" [Accused of murder], "Das Killer-Gen" [The killer-gene]; "Blutrünstige Rivalen" [Bloodthirsty rivals]) (references in Sommer 1996). But then, purging language and employing euphemisms will not in itself foster desired political change. It may, on the contrary, just cover up fields of conflict. Moreover, any paradigm can be used to incite a war – prime examples being such diverse ideologies as the Christian doctrine to love one's neighbor, Buddhist belief in the vanity of life, or Marxist utopias of equality. I cannot see what harm talk about infanticide has done, but the fear that it *could* certainly generates much of the heat of the debate.

I will demonstrate how transgressions from "Ought to Is" and "Is to Ought" are committed by all parties involved: defenders and critics from within and outside the paradigm.

Nature revisited

The founding fathers of classical ethology – Nikolaas Tinbergen, Karl Ritter von Frisch, and Konrad Lorenz – did not offend the public with gory stories about a brute natural world. Non-human animals, if anything, were the better humans: they acted for the good of the group and did not kill each other with the ease that humans exhibit. Konrad Lorenz made such cultural pessimism explicit in *Das sogenannte Böse* (*On Aggression*, but literally translated as "The so-called Evil", 1963; see also Lorenz 1955). The Austrian ethologist embodied a modern version of Jean Jacques Rousseau, and the opening sentence from *Émile ou de l'education* (1762) could have been written by Lorenz himself: "Everything is good as it comes from the hands of the maker of things; everything perverts under the hands of man".

Such romanticism fueled 19th century portrayals of indigenous peoples as "noble savages", living in peace with themselves and their environment. Visions of politically and ecologically correct *Naturvölker* (nature peoples) are still nurtured by well-meaning Green idealists, particularly in continental Europe. One can easily criticize such back-to-nature missionaries by, for example, pointing out that the rates of homicide in the Highlands of Papua New Guinea are several times higher than in the streets of Los Angeles (Daly & Wilson 1988), and that native Americans had all but wiped out large game before European settlers arrived (Kay 1994).

The good-for-the-whole paradigm in animal behavior came to a rapid demise when field studies such as those on lions, langurs and chimpanzees reported killings amongst conspecifics (Hrdy 1974; Bertram 1975; Goodall 1986). The parallel rise of gene-centered sociobiology and modern behavioral ecology in the 1970s and 1980s was fostered by eloquent treatises accessible to the general public (Wilson 1975; Dawkins 1976). Efforts to debunk romanticism and group selectionism coincided with an unavoidable emphasis on "nature red in tooth and claw". Killings among conspecifics were compatible with the revised interpretations of social evolution, since animals were understood to maximize their individual reproductive success without taking the good of the group, population or species into account. Seemingly "nice" behaviors were also reinterpreted as only phenotypically altruistic. Maternal care, therefore, became genetically as selfish as cooperation amongst kin or the exchange of "reciprocal altruism" amongst non-relatives. Sociobiology thus equalized, categorizing both brute *and* cute stuff as "selfish", "deceptive" or "spiteful".

Frans de Waal (an ethologist of Dutch origin with perhaps a soft spot for classical ethology *sensu* Tinbergen) has complained that efforts to purge the study of animal behavior from group-selectionist hues has rendered it almost unacceptable to speak about "friendly" interactions. Purist sociobiologists would rather have him talk about "affiliative" interactions and relabel a "reconciliation sealed with a kiss" amongst apes as "postconflict interaction involving mouth-to-mouth contact" (de Waal 1996: 18f). De Waal tends to dwell on the niceties of animal societies – "Survival of the kindest" (1998) – and is perhaps in danger of neglecting a chicken and egg problem: that aggression is not caused by reconciliation but that reconciliation is caused by aggression, which is hence always *a priori*. Still, he rightly identifies the exaggerated swing of the pendulum to a behavioral ecology devoid of group-selectionist interpretations (not withstanding the fact that intergroup competition may sometimes select for group-benefiting behaviors; cf. Sober & Wilson 1998).

Despite de Waal's complaint, which some might view as merely semantic, there is still fundamentalistic resistance in the scientific community to the idea that animals act selfishly – particularly when it comes to infanticide.

Non-adaptationist explanations: critique from within the paradigm

Outspoken critics of adaptationist theories about infanticide are nowadays rare among behavioral ecologists. Still, their voices are clearly heard because they receive disproportionate attention concordant with the status of a minority. Arguments and counterarguments are often based upon particulars of reports and alleged data errors. As important as details are, I will also highlight general strings of argumentation.

Adaptationist hypotheses are typically debunked by declaring that favorable evidence distorts what goes on in nature. This attempt is two pronged: one can question the database or declare that infant-killings are maladaptive. Both positions have been invoked most prominently by

Phyllis (Jay) Dolhinow – the pioneering researcher on wild langurs (Jay 1963) – and her students and co-workers at the University of California at Berkeley (e.g., Dolhinow 1977; Curtin & Dolhinow 1979; Boggess 1984; Fuentes 1999). This group has recently been joined by Robert Sussman from the Washington University in St Louis. He and his student Thad Bartlett plus coworker James Cheverud are responsible for two papers (Bartlett *et al.* 1993; Sussman *et al.* 1995) and an accompanying press release (which resembles an anti-adaptationist pamphlet) proclaiming "Infanticide more myth than reality" (Aronson 1995). In 1998, Sussman became editor of *American Anthropologist*, which had published his original retort to the sexual selection hypothesis as applied to non-human primates. The rejectionist platform was broadened when the journal soon thereafter carried an article entitled "Infanticide by male lions hypothesis: a fallacy influencing research into human behavior" (Dagg 1999a). The paper listed "faulty science" as a keyword.

Such publications undermine the credibility of certain behavioral researchers, including myself. The primates article therefore became the subject of a rebuttal (Hrdy *et al.* 1995); the lion article provoked a letter, signed by 17 defenders of adaptationist explanations, sent to Sussman in his position as journal editor (Silk & Stanford 1999).

This development can be viewed as testimony to scientific progress that lives from the exchange of arguments and counterarguments. However, scientists do not simply sit down and weigh evidence. The specific academic climate in which we were brought up rewards or punishes specific opinions, and our views therefore tend to be biased. Not denying self-interest, I will take issue with Sussman and colleagues – since they explicitly referred to my own work (see also van Schaik, Chapter 2). The exercise shall demonstrate how evidence is inflated if we sympathize with explanations and how evidence is downplayed if we dislike explanations.

> A "careful" literature examination found 48 "observed" infanticides amongst wild primates (Bartlett *et al.* 1993: 960) – but failed to list criteria for what counts as "confirmed". For example, included is the fate of little langur M1.4 at Jodhpur, India, whose serious wounding I witnessed and who later disappeared. However, other such circumstantial evidence is often excluded from the sample. The authors' tally for Jodhpur thus adds up to 13 infanticides – whereas I counted more than 50 (Sommer 1994). Similarly, only 3 recognized infanticides among mountain gorillas compare with 17

or so cases considered by other researchers (Fossey 1984; Watts 1989). What counts as "unequivocal" evidence is obviously largely in the eye of the beholder.

Captive primates are only included selectively. Killings that allegedly refute the sexual selection theory are cited (Bartlett *et al.* 1993: 978), while supporting cases are ignored (e.g., Böer & Sommer 1992).

The review states that "many instances of infant killings are confounded by unnatural conditions", citing the killing of resident males by humans or the darting of a mother preceding an infanticide (Bartlett *et al.* 1993: 981). Alternatively, one could value such cases as strong support for adaptationist explanations. In fact, an early critic of the sexual selection theory lamented the lack of (experimental) manipulation of group structures (Schubert 1982). In fact, the authors may have ethical concerns about lethal experiments (as do I and other researchers, e.g., Hrdy 1982: 248). But then, evidence in which field researchers protected infants from attacks is also excluded (Bartlett *et al.* 1993: 985).

Some chimpanzee infanticides are omitted "because of the poor observation conditions" (Bartlett *et al.* 1993: 985). However, study sites with good observations are labeled as "urban-like setting" (P. Dolhinow, cited in Aronson 1995: 2). Similarly, habitats without predators count as "disturbed". But when predators abound, they, rather than new incoming primate males, are considered to be the infant-killers (Bartlett *et al.* 1993: 968).

My observation of a dog snatching an infant langur that fell from a tree after being seriously wounded by a new male (Sommer 1987: 178) begs for the simple interpretation that the attacker caused the death. Instead, my description is rated "significant because it highlights the possible role of feral dogs in the deaths of infant langurs" (Bartlett *et al.* 1993: 968). One could as well consider the role of crows that also scavenge on dying infants.

The authors stress that "the most salient characteristic of the infanticidal episodes . . . is the generalized, overt aggression exhibited by adult males" (Bartlett *et al.* 1993: 984). Moreover, males suffering injuries during invasions into troops provide evidence that "infants were not the sole attack target" (R. W. Sussman, cited in Aronson 1995: 3). However, this is exactly predicted by the sexual selection hypothesis, according to which infant-killings occur in connection with aggressive competition amongst males for access to females. More importantly, how can a male kill an infant without being aggressive? The critics demand non-aggressive aggression. In addition, it looks as if they confuse a proximate mechanism (aggression) with an ultimate function.

The authors of the review suggest that "infants often place themselves in danger by their own actions" – such as clinging to their mothers during attacks (Sussman *et al.* 1995: 149). However, how likely is it that unweaned infants will *not* cling to their mothers when aggressive males approach? I witnessed time and again that males singled out infants, even wrenching them from their mother's breast. Female defenders hardly ever suffered a scratch, even if they wrestled with the highly aggressive male – presumably, because females are potential mates (see photos in Sommer 1987).

The review concludes that infanticide amongst wild primates failed "to support the interpretation of infanticide as a primatewide adaptive complex"; moreover, of the 48 infanticides that were accepted as "observed", 88% were allegedly "not compatible" with the sexual selection hypothesis (Bartlett *et al.* 1993: 958; Sussman *et al.* 1995: 149; a colporteur is Dagg 1999b: 20). This is gravely misleading. Incompatible would be cases in which fathers killed their own offspring or when males killed weaned offspring. However, information about the genetic relationship between killer and victim is in many cases simply incomplete. Such cases do not at all contradict the hypothesis.

Using such distorted evidence, the anti-adaptationist group condemns the widespread acceptance of the sexual selection hypothesis as "academic mythology" (Bartlett *et al.* 1993: 984) or, even bolder, as "myth of legendary proportions" (P. Dolhinow, cited in Aronson 1995: 2). Similar statements are reiterated by the major European pocket of resistance against adaptationist explanations: Austrian ethologist Irenäus Eibl-Eibesfeldt, a loyal disciple of his academic mentor Konrad Lorenz (Eibl-Eibesfeldt 1984, 1997). Early in his career Eibl-Eibesfeldt wrote an article with the programmatic title "Warum sich Tiere nicht töten" (Why animals don't kill each other; Eibl-Eibesfeldt 1965). Ironically, he initially referred to the critique of infanticide by my own supervisor (Vogel 1979) in concluding that "an often cited example for the recklessness of genetic selfishness turns out to be the product of premature conclusions" (Eibl-Eibesfeldt 1984: 125; translated by V.S.). This statement was reiterated 14 years later, citing the review conceived at Washington University as additional evidence (Eibl-Eibesfeldt 1997: 142).

Eibl-Eibesfeldt is uneasy about the occurrence of infanticide because it contradicts the basic assumption of classical ethology that an "aggression drive" – essentially an adaptive "instinct" – is bridled by an "inhibition to kill conspecifics" that ensures that the species survives. The frequent killings amongst humans cannot easily be reconciled with this concept. The term "pseudospeciation" is therefore introduced to explain why homicide is not inhibited: "de-humanizing" does the trick. Interestingly, Jane Goodall borrowed this idea, stating that wild chimpanzees "dechimpized" their opponents before killing them (Goodall 1986: 532). Obviously, "pseudospeciation" rests on the postulate of a "killing-inhibition". Both terms are as obsolete as the classic concept of "instincts" (see Zippelius 1992). Otherwise, we would need terms such as "de-lionization", "de-langurization", and "de-mouseation." Eibl-Eibesfeldt muses that infanticide is a pathology: chimpanzees (and humans) have perhaps undergone rapid brain expansion, during which some fine-tuned control mechanisms were lost (Eibl-Eibesfeldt 1997: 143).

A similar "loose screw argument" was initially employed by Dolhinow (1977). Infanticide amongst langurs was considered to be "aberrant", caused by the stressful conditions at "atypical" sites (see also Aronson 1995: 2). I believe this competing hypothesis has outlived its justification for the following reasons. (1) I could never detect a hint of "pathological" behavior during my own observations of langur infanticide; the killers were apparently healthy individuals acting under highly predictable circumstances (Sommer 1987). (2) Langur infanticide has meanwhile been observed at sites with minimal human influence. Provisioning in particular – believed by Dolhinow to cause "crowding" – is absent from, for example, Kanha, India (Newton 1986), and Ramnagar, Nepal (Borries *et al.* 1999b). (3) Other primate species commit infanticide in the wild at undisturbed sites (for a review, see Bartlett *et al.* 1993). (4) Current DNA analyses for Ramnagar langurs show that male attackers were not related to their victims (Borries *et al.* 1999b).

The interpretation of infanticide as a pathology has its roots in typological reasoning. Animals are believed to have a species-specific repertoire of behavior. Those who deviate from such "universals" are consequently "deviant", "maladaptive", "pathological" or "abnormal". "Overcrowding" (Curtin & Dolhinow 1979) is a corresponding misnomer since no God-given gold standard for population density exists. Animals will reproduce rapidly until a habitat's carrying capacity is reached. Different behaviours might be employed under low and high densities and some individuals will do better than others (cf. Moore 1999). One cannot have it both ways and rightly emphasize that "one genotype can produce a wealth of different phenotypes" (Dolhinow 1999: 194) while at the same time hang on to the dichotomy of "normal" versus "atypical" (Dolhinow 1999: 195). It is therefore quite peculiar if not self-contradictory that Dolhinow (1994) tends to stress the problematic "reification" of social systems, pointing towards the tremendous "intraspecies behavioral variability" of primates. This resembles modern behavioral ecology, which has done away with the emphasis on "uniformity" and the labeling of alleged "asocial" behavior as "sick". Instead, selection is believed to produce individuals who are capable of strategic and tactic responses, resulting in behavioral polymorphisms and "plasticity in primate societies" (Fuentes 1999: 183). A behavior such as infanticide is therefore not necessarily an all-or-none phenomenon, but may or may not occur – depending on varying socioecological circumstances. Such a theorem has room for non-conformist life trajectories.

The British ethnologist Alfred Reginald Radcliffe-Brown (1881–1955) helped to pave the way to (since outdated) group-selectionist ideas. He maintained that the health of a society could be measured by how well its members were integrated, societal structure being "like that of the organic structure of a living body" (Radcliffe-Brown 1952, cited in Jay 1963: 224). The analogy to body cells suggests that the individual within society has little value. Dolhinow applied this to societies of non-human primates where age and sex classes fulfill roles to support the "functionally integrated structure". Variability in behavior may lead to "temporary alterations in the 'normal' structure" but such "disequilibrium" will soon give way to the original "equilibrium" (Jay 1963: 239). This catechism of harmonious togetherness knows no place for acts that do not foster group cohesion.

Tragically, Nazi ideology was built on similar convictions, and Konrad Lorenz himself lent pseudoscientific credibility to it. He, who held a degree in medical sciences, proposed to eradicate "asocial" members from the "supra-individual organism" of the Deutsches Reich, just as a doctor has to cut out cancerous cells from individual bodies (Lorenz 1940, cited in Müller-Hill 1984). There is certainly no straight line from the Nazis via Lorenz to the rejection of the idea that so-called "selfish" behavior is pathological. However, it is worth noting that group-selectionist (= systemic) arguments can be as easily perverted into fascist currency as can gene-centered (= individualistic) views (cf. Segerstråle 1992, 2000).

Rather unspecified political fears were expressed by Phyllis Dolhinow: "You are playing with fire when you use such a loaded word" as infanticide (cited in Aronson 1995: 3). She stops short of transforming into an outside-paradigm critic; but Anne Innis Dagg (1999a) in her attack on adaptationist explanations clearly does become one. The sexual selection hypothesis "resonates with Western culture in which many people accept male dominance and aggression and condone in part the control of female sexuality by men". She commits the moralistic fallacy by stating how the world ought to be: "There is little hope of changing society if social behavior is considered as a genetic trait". Dagg is not very specific about which of the multitude of existing societies she would want to change and how and why, but her agenda is obviously an anti-conservative and progressive Western platform: "Sociobiology clearly treads on dangerous ground" and "locates us in a biological fatalism that a correct cultural interpretation would avoid" (Dagg 1999a: 940, 947–8). Dagg (1999a: 947) categorizes previous research about infanticide as "not good science". The moral flavor of her "good" is the subject of the following section.

Science is politics: critique from outside the paradigm

The modern evolutionary paradigm maintains that the structure of animal societies is delicately balanced between cooperation and competition – with infanticide as one consequence. Is this really a hard and fast "objective" approach to the study of nature? Scientists may believe that their methods safeguard them from being subjective. However, topics certainly become fashionable (or unfashionable) with the prevailing *Zeitgeist*.

The theory of evolution is linked to the 19th century mantra of "progress", fostered by Victorian optimism after the so-called civilized nations had gone through a period of "enlightenment".

Darwin did not dwell on the selective forces reflected in the morphology of genitals – probably muffled by the mores of Victorian England. It took a century, and perhaps the so-called "sexual revolution", until genital sexual selection became a recognized area of research (cf. Dixson 1998).

"Overcrowding" was a buzzword in the 1960s (Calhoun 1962) when fears about human population growth flared up. Early reports about infanticide amongst monkeys were also linked to this concept (for critique of underlying assumptions, see Moore 1999).

"Tool-making" defined early humans in much of 19th and early 20th century palaeoanthropology – in all likelihood a reflection of the abundant technology associated with the industrial revolution. Likewise, theories of "Machiavellian intelligence" as a means to manipulate others (Byrne & Whiten 1988) flourished when information technology became widely available.

Still, average scientists tend to be naive positivistic empiricists, convinced that "results" describe "reality" more or less objectively. Disciplines more open to meta-theories such as philosophy or theology point out that all measurement depends on categories. Otherwise, we would have nothing to quantify. But do categories really exist or are they in some ways social or linguistic constructs? Consider, for example, "species". The borders of this category are hard to define in a longitudinal phylogenetic perspective, but even cross-sectional perspectives reveal blurred edges such as interbreeding with other "species". Categories are certainly useful but the heuristic importance is not in itself proof that such entities exist per se. The philosophical tradition that argues in favor of the existence of categories has its roots in Plato's teaching about "ideas": perfect ["ideal"!] "horsehood" or "beautihood" exists independent from the imperfect specimen of horses or beautiful beings we see here on earth. This tradition is, somewhat confusingly, not called "idealism" but "realism" – because the categories are believed to be real entities. The opposing school according to which categories are simply "names" (Latin *nomen*) is called "nominalism".

If reification troubles physics (does "gravity" exist, or an "atom"?) then the conversion of abstractions into concrete entities poses even more of a problem for biological sciences (is "intelligence" or "race" something concrete?; cf. Segerstråle 1992). The debate about infanticide is heavily infested with essentialized terms such as "species", "population", "instinct", "social system", "selection". Thus, even natural scientists should warm to the idea that we not only define the borders of the world by naming the parts we find in our environment, but that we in certain ways also "create" worlds.

This has implications for the judgment about the vigorous attack on the *naïveté* of natural sciences that evolved in the late 1980s from the left-leaning postmodern movement (cf. Haraway 1989). For scientists, the choice and usage of certain categories to "foreground" (Haraway 1989: 311) could simply be a matter of heuristics that employ reductionism and simplification. However, postmodern outside-paradigm critics assume that such choices are contaminated with politics. For example, evolutionary biologists who believe that they can *reconstruct* the natural history of infanticide can be *de-constructed* as promoting male violence against females. To unmask hidden agendas is to teach that the truths about the world are moral truths. Scientists may postulate that they try to distinguish true from false, but according to postmodern critique any construction of reality is a decision between good and evil (for the following, cf. Cartmill 1998).

To stress that humans are a specific kind of animal, that they are influenced by genetic information and that there was no right or wrong for the billions of years it took life forms to evolve on earth is a deeply suspicious notion for postmodernists: It restricts the (allegedly) unlimited freedom of humans to shape their destiny. Humans are believed by postmodernists to be utterly different from other animals – an idea that is not easily compatible with that of evolution. Such a *Weltanschauung* has been labeled "secular creationism" because the dislike for evolutionary theory is similar to that found in religious creationism where evolution is unacceptable because it "demotes" humans to the status of beasts. The Christian Right ideology of creationism is virtually endemic to North America. In contrast, the multicultural Leftist ruminations have not only infiltrated Europe but were in large part conceived in France (e.g., Derrida 1970; cf. the counterattack mounted by Sokal & Bricmont 1998).

As diverse as these secular and religious discourses may be, they are united in their dislike for Darwinism and follow the moralistic fallacy. Scientists may find the accusations entirely silly, but the basic idea of a feedback in which science is influenced by and itself influences the given contemporary political climate is not unwarranted. Thus, even if science does not have an explicit political agenda, it may have one implicitly. I will now evaluate some typical accusations raised by postmodernists.

Accusation: science is sexist

Postmodernists argue that the patriarchal structure of science places rationality above feelings and thus facts above values, rendering it sexist. Behavioral ecology should not have much of a problem with this because women and the particular views of women, especially on male-biased interpretations, became deeply influential in the development of this field. For example, the orthodox concept that male mammals are promiscuous (because their reproduction is limited by access to fertile females) and that females are monogamous (because they do not gain anything from copulations with multiple partners) has been replaced with a whole repertoire of reproductive strategies that have little to do with the idea that females are "passive" (Hrdy 1981). Thus females may be inclined to seek extra-pair copulations in order to accrue better genes, to elicit additional parental investment or to confuse paternity in order to minimize the risk of (*sic!*) infanticide.

One consequence of these discoveries is the "re-education" of male researchers who will now readily admit the nastiness of the reproductive strategies of their own "demonic" sex (Wrangham & Peterson 1996). But male-ist *mea culpa* might be unnecessary because notable "Darwinian feminists" such as Sarah Blaffer Hrdy, Barbara Smuts or Patricia Gowaty disagree with the postmodern paradigm by arguing that females are not innately "better" than males. They merely compete by other means – such as female choice or inciting male–male competition (Gowaty 1997a). Females will even tip the balance of physical power if they can get away with it – as is illustrated by female coalitionary dominance over males in perhaps our closest living relatives, the bonobos (Parish 1996). Finally, even infanticide committed by *females* became a resilient feature (Digby, Chapter 17).

Nevertheless, Haraway warns that to center debates on the biological meanings of infanticide "could be dangerous territory on which to build feminist approaches to science" because "women rarely control the field structure of interpretation" (Haraway 1989: 311, 418). This brings us to the ultimate "evil" of behavioral ecology: competition.

Accusation: science is capitalistic

Behavioral ecologists are most easily assailed in this arena given that game-theoretical approaches and the method of cost–benefit analysis are direct descendants from economics (cf. Maynard Smith 1982). Centers such as that for Economic Learning and Social Evolution at University College London evidence this basic link. One could even subsume all of sociobiology and behavioral ecology as a subdiscipline of economics – although the primatologist stumbling through a muddy rain forest hoping to glimpse who attacked whom does not easily pass as a stockbroker. Still, sociobiologist fieldworkers are likely to come up with a net-benefit interpretation for the observation. Undoubtedly, sociobiological arguments about infanticide are "firmly anchored in the narrative of genetic investment" (Haraway 1989: 430).

The German sociologist Max Weber (1864–1920) connected capitalism with certain religious traditions. He drew attention to the Christian concept of predestination fostered by the Swiss reformatory Johannes Calvin (1509–1564). Calvin disagreed with the Catholic church on how to obtain salvation. Catholicism taught the principle of "ut . . . ut", which states that a believer could acquire eternal life through a mixture of God's grace *and* good deeds. Calvin rejected this idea – for God would not be absolutely free if He allowed Himself to be influenced by the virtuous deeds of His creatures. Thus the fate of humans was already predestined by God's inexplicable decisions even before birth: for either eternal salvation or eternal damnation. It did not matter what one did. Such conviction came in handy for early capitalists who argued that their own riches were testimony of God's choice; helping the poor would mean to revolt against the divine plan. Such thinking was fertile soil for the social Darwinism of the 19th and 20th century, except that God's will was substituted by the selection processes of evolution (cf. Mühlmann 1984). Evolutionary biologists would be naive to assume that their ideas about mechanisms of adaptation are not at least subtle extensions of such histories of intellect.

Accusation: science is oppressive

Postmodernists state that science oppressively demands that we argue in an approved way and that science is thus part of a conservative strategy to suppress political progress. One can raise the counterargument that scientists span a broad range of political thought or even conclude that sociobiology is a Communist conspiracy. After all, J. B. S. Haldane and John Maynard Smith were members of the British Communist Party, Robert Trivers was involved in radical Black politics, and "E. O. Wilson and most other leading sociobiologists are left-center liberals or social democrats" (van den Berghe 1981: 406). However, facts will matter little because the postmodern critique assumes an "inherent connection between political motivation and bad science" (Segerstråle 1992: 203). Any meaningful dialogue ends if postmodern critique voices such totalitarian views.

Interestingly, argued from within the paradigm, science is designed to be non-oppressive and open to heretics. Ideally, and often factually, those are rewarded who overthrow a reigning paradigm by convincing their peers that a competing hypothesis offers better explanatory value. Such a struggle is reflected in the exchange of arguments between adaptationist and non-adaptationist explanations of infanticide (data artifact, pathology, by-product). The latter are notoriously annoying, upsetting, and unreasonable for me (and others). Still, my opponents have – and should have! – the freedom to criticize my work, which will hopefully spur me to deliver stronger arguments or new data.

However, deriving testable predictions from competing hypotheses has its gray zones. For example, the often reiterated Popperian theorem that the possibility of "falsification" is the cornerstone of proper scientific method is limited. For instance, one cannot falsify the assumption that something (God perhaps, or the death star) does *not* exist – if it *really* does not exist – because the falsification would have to show its existence. Non-adaptationists may at times feel that they are trapped in a similar circle. For example, Phyllis Dolhinow (pers. comm.) complained that journals tend not to accept articles that provide "negative evidence", i.e., reports on resident male changes in primates groups that did not lead to infant-killings.

Moreover, hypotheses multiply once a paradigm "goes baroque" (typically, before it "goes broke". . .). Behavioral ecology may be approaching this era. Ironically, it is more difficult to reject certain hypotheses once the empirical basis is broadened, because living systems are complex. For instance, half a dozen explanations are now available for why infanticide might be adaptive, and one can almost never go wrong with a reference to resource competition. In addition, infanticide has advanced to a mega-paradigm in that the risk of infant-killing is contemplated to be the very reason for male–female associations itself (van Schaik & Dunbar 1990; van Schaik & Kappeler 1997). Consequently, and somewhat paradoxically, infanticide is relatively rare or may even have become extinct because the severe threat bred sophisticated counterstrategies, as perhaps happened in gibbons (Reichard & Sommer 1997) and bonobos (Parish 1996). Similarly, thousands of hours of focal animal sampling in downtown London may never yield a single datum point about a pedestrian killed by a car; still, nearly all movements of *Homo sapiens* in this habitat reflect avoidance of vehicles. We are therefore advised to not confuse rates of mortality with the risk of being killed (Dunbar 1988). So it seems that the study of infanticide is slowly evolving into a study of events that do – more often than not – *not* happen.

Amongst the fallibles

The confrontation with outside-paradigm critics does not lend itself to solutions because the study of nature cannot be free from subjective constraints. In physics, Heisenberg's uncertainty principle states that one cannot observe something without influencing its path because the minimum amount of energy needed for a measurement is a photon, which will inevitably influence the very thing it is supposed to measure. The measured entity will, by the same token, influence the measurer. Thus "objectivity is a subject's delusion that observing can be done without him" (Heinz von Foerster, cited in von Glasersfeld 1992: 31). Our epistomology is left with, at best, a "hypothetical realism" (Lorenz 1973). As a consequence, natural scientists have no monopoly on what is a correct world-view. By the same token, postmodernist schools can be light-handedly caught in their own de-constructionist nets: Time dependency is definitely characteristic of late 20th century agendas on political and social equality. To declare that one holds the holy grail of "a correct cultural interpretation" (Dagg 1999a: 947f) is self-righteous and approaches totalitarianism.

I do not hesitate to conclude that natural scientists, too, are not immune against the two fallacies. We commit moralistic fallacies with transition from "Ought to Is" when formulating hypotheses: the hypothetico-deductive method cannot work without *pre hoc* categories, and these tend to be derived from a capitalistic cost/benefit approach to nature. Similarly, we commit naturalistic fallacies and transgress from "Is to Ought" when we like or dislike what study-animals do. Who could remain unmoved if our closest primate relatives engage in conspecific killings? Understandably, primatologists sometimes intervened in infanticidal attacks (cf. Goodall 1986). The problem is most obvious with respect to the widespread practice of child abuse and infanticide amongst humans, since few behavioral ecologists will *not* wish this pattern to decrease. Such moral agenda is even employed to justify the study of infanticide: "If you know these things better, you know what to do, take certain measures, counsel people. It arms us" (Carel van Schaik, quoted in Zimmer 1996: 78). The position resembles that of Darwinian pioneer Thomas Henry Huxley, who in 1888 proclaimed that humans could not learn anything from nature except that it is full of evil. The "good natured" heritage that de Waal (1996) stressed is just the flipside of this coin, which will never become outdated as long as humans interact. Consequently, we probably cannot develop rational ethical choices without recurring to nature (Liedtke 1999).

Does this mean that empiricist approaches to infanticide are futile? Outside-paradigm critics may think so, whereas I do not. Behavioral ecology reflects a certain type of culture to which members of the scientific community adhere, and this set of rules produces impressive results for those who "believe" in it ("emic perspective"). For me, this is enough justification for my work. We cannot – and should not! – expect a synthesis or reconciliation with a paradigm that operates under different rules ("etic perspective").

Perhaps, there can one day be the "new holism" E. O. Wilson (1975: 7) hoped for and still dreams about (Wilson 1998) – enabled by evolutionary theory. In the meantime, I at least, do not want my world to be stirred into a gigantic homogenization of paradigms. My desire is probably evidence for the trend to base ethics on esthetics. The powerful proximate mechanisms of my neurochemistry make me cherish a world full of beautiful biodiversity – and mind-boggling tradidiversity. Live, let live and – one hopes – be allowed to live: a wisdom strangely obtained from the study of infanticide.

The within-paradigm critique of the study of infanticide is a conflict that in principle can be resolved. Opponents subscribe to similar if not identical sets of rules that include careful comparison or experiment. We will have to design conflicting hypotheses in such a way that their predictions will become mutually exclusive. This is relatively easily said and done. Alas, the hard part is to test the predictions under natural conditions because observations leave much room for subjective interpretation: one observer's "goal-directed attack" might be another's "generalized aggression". Similarly, DNA analyses for langur monkeys at Ramnagar, Nepal (Borries *et al.* 1999b), that exclude attackers of infants as fathers can be interpreted as supportive for the sexual selection hypothesis. Alternatively, one can raise the bars and reject the sample size of this study as too small or point out the lack of evidence for other langur populations or other primate species. In any case, we should not fool ourselves that in the end certain hypotheses can be proven or – à la Dagg (1999a: 947) – "disproven". Like it or not, we will have to resign ourselves to "plausible" and "implausible" explanations.

Yet, on the whole, I am optimistic about this process – as much as I may be upset about notorious counterarguments. But this is probably true for my opponents as well. I view the science I conduct as a sports match – certainly not free from emotional and economic involvement, but subject to mutually agreed rules. My arguments may ultimately succumb. However, until then, I will defend my conviction with passion – in this case the belief that the pattern of infanticide has been brought about by selective forces.

2

Infanticide by male primates: the sexual selection hypothesis revisited

Introduction

The aim of this chapter is to evaluate the evidence for the sexual selection hypothesis for infanticide by male primates, in relation to other hypotheses for this phenomenon. The sexual selection hypothesis emerges largely vindicated. The second part of the chapter then sets the stage for the rest of this book by summarizing observed behavior during episodes of infanticide, the evidence for the decision rules used by males, and the effects of social organization on rates of infanticide by males, and by pointing out some of the remaining difficulties with the hypothesis in its present form.

Critics have repeatedly pointed out that the evidence for infanticide in primates is often circumstantial at best, and that researchers may have jumped to conclusions on the basis of premature extrapolation from snippets of observations (Boggess 1979, 1984; Bartlett *et al.* 1993; Sussman *et al.* 1995). This makes it difficult to critically evaluate the hypothesis that a male committing infanticide will on average gain reproductively by doing so. Of course, it has made it equally difficult to evaluate other hypotheses. In this chapter, I therefore evaluate the sexual selection hypothesis for infanticide by males in non-human primates by a careful review of the most appropriate sources of evidence: directly observed cases in wild primates.

The conditions in which male infanticide are expected to raise the male's fitness have been reviewed extensively before (Hausfater & Hrdy 1984). Infanticide will be an adaptive male reproductive strategy if: (1) the probability that the male had sired the infant(s) was zero or close to zero; (2) the mother can be fertilized earlier than if the infant(s) had lived; and

(3) the infanticidal male has an increased probability of siring the next infant(s) relative to the current offspring (Hrdy 1979; Hrdy et al. 1995). All three conditions must be met simultaneously. These predictions seem straightforward enough, but there is widespread misunderstanding (e.g., Bartlett et al. 1993; Agoramoorthy & Rudran 1995). In some cases, critics tested incorrect predictions, whereas in others they rejected the hypothesis because the expected outcome sometimes did not occur (e.g., when a new male is ousted again before he can mate). Here, I develop these predictions in somewhat more detail in order to assess whether such observations are exceptions to a trend that otherwise overwhelmingly supports the sexual selection hypothesis or whether they are so common as to confound any pattern.

As with other forms of behavior, it is useful to specify when the benefits of the behavior exceed its costs, and therefore when it is favored by natural selection. Although one does not usually need to specify the genetic mechanism for such an analysis to be productive (Grafen 1991), it is also important to consider the decision rules used by the organism.

The benefits (B) of male infanticide are the time gained relative to not committing infanticide, corrected for the change in probability of paternity. Assuming that a male has access to a particular female, and will only kill infants upon assumption of his tenure because he will gain subsequent reproductive access to the female, a male's total reproductive lifespan, T, can be made up of 1 birth following infanticide (i) and k births not following infanticide (n), after time intervals t_i and t_n, respectively. Therefore, $T = t_i + kt_n$.

If P_n is the probability of siring the next infant without committing infanticide after period t_n, and if the male's total reproductive lifespan is T, the male's benefit would be

$$B_n = P_n(T/t_n)$$

Alternatively, if P_i is the probability of siring the next infant following infanticide, i.e., after period t_i, and if the probability that the infant killed is actually sired by the male is p, then the benefit of infanticide is

$$B_i = (P_i - p) + [(T - t_i)/t_n]P_n$$

The male's net benefit from committing infanticide upon starting reproductive tenure in a group is therefore

$$B_i - B_n = P_n(-t_i/t_n) + P_i - p$$

If, for simplicity, we assume that $P_n = P_i = P$ (i.e., the male's residence status or dominance rank do not change in the time period gained by the infanticide), the net benefit will be positive if

$$[(t_n - t_i)/t_n]P > p \qquad (2.1)$$

It is therefore almost inevitable that a male committing infanticide benefits from doing so, since all it really requires is that there is a reduction in interbirth interval $(t_n - t_i > 0)$, given that $P \gg p$.

In deriving prediction (2.1), however, I have made some additional assumptions. First, I assumed that a female is willing to subsequently mate with the male who killed her offspring. Both theory and observation suggest that the potential cost to males of females' refusal to mate with them is negligible (see Discussion). Second, I assumed that there are no other costs, for example in finding the infant or in killing the infant. In anthropoid and many lemuroid primates where mothers carry their infants around, the costs of finding the infant can usually be ignored. However, killing may be costly if the male is injured while overcoming the defenses of the mother and possibly her relatives or the sire. This may reduce his tenure (T) if the injuries sustained make him less likely to defend his new position against male competitors, or if the females subsequently succeed in evicting him from the group. (Possible injuries sustained by the female need not be considered separately because they would effectively serve to lengthen interbirth intervals and are thus incorporated into the male's time advantage.) Hence, whether or not the above prediction is correct depends on the risk of injury for the perpetrator.

This formulation allows us to examine the predictions more closely, and also to evaluate interpretations by others. First, it has often been taken to mean that males must subsequently have sole access to the female and therefore sire her next offspring; in other words, that infanticide should occur only in one-male groups (e.g., Boggess 1979; Camperio Ciani 1984). However, note that while $P > p$ is required, it is not necessary that $P = 1$. Second, it is sometimes claimed that the male must be an outsider who has never mated with the female before. However, it is not required that $p = 0$ (although the issue of how males assess these probabilities needs to be discussed, see below). Third, a critical prediction is a reduction of the mean interbirth interval. In species with the appropriate life history this prediction is almost inevitably met (see van Schaik, Chapter 3), on average, unless infanticide causes a serious delay in fertilization due to female injury. It has, however, been predicted that infanticide should not be found where reproduction is extremely seasonal (Hrdy 1979; Hausfater 1984; Hrdy & Hausfater 1984; Leland *et al.* 1984; Hiraiwa-Hasegawa 1988) because this would not lead to reduced interbirth intervals. Below, I examine effects of infanticide on interbirth intervals in seasonal breeders. Fourth, it has been argued that where males are the same size as, or even smaller than, females, infanticide would be rare or absent. This, too, is an empirical issue. The profitability of infanticide in monomorphic species would be compromised if males were likely to sustain serious injuries.

As with each behavioral adaptation involving decision-making, it is important to remember that these predictions assume perfect information. In practice, of course, males must behaviorally estimate P_i and p, the approximate age of the infant(s) and possibly weigh the risks in light of the presence of defenders. Thus the male must assess each situation and decide on the attack. As a result, a perfect fit with the predictions need not be found in every single case. Rather, the outcome should be adaptive when averaged over multiple cases. Where it is found that the costs for the males are low, there will be less selection on strict assessment of these parameters (especially P and infant age) by the males, although large errors in estimating p are costly.

A cost not considered so far is the possibility that the infant was sired by a male relative. In this case, the net benefit becomes

$$[(t_n - t_i)/t_n]P > p + (rP_r),$$

where r is the relatedness between perpetrator and likely sire, and P_r the probability that this relative actually sired the infant. This increase in cost will make infanticide less profitable, although not necessarily enough to select against it.

There is never a net benefit for the female to have her infant killed by a male, although she may have a slight gain when her next infant is a son and has inherited the infanticidal tendency, and when she lives in an otherwise non-infanticidal population. This benefit, however, will disappear once infanticide is widespread.

We now test these predictions with observations from wild primates.

Testing the sexual selection hypothesis

Observed cases of infanticide in the wild

Some of the most vigorous criticism of the sexual selection hypothesis concerned the tendency to overgeneralize on the basis of incomplete observations (Bartlett *et al.* 1993). By including only cases where the attacks on the infants were directly observed (unlike Bartlett *et al.* 1993), we avoid interpreting as infanticide events that were caused by different processes. By limiting ourselves to wild populations, we also avoid the inclusion of possible behavioral artifacts produced by captivity.

Struhsaker & Leland (1987) presented a table detailing the 24 cases of infanticide reported in the literature until 1985 where the infanticide had actually been observed in a wild population. In Table 2.1, I use their table and have added all cases published since then known to me by the end of 1998, using the same criteria. However, I eliminated six cases (Struhsaker & Leland's cases 2 through 7), because they referred to hanuman langurs in Jodhpur. Instead, Table 2.1 includes the 12 directly observed cases of infanticide presented by Sommer (1994) in an exhaustive analysis of all observed and inferred cases of male infanticide among Jodhpur langurs (excluding one case of killing of a 4-year-old juvenile, killed under different conditions). Thus we have a total of 55 cases of observed male infanticides fulfilling the criteria of direct observation in a wild population.

Behavioral contexts

Angst & Thommen (1977) suggested that cases of infanticide had in common a context in which a reproductively capable male came into a position of top dominance, either by being placed into a group by humans, by immigrating into it and defeating or ousting the resident male, or by rising in rank within a group. The majority of reported observed cases of male infanticide in natural populations compiled in Table 2.1 comfortably fall within this category (46/54: 85%). In numerous captive situations (see Table 2.2), male infanticide also takes place in accordance with Angst & Thommen's specified context. Many of the reports of direct observations show deliberate, and sometimes persistent, stalking and direct attacks by the male on the infant or the mother–infant pair (e.g., Butynski 1982; Busse & Gordon 1983; Zunino *et al.* 1986; Tarara 1987; Borries 1997), rather than generalized aggression aimed at a variety of individuals (see also Palombit *et al.*, Chapter 6). Borries (1997) noted that infanticidal males tend to attack easy targets; infants are more likely

Table 2.1. *Observed cases of infanticide and their contexts and consequences in wild non-human primates*

Species	Case No.	Source[a]	Context	Actor and victim related?	Age, sex victim	Subseq. mating by actor[b,c]	Male injured?	Female injured?
Alouatta caraya	1	Zunino et al. 1986	Takeover	N	2 wk, m	Probable	N	N
Alouatta seniculus	2	S & L 1987, case 18	Immigration	N	?	Possible	N?	N?
Alouatta seniculus	3	S & L 1987, case 19	Rank rise (inferred)	Possible	1 d, m	Y	N	N
Alouatta seniculus	4	Agoramoorthy & Rudran 1995	Immigrant (male 2)	N	9 mos, f	Y	N	N
Alouatta seniculus	5	Agoramoorthy & Rudran 1995	Immigrant (subad. 16)	N	1.5 mos, ?	N	N	N
Alouatta seniculus	6	Agoramoorthy & Rudran 1995	Immigrant (male 14)	N	1 d, ?	Y	N	N
Alouatta seniculus	7	Agoramoorthy & Rudran 1995	Immigrant (male 10)	N	1.5 mo, f	Y	N	N
Alouatta seniculus	8	Agoramoorthy & Rudran 1995	Immigrant (subad. 16)	N	2 d, ?	Y	N	Y
Alouatta seniculus	9	Izawa & Lozano 1991	Rank rise, immigrant	N	8 mo, m	Probable	Y	Y
Alouatta seniculus	10	Izawa & Lozano 1991, 1994	Immigrant	N	2 d, m	Possible	N	N
Alouatta seniculus	11	Izawa & Lozano 1991	Immigrant	N	<1 mo, f	Y	N	N
Cebus olivaceus	12	Valderrama et al. 1990	Breeding male	Likely	1–2 wk, m	Y	N	N
Cebus olivaceus	13	Valderrama et al. 1990	Rank rise, immigrant	Unlikely	1 d, m	Y	N	N
Cebus olivaceus	14	Valderrama et al. 1990	Second-ranking male	Unlikely	9 mos, f	Prob. not	N	N
Cercopithecus ascanius	15	S & L 1987, case 10	Takeover	N	7 d, ?	Probable	N	N
Cercopithecus ascanius	16	S & L 1987, case 11	Takeover	N	1 d, ?	Probable	N	N
Cercopithecus mitis	17	S & L 1987, case 12	Takeover	N	6 mos, f	N	N	N
Cercopithecus mitis	18	Fairgrieve 1995	Takeover	N	<3 wk, ?	Y	N	N
Macaca mulatta	19	Camperio Ciani 1984	Rank rise	Possible?	<1 yr, m	Possible	N	N
Papio anubis	20	S & L 1987, case 14	Immigration	N	6.4 mos, f	Y	N	N
Papio anubis	21	S & L 1987, case 15	Immigration	N	8.5 mos, ?	No data	N	Y
Papio anubis	22	S & L 1987, case 16	Immigration	N	7.2 mos, f	Prob. not	N	N
Papio anubis	23	S & L 1987, case 17	Fissioned natal group	N	8.4 mos, f	No data	N	N
Papio ursinus	24	S & L 1987, case 13	No data	N	8 mos, ?	No data	N	N
Papio ursinus	25	Tarara 1987	Rank rise, natal	No data	6 mos, f	N	N	N (Orphan)
Theropithecus gelada	26	Mori et al. 1997	Immigration	Unlikely	"Unweaned", ?	Y	N	Y
Colobus badius	27	S & L 1987, case 9	Rank rise, natal	N	1 mo, m	Y	N	N
Colobus guereza	28	D. Onderdonk unpublished data	Immigration	Unlikely	1 d, f	No data	N	N

	Species	No.	Reference	Context					
29	*Presbytis cristata*		S & L 1987, case 8	Rank rise, immigrant	N	9 d, ?	Possible	N	N
30	*Presbytis entellus*		S & L 1987, case 1	Takeover	N	11.6 mos, ?	Y	N	N
31	*Presbytis entellus*		Ross 1993	Takeover	N	2 mos, f	Probable	N	N
32	*Presbytis entellus*		Sommer 1994 case 1	Takeover	Unlikely	3.3 mos, ?	Y	N?	N?
33	*Presbytis entellus*		Sommer 1994 case 2	Takeover	Unlikely	3.4 mos, ?	Y	N?	N?
34	*Presbytis entellus*		Sommer 1994 case 4	Takeover	Unlikely	3.1 mos, ?	Y	N?	N?
35	*Presbytis entellus*		Sommer 1994 case 15	Takeover	N	3.3 mos, ?	N	N?	N?
36	*Presbytis entellus*		Sommer 1994 case 18	Takeover	N	3.4 mos, ?	Y	N?	Possibly
37	*Presbytis entellus*		Sommer 1994 case 19	Takeover	N	4.6 mos, ?	Y	N?	N?
38	*Presbytis entellus*		Sommer 1994 case 42	Takeover	N	1.2 mos, ?	Y	N?	N?
39	*Presbytis entellus*		Sommer 1994 case 49	Takeover	N	1.5 mos, ?	Y	N?	N?
40	*Presbytis entellus*		Sommer 1994 case 51	Takeover	N	0.4 mo, ?	Y	N?	N?
41	*Presbytis entellus*		Sommer 1994 case 79	Takeover	N	7.4 mos, ?	Possible	N?	N?
42	*Presbytis entellus*		Sommer 1994 case 89	Takeover	N	8 mos, ?	No data	N?	N?
43	*Presbytis entellus*		Sommer 1994 case 101	Takeover	Unlikely	0.3 mo, ?	Y	N?	N?
44	*Presbytis entellus*		Borries 1997	Immigration	N	18 mos, f	Y	N	N
45	*Propithecus diadema*		Wright 1995	Immigrant	N	<1 mo, m	N	N	N
46	*Propithecus diadema*		Wright 1995	Immigrant	N	1–2 mos, m	Probable	N	N
47	*Propithecus diadema*		Erhart & Overdorff 1998	After resident disappears	N	1 d, m	Probable	N	N
48	*Lemur catta*		Hood 1994	Intergroup encounter	Unlikely	9 d, ?	Prob. not	N	N
49	*Gorilla gorilla beringei*		S & L 1987, case 22	Intergroup encounter	N	11 mos, ?	Probable	N	N
50	*Gorilla gorilla beringei*		S & L 1987, case 23	Encounter with lone male	N	1 d, ?	N?	N	Y
51	*Gorilla gorilla beringei*		S & L 1987, case 24	Rank rise, immigrant	N	3 mos, ?	Y	N	N
52	*Pan troglodytes*		S & L 1987, case 20	Strange female met	N	21 mos, ?	N	N	Y
53	*Pan troglodytes*		S & L 1987, case 21	Strange female met	N	21 mos, f	N	N	Y
54	*Pan troglodytes*		Hamai et al. 1992	Within group	Unlikely	6 mos, m	Y	N	N
55	*Pan troglodytes*		Hamai et al. 1992	Within group	Unlikely	5 mos, m	No data	N	Y

Notes:

subad., subadult; N, no; Y, yes; wk, week; d, day; mo, month; m, male; f, female; prob., probably.
[a] S & L, Struhsaker & Leland 1987.
[b] Possible, no mating observed but male remained in group;
[c] Probable, no mating observed, but male was only male resident in group.

Table 2.2. *The distribution of infanticide (including attempts) by males among species of non-human primates*

Note: Species included if original authors observed or inferred infanticide by males from indirect evidence

Species	Wild (source of report)	Context[a]	Captivity (source of report)	Infanticide observed?	Context[a]
(Mirza coquereli)			(Stanger et al. 1995)	(Yes)	7
Lemur catta	Hood 1994	4	Pereira & Weiss 1991	Attempts	2
Eulemur fulvus			Pereira & Weiss 1991	No	1 or 2
Eulemur macaco			Pereira & Weiss 1991	No	1 or 2
Varecia variegata			Pereira & Weiss 1991	No	1 or 2
Propithecus diadema	Wright 1995	2			
(Tarsius bancanus)			(Roberts 1994)	(No)	7
Cebus capucinus	Fedigan 1993; Rose 1994	2			
Cebus olivaceus	Valderrama et al. 1990	3, 7			
Alouatta caraya	Zunino et al. 1985; Rumiz 1990	1			
Alouatta fusca	Galetti et al. 1994	1			
Alouatta palliata	Clarke et al. 1994	1, 3			
Alouatta seniculus	Crockett & Sekulic 1984; Izawa & Lozano 1991; Agoramoorthy & Rudran 1995	2, 3			
Cercocebus galeritus	Kinnaird 1990	1 or 2			
Cercocebus torquatus			Busse & Gordon 1983; Böer & Sommer 1992	Yes; Yes	3; 1
Cercopithecus aethiops	Andelman 1987	2	Fairbanks & McGuire 1987	Attempts	1
Cercopithecus ascanius	Struhsaker 1977	1			
Cercopithecus campbelli	Galat-Luong & Galat 1979	1			
Cercopithecus mitis	Butynski 1982; Fairgrieve 1995	1	Böer & Sommer 1992	Yes	1

Species		Context			
Macaca cyclopis	M. J. Hsu & G. G. Agoramoorthy, unpublished data	1 (2?)	Angst & Thommen 1977	Yes	1, 3
Macaca fascicularis	de Ruiter et al. 1994	3	Pallaud 1984	No	
Macaca fuscata	Kawai 1965; Tokida 1976	2			
Macaca mulatta	Camperio Ciani 1984	3	Angst & Thommen 1977; Lindburg 1971	No; No	1
Macaca nemestrina			Kyes et al. 1995	Yes?	2
Macaca silenus			Lindburg et al. 1989	No	1 or 2
Macaca sinica	Dittus 1988	?			
Macaca sylvanus			Angst & Thommen 1977	No	1
Mandrillus leucophaeus			Böer & Sommer 1992	Attempts	1
Papio anubis	Collins et al. 1984	2, 4			
Papio cynocephalus	Shopland 1982; Tarara 1987	3, 4			
Papio ursinus	Collins et al. 1984; Palombit et al. 1997	2; 2			
Papio hamadryas	Angst & Thommen 1977	1	Angst & Thommen 1977; Rijksen 1981	Yes; Yes	1; 7
Theropithecus gelada	Mori et al. 1997	2	Moos et al. 1985	No	2
Colobus badius	Struhsaker & Leland 1985; Marsh 1979b; Starin 1994	1, 2			
Colobus guereza	Oates 1977; D. Onderdonk, unpublished data	2			
Presbytis cristata	Wolf & Fleagle 1977	1			
Presbytis entellus	Sugiyama 1965; Hrdy 1974; Newton 1986; Sommer 1994	1			
Presbytis senex	Rudran 1973	1			
Presbytis thomasi	Steenbeck 1999b	1			
Gorilla gorilla	Watts 1989	1, 4			
Pan troglodytes	Goodall 1986; Hamai et al. 1992	5, 6			

Notes:

[a] Contexts: 1, male immigration with replacement or after male disappearance (or removal in captivity); 2, male immigration without male replacement; 3, rise in rank of resident male; 4, attack between groups; 5, following female immigration; 6, attack by top-ranking males; 7, attacks by likely father.

to be attacked when they are not on their mother and especially when they are already injured or otherwise handicapped. This issue is discussed in more detail when evaluating other hypotheses.

Infanticide has also been observed in some other contexts. First, in several species, male infanticide is observed between groups (cf. Table 2.1): ringtailed lemurs, savanna baboons (see Shopland 1982; Collins *et al.* 1984), vervet monkeys (see Cheney *et al.* 1988), and chimpanzees (see below). In Thomas's langurs, group-living males occasionally leave their group to stalk other groups and ambush them, targeting infants; in this species, infanticide has been directly observed only in this context (Sterck 1997; Steenbeek, Chapter 7). Second, in chimpanzees, infanticide is also observed within communities in which male rank relations were stable. These two contexts are not necessarily inconsistent with the sexual selection hypothesis, because females could transfer after or before the infanticide, respectively, but clearly additional information is required to evaluate these contexts. Finally (not in Table 2.1), infant baboons may (rarely) be killed during escalated fights among males, in which they participate as "social tools", being carried into the encounter by one of the males (Collins *et al.* 1984; see also Paul *et al.*, Chapter 12). I discuss them again below.

Killers' relatedness to infants

Closely related to the question of context, and critical for the sexual selection hypothesis, is whether males kill infants unlikely to be sired by themselves. In 51 of 54 relevant cases (94%), it is impossible or unlikely that the infanticidal male killed his own offspring, because he was not present in the group during the mother's conception period, or if present was not seen to have mated or was unlikely to have been sexually active. The chimpanzee infants killed by males within the same community, had probably not been sired by these males, who were the dominant males of the group (Hamai *et al.* 1992). The three exceptions refer to two cases where males presumably (case 3) or certainly (case 19, alpha male died) rose in rank, and one case (case 12) in which the breeding male may have killed his own offspring. Thus, a genetic relationship between the infant and the infanticidal male is impossible or unlikely in 94% to 98% of the observed cases. In Sommer's (1994) large sample of 55 observed and presumed infanticides in hanuman langurs, a genetic relationship was impossible or unlikely in at least 95% of cases. Using DNA analysis, Borries *et al.* (1999b) showed that in none of the 16 cases of infanticide did the male kill his own offspring. Hence, in the overwhelming majority of cases, males did not kill their own offspring.

The potentially infanticidal male may be related to the sire. Although this situation is probably rare, in howler monkeys male relatives tend to form long-term alliances (Pope 1990) in which only the dominant male mates and sires offspring (Crockett & Sekulic 1984; Pope 1990). In such situations, previously subordinate males may become dominant. Similarly, in langurs and gorillas, sons may mature into the natal groups and could potentially take over. In gorillas, internal takeovers have been largely peaceful (Robbins 1995), but in howlers there are several cases of infanticide by males who rose in rank inside the group and were probably related to the sire (Crockett & Sekulic 1984; Clarke et al. 1994). Although infanticide may be selected against in this context, it need not be. In the red howler case, for instance, if the males are half-siblings and the reduction in interbirth interval is 32% (see below), there is still a benefit if there are no additional costs $((0.32 \times 1) > (0.25 \times 1))$.

Weaning status of infants

As is evident from Table 2.1, infants are usually killed at an age well below the average age at weaning for the species or population in question. However, it is impossible to prove this in individual cases, because within species the variation in the age at which infants are weaned can be appreciable. In some well-studied populations, the data suggest that older infants are progressively less likely to be killed in takeovers, as expected by the sexual selection hypothesis (Fig. 2.1).

The critical issue remains whether estrus is speeded up, and this can even happen when the infant has been weaned. For instance, Böer & Sommer (1992) reported on a captive male white-collared mangabey killing two small juveniles, 24 and 19 months old. Although the infants were weaned, their mothers had not yet returned to estrus and one did so 22 days after the attack, although the other took another 159 days. Borries (1997) noted that 92% of the females whose infants were killed were not sexually active before the event.

If the immatures killed by males were often juveniles, this should lead us to refute the sexual selection hypothesis. There are very few reports of the killing of juveniles. In a quantitative analysis of all 55 cases in which immature hanuman langurs were injured or killed by adult males, only two were weaned, i.e., juveniles. However, unlike all other attacks, the attacks on these (male) juveniles did not take place during takeover

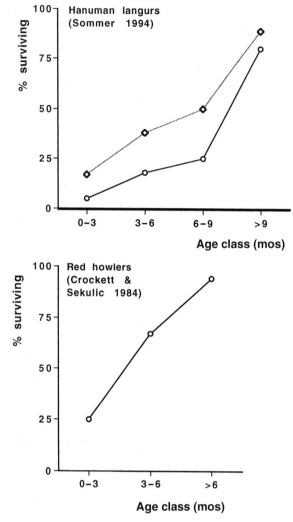

Figure 2.1. The probability that immature hanuman langurs (*Presbytis entellus*) and
red howlers (*Alouatta seniculus*) survive the presence of an infanticidal adult
male as a function of their age (after Sommer (1994) and Crockett & Sekulic
(1984)). In the langur panel (*top*) the upper curve refers to cases where attacks or
infanticides were witnessed; the lower curve also includes cases where
infanticide was considered likely or was presumed. In the howlers, all "very
probable" infanticides were included.

periods (Sommer 1994). In another study with a large sample, Borries (1997) noted that all 35 observed male attacks targeted infants.

Speeding up return to ovarian cycling

In general, primate females do not undergo ovarian cycles and are therefore not sexually active until near the end of lactation (Altmann *et al.* 1978; van Schaik 2000). However, the death of a dependent infant due to infanticide often causes a return to female sexual receptivity. It is obviously in the interest of the female as well that this return is rapid; females usually return to receptivity within a month, but sometimes as fast as less than 2 weeks in red colobus (Struhsaker & Leland 1985), 5 days in mantled howlers (Clarke 1983), 4 days in hanuman langurs (Sommer 1994), and 3 days in red howlers (Agoramoorthy & Rudran 1995) and captive blue monkeys (Böer & Sommer 1992). (Interestingly, return to ovarian cycling seems to be fastest in the species subject to the highest rates of infanticide.) When healthy infants of healthy mothers are killed by males, one would therefore expect a shortening of the current interbirth interval.

A conservative test of this proposition is to compare the observed interbirth interval following infanticide with the mean interbirth interval in the population (excluding infanticides, but including infants death due to other causes: cf. Bartlett *et al.* 1993). The directly observed cases of Table 2.1 suggest that infants are killed well before weaning, and the results of most other studies are also consistent with this proposal. However, for a reliable estimate, we need studies with larger sample sizes. These studies yield a reduction of 26% in provisioned hanuman langurs (Sommer 1994; excluding "three questionable cases"), of 25% in wild hanuman langurs (Borries 1997), and of 32% in red howler monkeys (Crockett & Sekulic 1984). The similarity in the langur estimates is especially interesting, since mean interbirth intervals differ strongly between the two populations (16.7 vs. 29 months, respectively).

Because the gain in interbirth interval decreases as the age at death of the infant increases, one can calculate the age beyond which reproduction is no longer speeded up. For the provisioned Jodhpur hanuman langurs this is about 6 months (Sommer 1994); for those at Ramnagar, Nepal, this is about 16 months (Borries 1997); for the red howlers of Hato Masaguaral this age is about 9 months (Crockett & Sekulic 1984). As estimated by the authors, 67% and 88%, respectively, of infant langurs were killed below these critical ages, and 100% in the howlers. Thus the correspondence between vulnerability to infancide and effect on female interbirth interval, while not perfect (cf. Agoramoorthy & Rudran 1995), is nonetheless reasonably strong.

In highly seasonal breeders, infanticide might not produce a reduced interbirth interval. Infanticide by males has been observed in several strictly seasonal breeders, such as ringtailed lemurs (Pereira & Weiss 1991; Hood 1994), Milne-Edwards' sifakas (Wright 1995; Erhart & Overdorff 1998), and some populations of blue monkeys (M. Macleod, unpublished data) and hanuman langurs (Newton 1986; Borries 1997). The obvious explanation is that females do not give birth every year, as in the sifaka and the langurs (and in many other species not listed in Table 2.1; cf. Bartos & Madlafousek 1994). Among Ramnagar langurs, with an interbirth interval of almost 2.5 years, Borries (1997) showed that if an infant langur is killed in its first year, but too late for the female to conceive again, there is still a reduced interbirth interval. M. Macleod (unpublished data) reaches the same conclusion for blue monkeys. However, even when births take place each year, as in provisioned ringtailed lemurs, it is still possible that infanticide is favored by natural selection if the loss of an infant improves the survival prospects of the subsequent infant. This question needs more attention, because it is also possible that males continue to commit infanticide even during periods or in conditions (e.g., *ad libitum* food supply in captivity) when birth rates are so high that they would not derive benefits from doing so.

Subsequent sexual access to mother

In general, the perpetrating males remain members of the social group and are thus potential fathers of the females' subsequent offspring, and many studies actually report matings between the infanticidal male and the females whose infants they had killed (Table 2.1). For the 49 cases with relevant data, males gain certain, probable or possible mating access in 25, 8 and 5, or a total of 38 cases (78%; see also Palombit *et al.*, Chapter 6). The exceptions are of interest. In cases 5 and 17, the infanticidal male was evicted by a new male before the female returned to receptivity; in case 45 he was evicted by the females, and in cases 25 and 35 no mating was possible because the infant was an orphan or the female disappeared soon after the infanticide. In case 14, the male wedge-capped capuchin who killed the infant was a beta male, and unlikely to sire infants. However, when he committed the infanticide, the alpha male was ill, and the beta male was the most likely to have replaced him had he not recovered (Valderrama *et al.* 1990). Hence, most of the exceptions to the rule that the male gets mating access are nonetheless consistent with the males' assessment of their chances of future matings postulated by the sexual selection hypothesis.

However, in at least four cases, subsequent mating was unlikely because the infant belonged to another group and neither the male nor the female transferred soon after the infanticide. Although some examples of infanticide in other groups are consistent with the sexual selection hypothesis because they may induce females to transfer to the perpetrating male, there are many exceptions to this rule, and better understanding of infanticide in this context is needed (see Discussion).

To sum up, in at least 78% of cases the male did get mating access to the deceased infants' mothers, although this access was not always exclusive. Sommer (1994) estimated that male hanuman langurs gained mating access in about 96% of cases, although they were present during the subsequent conception, and presumably sired the infant in only about 76% of cases. However, in another population, Borries *et al.* (1999b) demonstrated that infanticidal males were the most likely sire of the next offspring.

Injuries

In some cases, animals are injured during the infanticidal attacks. If the male is injured, this might reduce the likelihood that he retains tenure of the group, and thus reduce or negate his reproductive advantage. In the survey, whenever detailed descriptions of the events were provided, but no mention was made of injuries, it was assumed that injuries did not occur. In only one case did the male sustain an injury, a 4 cm laceration on a foreleg in a red howler monkey (Izawa & Lozano 1991), which may have contributed to the subsequent immigration of another male. Thus, overall, the risk of injury to males is surprisingly small despite the sometimes vehement counterattacks to which they are subjected. This suggests that males carefully select victims and contexts in which risk of injury is minimal.

Female injury is relatively more common, both to the mothers and to other females defending the infants (examples: Galat-Luong & Galat 1979; Izawa & Lozano 1991; M. A. van Noordwijk & C. P. van Schaik, unpublished data). Most of these injuries were superficial. However, there were also serious injuries in gorillas and chimpanzees, and recovery from these injuries may increase the interbirth interval and thus negate the benefit to the male of killing the infant. The case of female injury in gorillas was unusual, in that the female with a newborn infant herself rushed to attack a strange silverback involved in a prolonged series of displays and charges with the resident silverback, and was struck by this male. The infant died within a day, the female went missing after a week and was presumed dead (Fossey 1984). Chimpanzee females attacked by individuals from other communities often sustain serious injuries, as well as losing their infants (Goodall 1986), but these attacks are not always made by males only and do not seem to fit the sexual selection hypothesis (see Discussion). Thus, apart from the chimpanzee cases, female injuries seem to be relatively uncommon and minor.

Remarkably, there seems to be no effect of size dimorphism on either incidence or injuries, since infanticide is also seen in a variety of species in which males are the same size as or even marginally smaller than females (especially lemuroids, but also some langurs; see Tables 2.1 and 2.2). This, again, suggests that males carefully select situations of low risk and females tend to avoid excessive risks during the defense. In langurs, for instance, the most vehement defenders of infants tend to be aging female relatives, for whom injuries would carry a less serious fitness cost.

Conclusion

The cases laid out in Table 2.1 generally support the sexual selection hypothesis. Although exceptions occur, they tend to be consistent with the decision-making that would under normal circumstances lead to adaptive outcomes, with the obvious exception of between-group infanticide in chimpanzees and perhaps other species. Perfect fit with expectations is not expected because animals have neither perfect information nor perfect decision rules, but on average a net benefit should apply. If we are willing to pool estimates and create an average primate, we have a reasonable sample size to estimate the various parameters of prediction (equation 2.1). Table 2.1 and the other data reviewed here suggest a mean time gain $((t_n - t_i)/t_n)$ of 0.25 (although time gain estimates also included suspected infanticides, they should not be systematically biased), a mean p of 0.02 to 0.04, a value of P less than 0.8 but probably exceeding 0.4, and a small, though unknown, cost due to injury (injury may result in loss of tenure, but this need not be permanent).

With these estimates, equation 2.1 suggests that the costs of injury may be quite appreciable before the benefit (between 0.1 and 0.2 infants per female to whom access is obtained) is outweighed by the risk of killing one's own offspring (0.02 to 0.04). These values make male infanticide highly advantageous for the perpetrators, and may also explain why males may respond to small infants that they certainly did not sire ($p = 0$) with violent attacks so long as there is some, if small, prospect of future mating with the infant's mother, provided they can limit the risk of injury. Moreover, with the exception of between-group infanticide (especially in chimpanzees: see below), the exceptions to the predictions do not fall into a distinct pattern but rather seem to reflect predictable imprecision in the male decision-making process of a low-cost behavior or unpredictable environmental events. On the whole, therefore, these data provide strong support for the hypothesis that males commit infanticide as a reproductive strategy. This conclusion is also supported by a wealth of studies of captive groups (e.g., Angst & Thommen 1977; Pallaud 1984; Böer & Sommer 1992; Kyes *et al.* 1995), and by detailed recent field studies (Borries & Koenig, Chapter 5; Palombit *et al.*, Chapter 6).

Other hypotheses

Social pathology

It has often been argued by critics that infanticide by males is not the result of any selected behavior patterns, but rather a consequence of human disturbance, either directly or because of unusually high densities (crowding), to which males respond with violence (summarized by Boggess 1979). This hypothesis has received extensive discussion (see various chapters in Hausfater & Hrdy 1984) and is now generally abandoned. First, infanticide in langurs, which sparked the debate, is not limited to situations of unusual crowding (Newton 1986). Likewise, no unusual crowding occurs in most other observed or inferred cases of infanticide in other species (Tables 2.1, 2.2). Second, a comparison of blue monkey populations by Butynski (1982, 1990) showed that infanticide rates were far higher in the low-densitiy population than in the high-density population, probably because intruder pressure and thus takeover rates were higher in the former.

This hypothesis also suggests that the males are psychologically disturbed and thus moved toward maladaptive behavior. This suggestion is incompatible with observations. First, males are often selective in their attacks (e.g., a red colobus male did not attack the vulnerable infant of his mother: Leland *et al.* 1984). Second, the infanticidal males later turn into strong protectors of the infants they themselves have sired (e.g., long-tailed macaques: van Noordwijk & van Schaik 1988; de Ruiter *et al.* 1994; drills: Böer & Sommer 1992; sooty mangabeys: Böer & Sommer 1992;

Busse & Gordon 1983; blue monkeys: Böer & Sommer 1992; Fairgrieve 1995). Third, the presence of elaborate evolved female and male counter-strategies (see below) is incompatible with the recent emergence of aberrant male behavior.

Some reports from captivity, however, suggest pathological killing. In captivity, hamadryas baboon males remarkably often kill their own offspring (Zuckerman 1932/1981; Angst & Thommen 1977; Rijksen 1981). Often, these males aim prolonged abusive behavior at the infants, rather than a single attack. These two features suggest that a special explanation is required, possibly relating to the uniquely derived ritualized herding and "punishing" behaviors of females. Spijkerman *et al.* (1990) reported a similar example for a captive male chimpanzee. Detailed analysis of such cases may lead to better insights into the decision rules used by the males.

A more subtle form of this hypothesis was proposed by Boggess (1984): the disturbance and crowding caused a shift in male change from gradual to sudden takeovers. It is a historic irony that, while the hypothesis has generally been discarded, this particular point has now been confirmed (*pace* Hrdy 1979). Sterck (1998) found that rates of infanticide in langurs are much higher in situations of serious human disturbance, and that the causal variable is variation in the takeover rate (see also Steenbeek, Chapter 7). Thus, while infanticide in all circumstances among langurs remains consistent with the sexual selection hypothesis, the rate is much increased due to human disturbance. However, this does not make infanticide pathological behavior.

Yet another form of the hypothesis, implied by Bartlett *et al.* (1993), is that infanticide by males is common only among langurs. It would therefore reflect a peculiarity of langur biology or the artificial ecological conditions in which the majority of langurs are found, rather than a general selective advantage for male reproduction. However, even by the conservative accounting of Table 2.1, infanticide has now been directly observed in wild populations of 17 species of non-human primate. If less conservative criteria are applied, allowing published credible reports of male infanticide in non-human primates, both directly observed and strongly indicated, many more species yield direct or indirect evidence of infanticide by males: 38 at the last count (Table 2.2). It is reported for four of the six major primate radiations (lemurs, platyrhines, cercopithecoids and apes; not in lorisids and tarsiers). Thus infanticide is not a taxonomically isolated peculiarity.

Infanticide as a by-product of male aggression

More recently, the social pathology hypothesis has been reformulated. The new version suggests that infanticide is an artifactual outcome of generally increased aggression rates and intensity accompanying the immigration of new males or instability in the ranks of resident adult males (Bartlett *et al.* 1993). Thus the attacks are thought to be aimed at the mothers rather than the infants, perhaps related to efforts to make the female mate with the male (Boggess 1984) or induce her to transfer (Fossey 1984). Hence the infants are killed inadvertently. Aggression by males aimed at females, probably in order to increase their mating access, has been documented in a variety of primate species (Smuts & Smuts 1993). Several studies do indeed show remarkably tenacious and vehement aggression by males toward females, in gorillas (Fossey 1984), hanuman langurs (Sugiyama 1965; Hrdy 1977; Ross 1993), and savanna baboons (Collins *et al.* 1984), although such behavior is not the rule even in these species. Hence, it is possible that at least some cases of infanticide are a by-product of aggression aimed at females.

Nonetheless, the by-product hypothesis is not a viable alternative for most observations of infanticide by males. First, it would not predict that females with infants are attacked any more heavily than females without infants; indeed, the opposite could be argued because females without infants are more likely to be receptive. However, although Ross (1993) reported that the infanticidal male also harassed a female without an infant, the great majority of male attacks is aimed at females with infants. Second, in most cases reviewed here, the attacks on infants are often highly directed and not preceded by episodes of intense aggression. Indeed, often there was no aggression at all before the sudden and seemingly unprovoked attack by the male (e.g., mangabeys, Busse & Gordon 1983; sifaka, Erhart & Overdorff 1998; langurs, Sommer 1994, Borries 1997; gelada, Moos *et al.* 1985, Mori *et al.* 1997; baboons, Palombit *et al.*, Chapter 6). Third, attacks on lone infants are not uncommon (e.g., vervet monkeys, Cheney *et al.* 1988; ringtailed lemurs, Hood 1994; hanuman langurs, Borries 1997). Fourth, in many cases the attacks were unprovoked and during periods of low agonism. Some reports emphasize that males immediately stop attacking the female after the infant has been eliminated (e.g., Butynski 1982; but see Fossey 1984), thus supporting the notion that it is the infant, not the female, that arouses the male's aggression. Indeed, in some cases, aggression tends to be higher following the infanticide when mothers who lost their infants, and sometimes their female kin, continue to attack the perpetrator (e.g., Wright 1995; Erhart & Overdorff 1998; D. Onderdonk, unpublished data). Finally, even in those cases where the aggression seems to be directed mainly at the females and even continues after the infant's demise, the consequences remain the same: the female's reproduction is speeded up and the male gains sexual access. Thus natural selection would still reward this form of male behavior.

Another version of this alternative simply states that the general mayhem accompanying male takeovers or escalated male conflicts for dominance in a group will disproportionately injure or kill the most vulnerable group members: infants (Bartlett *et al.* 1993; Sussman *et al.* 1995). The observations listed above are also incompatible with this version, but another prediction can be developed. If general aggressiveness were the cause of infanticide, the fairly exclusive focus on infants would not be expected. Being smaller than juveniles, infants may seem to be more easily killed, but they are also more closely guarded by their mothers, whereas juveniles tend to be moving around independently, often associating most with other juveniles. Yet, in hanuman langurs and red howler monkeys, risk is inversely related to age of the infant (Fig. 2.1), and thus with its tendency to move about. Indeed, in Sommer's (1994) study, only two cases of attacks on juveniles have been reported, and these took place 1 year after the takeover, well after the male had stopped attacking infants.

In contrast to the by-product hypothesis, the sexual selection hypothesis assumes that aggression and infanticide are separate behavioral systems. Perhaps the best evidence for this comes from studies of mice, where exposure to testosterone *in utero* appears to have a sensitizing effect on the neural areas mediating male–male aggression, but a desensitizing effect on the neural areas mediating infanticide (vom Saal 1984: 419). This suggests that, although some infanticide may in fact be by-products of aggressive thrashing about, the kind of goal-directed infanticidal behavior often described is best understood as a motivational system separate from general aggressiveness.

Exploitation: cannibalism

Although exploitation of the infant as food is not often mentioned as a possible benefit for the infanticidal males in primates, there are a few cases in which the infant is actually consumed by the male after the killing: in baboons (Saayman 1971; Collins *et al.* 1984; Palombit *et al.*,

Chapter 6), redtailed monkeys (Leland *et al.* 1984), blue monkeys (Fairgrieve 1995; M. Macleod, unpublished data), and especially chimpanzees (Hamai *et al.* 1992). Hence, exploitation benefits do not explain infanticide by males in the majority of species.

In infant-caching nocturnal prosimians, both males and females, including mothers, sometimes kill the relatively altricial infants, especially newborns, and cannibalism is not uncommon (dwarf lemurs, Stanger *et al.* 1995; tarsiers, Roberts 1994; and especially galagos, Hiraiwa-Hasegawa 1992). Since these species are often carnivorous, this may be true cannibalism. However, since most observations concern captive animals, and mothers may be provoked to eat their young by "stress" (Izard & Simons 1986), it is difficult to provide a functional interpretation for this behavior at this stage. Note that there is no evidence for sexually selected infanticide in nocturnal prosimians.

The question is whether, in the cannibalistic species, exploitation provides a better explanation than reproductive benefits. This is unlikely for most species. First, adult females should also cannibalize, but with the exception of gorillas (one case; Fossey 1984) and chimpanzees (three cases; Goodall 1977, 1986) females have not been observed to consume infants (in clear contrast to many rodents, see Blumstein, Chapter 8). Second, juveniles are neither killed nor cannibalized. Third, not all infants are cannibalized by the males: it is very rare in baboons and not universal in blue monkeys and chimpanzees. Finally, few of the primate species hunting vertebrate prey engage in cannibalism (Hiraiwa-Hasegawa 1992). In conclusion, then, because in the species with cannibalism males still tend to kill unrelated infants, and reap the reproductive benefits of infanticide, such as earlier return to receptivity, cannibalism provides an additional rather than an alternative benefit for infanticide by males.

Nonetheless, the puzzling cases observed among chimpanzees deserve further discussion (see below), because infants were treated as prey (especially in the instances of within-group killing), the social scenes (especially in the instances of between-group killing), including begging and food sharing, resembled those seen following hunts of allospecific prey, and this kind of killing peaked during hunting seasons (Hamai *et al.* 1992; cf. Hiraiwa-Hasegawa 1992).

Resource competition
Rudran (1973) and Agoramoorthy & Rudran (1995) have suggested that infanticide will, on average, result in increased access to limiting

resources, especially food, for the killer and his descendants. Many of its predictions overlap with those of the sexual selection hypothesis, but it also makes separate ones. First, juveniles should be killed as well, perhaps even preferentially, since they consume more resources (Hrdy 1979) and can be easily killed. Second, females should be expected to commit infanticide of unrelated infants more often (cf. Hrdy 1979; Sherman 1981; Digby, Chapter 17). Clearly, both predictions are not met in the studies reviewed above. (A third prediction is that the infants sired by the infanticidal male should have improved growth and survival due to the elimination of competitors. Analysis by C. M. Crockett & C. H. Janson (unpublished data) of the same red howler population as studied by Agoramoorthy & Rudran (1995) did not support this prediction.) Finally, Butynski (1982) compared infanticide between two populations of blue monkeys, and found that it was far more common in the population living at lower density in the richer habitat. Thus resource competition is unlikely to provide a widely applicable explanation for infanticide by males.

Adoption avoidance

One benefit of male infanticide not considered so far for primates is the avoidance of misdirected male parental care (Sherman 1981; Elwood & Ostermeyer 1984). Infanticide by males is expected to be more likely where males show extensive parental care, even in the absence of effects on female reproduction. Among primates, misdirected paternal care is not likely to lead to infanticide, because, paradoxically, in those species with the most extensive male care for infants, the callitrichids, paternity is a poor predictor of the amount of male care (Smuts & Gubernick 1992; van Schaik & Paul 1996–7). Moreover, those are the very species in which the tendency toward sexually selected infanticide is absent because of the relatively short lactation of the females and the incidence of postpartum mating (see van Schaik, Chapter 3).

Elimination of future rivals

In hanuman langurs and chimpanzees, male infants are killed preferentially. If it can be shown that this bias is not due to sex differences in the infants' behavior, a separate explanation is required. It is possible that infanticide may (additionally) serve to eliminate future competitors for access to mates for the resident male(s) or for his/their offspring (Hiraiwa-Hasegawa & Hasegawa 1994).

Langur males taking over groups show a significant bias toward killing male infants (Sommer 1994). In groups of various langur species, young males maturing in the groups join the males in loud calling and in defense against neighboring groups (e.g., Thomas's langur: Steenbeek, Chapter 7). If these maturing males were unrelated, however, their agonistic initiatives might at one point be directed against the group's resident adult male. It is therefore not surprising that among langurs, when the group's resident male leaves, the juvenile males all go with him, whereas the juvenile females tend to stay (e.g., Hrdy 1977; Steenbeek 1999b). Langur males also direct severe aggression at any non-related juvenile males that happen to have remained in the group after takeover (Sommer 1994). Male mantled howlers may also preferentially kill male infants (Clarke 1983); their social system also consists of one-male groups with takeovers.

Among chimpanzees, infanticide by males occurs in two different contexts: between groups and within groups (Goodall 1977, 1986; Hamai *et al.* 1992), and may have different functions. Killing infants within groups is consistent with sexual selection. Critics have argued that the lack of social instability among the males argued against this hypothesis. However, the facts are consistent with all the elements of the sexual selection hypothesis (see Hamai *et al.* 1992). The victims' mothers tend to be primiparous females who mated little if at all with the top-ranking males in the community during their conception period, but instead mated with adolescent or low-ranking ones. The killers were parties of top-ranking males who also gained subsequent sexual access to the females. Loss of infant is followed by strongly reduced interbirth intervals (Goodall 1986; Hamai *et al.* 1992). Unusual elements, however, were that all the infants killed were eaten and that there were always multiple males present.

The infanticides in between-group contexts among chimpanzees are probably not consistent with the sexual selection hypothesis. First, several adult males (sometimes accompanied by females without infants) viciously attacked females with infants; the attacks were clearly aimed at the females (who sustained serious injuries) and not their infants, some of whom escaped death (Goodall 1977, 1986). Second, in none of the cases observed did these females subsequently transfer. However, if many females have infants sired by outside males (cf. Gagneux *et al.* 1997), it is possible that the attacks made these females more willing to mate with the attackers in the future.

In both contexts, the infants killed were almost exclusively males

(Hamai *et al.* 1992; Hiraiwa-Hasegawa & Hasegawa 1994). Since the attacks are by large parties on much smaller parties or individuals with dependent offspring, little risk is involved. Killing the male infants of a neighboring community may strengthen the position of the killers' sons in the balance of power with the neighboring community, and may even allow expansion of their territory (Hiraiwa-Hasegawa & Hasegawa 1994). The recently immigrated peripheral females who tend to inhabit the overlap areas (Hamai *et al.* 1992) are also likely to have infants sired by males from neighboring communities, or by adolescent males inside the community. Male infants sired by males from outside the community may be less valuable as future allies when adult, either because relationships with unrelated males are less supportive (cf. Struhsaker & Leland 1985; Pope 1990) or because choosing closer relatives as allies brings greater inclusive fitness returns (male community members tend to be relatives: Morin *et al.* 1994). Consistent with this hypothesis is that immigration of juvenile males is quite rare, but that the two male juvenile immigrants observed in Mahale both ended up getting killed by resident males (Hamai *et al.* 1992). The two different contexts of infanticide by male chimpanzees might thus both serve this function of elimination of unrelated males, which is most likely to be adaptive in a patrilineal society. However, more observations and paternity estimates are needed before this hypothesis can be considered as supported.

Conclusion

The cases reviewed here are largely or entirely consistent with the sexual selection hypothesis. The social pathology or resource competition hypotheses do not explain the majority of cases of infanticide in primates. The by-product hypothesis may explain some cases not consistent with the sexual selection hypothesis but is not a general explanation for the phenomenon. Exploitation or cannibalism may provide an added incentive in chimpanzees, and elimination of future rivals may produce additional benefits in chimpanzees and langurs. Only the sexual selection hypothesis can explain the majority of directly observed cases among wild primates, many examples of infanticide by males in captive settings, the taxonomic distribution among both primates and non-primates of infanticide by males in relation to female reproductive life history, and the patterns of infanticide by males in non-primate mammals: lions (Pusey & Packer 1994) and rodents (Brooks 1984). Hence, in the rest of this book it is assumed that the taxonomically most widespread benefits of infanticide by mammalian males, and the most common benefits within species, are male reproductive benefits. In the rest of this chapter, I examine selected aspects of the sexual selection hypothesis in more detail.

Defenders of infants

Not all infants are killed in situations of acute infanticide risk. In red howlers, of 60 infants in the vulnerable age-bracket during takeovers or male rank reversals, 41 (68%) survived, 7 (12%) with injuries (Crockett & Sekulic 1984; their Tables II–IV). In mantled howlers, 60% of infants (N = 15) survived takeover periods (Clarke 1983). In the hanuman langurs of Jodhpur, of the 112 infants present during male takeovers or born soon after, 71 (63%) survived, of which 14 (13%) had injuries (Sommer 1994: 159); in those of Ramnagar, 37 infants were present during male immigrations, of which 29 survived (78%), but 10 (27%) sustained injuries (Borries 1997). The high survival is due in part to defense by mothers, other females, and resident males. For this section, we draw not only on the cases assembled in Table 2.1 but on all relevant observations reported in the literature.

Defense by females
There are three times when females could attack infanticidal males: when the male appears and can be judged to be infanticidal, when the male attacks the infant, and after the male has attacked and perhaps killed the infant.

Reports of females attacking males before they have done any harm to infants are relatively rare, because in most cases this role is played by the group's male(s). However, langur females occasionally attack intruding males (Sugiyama 1966; Mohnot 1971; Steenbeek 1996). In some species, females may have control over whether strange males can immigrate into the group by attacking them, often in coalitions (Fairbanks & McGuire 1987; Smuts 1987b; Starin 1994).

Attacks on infants or mother–infant pairs by infanticidal males almost invariably provoke vehement counterattacks by females, alone or in coalitions, in virtually all studies reviewed here. Borries (1997), for instance, reported that adult females (usually both mothers and others) defended attacked infants in 88% of the observed attacks. Persistent female counterattacks may serve merely to delay infanticide (hanuman langurs, Mohnot 1971; Hrdy 1977), but the delay may be long enough for the male to be ousted by another male (blue monkeys, Butynski 1982), for the females to evict the male (after 2 months, Milne-Edwards' sifaka, Wright 1995), or for the infants to have grown out of the vulnerable period (Steenbeek 1996). Sometimes, the female counterattacks continue after the death of the infant (e.g., Zunino *et al.* 1986; Valderrama *et al.* 1990), but this might have given temporarily stunned or immobilized infants enough time to recover. This is why so many infants survive the vulnerable period, even though they may get injured. That defense by females is not more effective is because it is a very unequal contest, where the would-be infanticidal male often is seen to follow the mother–infant pair for periods of hours, days or even weeks (e.g., Butynski 1982; Wright 1995), and can simply wait for a brief moment of female inattention to attack the infant.

In some other cases, however, females do not seem to put up an effective fight. Clarke (1983) reported that a female mantled howler simply abandoned her infant, and ignored its distress calls, following a male takeover, and mated with the new male; the infant soon died. Böer & Sommer (1992) report that a captive female blue monkey, after 10 minutes of intensive chasing, pushed her infant toward the attacking male (see also Steenbeek, Chapter 7). Females may also not receive any support from female relatives (Fairgrieve 1995). It is possible that, in such cases, females assess their risk of injury as being very high. Thus, Starin (1994) noted that female coalitions (with a male) were effective in one population of red colobus without sexual dimorphism, whereas only males defended infants in a highly dimorphic population. Unfortunately, insufficient quantitative data are available to test the impact of these factors.

The prevalent response of vulnerable females to the presence of potentially infanticidal males is avoidance. Groups of females that lost their male avoided contact with other groups or males, thus delaying immigration of new males, sometimes successfully (Steenbeek 1996). After male immigration, mothers with infants often stayed away from new males (e.g., hanuman langurs, Sommer 1994; baboons, Collins *et al.* 1984). In captive vervets, mothers of young infants restrained their infants more when strange males were present (Fairbanks & McGuire 1987). Likewise, a few studies report that females with vulnerable infants avoid getting involved in between-group conflicts (red howlers, Sekulic 1983b; gibbons, Mitani 1987; red colobus, Starin 1994). Occasionally, this avoidance goes so far as to become emigration from the group (red colobus, Marsh 1979b; Thomas's langurs, Sterck 1997).

Defense by males

It was noted above that the disappearance or incapacitation of the likely sire is the most common context for infanticide, suggesting that these males play a major role in infant defense, directly by active defense of the infants when under attack, or indirectly by preventing immigration or takeover by other males. This role of males has been confirmed experimentally, thus eliminating confounding factors, in the field (Sugiyama 1966; Kummer *et al.* 1974; Galat-Luong & Galat 1979), and on numerous occasions in captivity (e.g., Angst & Thommen 1977; Gomendio & Colmenares 1989). Most studies also report that possible and likely sires, if they are still present, actively defend the infants against infanticidal attacks (though not always successfully), alone or in coalition (see also Palombit *et al.*, Chapter 6). Borries (1997) quantified this, reporting that adult males, always putative sires, defended infants in 72% of the known attacks, and seemed to do so more effectively than females.

The male role in defense is not surprising. In most cases, likely sires have a higher average relatedness to the infants than female relatives of the mother, they have more favorable time budgets, and can therefore spend more time vigilant for infanticidal males (e.g., Rose & Fedigan 1995; cf. Butynski 1982), and their greater ability to inflict injuries on the attacking males may make infanticidal attacks too costly when a healthy and powerful male is associating with the female. Moreover, since males reach top rank in multimale groups or tenure in one-male groups only once in their lifetime (cf. Rajpurohit & Mohnot 1988; van Noordwijk & van Schaik 1988), selection would favor their taking greater risks in defending their infants (provided their likelihood of paternity is high enough) than the challengers, whose reproductive career lies largely ahead and would thus incur serious fitness reductions from sustaining debilitating injuries.

Females may actively support the protector male and thus discourage takeovers or immigrations. Females with vulnerable infants support the likely sire against the intruder or challenger, whereas they do not support, or even affiliate with, the new males, when they are non-lactating or pregnant (longtailed macaques, M. A. van Noordwijk & C. P. van Schaik, unpublished data; red howlers, Sekulic 1983b). When immigration is not prevented, in a few species, lactating or highly pregnant infants leave with the former resident male (langurs, Sugiyama 1967, Rudran 1973, Sommer 1994; gorillas, Watts 1989; see also Sterck & Korstjens, Chapter 13).

There are, nonetheless, a few puzzling cases of lack of defense by males, in red howlers (Izawa & Lozano 1991) and sifaka (Wright 1995). In these cases, it is possible (though not likely) that the males were not the infant's sire.

Male decision rules

The predictions developed in the first section suggest that males need to estimate their chances of paternity (P and p), the size of the infant (and thus their time gain) and the risks involved in infanticidal attacks. Here I review what mechanisms males may use to arrive at their decisions.

Paternity

Male mammals cannot recognize infants as kin (König 1989; Elwood & Kennedy 1994), and must therefore base assessment of the likelihood of paternity entirely on mating history (but see Treves, Chapter 10). No deliberate experiments have been done with primates, as opposed to rodents (vom Saal 1984; Perrigo & vom Saal 1994). In rodents, there is evidence for three rules: the insufficiency of mere association, the necessity of mating, and the disturbing effect of separation during pregnancy. In mice, males who have ejaculated with the female and hence are most likely to be fathers of the infants they encounter are inhibited from killing them. The inhibition gradually gets stronger after mating, peaks at roughly one mouse gestation period after mating (three weeks), and disappears again at what would normally be the time of weaning. The extraordinary shift from paternal or benevolent to infanticidal responses to infants is apparently calibrated to photoperiodic cues (number of light/dark cycles experienced) rather than absolute time elapsed since ejaculation (Perrigio & vom Saal 1994: 366). In gerbils and mice, separation of the male and the female during gestation makes males far more likely to commit infanticide on their own offspring (Elwood & Kennedy 1994). This response is adaptive if separation implies that females have mated with other males. However, not only is there variation in details of these mechanisms among species, as expected, but there is even variation within species (Gubernick *et al.* 1994).

In primates, we must rely on observational data to provide circumstantial evidence for these rules. The numerous examples of males killing infants of females with whom they had long been familiar, for example because they are in their natal group or immigrated months or years earlier, demonstrates that mere familiarity is not a decision rule of the male. Mating history plays a decisive role. For instance, when resident males take over dominance in howler groups, they attack infants, because they never mate with the females and sire offspring unless dominant (Crockett & Sekulic 1984; Pope 1990). Merely having mated, however, is not enough, and both the relative frequency of mating and its timing probably play a role. Hasegawa (1989) listed several cases in chimpanzees where a male was known to have mated with a female but yet killed her infant. In none of these cases, however, was the male known to be the most likely father who mated during the presumed period of conception. Thus "experience of copulations cannot restrain the male from killing his partner's infant" (Hasegawa 1989: 101).

As to timing relative to conception, male primates do not seem to have the clocks that some male rodents do, which opens the possibility for females to manipulate paternity estimates by males by facultative sexual activity during pregnancy (Hrdy 1979; Hrdy & Whitten 1987; van Schaik *et al.* 1999). However, the effectiveness of these matings in preventing infanticidal attacks decreases as pregnancy progresses (probably because females become less attractive, although a real clock mechanism cannot be excluded). Studies with reasonable sample sizes or infants born late after takeover suggest that matings during pregnancy are effective in preventing infanticide up to about 80 days before parturition in red colobus (Struhsaker & Leland 1985), about 60 days in hanuman langurs (Sommer 1994), about 45 days in blue monkeys (Fairgrieve 1995), and about 40 days in sooty mangabeys (Busse & Gordon 1983). Obviously, experiments would provide greater control and thus permit more refined estimates, especially since some observations suggest that matings even rather early in pregnancy are not effective in preventing infanticidal attacks (Borries *et al.* 1999a).

The effect of separation is probably also found in primates. Separation may account for two cases where the male possibly killed his own offspring because they were not with the female for much of the gestation and birth (red howlers, Crocket & Sekulic 1984; chimpanzees, Kawanaka 1981). There are many examples from captivity. Gust *et al.* (1995) reported that a dominant male sooty mangabey who had been absent from his group during a few months, attacked an infant born during his absence from the group (even though it was ascertained that he had sired it). Bingham & Hahn (1974) noted that a female gorilla, removed from her group 3 weeks before giving birth, was attacked by the likely sire when she was reintroduced 5 days after birth. Eisenberg (1981) reported on infanticide attempts by male brown spider monkeys and white-handed gibbons on their own offspring, after they had been separated during part of the gestation. In many solitary mammals, where separation is frequent, females aggressively keep their mates at a distance some time before parturition (gerbils, Elwood & Ostermeyer 1984; tarsiers, Roberts 1994).

Other parameters

In addition to paternity estimates, males need more information before they can make adaptive decisions. First, future mating prospects must play a role. In the great majority of cases, males attacking infants are newly dominant males or males about to become dominant (and thus already likely to be confident of future paternity). The major causal role played by testosterone in this respect has been demonstrated in rodents, along with various other experiential and genetic factors (Svare *et al.* 1984; vom Saal 1984). Second, the age of the infant is a factor. In many species, young infants have a natal coat, and after they lose this coat may have a lower stimulus value in inciting male attacks. This, of course, raises the question of why such natal coats exist at all (see Treves, Chapter 10). Finally, males must evaluate the immediate risks. Although there are virtually no data on risk assessment, some evaluation must take place. Some of the more bizarre cases of infanticide involve situations of low risk because the infant was unguarded by either the mother or males.

Discussion: remaining problems

Although the sexual selection hypothesis currently provides the most adequate explanation for most cases of infanticide by male primates, some questions remain. These are discussed below.

Why infanticide between groups?

The killing of infants belonging to other groups is not easily explained by the sexual selection hypothesis. The infants will generally be unrelated to the perpetrator, and the female's receptivity will be advanced. However, for infanticide to be adaptive, the male should have good prospects of gaining sexual access to her. There are various possibilities. First, between-group fertilizations may be common. For instance, about half the infants in one chimpanzee community were sired by extra-group males (Gagneux *et al.* 1997). Second, in gorillas and Thomas's langurs, between-group infanticide may considerably increase the probability that the infant's mother will subsequently transfer into the perpetrator's group (Fossey 1984; Watts 1989; Sterck 1997; Steenbeek, Chapter 7). Females may thus use a male's ability to kill infants as an honest indicator of male quality, since this also predicts his ability to prevent such attacks from the outside. Third, infanticide may increase the probability that the perpetrator can immigrate into the female's group. Perhaps a male who has lost one or more infants in his own group is more easily ousted by an invading male because his incentives for staying are reduced. This scenario is plausible if a male reaches the end of his tenure and the benefits of staying on are mainly those of protecting the current infants rather than siring new ones, but there is no evidence for it. Finally, and least likely, the male may benefit in his own group, because of strong female mating preferences (Hood 1994). Females may choose mates on their ability to defend infants, and, if the ability to protect an infant is correlated with that to kill one, the chances of being preferred by females rise for infanticidal males.

There is virtually no evidence to evaluate these possibilities, and many observations are needed to distinguish each from the simpler alternative that males with a high motivation to kill infants in their own group respond to any infant in the vulnerable age category and in a situation enabling risk-free successful attack act in the same way, by killing it. For instance, the case in ringtailed lemurs reported by Hood (1994) refers to an infant left behind when his group was chased away during a violent between-group conflict.

The genetics of infanticide by males

Although the criticism that infanticide by males is too rare to produce substantial selective benefits (Sussman *et al.* 1995) is refuted by its common occurrence in the appropriate conditions, questions as to its genetic basis and its universality in the species with infanticide by males remain. These questions are difficult to answer without experiments, especially since the behavior is clearly conditional. However, such evidence is available for rodents. In mice, the existence of strain differences in infanticidal tendencies strongly supports the assumption that genes are involved (Svare *et al.* 1984; Perrigo & vom Saal 1994). But note that a genetic basis does not imply absence of environmental influences. In male mice, infanticide is clearly conditional, influenced by social status (e.g., dominant versus subordinate), reproductive status (particularly whether or not the male has mated during the past 3 weeks), and seemingly random developmental factors (such as intrauterine position).

It is possible that, as in laboratory mice, the tendency to attempt infanticide in the appropriate social conditions is not universal in the primate species in which infanticide by males occurs. However, infanticide is probably the evolutionarily stable strategy (ESS) for males. Calculations by Chapman & Hausfater (1979; Hausfater 1984) suggested that infanticide can always invade a population where males are non-infanticidal in the social and demographic conditions shown by langurs. However, these calculations also suggested that infanticide need not spread to fixation, because at certain tenure lengths, when an infanticidal male is followed by another infanticidal male his own offspring will not yet be weaned, and will thus be killed, canceling any reproductive advantage of the infanticidal tendency. But this result is found because fixed mean tenure lengths are used in the calculations. If we assume that tenure lengths will have high variance, then the average reproductive success for infanticidal males will be higher. Hence, male infanticide is probably the ESS and will be a chronic threat to the females.

In the absence of experiments, no confirmation of this result is possible, but the observational evidence suggests that the infanticidal tendency is very common, if not universal. First, if some males kill no offspring at all following takeovers, there are plausible explanations for these exceptions (e.g., earlier residence or presence of close relatives, Sommer 1994). Second, infanticide is very reliably provoked experimentally even in species where it is rare in the wild (e.g., longtailed macaques, Angst & Thommen 1977, Pallaud 1984; rhesus, Carpenter 1942, Angst & Thommen 1977; pig-tailed macaques, Kyes et al. 1995; hamadryas, Gomendio & Colmenares 1989).

The rarity of infanticide in some species raises the question of how natural selection can maintain a tendency if it is so rarely expressed. One possibility is that we may underestimate the rate of infanticide in these species. Certainly, in most of the species listed in Tables 2.1 and 2.2, the average male will find himself in a situation in which infanticide is an option at least once in his lifetime. Perhaps the demographic conditions in which we study these species (e.g., the extremely large free-ranging provisioned groups of several macaques) suppress the rates of infanticide relative to the levels in the wild (cf. Kawai 1963; Tokida 1976). Another possibility is that, even in the species with a very low rate, the rate may be high enough. Other traits, such as resistance against specific diseases, are also selected for only sporadically and maintained by these occasional spikes of strong selection despite long periods of weak or no positive selection (see Endler 1986).

Should females resist?

By the same token, females should often enough encounter situations in which their infant is at risk of infanticidal attacks to provide powerful selective pressures for counterstrategies. It has been suggested that infanticide is prevalent because males benefit more from committing infanticide than females would lose from being subjected to it (Hausfater 1984; Hiraiwa-Hasegawa 1988). This conclusion ignores that the outcome of sexual conflict not only depends on the selective advantages of the different players but is also strongly influenced by power. With respect to infanticide there is a distinct power asymmetry, with males being stronger, not encumbered by the infant, and able to bide their time until the opportunity for relatively risk-free attack arises.

It has also been suggested that females would benefit from losing their infants because their sons would inherit the advantageous infanticidal trait that would enhance their fitness (Itô 1987, cited in Yamamura et al. 1990). First, this is only true when the trait is still rare and would thus only explain its initial spread and not its maintenance once it is common. Second, in models this benefit arises only if one assumes unrealistically strong linkage disequilibrium between the infanticidal trait expressed in males and the lack of resistance expressed in females and a very small effect of infanticide on female fitness (Yamamura et al. 1990). Intuitively, individual females are expected to benefit most if they mate with such males but not lose their infants to them.

The obvious conclusion is that, in slowly reproducing taxa such as primates, the death of an infant represents such a large proportion of a female's lifetime reproductive output that one would expect considerable selective advantages for effective female counterstrategies, provided their costs are not prohibitive (Hausfater 1984). Accordingly, much of this book is devoted to exploring the pervasiveness of these counterstrategies, which may involve immediate social behavior such as active defense or emigration, association patterns with protective males or females, parental decisions such as premature weaning or resorption and abortion of embryos, as well as reproductive physiology that serve to modify mating behavior during ovarian cycles or at times of infertility.

Female counterstrategies could be also be punitive (Hrdy 1979; Hausfater 1984), i.e., females could easily eliminate infanticide by simply refusing or postponing mating with infanticidal males, the so-called penalty strategy. This is unlikely to be an ESS because (1) postponing mating with infanticidal males in a one-male setting is costly to females and will be the ESS only under highly unusual demographic conditions (Hausfater 1984); and (2) once infanticide is present, selection of non-infanticidal males as mates in either one-male or multimale groups would usually mean selection of less effective infant protectors. Thus only non-punitive counterstrategies can realistically be expected and they will be explored elsewhere in this book (cf. Hrdy 1979; Smuts & Smuts 1993; van Schaik 1996; Ebensperger 1998a).

Acknowledgments

The review presented here was started while the author was a visitor at the Deutsches Primatenzentrum as Preisträger of the Alexander von Humboldt foundation. I thank Drs J. Ganzhorn, P. Kappeler and H.-J. Kuhn for making this stay possible; Miya Hamai for references of accounts of infanticide in Japanese macaques; Daphne Onderdonk, Minna Hsu and Mairi Macleod for sharing unpublished manuscripts; Carola Borries, Peter Kappeler, Andreas Koenig, Maria van Noordwijk, Signe Preuschoft and Volker Sommer for valuable discussion; and Charlie Janson for helping to formulate the predictions.

3

Vulnerability to infanticide by males: patterns among mammals

Introduction

The sexual selection hypothesis for the selective advantages of infanticide by males requires that certain conditions be met. Provided the male is able to locate the infant, in order to derive reproductive benefits from infanticide he must be able to kill it with limited costs, the female must resume ovarian cycling earlier or produce more offspring than she would do otherwise, and he must gain mating access to the female when she resumes cycling (Hrdy 1979; van Schaik, Chapter 2). Whether or not these conditions are met depends on life style and life history. Life style variables are the location of the infants relative to the female, the presence or absence of hiding places for infants, and the degree of predictability of female spatial position in territories. Life history variables include the degree of infant precociality, and hence their ability to escape from attacking males, and the speed of female reproduction, i.e., their ability to be pregnant and lactating at the same time.

In this chapter, I examine whether infanticide by males is concentrated in species with the expected female life history. Infanticide by males is most advantageous where lactation is long relative to gestation. In such species, postpartum mating and early pregnancy are impossible because this would produce two sets of young of different ages, different needs and different competitive power for access to milk. It is therefore expected that where lactation (L) lasts longer than gestation (G), females are forced to have lactational amenorrhea, which forecloses the option of postpartum mating and thus makes them much more vulnerable to infanticide by males (for details, see van Schaik 2000). In accordance with this expectation, there is a tight relationship between L/G and the

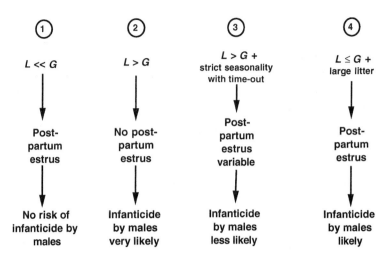

Figure 3.1. The life history hypothesis for infanticide by males. Case 1 (presumed ancestral state): short lactation (*L*) and long gestation (*G*); no vulnerability to infanticide by males. Case 2: long lactation and short gestation; vulnerability to infanticide. Case 3: despite relatively long lactation, low infanticide risk because of seasonal breeding and reproductive dormancy (provided that breeding effort in 1 year does not affect effort in subsequent year). Case 4: despite relatively short lactation, infanticide risk is high because size of current litter negatively affects size of subsequent litter.

presence of postpartum mating: where gestation lasts longer than lactation most species of mammals have postpartum amenorrhea. This is indeed found among mammals (van Schaik 2000). Animals with $L>G$ will tend to be at risk of infanticide by males.

Figure 3.1 illustrates the logic of this argument. Case 1 shows the ancestral situation. Infanticide by males is unlikely in species in which females have a fast life history and undergo postpartum estrus. Thus the female can combine being pregnant and caring for the current dependent offspring. Case 2 is where lactation exceeds gestation and vulnerability to infanticide is high. Figure 3.1 also illustrates two complications. Case 3 depicts the situation where there is only a single breeding event in a year, and this event is seasonal, and where females give birth each year to the same-sized litter regardless of the breeding effort in the previous year. In such a case, postpartum estrus may or may not occur, depending on gestation length and the presence of delayed implantation, but infanticide is not expected, since there is no reproductive advantage to the male. Note, however, that in many cases of seasonal breeding, there may still be an advantage, if more breeding cycles can be completed within a breeding season, or if breeding effort in the current year affects the effort in the subsequent year. Case 4 generalizes this latter problem, showing that where litters are large, even if lactation is relatively short, loss of part or all of the litter will increase the size of the subsequent litter, or even speed up the gestation of the next litter (e.g., Elwood & Ostermeyer 1984). Thus, while animals with $L/G < 1$ will generally be at far lower risk, large litter size may undo some or all of this advantage. In the broad-sweep test that follows, I am forced to ignore most of the complications of cases 3 and 4. The predictions are tested at different taxonomic levels.

Vulnerability: variation across mammals

Lactation/gestation ratios

One way to test this prediction is to examine patterns across mammals. Although a thorough survey of the taxonomic distribution of infanticide by male mammals has yet to be undertaken (for an attempt, cf. Ebensperger 1998a), and although observation biases are likely to affect the available data, the known cases of male infanticide are clearly concentrated in a few mammalian orders: primates (van Schaik, Chapter 2), carnivores (Packer & Pusey 1984; Ebensperger 1998a) and rodents (Brooks 1984; Hoogland 1995; Ebensperger 1998a; Blumstein, Chapter 8). Outside these three orders, there are scattered records (see below), although infanticide may be common in some taxa. Thus, in bottle-nosed dolphins, recent observations of stranded infants strongly indicate infanticide, most likely by males (Patterson *et al.* 1998). It is suspected that infanticide occurs in many other dolphins and porpoises.

For this analysis, I used a dataset containing 587 species of eutherian mammals. Initially, data on gestation and lactation were taken from the ecological and life history literature (sources listed in van Schaik 2000). Subsequent entries were obtained from the much larger compilation of Hayssen *et al.* (1993). The L/G ratios have limited accuracy and reliability. Gestation is defined here as the interval from implantation to birth. Variation in gestation estimates may be due to variation in litter size and ecological conditions. Bias in gestation may arise due to unrecognized and often variable delayed implantation; the length of the delay often depends on litter size (e.g., Daniel 1970). Hence, whenever available, estimates corrected for delayed implantation were used[1]. Lactation can be highly variable depending on ecological conditions, and since weaning

---

1. Species with obligatory and fixed delayed implantation are overwhelmingly seasonal, and tend not to be vulnerable to infanticide (e.g., Mead 1989; Hayssen *et al.* 1993).

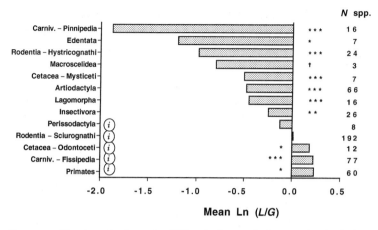

Figure 3.2. Mean lactation/gestation (L/G) ratios for taxa (orders or suborders) of non-volant mammals (data from van Schaik 2000, supplemented with those of Hayssen *et al.* 1993). Number of species in sample indicated by N spp.; taxa in which deliberate infanticide by males has been recorded are labeled with *i*. *t*-tests were used to test against equal length of lactation and gestation (*** $P<$ 0.001; ** $P<$0.01; * $P<$0.05; † $P<$0.10).

tends to be gradual estimates are often crude. Nonetheless, if robust trends emerge, they are unlikely to be due to such unbiased noise.

Figure 3.2 gives the mean L/G ratios for a variety of higher mammalian taxa (where taxonomic boundaries were sometimes adjusted to reflect major life style differences: pinniped versus fissiped carnivores, odontocete versus mysticete cetaceans, sciurognath versus hystricognath rodents). A crude analysis of variance of species values of ln (L/G) among these units show highly significant variation (Kruskal–Wallis test because variance of various taxa is heteroscedastic, H [19] = 246.6, P<0.0001; or if taxa with one or two species are removed: H [13] = 239.6, P<0.0001).

Taxonomic distribution of infanticide records

By the conservative accounting of Chapter 2 (see Table 2.1), infanticide has now been directly observed in wild populations of 17 species of non-human primate. This conservative approach was taken in response to criticism suggesting that fieldworkers had uncritically ascribed infant deaths to infanticide. However, a less conservative accounting is required if we are to detect effects of taxonomy or ecological, life history or social variables correlated with taxonomy. Thus, Table 2.2 furnishes a complete overview of published credible reports of male infanticide in non-human primates, both directly observed and strongly indicated. This table indicates that infanticide by males is now witnessed or strongly suspected in 39 primate species, in 9 of which the only reports are from captivity. It is reported for four of the six major primate radiations (lemurs, platyrhines, cercopithecoids, and apes; not in lorisids and tarsiers).

Even in a well-studied group such as primates, the taxonomic distributions of infanticide are likely to be incompletely known. There is no confident way to distinguish between absence of evidence and evidence of absence. For instance, the absence of records for species in which small numbers of infants are observed at a time, such as nocturnal species or spider monkeys, gibbons and orangutans, could well be due to the low expected frequency. However, other patterns cannot be dismissed so easily. Among the platyrhines, infanticide by males is not reported for the callitrichids, even though they are kept in numerous facilities. It is not even observed when unrelated infants are placed into groups containing an adult male (e.g., Wamboldt *et al.* 1988; I. Kuderling, pers. comm.). Likewise, virtually no cases of infanticide have been reported from the intensively studied provisioned groups of several macaque species (*Macaca fuscata, mulatta, sylvanus*) once these groups were well established. Thus the absence of records of infanticide by males in primates may well reflect either possibility: true absence, or no evidence.

Table 3.1 compiles the non-primate mammals for which infanticide by males as a result of deliberate attacks on unrelated infants has been recorded. In most non-primate mammals, it is generally more difficult to argue that the absence of records of infanticide by males reflects the absence of infanticide rather than the absence of solid information. Nonetheless, infanticide by males is relatively common in carnivores and sciurognath rodents. It is also suspected to be common in cetaceans. A few cases of infanticide by males have been reported for equids (a very well-studied group, Feh 1999; cf. van Schaik 2000). Artiodactyls show little evidence of infanticide by males, apart from a single (clearly exceptional) case in red deer, and a series of well-documented cases for hippopotamuses. The observations of infanticide by males in pinnipeds were not included, because they can all be ascribed to accidents rather than to targeted attacks (LeBoeuf & Campagna 1994). Thus, especially non-primates will yield many false negatives (no infanticide recorded although it occurs in a given species). However, since false positives are unlikely, we should be able to get a first, if conservative, impression by comparing species with and without recorded infanticide by males.

Table 3.1. *Non-primate mammal species in which infanticide by males has been reported in a context consistent with the sexual selection hypothesis[a]. I distinguish between certain cases, and cases where circumstantial evidence supports the interpretation[b]*

Order	Species	Captive/ Wild	Source
Artiodactyla	*Cervus elaphus*	W	Ebensperger (1998a)
	Hippopotamus amphibius	W	Lewison (1998)
Carnivora – Fissipedia	*Crocuta crocuta*	W-prob.	Ebensperger (1998a)
	Enhydra lutris	C	Ebensperger (1998a)
	Felis catus	W	Ebensperger (1998a)
	Felis concolor	W	Ebensperger (1998a)
	Felis pardalis	W	Ebensperger (1998a)
	Lynx canadensis	W	Kitchener (1991)
	Lynx rufus	W-prob.	Ryden (1981)
	Mephitis mephitis	W	Ebensperger (1998a)
	Nasua nasua	W	Ebensperger (1998a)
	Panthera leo	W	Ebensperger (1998a)
	Panthera pardus	W	Ebensperger (1998a)
	Panthera tigris	W	Ebensperger (1998a)
	Ursus americanus	W	Ebensperger (1998a)
	Ursus arctos	W	Ebensperger (1998a)
	Ursus maritimus	W	Ebensperger (1998a)
Cetacea	*Tursiops truncatus*	W-prob.	Patterson *et al.* (1998)
Perissodactyla	*Equus caballos*	C	Duncan (1982)
	Equus przewalskii	C	Ryder & Massena (1988)
	Equus zebra	C	Penzhorn (1984)
Rodentia – Sciurognathi	*Acomys cahirinus*	C	Blumstein (Chapter 8)
	Apodemus sylvaticus	C	Blumstein (Chapter 8)
	Clethrionomys glareolus	W	Blumstein (Chapter 8)
	Cynomys ludovicianus	W	Blumstein (Chapter 8)
	Dicrostonyx groenlandicus	C	Blumstein (Chapter 8)
	Lasiopodomys brandti	C	Blumstein (Chapter 8)
	Marmota caudata	W	Blumstein (Chapter 8)
	Marmota marmota	W	Blumstein (Chapter 8)
	Meriones unguiculatus	C	Blumstein (Chapter 8)
	Microtus arvalis	C	Heise & Lippke (1997)
	Microtus californicus	W	Blumstein (Chapter 8)
	Microtus pennsylvanicus	W	Blumstein (Chapter 8)
	Mus musculus	C	Blumstein (Chapter 8)
	Paraxerus cepapi	W	Blumstein (Chapter 8)
	Peromyscus californicus	C	Blumstein (Chapter 8)
	Peromyscus leucopus	W	Blumstein (Chapter 8)
	Peromyscus maniculatus	C	Blumstein (Chapter 8)
	Phodopus sungorus	C	Blumstein (Chapter 8)
	Rattus norvegicus	C	Blumstein (Chapter 8)
	Spermophilus beecheyi	W	Blumstein (Chapter 8)
	Spermophilus columbianus	W	Blumstein (Chapter 8)
	Spermophilus franklinii	W	Blumstein (Chapter 8)

Table 3.1. (*cont.*)

Order	Species	Captive/ Wild	Source
	Spermophilus parryi	W	Blumstein (Chapter 8)
	Spermophilus townsendii	W	Blumstein (Chapter 8)
	Spermophilus tridecemlineatus	W	Blumstein (Chapter 8)

Notes:
W, wild; prob., probable; C, captive.
[a] i.e., infanticide by males unlikely to have sired the offspring kill the offspring and are likely to subsequently mate with the mother; especially in predators, cannibalism (killing and eating unrelated dependent offspring) alone is insufficient if it was not in a context with possible subsequent mating access to the mother.
[b] For brevity, we refer to careful reviews (Ebensperger 1998a; Blumstein, Chapter 8) rather than original sources, but for non-rodents original sources were consulted where context was open to doubt (as well as Packer & Pusey 1984). Where data are available from both the wild and captivity, wild data take precedence.

Remarkably, the higher taxa with positive values of ln (L/G), and thus relatively long lactation periods, are exactly the ones for which infanticide by males is thought to be common (Fig. 3.2). Although this is of course not a formal test, the broad taxonomic pattern emerging in life history and infanticide is fully consistent with the expectations based on a sexual selection model for male infanticide. We can do more detailed tests for species known to have male infanticide.

L/G and infanticide

The sexual selection hypothesis predicts a correlation between L/G of a species and its vulnerability to infanticide. There are two ways to test this prediction. First, if we are willing to assume that the absence of evidence for infanticide by males reflects evidence of its absence, we can directly compare the species with and without male infanticide with respect to their L/G values. Although the data are poor, even for primates, this test is conservative. In effect, it classifies species as non-infanticidal if we have no information. However, if the species with no recorded infanticide are a biased subset with respect to life history, and further study would reveal infanticide by males in these species, a second test may be deemed more appropriate, in which we test whether the observed L/G distribution is significantly greater than 1. This test, too, is conservative because an L/G ratio of slightly less than 1 may not eliminate infanticide risk to zero, and because species with large litters may be vulnerable if $L < G$.

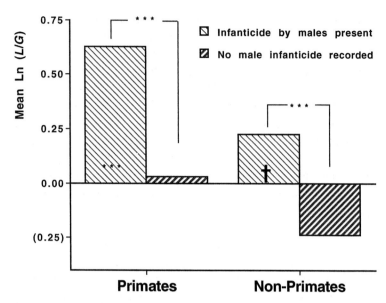

Figure 3.3. A comparison of lactation/gestation (*L/G*) ratios of species in which
infanticide by males is recorded with those in which such infanticide is not
recorded (absent or unknown), for primates (left) or all non-primate mammals
(right). Statistical tests refer to the difference between species with and
without known infanticide by males, and to difference between *L/G* ratio and
equal *L* and *G* (i.e., ln (*G/L*) = 0) for the species with male infanticide. (*** *P* <
0.001; * : *P* < 0.05; † : *P* < 0.10).

For all mammals with data for *L* and *G* (*N* = 587), mean ln(*L/G*) =
−0.199, which is significantly below 0 (*t* = 7.25, df = 586, *P* < 0.0001), indicating that the average mammal has a gestation period longer than lactation. For the 20 primate species with data on *L/G* for which infanticide by
males is recorded, mean ln(*L/G*) = +0.627, well above the mean ln(*L/G*) =
+0.031 for the 40 species with no recorded infanticide by males (*t* = 3.45,
df = 58, *P* = 0.001; Fig. 3.3). The mean ln(*L/G*) for species with recorded
infanticide is also well above zero (*t* = 4.89, df = 19, *P* < 0.0001), as predicted by the life history argument of the sexual selection hypothesis.
Among primates, life style and life history are clearly correlated (van
Schaik & Kappeler 1997; van Schaik 2000). Females of species that carry
their young at all times have relatively long gestation, whereas females
that cache their dependent infants or leave them in a nest have short lactation relative to gestation. Infant-carrying females of species with extensive allomaternal care (or communal breeders) have secondarily reduced
length of lactation, which also should make them less vulnerable to

infanticide. Hence, among primates, reports of infanticide are concentrated exclusively in species in which females carry their young but do not receive substantial help in rearing their young.

Similar patterns are found among non-primate mammals. As shown in Fig. 3.3, the mean $\ln(L/G)$ of these 35 species is also positive (0.225), much higher than the mean of -0.281 for all 492 remaining non-primate mammals ($t = 4.57$, df $= 525$, $P < 0.0001$). The mean $\ln(L/G)$ is also significantly greater than zero ($t = 2.33$, df $= 34$, $P < 0.05$), but it includes several species with $L < G$. Inspection shows that the species with relatively short lactation which are nonetheless subject to infanticide by males are rodents with large litters in which loss of one litter either speeds up female reproduction or tends to increase the size of the next litter (Elwood & Ostermeyer 1984).

These results may to some extent be spurious due to lack of phylogenetic independence of the values of related species. Use of phylogenetic contrasts (Harvey & Pagel 1991) to correct for this independence is impractical at this stage, because (1) the number of species involved is unusually large, (2) the phylogenetic relationships for this large taxonomic array are still unsettled, and (3) only changes in L/G that cross (not strictly defined) vulnerability classes should be included (or in other words, the effect of L/G on vulnerability is highly non-linear). Future analyses could be limited to orders and finer vulnerability estimates that include litter size and seasonality as well as L/G. As a preliminary sensitivity analysis, we can compare patterns in L/G ratios across monophyletic taxa. Table 3.2 presents the comparisons by taxon: five out of six have a higher mean L/G ratio for the species in which infanticide by males has been recorded than for the species in which it has not been recorded, and therefore may be absent. Two of these differences are statistically significant. For fissiped carnivores and primates with infanticide records, lactation is significantly longer than gestation. Moreover, in each taxon, the L/G ratios of species with infanticide by males are close to 1 or greater, as predicted.

Discussion

The compilation of cases of infanticide by males confirms the remarkable concentration of observations in just three orders. While some of this may be due to observational biases toward large diurnal animals that are non-fossorial, the overall pattern cannot be explained this way. Many large

Table 3.2. *Taxonomic breakdown of L/G ratios in species with and without reliable records of infanticide by males (IM) in a context consistent with the sexual selection hypothesis*

	N spp.	Mean ln (L/G): IM recorded	N spp.	Mean ln (L/G): No IM recorded	Different? t-test
Artiodactyla	2	−0.065	64	−0.498	NS
Carnivora – Fissipedia	13	+0.586*	64	+0.165	**
Cetacea – Odontoceti	1	+0.497	11	+0.153	NS
Perissodactyla	2	−0.205	6	−0.102	NS
Primates	20	+0.627***	40	+0.031	***
Rodentia – Sciurognathi	16	+0.034	177	+0.015	NS

Notes:
Tests in the species with recorded IM are against the null hypothesis of ln (L/G) = 0. The test in the last column is for differences in ln (L/G) of the two classes of species.
* $P < 0.05$; **, $P < 0.01$; ***, $P < 0.001$; NS, not significant.

artiodactyls are well studied, yet there are remarkably few examples of infanticide by males for this order. Likewise, many nocturnal and fossorial rodents have been studied, often in captivity, for evidence of infanticide by males. This taxonomic distribution shows that it is most common in taxa in which young are highly altricial at birth (sciurognath rodents, fissiped carnivores) or in which relatively precocial young grow and develop slowly (primates). If infanticide turns out to be common in cetaceans, they would fit the primate pattern.

The distribution at the ordinal level suggested a correlation with life history (Fig. 3.2). Detailed tests at the species level remain consistent with this impression. In both primates and non-primates, L/G ratios of species in which deliberate infanticidal attacks by males have been recorded are higher than those in species without such records. Because the absence of records does not necesarily imply the absence of infanticide, I also tested whether species with recorded infanticide by males are more likely to have relatively long lactation relative to gestation. In primates, this prediction was also confirmed. It was even confirmed in non-primates, in which larger litters are common and species can therefore be vulnerable to infanticide even if lactation is shorter than gestation. This whole pattern therefore strongly supports the sexual selection hypothesis for infanticide by males (Hrdy 1979; Hausfater & Hrdy 1984; van Schaik, Chapter 2), and is not favorable for especially the non-adaptive hypotheses that suggest that infanticide is a by-product of aggressive takeovers by males. Such takeovers occur in many radiations, not only the ones in which males commit infanticide, including many well-studied artiodactyls.

In some species, however, infanticide by males need not serve the function hypothesized by the sexual selection hypothesis, but rather may represent avoidance of misdirected paternal care or especially opportunistic cannibalism. In a broad overview such as has been attempted here, some functional hetereogeneity is difficult to avoid. A few observations of opportunistic cannibalism in carnivores in a context not related to male immigration or takeover (e.g., when the whole group moves) have not been included. Nonetheless, many carnivores and rodents readily eat meat, and can easily devour altricial infants when unrelated litters are encountered. The best way to distinguish between mere vulnerability due to alitriciality at birth or very slow development and vulnerability due to effects of killing on female reproduction is that, under the former hypothesis, both males and females should readily kill infants, whereas only males are expected to do so under the latter. The taxonomic coincidence between infanticide by males and females is poor in primates (Digby, Chapter 17), but better in rodents (Blumstein, Chapter 8). However, note that the reproductive benefits for males still obtain if infants are cannibalized, thus exerting additional selection on male tendencies to commit infanticide (and on female tendencies to reduce its risk).

In conclusion, infanticide by males is strongly concentrated in those species whose life history produces a reproductive benefit to the male, either due to a faster return to female fertility, or due to increased litter size when the current offspring are killed. On the other hand, it is rarer in species where killing the infant would make no difference in the female's return to estrus. These patterns therefore provide important support for the sexual selection hypothesis.

Part II

Infanticide by males: case studies

4

Infanticide in red howlers: female group size, male membership, and a possible link to folivory

Introduction

Food acquisition and predator avoidance are two major factors proposed to promote social grouping in animals (Krebs & Davies 1987). In the primate literature, these two factors have been presented as alternative hypotheses (Wrangham 1980; van Schaik 1983). Defense against potential infanticide by new males also favors social grouping in primates: females form permanent associations with protective males, including their dependent infants' fathers (Wrangham 1979; van Schaik & Dunbar 1990; Smuts & Smuts 1993; van Schaik & Kappeler 1993, 1997; Sterck *et al.*, 1997). The present volume originated from the recognition of the importance of infanticide avoidance in shaping primate social organization.

Factors favoring the formation of permanent social groups of females and males are not necessarily the same factors influencing the size of the groups. For example, many believe food competition to be the principal factor limiting primate group size (Janson 1988; van Schaik 1989; Isbell 1991; Janson & Goldsmith 1995). Upper group-size limits may be mediated by the maximum daily travel distance individuals can sustain as they forage in groups (Wrangham *et al.* 1993). In this chapter, we present evidence from red howler monkeys (*Alouatta seniculus*) that infanticide, too, may play a role in limiting group size in primates (see also Steenbeek, Chapter 7). When infanticide rates increase with the number of reproductive females, females may opt for dispersal, thus keeping total group size small. We propose that indirect evidence previously suggesting the role of food competition in limiting group size might actually reflect infanticide.

Van Schaik (1983) demonstrated an apparent reduction in female fecundity with increasing female group size consistent with food competition; this conclusion was based on infant/female ratios from censuses rather than actual birth rates. Reviews of primate foraging effort (Isbell 1991; Janson & Goldsmith 1995) indicated that larger primate groups within a given species and population tend to have longer day ranges, implying greater foraging costs. However, this trend was significant only for primates whose diets are largely frugivorous or clumped. Among leaf-eating primates, several species, including red howlers, show little or no increase in foraging effort as group size increases and yet, paradoxically, live in social groups smaller than most primates (Janson & Goldsmith 1995).

Perhaps predominantly folivorous primates are not able to increase foraging effort because low energy content or extractability of their diet limits maximum daily energy expenditure to low levels (Nagy & Milton 1979). If they do experience food competition but cannot increase foraging effort (e.g., daily travel distance) to compensate, we expect females to suffer reduced fecundity in larger groups, as suggested by van Schaik's (1983) analysis. Alternatively, some primate populations actually experiencing little food competition may be vulnerable to increased mortality of infants in larger groups due to factors unrelated to food, a pattern erroneously suggesting a relation between group size and female fecundity (Isbell 1991). These other factors might limit group size to a level below that associated with food competition. A plausible mechanism relating infant death rates to group size is infanticide.

Males in various primate species may kill infants after taking over the major reproductive position in a group (Crockett & Sekulic 1984; Leland et al. 1984; Struhsaker & Leland 1987; Newton 1988). If infanticides preferentially occur in groups with more females, the per capita rate of infant mortality could actually increase with group size. Birth rate itself might then be independent of group size or, in species where infant death results in shortened interbirth intervals, birth rate might actually increase, rather than decrease, with larger group size. In red howlers, total social group ("troop") size is directly related to the number of reproductive adult females (Crockett 1984), and this relation is probably the case for most polygynous primates. Analyses of group size therefore often focus on female group size.

The published data on the relation of infanticide to female group size are most compeling for lions (*Panthera leo*). Female reproductive success is greatest in prides with three to ten females, because those smaller and larger are more prone to male takeover and infanticide (Packer et al. 1988).

Maturing females are more likely to disperse if their natal pride would have exceeded ten females (Pusey & Packer 1987b). The advantages of cooperative hunting are diminished by food competition at large kills such that food intake in solitary lionesses and those living in small prides is not significantly different from that of those living in large prides (Packer 1986). Packer (1986) proposed that, because the costs of infanticide opposed those of food competition, lion sociality derived from mother lions and their daughters raising their cubs communally to defend them against infanticidal males.

Evidence from two primate species hints that the risk of male invasion and associated infanticides increases with the number of adult females. In geladas (*Theropithecus gelada*), larger harems (four to ten females) experienced male takeover at a much higher rate than small harems (Dunbar 1984b). Infanticide after gelada harem-male replacement has been observed in captivity (Moos *et al.* 1985) and was reported recently following male incursion in the wild (Mori *et al.* 1997). In hanuman langurs (*Presbytis* (*Semnopithecus*) *entellus*), infanticide rate was about three times as high in the larger of two closely monitored groups (Borries 1997; average female group sizes of 6.5 and 12.9, C. Borries, pers. comm.).

Here we examine demographic data from a population of red howler monkeys in which numerous male invasions and infanticides have been documented (Rudran 1979; Crockett & Sekulic 1984; Agoramoorthy & Rudran 1995). In the study population, adult female group size never exceeded four, and the resulting small average troop size had been attributed to inferred food competition (Crockett 1984). However, because of the apparent lack of relation between female group size and foraging effort in this species (Sekulic 1982b; Janson & Goldsmith 1995), we began to suspect infanticide to be a better explanation for small troop size in red howlers.

Infanticide risk may result from an interplay between numbers of females and males in a group. Larger female groups will be more attractive targets for incoming males only up to the point that the new dominant male can exclude other males and monopolize breeding. In general, male membership in polygynous groups increases significantly with the number of adult females (Andelman 1986; Mitani *et al.* 1996b). Breeding males might tolerate the presence of additional resident males if they helped to deter invaders. In *P.* (*S.*) *entellus*, one-male troops are more vulnerable to takeover (Newton 1986, but see Borries & Koenig, Chapter 5). Even when more than one adult male is resident in a red howler troop, only one does nearly all of the mating and usually sires all of the offspring (Sekulic 1983a; Pope 1990). We already knew that single adult males were less likely to invade red howler troops and succeed in replacing resident male(s) than coalitions of two invading males (Crockett & Sekulic 1984). When a single male was able to enter a troop, he usually (three out of four cases) was unable to expel the resident and instead co-resided with him (Crockett & Sekulic 1984). Other multimale red howler troops developed through maturation of natal males (Crockett & Eisenberg 1987). Because male membership of howler troops is variable (Crockett & Eisenberg 1987), we included the relation between the number of resident males and rates of incursion and infanticide in our analyses.

Here we evaluate the following predictions for red howlers:

1. Infanticide rates increase with increasing adult female group size whereas infant mortality from other causes is unrelated to female group size.
2. Troops with more males should experience lower rates of incursion and infanticide, at a given female group size.
3. If infanticide is related to female group size and is the major source of infant mortality, birth rate might increase with female group size because interbirth intervals are shortened after infant death.

In the discussion, we evaluate the generality of the findings including a possible link to folivory.

Methods

Study site and monitoring techniques

The field research was conducted at Hato Masaguaral, a cattle ranch and wildlife preserve in the *llanos* (plains) in north central Venezuela. Red howler troops were studied in two habitats, a woodland and a gallery forest (Crockett 1984, 1985, 1996; Crockett & Eisenberg 1987).

During two years at the study site, C.M.C. censused troops in both habitats in addition to conducting a detailed behavioral study of one troop in each habitat. Sekulic (1983a), who witnessed an infanticide, intensively observed four of the woodland troops during one of the study years. Most woodland troops were contacted at least monthly during the two years, having been identified and monitored previously (Rudran 1979). The gallery troops were located and individuals identified during the first year of C.M.C.'s study. A set of 24 woodland troops monitored for two years (February 1979 to February 1981) and ten gallery troops monitored for one year (February 1980 to February 1981) were followed closely enough to provide data on group size, birth rate and inferred infanticide rate for the present analysis (Table 4.1). Additional data on male incursions were available through December 1981.

During census contacts, individuals were identified using physical characteristics (e.g., scars, depigmentation patterns on soles of feet or nipples, sex, size) and, in some cases, ear tags. Troop lists carried into the field described identifying characteristics of individuals in words and drawings. Changes in troop membership – including births, immigrations, emigrations, deaths and disappearances – were noted. Because troops were small (mean size, eight to ten individuals), relatively cohesive, and in habitats with good visibility, troop members were rarely missed during census counts. Since troops never contained more than four adult females, maternal identity could be determined for nearly all infants. Census contacts took place approximately monthly, and most infants could be assigned a birth date to the nearest month. When changes in adult male membership occurred, a troop was sometimes contacted more than once a month. Census methods are described in more detail elsewhere (Crockett 1996).

Between October 1978 and December 1981, Sekulic (1983a) observed one infanticide (Fig. 4.1) and Crockett & Sekulic (1984) inferred 19 others when infants disappeared in conjunction with adult male incursions, replacements, or status changes (these incidents combined are "male changes"). Seven additional infants received severe injuries of a sort, suggesting that they had survived infanticide attempts. Infant mortality in stable troops (i.e., with no detected male changes) was very low. Infant mortality (death or disappearance before 1 year of age) was 31% in troops with documented male changes (19 out of 60 infants born) compared with only 9% in other troops (7 out of 78 infants born). In the eight most closely monitored troops with male changes, 14 infants disappeared within 1 month of nine changes (9 troop-months), compared with only two infant disappearances from these same troops at other times (246 troop-months) ($P < 0.001$). Thus, any infant disappearance in a troop with a recent history of new males entering, former resident males leaving, co-resident males fighting, or outside males attempting a takeover was considered to be infanticide. The infanticides were rated as very probable (14 cases), probable (2 cases), or questionable (3 cases), depending on how much was known about the male change events, and whether stalking of infants or injured infants were observed (Crockett & Sekulic 1984). During the

Table 4.1. *Red howler troops monitored, births, deaths, composition and corresponding male change and infanticide events*

Troop	Troop size 1/81	Mean AF	No. AF at infanticide	Obs. BR	No. infants born	No. infanticide	No. other deaths	Mean troop males	Mean males up to incursion	No. AM + SAM at incursion	Comment [a]
M51	9	2.50	3	0.60	3	1		2.00	2	2	Attempted incursion N, 12/80 (infanticide 14 and injury case d)
M52	12	3.00		0.83	5			3.00			
M53	9	2.88	3	0.87	5	2		2.00	2	2	Replacement A, 8/79: 2 males replaced 2 males (infanticides 1, 2 and injury case a)
M54	11	2.25		0.89	4			2.25			
M55	11	2.63		0.95	5		1	2.88			
M56	7	1.75		0.86	3			2.00	2	2	Replacement D, 1/81: 2 males replaced 2 males (1 dead); no infants observed in troop prior to invasion[b]
M57	6	2.00		0.75	3			1.00			
M61	10	2.63		0.76	4		1	1.96			
M62	9	2.00		1.00	4			1.50			
M63	11	3.00	3	0.83	5	1		2.75			Status change Q(infanticide case 18, 11/79)
M64	8	2.63		0.57	3			2.00			
M65	14	3.00		0.83	5			3.00			
M66	16	4.00		0.75	6			2.58			
M67	12	2.75		0.73	4			2.50			
M71	11	3.00	3	0.67	4	1		2.46	2.8	1	Status change V (infanticide 20, 10/79); overlap L (injury case c, 4/80)
M72	10	4.00	4	0.63	5	3		1.71		1[c]	Status change O resulting in death of 1 male (infanticides 15, 16, 17, 6–10/80); replacement E (no detected infanticides)
M73	9	2.00		0.75	3		1	2.00			
M74	11	3.13	3	0.96	6	2		2.33	2.65	1	Overlap K, 5/80 (infanticides 11, 12)
M75	9	2.00		0.75	3			1.17			

Troop											Comments
M76	9	3.00	3	0.83	5	3		1.63	1.33	2	Replacement C, 4//80: 2 males replaced 1 AM; SAM stayed (infanticide cases 3, 4, 5)
M77	11	3.00		1.00	6			2.33			
M78	9	2.00		0.50	2			1.38			
M79	11	2.38		0.63	3			1.50			
M58	13	3.00		0.67	4			2.17	1	1	Replacement B, 8//79: 2 males (+1 nearly SAM) replace 1 male (long interbirth intervals suggest undetected infanticides)[b]
G1	8	3.00		1.00	3			1.00			
G5	8	2.00		0.50	1			1.00			
G6	9	3.00		1.00	3			1.00		1[c]	Replacement H, 12/81: 2 males replaced 1 male (injury case b)[b]
G7	5	1.75		0.57	1			1.00			
G9	10	2.75	3	1.09	3	3		1.50	1	1	Replacement F, 11/80: 2 males replaced 1 male (infanticides 6, 7, 8)
G11	8	3.00		1.00	3		1	2.00			
G12	10	2.00		1.00	2			2.25			
G15	6	2.00		1.00	2			1.00			
G20	11	3.00		0.67	2			2.00			
G21	9	3.00	3	1.00	3	1		1.00	1	1 & 1[c]	Attempted incursion M, 2/81 (infanticide 13); Overlap J, 12/81 (different male from M)[b]

Notes:

M, Woodland troop monitored for 2 years; G, gallery troop monitored for 1 year; BR, birth rate; AF, adult female; AM, adult male; SAM, subadult male.

[a] Male change letters, infanticide case numbers and injury (infanticide attempt) case letters from Crockett & Sekulic (1984). Replacement: original male(s) gone. Overlap: invader co-resides with original male(s).

[b] Unrecorded births and infanticides possible because troop was not monitored in the few months immediately after incursion.

[c] Incursion occurring March–December 1981.

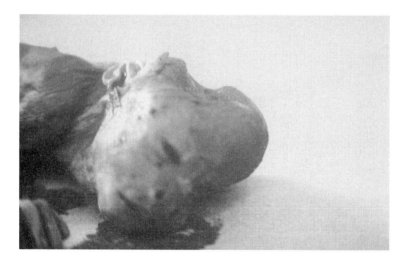

Figure 4.1. One-day old red howler victim of infanticide observed by Ranka Sekulic (1983a) (case 20, Table 4.1) showing crushed skull (photo by Robert D. Brooks; Zimmer 1996) and canine puncture wound in skull (photo by Ranka Sekulic; Sommer 1989b).

period analyzed here (a subset of the previous study), 1 observed, 13 very probable, 2 probable, and 1 questionable infanticides occurred (Table 4.1).

The mean age at death or disappearance for observed and inferred infanticide victims was 1.9 months ($N = 17$); the oldest was 8 months. For further details regarding monitoring of births and interbirth intervals,

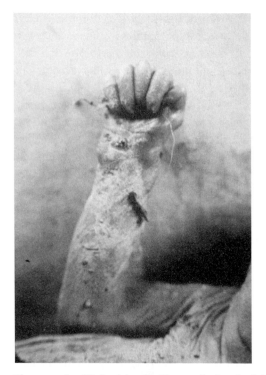

Figure 4.1. (*cont.*) Infant's hand holds strands of mother's fur (photo by Ranka Sekulic).

see Crockett & Rudran (1987a,b); for male changes and infanticides, see Crockett & Sekulic (1984) and Sekulic (1983a).

Measures

Birth rates were calculated over 2 years for woodland troops as the number of recorded births in a given troop divided by the number of "female-years" (total births/total months each adult female was present from February 1979 through January 1981 summed across all females in the troop per 12 months). Females were considered adult at the onset of the first pregnancy, approximately 4.5 years (Crockett & Pope 1993). Data were available for ten Gallery troops for only one complete year when a high proportion of primiparous births occurring in newly formed troops produced some birth rates exceeding 1.0 per year. Thus, parametric analyses involving birth rates were restricted to the woodland 2-year data set.

Infant mortality (death before one year of age) was dichotomized into "infanticide" or "other". The infanticides used in the present analysis include 17 cases previously attributed to infanticide (all those listed in Table II of Crockett & Sekulic 1984, except for cases 10 and 19 which were excluded because they occurred in woodland troops before or after the 2 year period analyzed, and case 9 which occurred in a gallery troop monitored for less than 1 year). The few remaining cases of infant disappearance were classified as "other" mortality. In the 24 woodland troops during the 2 years, 100 births occurred, 13 infant deaths were attributed to infanticide, and 2 "other" infant deaths were recorded. In the 10 gallery troops monitored for 1 year, there were 23 births, 4 infanticide deaths, and 1 "other" infant death. Although a few neonatal deaths might have been missed, we assume that they were proportionally attributable to infanticide or other causes. We calculated infanticide and "other" infant mortality rates for each troop individually by dividing the number of infants that died during the study period by the number born during the same period (mortality rate per infant born).

We defined an *infanticidal bout* as one or more infanticides or suspected infanticides following a given male change event. We used bouts in addition to individual infant deaths in some analyses because multiple deaths sometimes occurred in the same troop over a short period of time and thus are not likely to be statistically independent. Bouts were considered to be independent because they were associated with temporally separate male change events in the same troop or in different troops. Male changes included successful and attempted incursions by new males, and status changes occurring some time after a male immigrated but did not immediately replace the resident male (Crockett & Sekulic 1984). There were seven infanticidal bouts in woodland troops during two years and two bouts in gallery troops during 1 year. Some infanticidal bouts during the study followed incursions that happened prior to the study period, and not all incursions were followed by detected infanticides during the study period.

Female group size was the average number of females in a troop during the time period in question for comparisons involving birth and death rates, or the total number present at the time of an infanticidal bout.

Male group size: because subadult males of age 5–6 years approach adult weight (Crockett & Pope 1993) and may be capable of defending against adult male invasion, we used the average number of adult and subadult males for male group size. Because male incursions sometimes changed the number of males for troops that experienced incursions, we used the average male group size across the months prior to the incursion event for comparisons with infant mortality rates. For groups with no incursions (whether or not they experienced infanticide), we used the average number across the entire census period. For analysis of male group size and incursion incidence, we used the number of adult and subadult males at the most recent census prior to the incursion, usually one month. For correlations between female group size and male group size, we used the average number of males across the census period, including changes following incursions.

Statistical tests

Tests were computed with the SAS statistical package or Data Desk (Velleman 1997). All probabilities were two-tailed with an alpha level of $P = 0.05$. Because data from the two habitats differed in duration and variance estimates, parametric statistics and all analyses of birth rates were restricted to the woodland data set. Non-parametric Spearman correlations, with correction for ties (Zar 1984), were run on woodland alone and both habitats combined.

Results

Infanticide and female group size

Infanticide was more prevalent in troops with more adult females. Nine bouts of infanticide occurred in the woodland and gallery troops during the study period, all nine when troops contained three or four females. As the study troops contained three or four adult females only 61% of the time, the probability of infanticide bouts occurring only in the larger female groups by chance was small (binomial test, $P = 0.024$).

Only infant deaths attributed to infanticide were related to female group size. Infanticide rate per infant born in a troop correlated significantly with the average number of adult females ($r_s = 0.48, P < 0.02$, $N = 24$, woodland; Figure 4.2; $r_s = 0.41$, $P < 0.02$, $N = 34$, woodland and gallery). "Other" infant death rate did not ($r_s = -0.22$, NS, $N = 24$, woodland; $r_s = -0.08$, NS, $N = 34$, woodland and gallery).

The relation of number of males to incursion and infanticide

The number of group males was not significantly correlated with either infanticide rate ($r_s = -0.05$, NS, $N = 24$, woodland; $r_s = -0.19$, NS, $N = 34$,

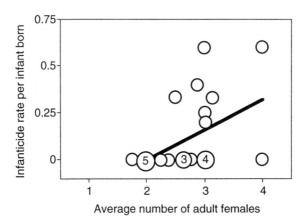

Figure 4.2. Deaths from infanticide per infant born versus number of females averaged over the entire 2-year study period (woodland data only). Numbers inside enlarged circles indicate number of troops with identical x and y values. The regression line is plotted for illustrative purposes only.

woodland and gallery) or "other" infant death rate ($r_s = 0.03$, NS, $N = 24$, woodland; $r_s = 0.22$, NS, $N = 34$, woodland and gallery). The number of adult females, adult and subadult males combined also was not significantly correlated with either class of infant mortality rate (woodland, $N = 24$: Infanticide, $r_s = 0.25$, NS; "other", $r_s = -0.05$, NS; both habitats, $N = 34$: Infanticide, $r_s = 0.14$, NS; "other", $r_s = 0.17$, NS).

Larger numbers of males might deter incursions by non-resident males and reduce the risk of infanticide. However, we found no effect of the average number of males on the rate of infanticide in the correlations reported above. Because female and male group sizes are significantly correlated ($r_s = 0.57$, $P < 0.01$, $N = 24$, woodland; $r_s = 0.40$, $P = 0.05$, $N = 34$, woodland and gallery), the potentially beneficial effect of larger male group size in reducing incursion rates may be confounded by larger female groups attracting more male incursions. We used logistic regression to predict the likelihood of an infanticidal bout occurring over the 2 year study period in a woodland troop as a function of both female and male numbers (up to the point of incursion, if any). We found a significant positive effect of mean number of adult females, as expected from the previous section's results, but the effect of males, while in the expected negative direction, was not significant. The regression equation is log (P(infanticidal bout)/(1 − P(infanticidal bout)) = 5.30 + 2.40 (mean number of females) − 1.09 (number of males). For the entire model, $\chi_2^2 = 6.54$, $P < 0.05$; effect of females, $\chi_1^2 = 4.15$, $P < 0.05$; effect of males, $\chi_1^2 = 1.65$, $P = 0.20$).

The weak effect of male group size may result because, whereas females associated with larger male groups may be safer from infanticidal bouts following male incursions, they may be more vulnerable to such bouts resulting from status changes among resident males. During the period March 1979 to December 1981, ten successful and two attempted male incursions were observed in the woodland and gallery troops, six associated with bouts of infanticide. Eight of the 12 incursions occurred in troops with only one male and four when troops had two males. The observed incursions occurred at significantly smaller male group sizes than those expected at random from the distribution of male group sizes during the study period. Study troops contained 1, 2, or 3–4 males for 29.2%, 46.6%, and 24.6% of the study period, producing expected values of 3.5, 5.6, and 2.95 incursions, respectively (chi-squared one-sample test, $\chi^2_2 = 9.17, P < 0.05$). Thus, larger male group size was associated with lower risk of incursion. However, the remaining three bouts of infanticide observed in the study occurred in the absence of recent incursions and were instead related to status changes among males. Such status changes were identified in two troops with two adult males and one subadult male and in one with two adult males. Infanticide risk from within is therefore not reduced in troops with more males.

Consequences of infanticide for fecundity of females

This population of red howlers had high infant survival, and infanticide was a major source of mortality. In the present sample, overall infant mortality was only 17% and of these infant deaths 85% were known or suspected infanticides. In a larger sample, over more years, total infant mortality was about 20% and at least 44% was observed or suspected infanticide (Crockett & Rudran 1987b). Similarly, in a red howler troop in a Colombian tropical rainforest, 8 out of 12 (67%) infants born between 1986 and 1993 fell victim to observed or suspected infanticide (Izawa & Lozano 1994).

In the Venezuelan red howlers, interbirth intervals were significantly shortened after infant death from an average of 17 to 10.5 months (Crockett & Sekulic 1984; Crockett & Rudran 1987b). Having shown that the risk of infanticide increases with the number of troop females, we tested whether this is reflected in birth rates that vary as a function of female group size. We calculated the Spearman rank correlation of the annual birth rate per woodland female (averaged over the 2 years) and the mean number of adult females ($r_s = 0.06$, NS, $N = 24$). Figure 4.3 illustrates that birth rates clearly do not increase with group size, in contrast

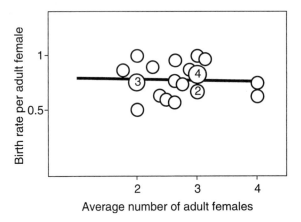

Figure 4.3. Birth rate per adult female per year versus the number of adult females in a group averaged over the entire 2-year study period (woodland data only). Numbers inside enlarged circles indicate number of troops with identical x and y values. The regression line is plotted for illustrative purposes only.

to what we expected if infanticide causes reduced interbirth intervals in larger female groups. This observed pattern might result from declining fecundity in larger groups produced by factors unrelated to infanticide. In other words, the expected positive slope in birth rate versus number of females is negated by factors producing a negative slope. The net effect is a slope approximating to zero.

To further explore this possibility, the average decrease in interbirth interval after infant death from 17 to 10.5 months was used to calculate the expected number of births in the absence of infanticide. If all the infants in a group survived, the observed birth rate would be about 40% less than if they had all died. We used the following formula to adjust the observed birth rate (OBR) in each woodland group (based on the entire 2-year study) by the inferred infanticide rate per birth (I) to produce the expected number of births in the absence of infanticide, the adjusted birth rate (ABR):

$$ABR = OBR\,(1 - 0.4I)$$

We acknowledge that a number of the infanticides occurred in the second year of the study and would thus have affected the birth rate only after the study finished. However, we assume that this bias is counteracted by the births that occurred in the first year of the study that could be attributed to infanticides that happened before the study began. Also, we

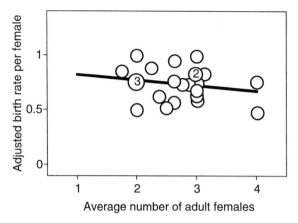

Figure 4.4. Adjusted birth rates per adult female per year versus the number of adult females in a group averaged over the entire 2-year study period (woodland data only). Adjusted birth rates are corrected for the decrease in interbirth interval that follows infanticides. Numbers inside enlarged circles indicate number of troops with identical *x* and *y* values. The regression line is plotted for illustrative purposes only.

have shown that the "other" infant death rate was very low and independent of female group size, so it was not figured into the expected value calculation.

Although the correlation of adjusted birth rate and female group size is nonsignificantly negative ($r_s = -0.11$, $N = 24$, NS; Fig. 4.4), it is tempting to suggest that birth rates, adjusted for the higher rate of infant mortality observed in larger female groups, are somewhat lower in larger groups.

Discussion

Our first prediction, that infanticide rate increases with increasing adult female group size whereas infant mortality from other causes is unrelated to female group size, was supported by data from Venezuelan red howler monkeys. Our second prediction was partially supported. Troops with more adult and subadult males did experience a lower rate of incursion by non-resident males. However, the rate of infanticide did not decline significantly with increasing numbers of males. This was because one third of the infanticidal bouts occurred in multimale troops that had not experienced recent incursions. Instead, these infanticides were associated with status changes between males that had co-resided for some time. The prediction that birth rates might increase with female group size was not supported. However, when troop birth rates were adjusted by the infanticide rate to compensate for average reductions in interbirth intervals after infant death, the correlation – although not significant – was somewhat more negative. This weak finding is consistent with (or at least, does not contradict) the hypothesis that, when the effects of infanticide are removed, larger group size may depress birth rate. Possibly, this may reflect a weak effect of food competition in larger troops.

The strongest conclusion from our analysis is that red howler females in larger groups incur a reproductive cost, assessed by increased infanticide rate. In some species, multimale troops may benefit females when larger female groups prevent single males from monopolizing them. Increasing female group size might actually lead to decreased risk of infanticide when decreased benefit or paternity confusion resulting from females mating with more than one male provides some protection against infanticide (Smuts & Smuts 1993). However, the present analysis indicates that, by being in a larger female group, the disadvantages to red howler females outweigh the advantages of attracting more, potentially protective, males. The risk of incursion is reduced when adult females associate with more adult and subadult males, but this benefit does not translate into a reduced risk of infanticide. Situations such as multimale troops resulting from a previous incursion of two males, or a single invader incapable of driving out the resident, or when a male matures in a troop in which the breeding male is not his father, are candidates for status changes (demoting or evicting the former breeding male) and associated infanticides (Crockett & Sekulic 1984).

Infanticide, then, provides an evolutionary hypothesis to explain the empirical observation that red howler monkeys are found in small troops characterized by no more than four adult females (Crockett & Eisenberg 1987). The primary proximate explanation is the emigration of maturing females that is directly related to the number of adult females already present in the natal troop (Crockett & Pope 1993). About 50% of maturing females disperse or disappear from troops with two adult females, compared with about 85% from those with three breeding females, 100% from four female troops, and none from the few troops with a single female after the death of a second. Furthermore, female emigration is costly in terms of delayed age at first breeding, dietary deficiencies while solitary, and probable increased mortality (Crockett & Pope 1993). Emigrant red howler females almost never succeed in entering existing troops and must form new ones if they are to breed successfully (Crockett 1984; Crockett & Pope 1993). Perhaps because emigration is a costly option, female red howlers appear to engage in aggressive competition for the limited positions (Crockett 1984; Crockett & Pope 1988).

Because total troop size is directly related to the number of reproductive females, the originally hypothesized ultimate explanation for the limited number of females in red howler groups was food competition (Crockett 1984). However, this hypothesis was based on theory (Wrangham 1979, 1980; Clutton-Brock et al. 1982), not on feeding data. Red howlers seem to fit the paradigm proposed by van Schaik (1989) of within-group scramble competition for food, although the relative importance of between-group contest (i.e., territorial defense) is uncertain (Crockett & Eisenberg 1987). Van Schaik (1989: 204) proposed that, if within-group scramble is the primary source of food competition, then "whenever a female finds herself in a group well above or below the optimal size she should be willing to move, provided that migration costs (m) are not prohibitively high. The value of m depends on the risk of predation outside a group, and the ease of settling in existing groups or of finding others with whom to found a new group." The usually low rates of female migration in mammals (Greenwood 1980) should then be the result of high migration costs relative to the costs of natal breeding. We interpret the red howler data as support for the hypothesis that infanticide-related costs of large female group size oppose the high costs of emigration.

The general model that we propose is illustrated in Figure 4.5. We hypothesize that increasing female group size leads to increased infanticide rate per infant born up to the point where the number of females cannot be monopolized by a single breeding male. As the number of breeding males increases, the advantage of infanticide and the ability of invaders to take over a troop decreases. Thus, at the largest female group sizes, the infanticide cost declines. On the other hand, lost reproductive potential through delayed conceptions and infant mortality due to food competition is expected to increase as group size increases in a within-group scramble species. At female group sizes where both infanticide risk and food competition are low, the benefits of female emigration are outweighed by the costs. For the red howler population studied, we hypothesize that the costs of infanticide outweigh the costs of emigration at female group sizes below that which food competition comes in to play.

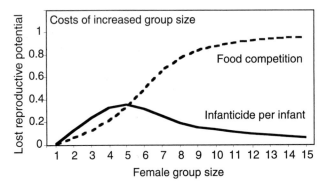

Figure 4.5. Hypothesized relationships between female group size, food competition (dashed line), and risk of infanticide (solid line) in red howlers. We suggest that food competition in red howlers limits the ability of females to increase group size to the point that it induces a multimale social structure with an associated large degree of paternity confusion and low incidence of infanticide. With female group size constrained, the increased risk of infanticide per infant born causes females to compete strongly to keep group size small.

We cannot say anything directly about food competition in red howlers. However, if food competition can be recognized by reduced birth rates in larger social groups (van Schaik 1983), then our analysis suggests that this effect is weak over the female group sizes examined (Fig. 4.3). Although at odds with the prevalence of negative relationships observed between numbers of infants per female and group size across primate species (van Schaik 1983), infants per female may not be a good reflection of true fecundity. Higher rates of infanticide also can reduce the ratios of infants per female disproportionately in larger groups. Thus, van Schaik's (1983) conclusion that food competition is the actual cause of reduced infant/female ratios in larger primate groups could be wrong. Indeed, all of the primate populations analyzed by van Schaik (1983) were of genera in which infanticide has been reported in the wild (*Propithecus, Alouatta, Macaca, Papio, Cercopithecus, Colobus, Presbytis, Semnopithecus, Trachypithecus*; we are following the colobine taxonomy of Oates *et al.* (1994; see van Schaik, Infanticide by males, this volume). Van Schaik (1983) did suggest that early infant mortality related to poor maternal condition was another hypothesis to account for the slope that he observed. When infanticide rates are selectively higher among groups with more females, as in the red howler study population, the direct cost to reproductive success may be through increased rates of infanticide rather than reductions in viable conceptions or thriving infants due to food competition.

Adjusting birth rates to compensate for the shorter interbirth intervals after infant death produced a negative but non-significant correlation between birth rate and female group size. The correlation was somewhat more negative than the median slope of infants-per-female versus group size reported in van Schaik's (1983) compilation. If this birth rate decrease in larger groups is a real effect, it might be due to food competition. However, it is more likely that female group sizes that increase vulnerability to infanticide in red howlers are associated with total troop sizes that are somewhat below those producing significant food competition (Fig. 4.5). Watts (1991b) reported rare harassment of immigrant females only in large gorilla (*Gorilla gorilla beringei*) groups, which otherwise appear to suffer little food competition (Watts 1985). The minimal evidence for food competition in the study population of red howlers is inconsistent with Rudran's (1979; Agoramoorthy & Rudran 1995) resource competition hypothesis that infanticide in red howlers is explained by the increased survival of the infanticidal male's future offspring by the elimination of food competitors. The sexual selection male reproductive strategy hypothesis is compellingly supported by the red howler data (Crockett & Sekulic 1984).

Given the strong reproductive costs of even rather small female group sizes, it is not clear why these howlers are social at all. Predation does not appear to be an important pressure, judging from the rather high frequency of solitary and extra-group animals in both this study site (6% of total population) (Crockett 1985) and in mantled howlers in Costa Rican dry forest (Glander 1992). Likewise, female howlers can survive many months as solitaries (Sekulic 1982a). However, they are unable to rear offspring successfully outside a troop. Continuous association by female howlers with a male may therefore be necessary for protection against infanticide by other males. The lower limit to red howler troop size might be a function of the minimum number of females with which a male would associate rather than abandon to take over a troop with more females. The evidence presented here suggests that infanticide may also contribute to the upper limit to social group size in red howlers. The relation between infanticide rate and number of females in red howler monkeys is the best explanation so far to account for the strikingly truncated female group sizes resulting from female emigration.

An important question is generality of the relation between female group size and infanticide. The hypothesis probably holds for the other howler species with small troop sizes like red howlers and for which infanticide also has been reported: *A. caraya*, *A. fusca*, and *A. pigra* (Zunino *et al.* 1986; Rumiz 1990; Galetti *et al.* 1994; Anonymous 1995). In a subsequent study of the Venezuelan red howlers, five more observed infanticides and eight more inferred cases of infanticide were reported (Agoramoorthy & Rudran 1995). Seven out of eight bouts of infanticide occurred in troops with three or four females and the eighth bout was in a troop with one subadult and two adult females. However, the distribution of female group sizes in the population as a whole was not reported.

Mantled howlers do not fit precisely into this scheme. *Alouatta palliata* troops tend to be larger than the other howler species. They can have up to 14 adult females and 6 adult males; male invasion and takeover, infanticide and female emigration all occur as well (Clarke & Glander 1984; Crockett & Eisenberg 1987; Glander 1992; Crockett & Pope 1993). Female mantled howlers are able to immigrate into existing troops (Glander 1992), whereas this is rare in red howlers (Crockett 1984). No data have been published relating female group size and infanticide rate, nor whether female mantled howlers tend to emigrate from larger troops and immigrate into smaller ones, as might be expected.

Mantled howler troops tend to have more males per troop than the other howler species and less sexual dimorphism in body size, perhaps reflecting lower male–male competition (Crockett & Eisenberg 1987). Possibly infanticide rates in mantled howlers decrease as male group size increases because the ability of males to monopolize females decreases. In *A. palliata*, copulation success increased with male rank, and the second and third ranking males in a three-male troop did mate but less than the most dominant (Jones 1981). Only the dominant male mated during the midpoint of the 2–4-day receptive period (Glander 1980). (Red howlers have receptive periods of similar length (Crockett & Sekulic 1982).) The estimated breeding synchrony (more than one female in estrus on the same day) is greater in mantled howlers than in red howlers (Nunn 1999b). It is feasible that with more males, monopolization of females might be reduced and paternity confusion increased. Paternity exclusion tests have been attempted for mantled howlers, but with little success due to low individual heterozygocity (Clarke *et al.* 1994; M. Clarke, pers. comm.). Maybe female mantled howlers have a different strategy for reducing infanticide risk by dispersing to troops with more adult males rather than emigrating from troops with more adult females. At present, the answer eludes us.

The black-and-white colobus (*Colobus guereza*) may turn out to have a pattern similar to that of the red howler. This species lives in small groups, usually with only one breeding male, and takeovers and infanticide have been inferred; some female emigration also occurs (Oates 1977; Dunbar 1987). The number of infants per female declined as female group size increased (Dunbar 1987). Although Dunbar (1987) attributed this to lower conception rates in females in larger groups, because there were also more males and more aggression, the intermittent nature of the field observations cannot rule out the possibility of undetected births and infanticide. Maximum troop sizes were very similar to the red howlers (fewer than 20 individuals), and black-and-white colobus troops never had more than six adult females (mostly two to four females) (Dunbar 1987).

Another species with notable similarity to red howlers is the Milne–Edwards sifaka (*Propithecus diadema edwardsi*). Males emigrate from natal groups at 5–6 years of age, enter nearby groups, and commit infanticide (Wright 1995). Females either breed natally or disperse, female group size is small (one or two females), and one maturing female's disappearance followed persistent aggression by the dominant adult female (Wright 1995). Wright (1995) speculated that female sifakas may disperse farther than males, a pattern documented for red howlers (Crockett & Pope 1993). The sifaka's small female group size may also be related to reducing infanticide risk.

Female emigration and infanticide occur in Thomas's langurs (*Presbytis thomasi*), a folivorous colobine species with small group size and a single breeding male–multifemale composition (Steenbeek 1996; Sterck 1997). Infanticide and infanticide attempts occur during male provocations, when a male from another bisexual group or an all-male band silently approaches and suddenly attacks a mother and infant (Steenbeek, Chapter 7). Infanticide avoidance appears to be the main factor influencing female dispersal in this species, because females appear to leave males that are no longer good protectors against infanticide (Sterck 1997, 1998; Steenbeek 1999a; Steenbeck, Chapter 7). Whereas the lower limit to Thomas's langur groups may be set by predation avoidance, the upper limit appears to be set by infanticide avoidance rather than feeding competition. Similar to red howlers, larger female groups of Thomas's langurs experience a higher rate of takeover, often accompanied by infanticide, such that female reproductive success is probably maximized in small groups (Steenbeek & van Schaik 2000).

Although, in geladas, troops with more females experienced a higher incursion rate, females rarely moved from one reproductive unit to another, and always within the same band (Dunbar 1984b). However, four females is the maximum number that a male can manage (e.g., regularly groom), and social fragmentation begins at female group sizes of five, the minimum size that males can take over successfully, partially due to a collective desertion by females (Dunbar 1984b). Thus the risk of takeover and infanticide may limit female group size in this species as well.

It may not be coincidental that the previous examples are species with predominantly folivorous diets. Perhaps a higher rate of infanticide in larger female groups will prove to be a particularly common pattern in leaf-eating primates. It has already been suggested that infanticide is more common in species with single breeding male social structures (Leland *et al.* 1984; Newton 1988). The folivorous genera in van Schaik's (1983) analysis tend to have more negative slopes of infants-per-female as female group size increases, indirectly suggesting that infant mortality – especially infanticide mortality – may be a penalty paid in larger groups (Mann–Whitney $U = 2$, 3 frugivores, 6 folivores, $P < 0.05$, 1-tailed). On the other hand, folivorous species with small group sizes showed only modest evidence of food competition as measured by day range (Janson & Goldsmith 1995). We expect group size in many of these folivorous species to be limited by strategies to reduce infanticide before the effects of food competition come into play.

We hypothesize that infanticide by males is particularly likely to favor smaller female group sizes in those species with the following characteristics. (1) One male can usually monopolize the breeding position, even if more than one mature male is present. This situation is especially likely to hold in species with little seasonality in breeding, as occurs in many folivores. Lower female overlap in estrous (lower breeding synchrony) is associated with fewer males across primate species (Nunn 1999b). Thus, whereas larger female groups are more attractive to males, leading to increased incursion rates, females do not benefit from forming larger female group sizes to attract even more mates and increase paternity confusion. At the group sizes where female overlap would favor enough males to achieve paternity confusion, food competition would be evident. (2) The presence of additional males may deter invasions but not the risk of infanticide from within. (3) Dietary habits, ecological conditions, and phylogeny favor female social relationships categorized as "dispersal-egalitarian" (Sterck *et al.* 1997). For example, females relying largely on non-clumped, folivorous food sources may experience little benefit from allying with kin to acquire or defend feeding sites. Thus the possibility of dispersal is enhanced, providing a proximate mechanism for limiting group size in response to infanticide.

Conclusion

We presented evidence that infanticide may limit group size in primates and suggested that indirect evidence previously cited as supporting the role of food competition in limiting group size might actually reflect infanticide. Demographic data from 34 troops of Venezuelan red howler monkeys (*Alouatta seniculus*) showed that infanticide rate increased with increasing adult female group size whereas infant mortality from other causes was unrelated to female group size. Birth rates were not significantly related to female group size. Troops with more adult and subadult males experienced a lower rate of incursion by nonresident males. However, the rate of infanticide did not decline with increasing numbers of males. This was because one third of the infanticidal bouts occurred in multimale troops that experienced status changes between males that had co-resided for some time. We suggest that infanticide-related costs of larger female group size oppose the high costs of emigration. Infanticide provides an evolutionary hypothesis to explain the empirical observation that red howler monkeys are found in small social groups characterized by no more than four adult females. In species in which infanticidal males are attracted to groups with more potential mates, in addition to favoring grouping because females must associate with protective males, infanticide may also limit group size. The effect seems to be more prevalent in folivorous species in which dietary effects create weak female social relationships, reducing costs of emigration, and where a single male can monopolize breeding.

Acknowledgments

This study would not have been possible without the commitment of Tomás Blohm and his family to conservation and their support of research at their ranch, Hato Masaguaral. The field research was funded by the Smithsonian Institution International Sciences Program, Friends of the National Zoo, National Geographic Society and Harry Frank Guggenheim Foundation. Thanks go to the Regional Primate Research

Center for incidental support (NIH Grant RR00166), editorial assistance by K. Elias, and literature search by the Primate Information Center. We are grateful to L. Isbell, L. Wolf, P. Waser, C. Borries, C. van Schaik and R. Steenbeek for helpful comments on this chapter. We thank R. Brooks and R. Sekulic for the use of their photographs. This is contribution no. 892 from the Graduate Program in Ecology and Evolution, State University of New York at Stony Brook.

5
_____

Infanticide in hanuman langurs: social organization, male migration, and weaning age

Introduction

Hanuman langurs (*Presbytis entellus*) have been among the first primate species where infanticide by adult males was observed and reported in the wild (at Dharwar, Sugiyama 1965; at Jodhpur, Mohnot 1971; at Abu, Hrdy 1974). Most cases were reported for hanuman langurs living in one-male groups (e.g., Sugiyama 1965; Mohnot 1971; Hrdy 1974, 1977; Newton 1986; Sommer 1994) and they usually took place after the resident male was replaced by a new immigrant. The occurrence of infanticide in hanuman langur multimale groups was first mentioned in 1980 by Ripley and has recently been documented in detail for a wild population at Ramnagar/Nepal (Borries 1997). Other studies on wild primate multimale groups confirmed that infanticide occurs in multimale groups (e.g., *Papio ursinus*, Busse & Hamilton 1981, see also Palombit *et al.* 1997; *Alouatta seniculus*, Crockett & Sekulic 1984; *Macaca fascicularis*, de Ruiter *et al.* 1994).

Generally, it is expected that the risk and frequency of infanticide should be lower in multimale groups as compared with one-male groups (e.g., Hrdy & Hausfater 1984; Leland *et al.* 1984; Altmann 1990; Newton & Dunbar 1994; Sommer 1994; van Schaik 1996). This is indeed supported by fewer infanticides in multimale groups of mountain gorillas (Robbins 1995), a low prevalence of infanticide in populations of hanuman langurs with a predominating multimale structure (Newton 1988; but see Sterck 1999) and low rates of infanticide in most baboon and macaque species (van Schaik, Chapter 2). The following factors have been proposed to reduce the risk of infanticide in multimale groups (e.g., Leland *et al.* 1984; Newton 1988; van Schaik 1996). (1) In multimale groups the sire may remain after the immigration of new males and is available as defender. In addition, due to promiscuous mating of females more than one male might assume to have fathered a particular infant and might defend it as well. Consequently infanticidal males may be less successful at committing infanticide (because of successful defense) or the frequency of attacks may be reduced (because of a higher risk for the attacking male). (2) In multimale groups several males compete for access to females, which may lead to a low reproductive skew amongst group males. Hence, a male faces a low probability of siring the next infant after infanticide. Consequently, the extent to which males benefit from infanticide should be low and the frequency of infanticide is expected to be low as well. (3) In one-male groups, reproductive careers of males are often restricted to a comparatively short period of time when they can monopolize a group of females. In contrast, overall male tenure may be longer in multimale groups. Males may even consecutively become members of different groups over their lifetime (e.g., Alberts & Altmann 1995; Borries 2000). Assuming a longer reproductive career of males in multimale societies infanticide might only rarely pay due to an unfavorable balance of costs and benefits. (4) Finally, in multimale groups female promiscuity and low mating skew may lead to a low paternity certainty so that infanticidal males risk killing own offspring. This may particularly apply to species where immigrant males preferentially attain low rank positions immediately after immigration and only gradually rise in rank later (e.g., van Noordwijk & van Schaik 1985).

In at least two multimale populations, however, infanticide is unexpectedly high. In particular, at the study site at Ramnagar, at least 31% of the infant mortality was attributed to infanticide (Borries 1997; cf. also 38% in *Papio ursinus*, Palombit *et al.* 1997). This unexpected result may be explained in two non-exclusive ways: first, even though reasonable, the four assumptions mentioned above may not hold. Second, the risk and frequency of infanticide may be influenced by other factors (other than social organization) that combine to override the effects of social organization. Most likely factors are male immigration rate and reproductive rate (e.g., Hrdy 1977; Chapman & Hausfater 1979; Leland *et al.* 1984; Packer *et al.* 1988; Crockett & Janson, Chapter 4). While frequent male immigrations will put infants more often at risk of being killed, a high reproductive rate will expose a higher number of infants to strange males.

Given these suggestions, in this chapter we summarize data on infanticide in a wild population of hanuman langurs living in multimale groups and examine the four assumptions (1–4; see above) suggested to lower the risk and frequency of infanticide in multimale groups. On the basis of these results we will model additional factors such as female group size, birth rate (i.e., gestation and pre-weaning period), immigration rate of new males, protection of infants, and number of infants exposed to new males in order to understand the interrelations between these factors and the occurrence and prevalence of infanticide in different social organizations.

Ramnagar: habitat, population, and methods

The langurs inhabit a semi-evergreen forest dominated by Sal trees (*Shorea robusta*) near the village of Ramnagar (Southern Nepal, 300 m above sea level, latitude 27° 44 N, longitude 84° 27 E, Podzuweit 1994; Koenig *et al.* 1997; Wesche 1997). The habitat undergoes strong seasonal changes and the langurs breed seasonally (conceptions, July through November, peak month August; births, January through June, peak month March; Koenig *et al.* 1997: 225). All adult members and group size developments were known for 18 groups. Seventy-two percent of the groups were multimale–multifemale, 28% one-male–multifemale. The mean group size was 18.3 individuals including 2.5 adult males and 6.6 adult females (census data, see below). Multimale groups averaged 20.5 members including 3.1 adult males and 7.2 adult females and were, thus, larger than one-male groups with 12.8 members including 1 adult male and 4.8 adult females. The number of adult males per group was significantly positively correlated with the number of adult females (Spearman rank order correlation coefficient for $N=176$, $r_s=0.56$, $P<0.001$; for $N=14_{\text{June 1993}}$, $r_s=0.76$, $P=0.002$). Females were generally philopatric (for exceptions, see Koenig *et al.* 1998) whereas males dispersed rather frequently. The mean tenure for non-natal males averaged 11.7 months (median 3.1 months) and most immigrations and emigrations occurred around the time of the mating season (Borries 2000). Most adult males were members of bisexual groups. Extra-group males were encountered singly, in duos or in temporary associations of up to five males.

Attacks on infants, (circumstantial evidence for) infanticides as well as group memberships, births, disappearances, and deaths were recorded in *ad libitum* sampling by 22 researchers (see Acknowledgments) from August 1990 through April 1996 during 37 291 hours spent in contact with the langurs. Most data are available for the three main study groups: P-troop (since August 1990, approximately 20 members), A-troop (since July 1991, approximately 10 members), and O-troop (since October 1992, approximately 30 members). Additional information for the other groups which were neighbors of the three main study groups was collected during census work performed every third month (June 1992 through March 1996) as well as during group encounters.

Paternity was determined via DNA analyses from feces by amplifying five informative microsatellite loci (Launhardt *et al.* 1998). If a male could be excluded by at least one microsatellite system he was treated as unrelated to the infant. The average paternity exclusion probability for all five loci combined was 88.8% (Launhardt 1998). For all infants attacked by males either all but one resident male or all resident males present during the conception period of the infant could be excluded from paternity. Males not present during conception but present during attacks could always be excluded as fathers. We could therefore define the following categories: *father* i.e., male not excluded from paternity and resident during the period when the respective infant was conceived; *possible sire*, i.e., male excluded from paternity but resident during the conception period; *non-father*, i.e., male excluded from paternity and not present during the conception period (Borries *et al.* 1999a).

The Ramnagar cases

Events
During the whole study period, 35 cases were observed belonging to the following categories: *attempted infanticide assumed* (N = 3), i.e., no attack on the infant was observed, infant was found wounded but survived, the injury could have been inflicted by the teeth of a male langur (small holes, sharp cuts, a distance of 3–4 cm between injuries) *and* the infant as well as other group members were afraid of, almost continuously watched, and avoided one or several of the group's males; *attempted infanticide* (N = 24), i.e., infant experienced non-fatal attack (clearly targeted approach of a male at high speed); *presumed infanticide* (N = 5), i.e., infant disappeared during the first two months after a new male immigrated into the group *and* the respective male was seen to chase females and/or infants prior and/or after the presumed event; *likely infanticide* (N = 2), i.e., infant died or disappeared after it had been wounded, circumstances as in "attempted infanticide assumed"; *witnessed infanticide* (N = 1), i.e., the male

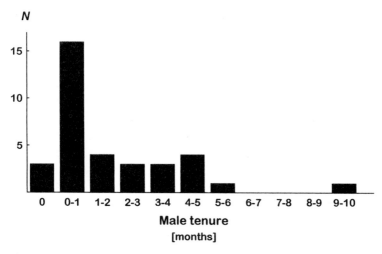

Figure 5.1. Male tenure at the time of attempted, likely, presumed or witnessed infanticide. The value 0 on the *x*-axis refers to three male residents attacking infants in neighboring groups. (For data, see Borries 1997, Table 1.)

was seen to attack and injure an infant that died from its injuries (cf., Borries 1997: 141f). These cases involved 18 individual infants from six groups and 11 individual adult males (for further details, see Table 1 in Borries 1997: 142).

Participants and paternities

All male aggressors were adult. In three cases the attackers were and remained residents in a neighboring group, in all other cases they resided in the same group as the infant they attacked. Male residents who attacked infants (32 cases) were on average in their third month of tenure (mean 2.5 months) with 50% of the males attacking during the first month (Figure 5.1). Only one male attacked in the 10th month of his tenure when he had thus already been resident for longer than the average gestation period (6.9 months at Ramnagar, C. Borries *et al.*, unpublished data). However, this male immigrated into the group only after the infant he is assumed to have killed later had been conceived. Compared with all other resident males, alpha males attacked infants significantly more often (47%, Figure 5.2; $\chi^2 = 16.6$, df$= 1$, $P < 0.001$). But still, males holding the rank positions two to five likewise attacked infants.

Infant victims were 10.0 months old on average when they were attacked (range 1 day to 21 months; Borries 1997: 143) and were not

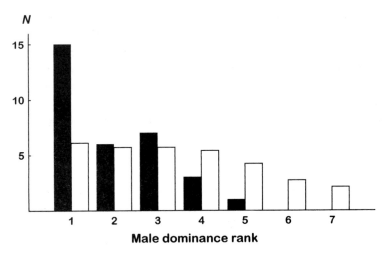

Figure 5.2. Dominance rank of males during attempted, likely, presumed or witnessed infanticide (black bars, observed cases; white bars, expected values; based on all males in the groups for all 32 events). (For data, see Borries 1997, Table 1.)

weaned. Note that the pre-weaning period at Ramnagar lasts from 19 to 32 months (mean 24.9 months, $N = 23$; C. Borries *et al.*, unpublished data). Most of the infants were alone when attacked, i.e., not in body contact with another langur (64% of 25 eyewitnessed cases, Borries 1997: 141) and those were significantly older than infants attacked while in body contact (age$_{alone}$, median 15.8 months, mean 15.2 months, range 4.6–20.3 months; age$_{body\ contact}$, median 1.2 months, mean 2.9 months, range 1 day to 16 months; Mann–Whitney U test, $U = 6.0$, $z = -3.76$, $P < 0.001$). Only infants attacked while alone were injured (19% versus 0%, Borries 1997: 142).

Infants were protected by other group members in all eyewitnessed cases. The mother, other adult females, immature females as well as adult males rushed to the site and chased the attacker or vocalized. Adult males defended in 65% of the cases (total $N = 17$, i.e., repeated attacks within one hour excluded here, Borries *et al.* 1999a) and only those males defended that had been residents at the time when the respective infant was conceived.

Genetic relationship between attacker and infant could be analyzed for 16 of 24 male–infant pairs involved in the 35 cases (Borries *et al.* 1999b). For the remaining pairs, DNA samples were incomplete. In all 16 cases the (presumed) attacker could be excluded from paternity. Protector–infant relationships could be determined in all cases. Only the father or possible sires (mean 1.9 males, range 1–4) defended infants while non-fathers (present in 88% of the cases; mean 2.6 males, range 0–5, attacker not counted) were never observed to defend (Borries *et al.* 1999a). In the course of the study, eight infants fell victim to (presumed/likely/witnessed) infanticide and four of the mothers delivered a subsequent infant. In all cases the presumed killer of the infant was the likely father of the subsequent infant (Borries *et al.* 1999b).

Impact

Fifty percent of the 52 infants born died before reaching 2 years of age, of which eight fell victim to infanticide. Infanticide, therefore, accounted for 31% of the infant mortality (Table 5.1). If only the three main study groups are considered for which the percentage of known causes of infant mortality was highest (i.e., 77%), infanticide accounted for 33% of all deaths and was the major cause of infant mortality. Twenty-two percent of infants died because their mother died or neglected them, 11% fell victim to predation and 11% died very early so they might not have been viable. The risk of infanticide, i.e., the percentage of all infants born who were (presumably/likely) killed by males was 15%, or 17% for the three main study groups. The frequency of presumed/likely/witnessed infanticide accounted for one event in 2.5 or 2.3 group-years respectively (Table 5.1).

Since infanticide only pays if females who lose an unweaned offspring will bear the next infant sooner, we analyzed the length of the interbirth interval in relation to the fate of the preceding infant. If an infant died or disappeared before it was finally weaned, the subsequent interbirth interval of the mother was significantly shorter compared to surviving infants (infant$_{did not survive}$, median 2.0 years, mean 1.6 years, range 1.0–2.0 years; infant$_{survived}$, median 3.0 years, mean 2.7 years, range 2.0–4.0 years; Mann–Whitney U test, $U = 76.5$, $z = -5.0$, $P < 0.001$; Figure 5.3). The particular age of unweaned infants does not seem to be decisive, although the shortest interbirth interval possible (i.e., 1 year) was only achieved if the infant died before completing its third month of life (Borries 1997: 144f).

Infanticide in a multimale population

In the langur population investigated, adult males living in multimale groups killed infants despite protector males (and females). Because

Table 5.1. *Life history variables, population characteristics, and infanticide for langurs at Ramnagar and Jodhpur. Infants were included, if they could have reached or reached 2 years of age (Ramnagar) or 1 year of age (Jodhpur) during the observation period*

	Ramnagar population (1) A B K O P S U X	Ramnagar main groups(2) A O P	Jodhpur main groups (3) B19 B21
Groups			
Mean female group size	7.0	7.8	9.8
Group years	20	14	21
Infants born	52	36	110
Infants died/disappeared	26	18	34–39 (4)
Presumed/likely infanticide	7	5	9
Witnessed infanticide	1	1	6
Infanticide as cause of infant mortality	31%	33%	38–44% (4)
Risk of infanticide	15%	17%	14%
Frequency of infanticide	Once in 2.5 group-years	Once in 2.3 group-years	Once in 1.4 group-years
Birth rate (1/interbirth interval) (5)		0.42	0.72
Weaning age (mos) (6)		Mean 24.9 Range 18.8–32.1	Mean 12.8 Range 11–15
Age of infants killed by males (mos) (7)		Mean 10 Range 0–21	Mean 4.1 Range 0–8.7
Male tenure (mos) (8)		Mean 11.7 Range 0–71+	Mean 26.5 Range 0–74

Notes:
(1) Borries 1997: 144, Table 3;
(2) A. Koenig & C. Borries (unpublished data);
(3) based on the original dataset and Sommer 1994: 188ff; unpublished data;
(4) five infants disappeared at an age of approximately 1 year during a gap in observations (exact age at disappearance unknown);
(5) Ramnagar, C. Borries *et al.* (unpublished data); Jodhpur, Sommer *et al.* 1992;
(6) Ramnagar, Borries *et al.* (unpublished data); Jodhpur, Rajpurohit & Mohnot 1991: 215;
(7) Ramnagar, Borries 1997: 143; Jodhpur, calculated from Sommer 1994: 188ff;
(8) Ramnagar, Borries 2000; Jodhpur, Sommer & Rajpurohit 1989: 300; for tenures ≥3 days, both populations.

males attacked or killed only infants they did not sire, and furthermore the death of an unweaned infant shortened the subsequent interbirth interval of the mother and the presumed killers sired the subsequent infants, infanticide seems to be a male reproductive strategy (Borries *et al.* 1999b). Considering the four assumptions for the prevalence and occurrence of infanticide in multimale populations, the Ramnagar study indicates the following:

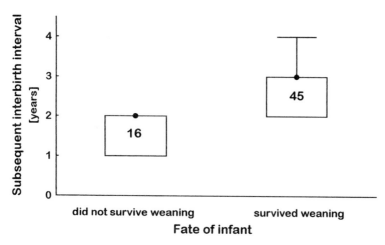

Figure 5.3. Median length of the subsequent interbirth interval for infants who died/disappeared before the weaning process was terminated and for infants who survived the weaning process (boxes = 25% to 75% interquartile range; whiskers = min–max; 61 infants of 6 groups; 43 group-years; values give the number of infants).

1. Adult resident males often (65%) defended infants against attacking males. Furthermore, during all witnessed attacks there always was at least one likely protector male present, on average almost two (Borries *et al.* 1999a). We do not know currently how many likely protector males on average remain in an infant's natal group until it is weaned, since the value given only refers to the number of males present during eye-witnessed attacks. However, the likelihood for attackers to be confronted with protector males seems to be quite high. Nevertheless, we assume the attacker's risk to be small, because we never saw a protector male injure an attacker, possibly because attackers in most cases held the alpha position (cf. Figure 5.2). Aggressors may even attempt to reduce their risks by preferentially targeting infants encountered alone so that they do not have to attack a caregiver as well, and possible protectors were at least not in body contact with the infant. In addition, hints accumulated that at Ramnagar attackers preferred handicapped infants as targets (with broken/paralyzed leg, unconscious after falling, Borries 1997: 141) thereby possibly increasing their chances for a rapid success. Thus we assume that, although in multimale groups attackers are facing a higher risk due to male protectors, this additional risk is not very high. Whether or not male protection actually reduces the risk of infanticide cannot be answered at the moment (but see below).

2. It seems reasonable to assume that in groups where several resident

males compete for access to females, the chances for a particular male to inseminate a particular female will be lower than in groups with a single male. Nevertheless, in all four cases documented for Ramnagar the presumed killer of an infant sired the subsequent infant of the mother (Borries *et al.* 1999b). The likelihood to benefit might, therefore, be higher than expected, which may be due to the fact that alpha males who were most likely to attack infants (Figure 5.2) on average sired significantly more infants than was expected (Launhardt 1998; Launhardt *et al.* 2000). Thus those males who attacked were mostly those who could expect rewards from infanticide. In addition, however, males up to rank 5 attempted infanticide even though their generally low reproductive success (Launhardt 1998) suggests much smaller chances to benefit from infanticide. Such a behavior of low-ranking males may be an additional hint for a low risk for attackers.

3. With an alpha male monopolizing most of the reproductive opportunities at a given time, the reproductive success of all other resident males is low (Launhardt 1998). Each male should therefore try to gain the alpha position and to hold it for as long as possible. However, alpha males can hold their position for just 1.5 mating seasons on average and we have no indication that a male can gain this position more than once in his life. It seems that at least the period with a reasonably high reproductive success is short. Later in life males may mainly help to sustain their past reproductive success via protection of progeny. Hence, the males in these multimale groups may, indeed, have a longer tenure in bisexual groups as compared to one-male populations. But the assumption that males living in multimale groups have a longer reproductive lifespan than males in one-male systems probably has to be modified, at least for Ramnagar.

4. Only infants not sired by the aggressor were attacked or killed. Aggressors were newly immigrants who attacked mainly in the first 6 months of their tenure (Figure 5.1) and thus during a period shorter than the shortest gestation period known for Ramnagar langurs (196 days, i.e., 6.4 months, C. Borries *et al.*, unpublished data). Infanticidal males did not risk killing their own offspring. A separate analysis of the temporal pattern of sexual behavior for attackers with the victims' mothers revealed that attackers possibly gain clues about paternity from the past sexual history (Borries *et al.* 1999a), a connection already suggested by, for example, Sugiyama (1965) and Hrdy (1974). Males even seem to be able to distinguish potentially fertile from pregnant females and only copulations with the former were taken as clues for paternity (Borries *et al.* 1999a). Thus female promiscuity seems to work as a counterstrategy against infanticide only in the long run: newly immigrant males who did not reside during a previous mating season

put infants at risk of being killed while those present during a mating season (but excluded from paternity) never attacked but defended infants. Nevertheless, a low risk of infanticide could result if newly immigrant males mostly attained low ranks. Given female promiscuity these males may not commit infanticide after gradually rising in rank to top dominance. At Ramnagar, however, 93% of rank changes were due to migration but not due to a gradual rise in rank and most alpha males gained their position immediately after immigration ($N =$ 10, mean tenure as alpha male 19.4 weeks). Those who did not gain the alpha position directly after immigration ($N = 3$) had very short tenures as alpha males (mean 2.3 weeks, maximum 3 weeks; Borries 2000).

In conclusion, none of the four predictions is supported by our data and it remains unlikely that these factors lower the risk and frequency of infanticide in the multimale groups of Ramnagar langurs. The fact remains, however, that infants were frequently defended by males and females. On the assumption that defense can prevent infanticide at least in some cases – all else being equal – one should still expect a lower risk and frequency of infanticide compared to one-male populations. But if we compare the results for Ramnagar with data from the langurs at Jodhpur, where bisexual groups are one-male and for which the available data basis is most comprehensive, a puzzling picture emerges (Table 5.1): (1) the frequency of infanticide is almost 80% higher in the one-male population; (2) the percentage of infants who fell victim to infanticide out of all infants who died before reaching the population-specific weaning age was slightly higher in the one-male population *but* (3) the risk of infanticide is identical at both study sites. A closer look at the population characteristics reveals striking differences between the two populations: at Ramnagar the birth rate is lower, the pre-weaning period is longer, and male tenure is shorter.

These data of the two best known hanuman langur populations suggest either that protector males at Ramnagar are unsuccessful when defending infants or that other factors override basic differences between one-male and multimale populations. Identifying and modeling the impact of such factors should lead to a better understanding of the interrelations between social organization and infanticide. In essence, this should help us to understand the occurrence and prevalence of infanticide in different populations or species and to control (statistically) for factors in within-population, cross-population and cross-species comparisons.

Infanticide and social organization: a modeling approach

In the following we explore how the variation of several factors acts on a predicted outcome of two measures of infanticide: the risk and the frequency. In doing so we do not consider (1) genetic dissimilarities between populations or species with regard to the trait of infanticide (see discussion in Hrdy 1979) or (2) social pathologies due to habitat disturbance (see discussion in Hrdy 1979) or (3) variables that may affect the observed frequency of infanticide, like observational conditions, length of observation, degree of habituation and group cohesiveness, which may also affect calculations of the risk of infanticide (see discussion in Borries 1997). Rather, we rely on an expected outcome, i.e., we suppose that males try to commit infanticide whenever there appears to be a chance for them to benefit, but their success and impact is modified by a small number of distinct variables.

Definitions

To judge the risk of infanticide various measures have been used, i.e., causes of mortality in percentage of dead infants (e.g., Struhsaker & Leland 1985; Crockett & Rudran 1987b; Newton 1987), the number of victims in relation to infants exposed to new males (e.g., Sommer 1987, 1994; Sterck 1998), and the number of victims in relation to all live-born infants during a particular time period (e.g., Borries 1997; Crockett & Janson, Chapter 4). Since it has been suggested that a lower risk of infanticide due to social organization would predict that females should favor living in multimale groups (e.g., Newton 1988; van Schaik 1996), we suggest that risk of infanticide should essentially be based on the third measure: it is the only measure which records the impact of infanticide on male and female fitness. We, hence, proceed on the following definition:

$$R(t) = N(t)/L(t) \qquad\qquad (5.1)$$

where $R(t)$ represents the risk of infanticide during the period t, $N(t)$ the number of infants killed by males during period t, $L(t)$ the number of live born infants during period t, t the observation period; to avoid any bias, t has to be adjusted, so that all infants included could have or have survived until mean weaning age (population specific weaning age) during the observation period.

Since earlier considerations on a link of social organization and infanticide were based on frequency (e.g., Leland *et al.* 1984), here we consider frequency of infanticide as well using the following definition:

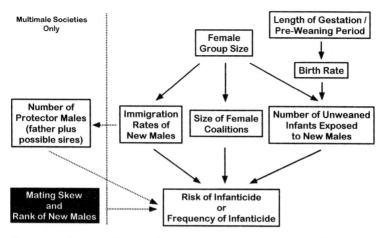

Figure 5.4. Scheme emphasizing the factors proposed to influence risk and frequency of infanticide (for definitions, see text). Dotted lines are applicable only in multimale societies. Factors in the black box were not modeled.

$$F(t) = N(t)/t \qquad\qquad (5.2)$$

where $F(t)$ is the frequency of infanticide during period t.

Influencing components

Given these definitions we consider at least four main factors to influence the risk and frequency of infanticide (Figure 5.4). Basically the number of unweaned infants exposed to new males is one factor that will determine the risk and frequency of infanticide. Obviously, this factor depends on female group size and birth rate, the latter depending almost entirely on the length of gestation and pre-weaning period. Pre-weaning period is essential, because late weaning puts infants at risk for a longer period (e.g., Borries 1997). In addition, it can be expected that male immigration rate is a modifying variable, i.e., the more often new males immigrate the more often infants will be at risk (e.g., Hrdy 1979; Leland *et al.* 1984). Male immigration rate itself may be modified by female group size (e.g., Packer *et al.* 1988; Crockett & Janson, Chapter 4). Apart from these two basic factors, a reduction may be expected owing to defense of infants by females and males. The risk and frequency of infanticide should depend on the size and success of female coalitions, which may be a function of female group size (e.g., Treves & Chapman 1996). In addition, in multi-male populations males who are still resident after a new male immigrated may protect infants and, hence, may lower the risk and frequency of infanticide (e.g., Hrdy 1979; Packer 1979; Busse & Hamilton 1981; Collins *et al.* 1984; Borries *et al.* 1999a). The number of male protectors available is supposed to depend on male migration pattern, which may depend on female group size.

The reader may note that we do not consider interrelations between mating behavior and dominance rank of new males (Figure 5.4, black box). Low-ranking immigrant males who later rise in rank may then commit infanticide provided that mating is skewed (Crockett & Janson, Chapter 4; Soltis *et al.* 2000). Risk and frequency of infanticide would be influenced by a combination of immigration rates of males attaining high dominance right after immigration and of rates with which low-ranking males attain top dominance. If, however, mating skew is low, males rising in rank may not commit infanticide (or at lower rates), because paternity is confused. Consequently, risk and frequency of infanticide would be influenced only by the immigration rate of males attaining high dominance. Since we currently see no way to model these factors we rely in the following on the mere immigration rate of new males (see also below).

Assumptions

In order graphically or mathematically to model the two measures we make three specific assumptions:

1. *Success rate:* To judge the impact of infanticide we assume that males who commit infanticide in different populations have the same impact regardless of fecundity. Consequently, males in different groups, populations or species have equal relative success rates. Essentially we ignore here a potential effect of safety in number for infants.
2. *Defense:* In a similar way we treat the effects of female and male defense. Owing to defense, females (and males) can prevent a fixed proportion of infanticidal attacks depending on the size of coalitions. Again we assume this measure to be independent of the number of infants exposed to infanticidal males. This assumption seems to be justified, since the success in preventing infanticide should be based on single encounters, i.e., given a particular number of defending individuals the likelihood of succeeding in defense should be similar across all similar events.
3. *Male immigration:* In one-male groups immigration of new males will lead to the replacement of the breeding male (i.e., takeover) and the loss of the only protecting male. With regard to multimale groups one has to differentiate between multimale societies where all adult males are replaced during immigration (e.g., lions; see Packer *et al.* 1988) or

populations where at least some males stay following the immigration of a new male. The former closely resemble the pattern of one-male groups and, hence, will be ignored in the following discussion. In multimale populations such as Ramnagar, immigrant males can be former residents re-immigrating or strange males (Borries 2000). As shown above, infants are at risk only from new males (or those absent during the period when the infants were conceived; e.g., Busse & Hamilton 1981; Borries *et al.* 1999a). And these new males, regardless of their rank, attack infants. Even though most of the attacks came from the new alpha males, we may ignore a potential rank effect, since in the Ramnagar population newly immigrating males mostly acquire a high rank (Borries 2000; see also Busse & Hamilton 1981). Consequently, we consider that the immigration of a new male into a multimale group puts infants basically at the same risk as infants in a one-male group following takeover. Hence a reduction in risk or frequency of infanticide would be entirely owing to defense.

Modeling the factors indicated in Figure 5.4 we will mostly refer to three hypothetical populations:

1. The first hypothetical population serves as general baseline (dotted lines, Figures 5.5, 5.6) where defense by either sex may occur but has no effect. Since we assume that success rates of infanticidal males are independent of fecundity, the risk of infanticide should be equal across different female group sizes and the frequency of infanticide should increase monotonously with female group size.

2. The second population is composed entirely of one-male groups where the joint defense of infants by female coalitions reduces the risk and frequency of infanticide (hatched lines, Figures 5.5, 5.6). An increasing size of female coalitions (depending on female group size) will be more successful in defense. However, at a certain point the success may no longer be improved by additional females. We assume here two arbitrary limits, i.e., females are increasingly successful in defense up to an intermediate group size and females can reduce the risk and frequency of infanticide by a maximum of 10% (compared to the baseline).

3. The third population is composed almost entirely of multimale–multifemale groups (solid lines, Figures 5.5, 5.6). For reasons of simplicity, we suppose that the number of males per group is positively correlated to female group size. Hence, smaller groups are more likely to contain only a single adult male while larger groups almost always contain multiple males. In contrast to one-male groups, multimale groups may contain some males who defend infants. Following our findings for Ramnagar langurs we assume that mating

(a)

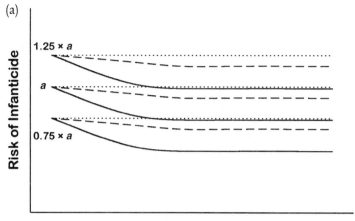

(b)

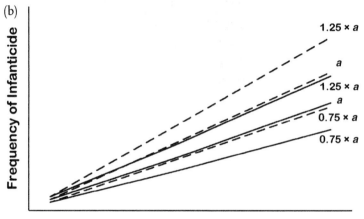

Figure 5.5. Predicted relationships between (a) risk and (b) frequency of infanticide and female group size for three hypothetical populations (dotted, baseline; hatched, one-male groups; solid, multimale) under three immigration regimes (0.75a, a, 1.25a). Immigration rates were held constant across female group size.

is promiscuous and fathers and possible sires protect infants accordingly. Analogous to female defense we assume that an increasing number of protector males will be more successful, males are increasingly successful in defense up to a certain number reached at an intermediate group size and males can reduce the risk and frequency of infanticide by a maximum of 20% (compared to the baseline). It is supposed that male protection adds to the protection of females.

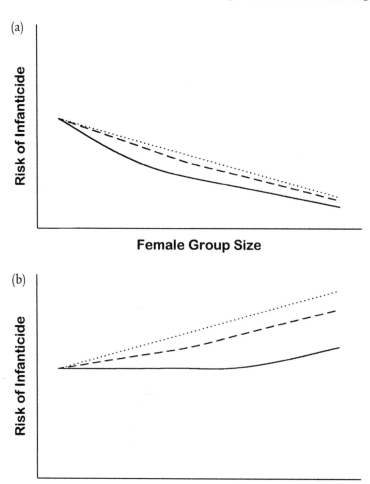

Figure 5.6. Expected risk of infanticide for three hypothetical populations (dotted, baseline; hatched, one-male groups; solid, multimale) under two immigration regimes. With increasing female group size male immigration rate is (a) decreasing and (b) increasing.

Male immigration and infanticide

Considering first that male immigration rates are constant across group size, defense of infants should reduce the risk of infanticide (Figure 5.5a). And provided that male defense is successful, it is to be expected that the risk of infanticide should be lowest in multimale populations. This is true, however, only if populations or species with similar immigration rates are compared. If overall immigration rates vary across populations,

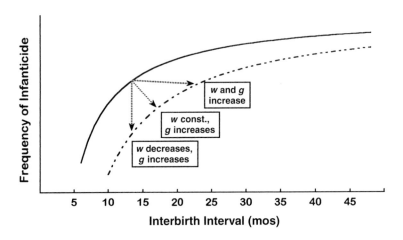

Figure 5.7. Influence of the length of the pre-weaning period (*w*) on the frequency of infanticide for two gestation periods (*g*; solid line, *w* increases at *g* = 5 months; hatched line, *w* increases at *g* = 9 months). Arrows indicate the influence of changes in the length of gestation (*g*) for three possible changes in *w* (for equations and further explanations, see text).

the risk of infanticide varies as well (Figure 5.5a). In a multimale population with an intermediate immigration rate (*a*) the risk of infanticide may be similar to a one-male population with lower immigration rates (0.75*a*). Given an even higher immigration rate (1.25*a*) in a multimale population, the risk of infanticide may be far higher than in a one-male population with low immigration rates (0.75*a*). In contrast, the assumed basic differences of social organizations are reinforced, if multimale populations have low immigration rates (0.75*a*) and one-male populations high rates (1.25*a*). Note that the immigration rates were modified only slightly, i.e., if we consider an intermediate immigration rate of "one new male per 24 months" (*a*) the proposed variation is 18 (0.75*a*) to 30 months (1.25*a*). Similar results are obtained for frequency (Figure 5.5b; baseline condition not shown). Depending on the immigration rates across populations, differences in the frequency of infanticide with regard to social organization may be diminished, reversed or reinforced.

Next we varied male immigration rates across group size and assumed that risk of infanticide parallels immigration rates (see Packer *et al.* 1988), for example if the frequency of immigrations decreases/increases with increasing group size so will the risk for the infants decrease/increase (Figure 5.6, dotted lines). Varying immigration rates should, however, affect the number of protector males. Under the condition of a reduced immigration rate with increasing group size (Figure 5.6a), replacement of protector males is reduced as well. A maximum reduction of the risk of infanticide should be reached at a group size smaller than intermediate. In contrast, increasing immigration rates across group size (Figure 5.6b) should lead to a faster replacement of protector males with increasing group size. Thus the maximum reduction of the risk of infanticide is reached only in larger groups. These two effects of immigration rates across group size reduce the expected differences between social organizations under decreasing immigration rates and increase the differences under increasing immigration rates. Moreover, in Figure 5.6 different costs and benefits of gregariousness become obvious. Under decreasing immigration rates large groups are beneficial, because the risk of infanticide becomes lower (Figure 5.6a). In contrast, in populations with an increasing immigration rate, infants in large groups face a considerably higher risk of being killed than infants in small groups (Figure 5.6b). Hence, in contrast to earlier assumptions (e.g., van Schaik 1996) infants in small groups, which are more likely to be one-male, may be at a lower risk than infants in large multimale groups (Crockett & Janson, Chapter 4).

Overall, the expected relationship between risk or frequency of infanticide and social organization is found only if immigration rates are similar.

The number of unweaned infants and infanticide

To exemplify the impact of life history variables and female group size on the number of unweaned infants (Figure 5.4), we proceed on the definitions for risk and frequency of infanticide (cf. equations 5.1 and 5.2). Frequency of infanticide was defined as the number of victims during a particular observation period. The number of victims at a given male immigration will then depend on the number of unweaned infants exposed to a new male modified by the success of this male

$$N = su \qquad\qquad (5.3)$$

where s is the success rate of infanticidal males and u the number of unweaned infants. The factor s describes the proportion of all infants actually killed and, hence, the interrelation between attacks by males and the success in defense by males and females. This factor includes immigration rates of males as well as variables such as mating skew and rank at immigration. Hence s delineates the impact of male demography, male reproductive strategies and male/female defense.

For convenience, we consider in the following the number of victims at a given male immigration, thus excluding the time factor. The number of unweaned infants at a given male immigration is given by the number of females multiplied by the proportion of time each female has an unweaned infant

$$u = fp \qquad (5.4)$$

where f is the number of females and p the proportion of time a female has an unweaned infant. To calculate p, we assume that conception takes place around weaning. Hence, the interbirth interval will be the sum of gestation and pre-weaning period

$$i = w + g \qquad (5.5)$$

where i is the nterbirth interval, w the pre-weaning period and g the gestation. Hence the proportion of time a female has an unweaned infant is given by

$$p = w/(w + g). \qquad (5.6)$$

Using equations 5.4 and 5.6 the number of victims can be calculated via

$$N = sfw/(w + g). \qquad (5.3a)$$

In addition to the sole impact of social organization (via male defense) equation 5.3a suggests three further influencing variables. Obviously female group size will monotonously decrease or increase the expected frequency of infanticide. The impact of gestation and pre-weaning period is outlined in Figure 5.7. We considered two values for gestation (5 and 9 months) and interbirth intervals to vary between 5 and 45 months, so that the length of the pre-weaning period would change accordingly. Basically, three major results can be deduced from Figure 5.7. First, if gestation increases from 5 to 9 months and the pre-weaning period remains constant, a lower frequency of infanticide should be expected in populations/species with a longer gestation. Second, if both pre-weaning period and gestation increase, the expected frequency should remain essentially constant, even if the interbirth interval increases. Finally, if the pre-weaning period increases at a constant gestation, the frequency of infanticide is expected to be higher where weaning occurs late.

In addition, it should be mentioned that the interbirth interval need not coincide with the sum of gestation and pre-weaning period. Females may conceive sometime after they weaned their infants. If so, equation 5.5 has to be modified to include this non-pregnant and non-lactating period, so

$$i = w + c + g \qquad (5.5a)$$

where c is the time during which a female is neither pregnant nor lactating. On the basis of equation 5.5a, equation 5.3a will change accordingly

$$N = sfw/(w + c + g). \qquad (5.3b)$$

Basically, the introduction of c or an increase of c across populations or species would lead to a decrease in the expected frequency of infanticide, since it increases the time without an unweaned infant. However, the outcome depends on the exact interplay of all three factors, w, c and g.

In conclusion, increases in female group size (always) and pre-weaning period (mostly) lead to an increase in the expected frequencies of infanticide. In contrast, extension of gestation and a non-pregnant non-lactating period will (mostly) lead to a reduction in the expected frequencies.

Considering the risk of infanticide we applied equation 5.3a to equation 5.1

$$R = [sfw/(w + g)]/L. \qquad (5.1a)$$

The number of live born infants L is given as the product of the number of females times birth rate

$$L = fb \qquad (5.7)$$

where b is the birth rate. Birth rate itself can be expressed as the reciprocal of the interbirth interval i, hence

$$b = 1/i. \qquad (5.8)$$

Applying equation 5.5 leads to

$$b = 1/(w + g). \qquad (5.8a)$$

Hence, risk of infanticide may be calculated via

$$R = [sfw/(w + g)]/[f \times 1/(w + g)] \qquad (5.1b)$$

which simplifies to

$$R = sw. \hspace{6cm} (5.1c)$$

Thus risk of infanticide essentially depends on only two factors: the factor s (see above) and the length of time a female needs to wean her infant. Risk of infanticide is directly proportional to the pre-weaning period, i.e., any increase or decrease in pre-weaning period should match a proportional increase or decrease in risk.

In conclusion, social organization does not predict risk and frequency of infanticide unless female group size and life history variables are taken into account. Particularly, the pre-weaning period may tremendously affect the risk and frequency.

Life history variables, population characteristics and infanticide

With regard to these suggestions we have to reconsider our findings. At Ramnagar the low observed frequency of infanticide and the comparatively high risk of infanticide may be related to several factors (Table 5.1): male tenures are short and the immigration rate of new males varies between one new male every 2 years (small groups) up to 2 new males every year (large groups; re-immigrating males not considered; A. Koenig & C. Borries, unpublished data). Birth rate is low, weaning is late and our main study groups contained on average 7.8 adult females. In comparison, the langur females at Jodhpur reproduce about twice as fast, wean their infants earlier, female group size is larger (20%) and male tenure is longer. At both sites gestation is rather similar (Ramnagar 209 days, Jodhpur 200 days) and females conceive approximately at the time when weaning takes place (Sommer *et al.* 1992; C. Borries *et al.*, unpublished data).

Hence, given the suggestions from our modeling approach and assuming that protector males at Ramnagar would be unsuccessful, one should expect a much higher risk at Ramnagar compared with Jodhpur. Frequency of infanticide should not be affected in a similar way, since female group size at Ramnagar is smaller and pre-weaning period is not linearly related to frequency of infanticide. One may, therefore, expect that frequency of infanticide should be similar or slightly higher at Ramnagar. In reality, however, risk is equal and frequency lower. Thus we suppose that risk as well as frequency are lower than expected and that this "low" risk and frequency of infanticide mirror the effect of multimale groups. If males would not successfully defend the infants, the risk at Ramnagar should by far exceed the risk at Jodhpur.

Overall, even though we did not explicitly mathematically model all factors, we hope it became clear that the interrelations between the different factors and their proposed influence may strongly affect the actual risk and frequency of infanticide in a given population. This remains true even if the success of protector males is higher than assumed here, since the expected effects of immigration, female group size, pre-weaning period, and gestation would still hold. Moreover, the variations applied were small and clearly in the range for extant primate populations and species. If all things were equal the predicted relationships between social organization and infanticide should uphold. Things are not equal, however, and we suggest that observed deviations from the expected pattern regarding infanticide and social organization are caused mainly by variation in population characteristics and life history variables. Even though the comparison of two populations is hardly a strong test, the data from Jodhpur and Ramnagar seem to show that (1) male defense may reduce the impact of infanticide, but that (2) the variation in other factors may override any effect of social organization. Future studies on relations between infanticide and social organization, therefore, require more precise, long-term data on the different factors in question and cross-species and cross-population comparisons must statistically control for these factors.

Acknowledgments

We would like to thank our Nepalese co-researchers Dr M. K. Giri, Dr P. Shresta and Professor S. C. Singh (Natural History Museum, Kathmandu), the Research Division of the Tribhuvan University (Kathmandu), and the Ministry of Education, Culture, and Social Welfare (His Majesty's Government, Kathmandu) for support, cooperation, and the permission to conduct the field study at Ramnagar. We are indebted to all participants of the Ramnagar Monkey Research Project for their contribution to the database (Hari Acharya, Ulrike Apelt, Ralf Armbrecht, Jan Beise, Hari Cchetri, Mukesh Chalise, Dud Bikram Ghale, Shiva Lama, Kristin Launhardt, Erik Mittra, Julia Nikolei, Julia Ostner, Doris Podzuweit, Yam Bahadur Rana, Oliver Schülke, Keshab Thapa, Sylvie

Thurnheer, Karsten Wesche, Paul Winkler and Thomas Ziegler). We thank Volker Sommer for the permission to reanalyze the data of the two main study groups of Jodhpur. Our sincere gratitude goes to Sarah Hrdy, Charlie Janson, Julia Ostner, Oliver Schülke and Carel van Schaik for their very useful comments on an earlier draft. We particularly thank Charlie Janson for introducing simplifications to our initial modeling attempt. Financial support of the field work by the Deutsche Forschungsgemeinschaft (DFG, grant nos. Vo124/19–1+2, Wi966/4–3; Institute of Anthropology, Göttingen; C.B.) and the Alexander von Humboldt-Stiftung (AvH, Bonn, V-3–FLF-1014527; A.K.) is gratefully acknowledged. The joint field study at Ramnagar further benefited from individual grants generously provided by the Alexander von Humboldt-Stiftung (AvH), the Deutsche Forschungsgemeinschaft (grant nos. Wi966/3–1+2 and He2699/1–1+2), the Deutscher Akademischer Austauschdienst (DAAD), the Deutsche Gesellschaft für Technische Zusammenarbeit (GTZ), and the Ernst-Stewner-Stiftung. During data collection C.B. was a member of the primatological group of late Professor C. Vogel and Dr P. Winkler (Institute of Anthropology, University of Göttingen, Germany).

RYNE A. PALOMBIT, DOROTHY L. CHENEY, JULIA FISCHER,
SARA JOHNSON, DREW RENDALL, ROBERT M. SEYFARTH, AND
JOAN B. SILK

6

Male infanticide and defense of infants in chacma baboons

Introduction

Sexually selected infanticide is relatively widespread among primates, but has been documented primarily in one-male–multifemale reproductive units, e.g., in guenons (*Cercopithecus* spp.) (Butynski 1982; Fairgrieve 1995), langurs (*Presbytis* spp.) (Hrdy 1974; Newton 1988; Sommer 1994), howler monkeys (*Alouatta* spp.) (Crockett & Sekulic 1984), and mountain gorillas (*Gorilla gorilla berengei*) (Fossey 1984; Watts 1989). Although male infanticide has been invoked as a selective force in multimale–multifemale groups, such as in macaques (*Macaca fascicularis*) (van Noordwijk 1985), capuchin monkeys (*Cebus olivaceus*) (O'Brien 1991), and chimpanzees (*Pan troglodytes*) (Smuts & Smuts 1993), it has rarely been observed in these species (e.g., Valderrama *et al.* 1990; Camperio Ciani 1984) or follows patterns partly inconsistent with Hrdy's (1974) sexual selection hypothesis (Hiraiwa-Hasegawa & Hasegawa 1994). Thus current data suggest that the presence of multiple males in a primate group discourages infant-killing by other males.

Relative to one-male groups, the presence of additional, reproductively active males may both dilute the benefits of infanticide and increase its costs. Exploitation of the reproductive opportunity created by infanticide depends upon the perpetrator's ability to monopolize subsequent fertilizations, which is a function of social variables such as the number of males in a group, the intensity of male–male mating competition, and the potential for effective mate guarding. In multimale groups, defense of infants by other resident males may increase the costs of infanticide, while mating competition may decrease the benefits. The relative contribution of these two processes in inhibiting infanticide in multi-male groups remains unclear.

Nevertheless, infanticide is a suggested reproductive strategy in several primate species living in multimale groups: the red colobus monkey (*Procolobus badius*) (Leland *et al.* 1984; Struhsaker & Leland 1985), the hanuman langur (*Presbytis entellus*) (Borries 1997; Borries & Koenig, Chapter 5), the red howler monkey (*Alouatta seniculus*) (Clarke & Glander 1984; Sekulic 1983a) and the subject of this chapter, the chacma baboon (*Papio cynocephalus ursinus*).

Chacma baboons of southern Africa live in large, multimale–multi-female groups of 60–80 individuals. On the basis of their observations of a population inhabiting the Moremi Game Reserve in the Okavango Delta, Botswana, Busse & Hamilton (1981) were the first to argue that infanticide functions as a male reproductive strategy. Alpha males achieve greater mating success than other males in the group (Bulger 1993), but the duration of alpha tenure is brief relative to female interbirth intervals, averaging well under 1 year (Hamilton & Bulger 1990). Thus a male that has recently immigrated into a group and attained the alpha position in the dominance hierarchy may increase sexual access to fertile females by killing infants, thereby terminating lactational amenorrhea in their mothers sooner than if infants survived to weaning.

Collins *et al.* (1984) summarized observations of infanticides over a 3-year period at Moremi. This dataset comprised four cases: two directly observed attacks and two inferred infanticides based on circumstantial evidence. Two episodes in which infants survived injuries inflicted by males were also described (see Appendix 6.1). These data confirmed that infant death accelerated the resumption of cycling in mothers, suggested that infanticide was more common in the chacma baboon than among East Africa conspecifics such as the olive baboon (*P. cynocephalus anubis*), and directed attention at the much shorter tenures of alpha males in the Moremi population as a possible cause of this interpopulation difference. Nevertheless, the patterning of infanticidal attacks among the Moremi chacma baboons was also inconsistent with Hrdy's sexual selection hypothesis in some cases. For example, in one instance the infant injured by a male was already weaned (and presumably no longer inhibiting cycling in its mother); in another case, the attacking male was a low-ranking immigrant unlikely to fertilize the infant's mother subsequently. The data set and accompanying conclusions have been questioned on these grounds, as well as for other reasons (e.g., tranquilization of a mother via darting preceded the one directly observed infanticidal attack by a known male).

There is a clear need to re-evaluate the question of sexually selected infanticide in this baboon population. First, the implications of the inconsistencies in the Collins *et al.* (1984) record, which constitute a relatively large proportion of their observations, can be resolved only through consideration of additional data. Second, the assertion of systematic intraspecific variation in rates of infanticide among savanna baboons requires further supportive evidence. The possible generality of infanticide in chacma baboons is suggested by one observed and one inferred infanticide reported recently for a population inhabiting the Drakensberg Mountains, South Africa (Weingrill 2000; Appendix 6.1). Third, male infanticide has been argued to be a crucial selective agent behind the evolution of many features of the chacma baboon social system, such as infant-carrying by males (Busse & Hamilton 1981), the "tail-raising" visual display of females (Busse 1984a), triadic interactions involving males and infants (Busse 1984b), copulation calls (O'Connell & Cowlishaw 1994), heterosexual "friendships" (Palombit *et al.* 1997), and female–female social interactions (Palombit *et al.*, 2000). These arguments all presuppose the existence of infanticide as a sexually selected male reproductive strategy.

The occurrence of infanticide in chacma baboons has naturally directed attention to possible counterstrategies of females. Palombit *et al.* (1997) argued that chacma baboon friendships represent such a strategy. When they give birth, females typically establish and actively maintain close bonds with particular males, who presumably will defend their infants. Observational and experimental data suggest that friendships perform an anti-infanticide function, but Palombit *et al.* (1997) provided only limited descriptions of male responses to actual infanticidal attacks.

In this chapter we update the record for the Moremi chacma baboons by summarizing observations of infanticide over a 6-year period. We use these new data to test two predictions of the sexual selection hypothesis. First, a male will kill infants sired by other males. Second, a male will mate with females whose infants he has killed. Finally, we further evaluate the argument that males protect the infants of their female friends by examining the responses of males to actual and potentially infanticidal attacks, and by considering the implications of male defense for the sexual selection hypothesis.

Background and methods

Study area and subjects

The study site is situated in the Moremi Game Reserve in northwestern Botswana (23° 02′ E, 19° 31′ S). The Okavango Delta is a seasonal wetland comprising grasslands and raised "islands" covered with trees and shrubs. Descriptions of the study area are provided by Tinley (1966), Buskirk *et al.* (1974), Hamilton *et al.* (1976) and Ross (1987).

We summarize patterns of infant attack and mortality for the fully habituated study group, C troop, observed continuously from July 1992 to August 1998. Over this 6-year period, 31 adult (parous) females and 22 non-natal adult males resided in the group. At any given time, however, the group of 60–70 individuals generally comprised 3–8 fully adult non-natal males, 3–7 natal adult males (≥8 years), 22–26 adult (cycling) females, and their immature offspring. Maternal relatedness was known for all females and natal males because of the long history of prior observation (1977–1991) by W. J. Hamilton III and colleagues (e.g., Hamilton *et al.* 1976; Busse & Hamilton 1981; Hamilton & Bulger 1992).

In chacma baboons, both males and females are organized into linear dominance hierarchies in which all females are subordinate to males (Bulger 1993). Subjects were assigned dominance rankings based on the direction of agonistic interactions, i.e., "supplants," "grimace" visual displays, or overt aggression. The average tenure of alpha males in C troop was 6.6 months (SD = 4.2, N = 9 males), excluding several periods when the dominance relationship between the current alpha male and a contemporary new immigrant male was ambiguous to experienced observers. This estimate of alpha tenure duration corresponds roughly with the short mean of 5–6 months reported previously for the same population (Collins *et al.* 1984; Hamilton & Bulger 1990). Although males may remain in their natal groups into adulthood and eventually attain alpha status (Bulger & Hamilton 1988; Hamilton & Bulger 1990), all but one of the alpha males during 1992–1998 were non-natal individuals.

Infant wounding in Moremi chacma baboons

We describe below the directly observed and inferred cases of infant wounding. In some instances, observers were alerted to an infanticidal attack already underway by outbursts of intense screaming. The screaming associated with infanticides impressed observers as highly distinctive by virtue of its high amplitude and its patterning, i.e., sustained calling and widespread participation by large numbers of individuals of both sexes, many of whom were not involved directly in the interaction (for similar accounts, see Collins *et al.* 1984; Tarara 1987). Unless otherwise noted, all infants appeared healthy and lacked ostensible signs of illness or debilitation prior to attacks upon them. The cases described below are summarized in Table 6.1.

Direct observations

Directly observed infanticidal attacks on infants were by two non-natal adult males that had immigrated into the study group and/or had attained alpha status relatively recently. Male WA immigrated in April 1991, and attained alpha status for the second time during his residency in June 1992. Male DG immigrated into the group in June 1994 and became alpha male within a week. One non-fatal attack by a third male was also directly observed. Male TM was born in the study group in September 1985, emigrated in April, 1995 (entering a neighboring group), returned to his natal group in April 1997, and attained alpha status in October 1997.

Case 1

On 21 October 1992, loud screaming by multiple baboons directed observers' attention to male WA, who was running through the group with 61-day-old infant BR in his mouth. This infant had been seen alive 15 minutes earlier. Adult male TN, who was the friend of BR's mother, chased male WA, giving typical male loud calls ("wahoos") and screams. Several females including infant BR's mother, her 3.75-year-old daughter, and two unrelated high-ranking females also ran screaming at WA, followed by several subadult natal males unrelated to the mother; however, none of these individuals approached male WA closely or intervened directly.

Case 2

This infanticidal episode on 8 February 1993 essentially followed the same pattern as case 1. Loud screaming erupted and male WA was seen running with infant NK in his mouth. The 112-day-old infant had been seen unharmed in the previous 10 minutes. The adult male TO, who was the friend of NK's mother, ran over wahooing, approached WA, and appeared about to engage WA in a fight, but then withdrew. No other baboons

Table 6.1. *Summary of observed and inferred attacks on infant chacma baboons (1992–1998)*

Case	Infant	Date	Age (days)	Sex	Circumstances	Observer
Direct observations						
1	BR	21 Oct. 1992	61	F	Outburst of loud screaming by multiple baboons occurred and adult male WA was seen running through center of group with dead infant BR in his mouth. Male friend of BR's mother rushed over wahooing and screaming, but did not engage WA in fight	D.L.C., R.M.S.
2	NK	8 Feb. 1993	112	M	Outburst of loud screaming by multiple baboons as adult male WA ran through group with infant in his mouth. Male friend of NK's mother rushed over, but did not engage WA in fight. Male WA ate infant	D.L.C., R.M.S.
3	AR	13 Sept. 1994	54	M	Adult male DG attacked at close range female BL carrying infant AR. No males came to the aid of the female (but few were present at the time). Male DG ate the infant	R.A.P.
4	NH	15 Sept. 1994	353	M	Outburst of loud screaming occurred, and adult male DG was seen running with screaming infant in his mouth, pursued by two other baboons. Male DG dropped the infant, already dead, and ran off	R.A.P.
5	RB	20 Oct. 1994	2	M	Outburst of loud screaming and adult male DG was seen running with infant in his mouth. It was not clear whether the infant was already dead. Other adult males, including male friend of RB's mother, were > 200 m away at the time of the attack	R.A.P.
6	CC	16 July 1997	280	F	While her mother was away at a vervet monkey kill site, CC was attacked and bitten twice in the torso by the adult male TM. The infant was retrieved by a distantly related, 8-year-old natal male. The infant became listless in ensuing days, but eventually recovered	D.L.C., R.M.S.
Circumstantial observations						
7	HI	13 Nov. 1997	29	M	During an intergroup interaction, an adult male of the study group (probably TM) attacked infant HI, biting him in the face and head. The infant died the next day	D.L.C., R.M.S.
8	JB	26 Aug. 1993	1	F	Outburst of screaming, chasing, and wahooing of adult male GL occurred. Infant JB was discovered dead with newly inflicted puncture wound to head	R.A.P.

9	HP	11 Oct. 1994	131	M	HP was seen alive and carried by his mother in early morning. When the group was recontacted in mid-morning, the mother was holding the dead body of HP, which had two bleeding puncture wounds in head	R.A.P.
10	MM	20 Aug. 1996	191	M	Outburst of screaming occurred. MM was discovered splattered with blood and with apparent canine punctures on left abdomen and left thigh, but still alive and carried by mother. He was dead the next morning	D.R.
11	SN	15 Nov. 1992	366	M	SN disappeared from the group during a violent, prolonged chase involving multiple baboons of both sexes, running about screaming and wahooing. Infant's mother was discovered the next day to have a wound on her left arm	D.L.C., R.M.S.
12	SU	19 Sept. 1994	156	F	SU was seen alive and carried by her mother in early morning. When the group was recontacted several hours later, SU was missing and adult male DG was found with blood on his face and hands, chewing on animal flesh	R.A.P.

Figure 6.1. An alpha male chacma baboon in the Moremi study group, minutes after he captured and killed an infant; case no. 3 (photo: Ryne Palombit).

entered the vicinity of male WA. Male WA ate the infant as male TO and NK's mother sat nearby. NK's mother was later found to have a new wound on her left arm.

Case 3

At 09:40 on 13 September 1994, the study group encountered lions while crossing a grassy field. Baboons fled, alarm calling, in either of two directions, and ascended the trees of wooded islands on opposite sides of this field. About 30 minutes later, the baboons returned to the ground and were relaxed despite the continued presence of lions 100–125 m away, which prevented reunion of the two subgroups that had formed during the initial flight. At 10:30, alpha male DG began wahooing and chasing females. He initially chased female LE (carrying her infant), and then female CD (whose infant was absent). When male DG came to within 20 m of primiparous female BL holding her 54-day-old infant AR, he terminated his chase of female CD and set upon BL. She ran away screaming, carrying her infant, but DG pursued her around an acacia bush, and when he emerged from the other side, he was already carrying AR in his mouth (Figure 6.1). Female BL continued to scream for about 30 seconds, along with several distant baboons. She then climbed into a small tree approximately 15 m from male DG, as he ate the infant. No baboons came to the aid of female BL, but this may have been partly because the subgroup she and DG were in, formed after the encounter with the lions, included only one of the other seven non-natal, adult males of the group. BL's male friend, a 9.5-year-old natal male who was second ranking in the hierarchy, was in the subgroup with BL, and had been last seen with her 30 minutes earlier. He failed to intervene against DG, though he appeared and sat 5 m from male DG soon after he had begun eating the infant.

Case 4

On 15 September 1994 intense screaming broke out and male DG was seen running with an infant in his mouth and pursued by two unidentified subadult males. At one point during the chase, DG stopped momentarily, sat down, and handled the infant, which was screaming and struggling to get loose. Male DG then resumed running with the infant in his mouth and, about a minute later, dropped its dead body and withdrew. The infant's wounds consisted of: (1) a laceration (6.5 cm $\times$ 5 cm) on the lower back exposing part of the pelvis and vertebral column; (2) minor lacerations on the top of the head and on left parietal region; (3) one bleeding puncture on the left arm above elbow; (4) one bleeding puncture on the leg, 5 cm above heel. The victim was the 353-day-old male infant NH.

Case 5

Female RS gave birth to her first infant RB on 18 October 1994. During this and the following day, the newborn infant grasped his mother weakly, prompting her to hold him in place on her ventrum with one hand when walking. On 19 October 1994, male DG made an apparent infanticidal attempt, but RS's male friend EG intervened successfully (see below). At 07:04 the next day intense screaming erupted and adult male DG was seen running with RS's neonate in his mouth. It was unclear whether the infant was still alive or already dead. DG ascended a tree and ate the infant. Male DG was unmolested by any of the group's adult males, all of whom, including the mother's friend EG, were several hundred meters away foraging on aquatic roots. Some of the intense screaming was by females that were also foraging down by the river and were too far away to have seen the attack transpiring on the other side of an intervening hill.

Case 6

On 16 July 1997, female infant CC was 150 m away from her mother, who was attempting to acquire a piece of a vervet monkey (*Cercopithecus aethiops*) that had just been killed and was being eaten by an adult male. Adult male TM was with most of the other adult males of the group at the kill site, but he then returned to 280-day-old infant CC's location, seized her, and inflicted two canine punctures to her torso. Eight-year-old natal male MS rushed forward immediately, retrieved the wounded infant, and carried her into a tree. After searching unsuccessfully for her infant, the mother moved off with the rest of the group. Male MS attempted to follow after the group several times, leaving CC temporarily, but he always returned to her, for she was unable to keep up with him. About an hour after the group had left them, MS picked up CC and carried her back to the group, where she was reunited with her mother. By 19 July, CC was listless and appeared to be suffering from an infection. By 22 July, however, she was recovering, and, in fact, survived TM's attack. Although he was a natal male, TM was maternally unrelated to the infant CC. Male MS was the second cousin of CC.

Summary

In two instances (cases 3 and 6) infanticides were essentially observed from start to finish. In one case (number 4), the onset of the attack was unobserved, but the male was seen to kill the live infant in his possession. In the other cases, observers saw a particular adult male running with an infant in his mouth immediately after outbursts of screaming by multiple baboons. In five of the six cases, the infanticidal male was the current alpha male; in the sixth instance (case 6), the male attained alpha rank approximately 2.5 months after his non-fatal attack on an infant. Infants attacked by males DG and TM were conceived prior to these males' immigration into the group; the infants fatally wounded by male WA were conceived prior to his rise to the alpha position.

Circumstantial observations

The first four cases concern infants that were discovered with fatal injuries.

Case 7

During an interaction with a neighboring group on 12 November 1997, an outburst of screaming occurred from C troop baboons far from the interaction area and members of the other group. Observers arriving immediately found that the female LX's 29-day-old infant HI had received bite wounds on the face and head, but was still alive. Adult males AP, AU and TM were present, but the first two males were unlikely candidates for infanticide, since LX had established friendships with them when she gave birth to HI. On the other hand, several adult females present were directing intense screaming at alpha male TM, which suggested that he may have been responsible for the injuries. The infant died the next day of its wounds.

Case 8

On the morning of her birth on 26 August 1993, the infant JB appeared normal and healthy as late as 11:40. An hour later, many baboons began suddenly screaming intensely and several unidentified individuals were seen chasing one another through dense vegetation. The current alpha male MK was the subject of a focal animal sample at the time and was uninvolved, but adult male GL, who had immigrated into the group 3 days earlier and who would later become alpha, was seen wahooing and running from other baboons. Female JN was located 10 minutes later, grunting at a pregnant, 14-year-old female who was holding her now-dead infant. The infant's condition was characterized by the following: (1) a slightly bleeding, circular puncture approximately 3.5 cm above the left eye; (2) an exposed patch of scalp on the top of head; (3) multiple minor lacerations under the left eye, on the right hand, and around the nostrils; and (4) blood draining from its nostrils. Over the ensuing 30 minutes, JN unsuccessfully attempted to retrieve her infant's body from the other female, and finally withdrew. The pregnant female later abandoned the dead infant.

Case 9

At 06:45 on 11 October 1994, the adult female SH was seen carrying her 131-day-old infant, HP. When the group was contacted 45 minutes later, SH was found holding the now dead body of HP. The infant had a bleeding puncture wound on the temporal region of each side of its head. The current alpha male DG had attacked this infant and its mother several times in the preceding weeks (see below).

Case 10

In the late afternoon of 20 August 1996, an outburst of intense screaming occurred. The 191-day-old male infant MM, which had been seem healthy that morning, was found slumped on the ground near its mother. He was still alive, but had a non-bleeding puncture wound in the left abdomen and on the left thigh, as well as blood on his hair. Besides occasionally lifting his head and vocalizing, the infant hardly moved for the rest of the day. Though still alive when the group entered its sleep tree that day, the infant was dead by 09:00 the next morning.

The last two cases concern disappearances of infants in circumstances suggestive of possible infanticide.

Case 11

On 15 November 1992, the 1-year-old male infant SN disappeared from the group during a violent, prolonged chase involving multiple baboons of both sexes. Many males were running about wahooing and chasing one another, and females were screaming at an intensity previously associated with observed infanticides. The interaction was observed from a distance, but the infant's mother was discovered the next day to have a wound on her left arm.

Case 12

When the study group was first sighted on the morning of 19 September 1994, the adult female SY was carrying her 156-day-old female infant, SU. When recontacted several hours later, the infant SU was missing and the alpha male DG was simultaneously found with blood on his face and hands, chewing on a piece of animal flesh. Male DG had been observed to kill two infants in the preceding 6 days, one of which he had eaten (see above), and he had attacked the infant SU 5 days earlier (see below).

Summary

Although observations were defined as "circumstantial" when an infanticidal attack was not directly observed, it is clear that in four of these six cases, infant death coincided with newly inflicted, fatal wounding. There were two sources of injuries of these kinds: conspecifics or predators (such as lions or leopards). It seems unlikely that these wounds were inflicted by predators because of the absence of alarm calling or any of the flight responses typical of a predator encounter. Instead, the screaming and chasing of multiple baboons were consistent with an attack by a conspecific. The other two cases were instances in which the disappearance of an infant coincided with conditions associated with other observed infanticides.

The distribution of apparently canine-inflicted puncture wounds on the head (cases 7, 8, 9), torso (cases 4, 6, 10), and limbs (case 10) suggests that infanticidal chacma baboons deliver fatal bites on any body area they can seize, in much the same way as male hanuman langurs do (G. E. King & D. Steklis, unpublished data).

Other non-fatal attacks on females and infants

Besides the cases of infanticide describe above, we also observed a number of unsuccessful attacks on lactating females by the alpha male. Many of these interactions took the form of protracted chases of mothers with infants that contrasted sharply with the usually brief chases of cycling or pregnant females. We describe below several examples of such attacks and then summarize these and additional observations in Table 6.2. These *ad libitum* data provide further information on the response of males – friend and non-friend – to potentially infanticidal attacks.

Female SH and her infant HP

At 08:20 on 14 September 1992, alpha male DG appeared suddenly, wahooing, near female SH and her male infant HP, who fled screaming into a tree. DG followed and knocked SH and her infant to the ground by vigorously shaking the terminal branch she was hanging from. He jumped down after her, resuming his chase, whereupon an outburst of intense screaming and other calls from nearby baboons occurred. SH's male friend NP, the non-friend adult male WA, as well as several females ran over, but only NP moved in closely to DG. NP engaged DG in a fight, screaming and fear-grimacing as DG hit him, during which time SH escaped with her infant and climbed into the interior of a dense camel-thorn acacia bush. Although DG pursued her, the density of branches and long thorns in the bush apparently prevented him from gaining access to her, and he eventually withdrew. Male NP and an adult sister of SH sat 5 m away from the acacia during this time.

On 22 September 1994, alpha male DG emerged from behind foliage and rushed at the infant HP, who was sitting near NP, the male friend of his mother. NP immediately grabbed the infant and, carrying him ventrally and screaming continuously, ran off pursued by DG. A non-friend adult male WA ran over from 25 m away, but DG immediately chased him off some distance and then resumed chasing NP and HP. Again screaming, NP carried the infant into a camelthorn acacia bush. Male DG followed, as did male WA and the infant's mother SH, both of whom sat outside the bush. DG then moved off.

Female AL and her infant AA

On 18 October 1994, alpha male DG chased the female AL and her infant AA from a wooded island into a grassy field. AL's male friend EG and her 9-year-old son ran over immediately after she began screaming. EG

Table 6.2. Summary of non-fatal attacks on females and infants by alpha male DG

Date	Target of attack	Response of mother's adult male friend	Response of other individuals
14 Sept. 1994[a]	Female SH and Infant HP	Friend NP ran over immediately; fought male DG, hitting, screaming, and fear-grimacing to him; then sat below acacia bush SH and HP had fled into until DG withdrew	Non-friend adult male WA and several females including the adult sister of SH ran over; these females and other females nearby screamed; male WA withdrew, but SH's sister remained with NP under the bush until the attacker DG withdrew
14 Sept. 1994	Female SH and Infant HP	Friend NP ran over and sat under the acacia tree SH and HP had fled into	An adult female maternally unrelated to SH accompanied NP, screaming, to the base of the acacia tree SH and HP had fled into
22 Sept. 1994[a]	Infant HP	Friend NP grabbed infant HP as DG rushed at it, and carried it ventrally while screaming and fleeing; NP entered an acacia bush with infant and remained there until DG withdrew	Non-friend adult male WA ran over, was chased off by DG, returned, and sat nearby; infant HP's mother SH accompanied WA
18 Oct. 1994[a]	Female AL and Infant AA	Friend EG ran over, fought DG, biting and hitting him; then chased DG several times; then sat near female and infant	9-year-old son of AL ran over immediately, but did not approach attacker
7 Nov. 1994	Female AL and Infant AA	AL with AA ran onto wooded island pursued by attacker DG; when recontacted 10 min later, (new) male friend TO was sitting below the tree AL and AA had fled into	None
8 March 1995	Female AL and Infant AA	Friend TO ran over, threatened DG visually, then gave roars and threat vocalizations; DG hesitated and withdrew, whereupon AL and AA descended and ran off with male TO	None
30 July 1994	Female BT and Infant BZ	Friend WA ran over, sat under the tree BT had fled into screaming, with BZ; then grabbed a nearby 19-month-old	Non-friend adult male GL ran over and stood under the tree BT and BZ fled into

Date	Subjects	Description	Outcome
		juvenile (unrelated to BT) and roared while looking at attacker DG	
15 Sept. 1994	Female BT and Infant BZ	Friend WA ran over to within 15 m of acacia tree BT and BZ fled into; DG terminated chase of BT; BT descended with her infant and ran to WA	None
27 Oct. 1994	Female JN and Infant JK	Friend WA followed non-friend TO and sat under tree JN and JK had ascended; JN climbed down, joined WA, and followed him closely for 15 min	Non-friend adult male TO immediately ran over and sat under tree JN had climbed into
20 Jan. 1995[a]	Infant JK	Friend WA repeatedly threatened, hit, and grunted at infant JK when it approached alpha male DG and until it withdrew to its mother	None
14 Sept. 1994[a]	Infant SU	Friend GL ran over, chased DG; then sat below acacia tree into which SU's sister had carried it; while sitting under this tree, male GL grunted and looked at attacker DG and female holding infant	SU's 5-year-old sister grabbed infant, carried it pursued by attacking male into an acacia tree; SU's mother SY accompanied friend GL, screaming, and sat below the tree SU and her sister entered
19 Oct. 1994	Female RS and Infant RB	Male EG ran over, sat below tree RS and RB had escaped into; after 7 min, approached and presented genital region to alpha male DG, grunting; then sat near RS and RB in tree, barking occasionally.	None

Notes:
[a] These cases are described in greater detail in the text.

charged directly at DG, and both males slapped and bit at one another, though no obvious injuries were inflicted. DG then ran off wahooing and EG chased after him, but this failed to prevent DG from resuming his attack on screaming female AL and her infant. Again, EG engaged DG in a fight, which ended as DG ran away pursued by EG, though both males moved noticeably slower, the distance between them increased, and DG's rate of wahooing declined. After having chased DG in a wide arc, male friend EG moved off, whereupon DG rushed with renewed speed at a knoll of tall, dense grass within which female AL and her infant had remained in silence during the last male–male fight. She ran screaming from the grass carrying her infant, and DG chased them up a tree. Male friend EG immediately returned, climbed into the same tree, whereupon DG switched to chasing a young juvenile at the top of the tree. Thereafter, he sat stationary several meters from AL, AA and EG, while EG occasionally yawned. Several minutes later DG descended the tree and moved off.

Female JN and her infant JK

Though not involving overt aggression, an episode of 20 January 1995 suggested a potential threat to an infant. Infanticidal male DG moved to within 5 m of the 90-day-old infant JK, whereupon she approached to within 1 m of him twice in the space of a minute. Each time she did so, JN's male friend WA approached JK grunting, threatened her, and hit her, which caused her to scream and run back to her mother JN sitting 5 m away. When JK once again approached DG 4 min later, male WA grunted at her from several meters away and the infant immediately ran to her mother. Male DG then moved off.

Female SY and her infant SU

On 14 September 1994, DG appeared from behind a bush rushing silently toward the female infant SU, who was sitting on the ground. The mother SY was absent from the immediate vicinity, but her other 5-year-old daughter SR was nearby. SR grabbed her infant sister and fled, without screaming, pursued by DG. Having seen this interaction from a distance, SY ran forward, screaming continuously. Immediately, SY's male friend GL ran from 35 m away and chased after DG, who pursued SR and SU into a small acacia tree. GL arrived and sat below the tree, grunting continuously, glancing alternately at the female SR (holding SU ventrally) and at male DG. After about a minute, DG descended the tree and moved off.

Summary

Male friends interceded defensively in all 12 cases where the alpha male attacked or threatened infants but failed to injure them. Actual or potentially protective intervention was also observed by non-natal, non-friend males in four cases (33%), by subadult males once (8%), and by females (other than the mother) in three instances (25%). Thus, in seven cases (58%), baboons other than the male friend and mother became directly involved in the attack, but only male friends were ever observed to chase, threaten, physically attack or present to the alpha male, to carry infants in retreat, or to give roars, threat vocalizations, or wahoos.

Infant mortality

During the 6 year study period, 79 infants were born, of which one was stillborn, and 30 died or disappeared before the age of 15 months; two live infants were less than 1-year-old at the end of the study period. Thus, depending on the fates of these two infants, mortality among live-born infants was between 38% and 41%, which is equivalent to previously reported rates of 45% ($N = 38$ infants) for C troop at a comparable group size, and of 39% ($N = 28$) for a neighboring troop (Bulger & Hamilton 1987). Mortality rates, however, varied considerably over time. For example, during the 2 year period from July 1993 to July 1995, 76% of infants died ($N = 21$). Such high levels of mortality are comparable with those observed in a population of desert chacma baboons (Brain 1992).

Sources of mortality for live-born infants are summarized in Table 6.3. Conservative estimates of mortality due to infanticide range from a frequency of 31% (if only observed cases of attacks are used) to 37% (if all 11 deaths described in Table 6.1 are considered). If disappearances of apparently healthy infants are included in this estimate, up to 76% of infant deaths may have been due to infanticide. Whichever estimate is used, infanticide is clearly a significant cause of infant mortality in this population. Its frequency is comparable with, if not higher than, that observed in mountain gorillas (*Gorilla gorilla berengei*) (37%; Watts 1989), howler monkeys (*Alouatta palliata*) (40%; Clarke & Glander 1984), and lions (*Panthera leo*) (27%; Pusey & Packer 1994).

Sex ratios among newborns and dead infants did not depart significantly from a 1:1 ratio. Of the 79 infants born, 42 were female, 35 were male and 2 were of unknown sex. Of the 29 infants that died, 18 were male, 14 were female, and 1 was of unknown sex. The mean age of infants

Table 6.3. *Mortality among live-born infants at Moremi prior to 15 months of age (1992–1998)*

Cause of mortality	N	%
Disease/debilitation	5	16
Disappear with mother[a]	2	7
Disappear (with no prior sign of debilitation)	12	40
Infanticide	11	37
Total	30	100

Notes:
[a] One case of infant disappearance with mother appears to have been due to leopard predation based on the presence of baboon hair, drag marks, and leopard spoor at the sleep tree site on the morning these individuals were discovered missing.

that were killed was 132.4 days (SD $=128$, N$=11$), which was statistically indistinguishable from infants that died of other causes or disappeared (Mean $\pm$SD $=103.5$ (±102.5) days, N$=18$) (Mann–Whitney U-test, $U=90.5$, $N_1=11$, $N_2=18$, $P>0.10$), but was much less than the age of surviving infants at the time their mothers conceived their next infants (Mean $\pm$ SD $=710$ (±148) days; $U=0$, $N_1=11$, $N_2=26$, $P<0.001$).

Infant births and deaths were distributed similarly across maternal age categories ($\chi^2=3.34$, df$=3$, $P>0.10$), suggesting equivalent rates of mortality among females of varying ages. Of the 11 cases of suspected and inferred infanticide in Table 6.1, six involved primiparous mothers and six involved multiparous mothers. Given that there were by definition fewer primiparous females in the group (N$=20$ live births) than multiparous females (N$=58$ live births), this suggests that primiparous female chacma baboons are more vulnerable to infanticide, as occurs among some birds breeding for the first time (Palombit 1999).

Variability in male infanticidal behavior

Alpha males differed considerably in infanticidal behavior. Three of the eight males that became alpha during the study period accounted for all observed infanticides, and were particularly implicated in most circumstantial cases. Moreover, two correlations among the data of Table 6.4 suggest that infant mortality varied with alpha male identity. First, levels of mortality among infants born before and after takeover were positively associated with one another across males ($r= +0.74$, $P<0.05$, N$=8$).

Table 6.4. *Mortality following male attainment of alpha status ("takeover")
among infants already present in the study group and among infants born
thereafter*[a]

Alpha Male[b]	Alpha tenure (mos)	Mortality of infants alive at takeover (%)[c]	N	Mortality of infants born after takeover (%)[d]	N
WA	8	33	3	25	8
ZR	3	9	11	0	2
MK	1	0	10	0[e]	4
WA	3	0	5	–	0
GL	8	0	6	0	1
DG	11	83	6	60	10
AU	6	0	4	6	18
AP	14	7	15	50	4
TM	5	0	9	13	8
AP	Current (>6)	0	6	–	0
Mean, SD		13%, 27	75	22%, 23	55

Notes:

[a] "Mortality" is defined here as death due to injuries (Table 6.1) or overnight
disappearance *without* prior indication of illness.

[b] Alpha males are listed in chronological order.

[c] For males that attained alpha status more than once (WA, AP), infants that they may
have sired (i.e., that were conceived during a previous alpha tenure) are excluded.

[d] Infants born later than one gestation period (5.5 months) into the tenure of a
particular alpha male are excluded, since they are likely to have been sired by that male.

[e] Circumstantial infanticide case 8 is not attributed to male MK, since he was
unambiguously uninvolved in the attack (see the text).

Second, mortality among infants born after a takeover increased with the
duration of an alpha male's tenure ($r = +0.83$, $P = 0.01$, $N = 8$). These data
indicate that attributes of alpha males or conditions of their tenure may
explain some of the variation in infant mortality rates.

The sexual selection hypothesis

The sexual selection hypothesis predicts that males are unrelated to the
infants they kill. In the six directly observed cases, the alpha male was
unlikely to be related to the infant he attacked because it had been conceived prior to his immigration into the group (males DG and TM) or
while he was lower ranking (male WA). Although the identity of the
infanticidal individual was unknown in the six other attacks, in all cases

the current alpha male was a recent immigrant who had been absent from the group when the infant was conceived. In one instance (case 8), the suspected perpetrator was a new immigrant who shortly became alpha male.

A distinction is sometimes made between the killing of infants already present in the group and "prospective infanticide", in which a male kills infants conceived prior to his attainment of alpha status but born soon afterwards. The latter suggests that males must somehow evaluate paternity, but the former variant is assumed in economic models of infanticide (Chapman & Hausfater 1979). Prospective infanticide is known in hanuman langurs (*Presbytis entellus*), lions (*Panthera leo*), and tropical house wrens (*Troglodytes aedon*) (Freed 1987), but nevertheless occurs rarely. For example, in hanuman langurs, 66% of infants born after male takeover were unharmed, compared to 44% of those infants already present when a new male entered the group (Sommer 1994).

Prospective infanticide, however, was common in the C troop chacma baboons. In five of the six (83%) directly observed cases, the infant attacked by the alpha male had been born after he had achieved alpha status but before he had copulated with their mothers. More generally, mortality due to wounding or sudden disappearances (unrelated to illness or predation) was roughly equivalent, if not more common, for infants born after a new male became alpha (22%) than for those born before (13%) (Table 6.4).

The sexual selection hypothesis predicts that an infanticidal male will mate with the female whose infant he kills. Female chacma baboons resumed cycling within 1–3 months of their infants' deaths, in accordance with numerous data demonstrating that loss of an infant significantly accelerates resumption of cycling in savanna baboons (Altmann *et al.* 1978; Collins *et al.* 1984). In all five directly observed cases of infanticide, the attacking male copulated with the mother when she resumed cycling (Table 6.5). Moreover, in four of these five cases, the infanticidal male was the alpha individual both at the time of the attack and subsequently, when the mother conceived her next infant. In the fifth instance, the attacking male's alpha position was ambiguous with respect to a new immigrant. Given that alpha males in the study group generally achieve greater mating success with females around the time of ovulation (Bulger 1993), these data support the hypothesis that infanticide increases male sexual access to fertile females.

The identity of infanticidal males was unverified in the circumstantial

Table 6.5. *Reproductive patterns following observed and suspected infanticides*

Case[a]	**Known** or *suspected* infanticidal male[b]	Did infanticidal male copulate with mother when she resumed cycling?	Alpha male when mother conceived her *next* infant
1	**WA**	Yes	WA
2	**WA**	Yes	Ambiguous alpha: WA or new immigrant NP
3	**DG**	Yes	DG
4	**DG**	Yes	DG
5	**DG**	Yes	DG
7	*TM*	Yes	AP
8	*GL*	Yes	GL
9	*DG*	Yes	Ambiguous alpha: DG or new immigrant AU
10	*AP*	Not applicable[c]	Not applicable[c]
11	*WA*	Yes	Ambiguous alpha: WA or new immigrant NP
12	*DG*	Yes	DG

Notes:

[a] Case numbers are from Table 6.1. Case 6 is excluded because the injured infant survived.

[b] Known male attackers in directly observed infanticides are indicated by **boldface**. Suspected infanticidal males in circumstantial attacks are indicated by *italics*. In all cases of the latter except case 8, the suspected infanticidal male is operationally defined as the current alpha male. In case 8, the current alpha male was uninvolved in the infanticidal attack, but new immigrant male GL was observed wahooing and running at the time (see text).

[c] The female died shortly after her infant was killed.

cases, but in only one of five cases did the alpha male at the time of the infant's death unambiguously fall in rank before the mother subsequently conceived her next infant (Table 6.5). Even without positive identification of the attacker, this pattern is striking in light of the short tenure of alpha males in general.

Other hypotheses for male infanticide are less supported. The possibility that males killed infants in order to exploit them nutritionally is suggested by partial consumption of infants' bodies in three of the five directly observed infanticides. This finding by itself does not invalidate the sexual selection hypothesis, which makes no explicit predictions concerning what males will do with the bodies of infants they have killed. Moreover, orphaned infant chacma baboons were unharmed by resident adult males even though they presumably were especially vulnerable to cannibalism at that time ($N = 2$ orphans during this study; $N = 9$ reported by Hamilton *et al.* 1982).

The hypothesis that the killing of infants is a largely an accidental side effect of generalized inter- and intrasexual aggression (Bartlett *et al.* 1993) receives only limited support, primarily from two inferred infanticides. In case 11, an infant disappeared during a prolonged intragroup aggressive interaction involving multiple baboons. In case 7 an infant was mortally wounded during an intergroup encounter, in which group arousal and aggression are indeed generally heightened. This case does not necessarily disqualify the sexual selection hypothesis, however, since it remains to be determined whether this context provides infanticidal males with opportunities to kill infants. The actual onsets of infanticidal attacks were unobserved in most cases, but it is important to note that the intense screaming generally associated with them usually erupted spontaneously from individuals otherwise silent and relaxed in the immediately preceding moments. These outbursts of screaming were thus not part of escalating aggression, but appeared to be the vocal response to an apparently precipitous infanticidal attack already in progress or, in fact, accomplished.

In summary, the sexual selection hypothesis best accounts for the currently available data on infant-killing in the Moremi population.

Infanticide as a facultative strategy

Theoretical models (e.g., Chapman & Hausfater 1979; Hausfater *et al.* 1981) have assumed that "infanticide" versus "non-infanticide" may be a pure evolutionarily stable strategy (ESS) pursued by two different types of male or a mixed strategy pursued by individuals at some probabilities p and q, respectively. The data for Moremi baboons are few, but the fact that three of eight alpha males accounted for all observed woundings of infants, and were also implicated in most of the suspected infanticides, suggests that males vary in infanticidal behavior.

If some males commit infanticide at significantly higher rates than others, an important goal of future research is to discover why. Existing data can only suggest potentially important variables meriting further study. Reduced opportunity for reproduction, as reflected by the ratio of estrous to anestrous females, is one of these. For example, male DG attained alpha rank at a time when only one of the group's 18 females was

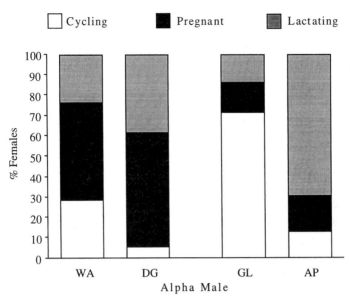

Figure 6.2. Distribution of reproductive conditions of resident females chacma baboons at the time of attainment of alpha position. Males DG and WA were responsible for a relatively large number of infanticides, whereas few infanticides were attributed to males GL and AP.

cycling. More generally, available data show that anestrous females, particularly pregnant females, were more abundant when males DG and WA each became alpha and embarked upon a series of infanticidal attacks, than when the males GL and AP became alpha and committed few, if any, infanticides (Figure 6.2). Because of long interbirth intervals and short alpha tenures, females that are pregnant at the time a new male becomes alpha are especially unlikely ever to become available as fertile mates to him, unless their infants subsequently die *in utero* or postnatally. Another possible reason for variation in infant-killing by males is the number of adult males in the group, since infanticide may be more costly in groups with more adult males. The case of male DG opposes this hypothesis, however, at least in its simplest version. He achieved alpha status at a time when male membership of the study group was at its highest level, including seven non-natal adult males and eight natal adult males (≥ 9 years old); the abundance of males failed to prevent male DG from killing a large number of infants. Finally, currently unidentified attributes of individual males may also contribute to variation in infanticidal behavior upon reaching alpha rank.

Clearly, more data are necessary to elaborate the causes of variation in infanticidal behavior of male chacma baboons. These data additionally promise to clarify the source of interpopulation differences, i.e., the notably lower incidence of infanticide in East African savanna baboons. At present, our data endorse the conclusion of Collins *et al.* (1984) that the brevity of alpha male tenure in Moremi chacma baboons (<7 months) is a likely proximate cause of infanticide in this population. The success of alpha males in monopolizing copulations with fertile females, which Bulger (1993) described as "the most striking aspect of male mating patterns" in Moremi chacma baboons, may further distinguish them from East African conspecifics (e.g., Bercovitch 1987; Altmann *et al.* 1988, 1996), although direct comparative data are needed to determine this. If so, male chacma baboons may be more likely to reap the reproductive benefits of infanticide. The reasons for the shorter tenures and possibly greater mating exclusivity of alpha male chacma baboons, however, remain obscure.

Friendships and male defense of infants

Palombit *et al.* (1997) argued that heterosexual friendships in chacma baboons are female counteradaptations to deter male infanticide. There were few opportunities to observe the responses of males to actual or potentially infanticidal attacks, although some preliminary patterns emerge. A protective role of male friends is suggested by their direct intervention in all attacks in which infants escaped injury, and their absence in two-thirds of the attacks in which infants were severely or fatally wounded. Male friends do not always act alone. In over half of attacks on infants, other individuals became involved in the interaction, including maternal relatives of either sex, and non-friend adult males.

The participation of both non-friend and friend males in defense is noteworthy because experimental data suggest that male friends are more predisposed than control males to aid females and dependent infants under attack (Palombit *et al.* 1997). There are at least two interpretations of this result. First, it should be kept in mind that it is sometimes difficult to identify "aiding" of females, let alone its relevant proximate causes, from the aroused running about and screaming of

Figure 6.3. A friendship in chacma baboons: a lactating female with her infant and an adult male (photo: Ryne Palombit).

multiple baboons following actual or potentially infanticidal attacks by alpha males. Indeed, it was precisely this problem, combined with the rarity of such attacks, that prompted Palombit *et al.* (1997) to use playback experiments to evaluate this question systematically under more controlled circumstances. Thus, interpretation of the *ad libitum* observations is complicated by the possibility that some baboons may not be "involved" in the sense implied here.

Alternatively, the few available data suggest that male friends invested more heavily than non-friends in defense of lactating females and infants under attack (Figure 6.3). This was variably expressed as reactions that were more direct and sustained, uniquely involving the protective carrying of targeted infants, initiating and maintaining close proximity to the aggressor alpha male, or engaging him with threats, vocalizations, appeasement gestures, chases, or, ultimately, fights. Non-friend males responded less conspicuously, but nevertheless actively, by rushing to the site of an attack and remaining there in a state of aroused vigilance. The possible benefit of their presence is suggested by the fact that their intervention occurred in one-third of attacks in which infants escaped injury but was absent in all cases where infants were wounded or killed (even though male friends defended in one-third of these). Participation of subadult males and females, however, was broadly similar in both injurious and non-injurious contexts (or, perhaps, slightly more common in cases where infants were seriously wounded). How non-friends specifically influence the outcome of attacks remains unclear, but deterrence may operate through elevated costs of attempted infanticide in the company of several adult males. The threat of more active defense by non-friends is implied, but in practice simply the presence of these males may be sufficient to modify alpha male aggression. Notably, however, non-friend males intervened in alpha male attacks on infants only when the mother's male friend interceded as well. In summary, the data from Moremi implicate the critical role played by male friends in protecting infants, but also hint at an important, though conditional, contribution to deterring infanticide by non-friend, adult males.

The hypothesis that emerges from the combined, if unequal, participation of both friend and non-friend males is that heterosexual friendships effectively deter infanticide partly because they facilitate communal defense of infants by several adult males. That is, a lactating female's male friend may or may not be capable of single-handedly preventing infanticide, but his immediate and substantive commitment to defending her may increase the probability of involvement by other males who otherwise eschew costly engagements with the alpha male on their own, and whose participation may be crucial in some circumstances. This hypothesis predicts that non-friend males' coalitionary support involves fewer costs (e.g., risk of injury) than the male friend's, and that they benefit from the interaction for reasons that may or may not be related to infant defense per se. For example, non-friend males may undermine the current alpha male's dominance status by supporting another male's aggressive confrontation with him. The importance of a particular male's defensive commitment is further corroborated by patterns of female competition for friends. When establishing friendships, lactating females routinely overlook several "unbonded" adult males available in favor of a male that already possesses a female friend, even though the ensuing competition with the rival female results in diminished sociospatial access to the male and a weakened friendship (Palombit *et al.* 2000). Some probability of paternity may be the crucial variable underlying female preferences for particular males, since 68% of friendships involved a male with whom the female was known to have copulated during the cycle in which she conceived her current infant (Palombit *et al.*, 1997) and since the strength of male response to playback of female screams is positively cor-related with male rank at the time the female conceived her infant (Palombit *et al.* 2000).

Conclusions

Although our data considerably augment the previous summary of Collins *et al.* (1984), they remain limited (e.g., are taken from a single group) and dictate a need for further testing of the conclusions offered here.

As in one-male primate groups, the social context of infanticide in chacma baboons is broadly related to replacement of the primary breed-ing male, in this case, the alpha male. Typically, the new alpha male in a chacma baboon group is also a recent immigrant, but this is not strictly necessary. Even a long-time resident of the group may attempt infanti-cide upon attaining (or resuming) the alpha position in the male hierar-chy. These were the two situations in which we observed infanticide. A third context for infanticide is possible in multimale groups and does not involve male replacement: if two females are simultaneously ovulating and the alpha male can sexually consort only with one of them, then the other female's infant might be vulnerable in future to infanticide by this same alpha male. Infanticide did not occur in this context, however, among Moremi chacma baboons.

In general, existing data support the sexual selection hypothesis more than alternatives. The systematic flexibility and ontogeny of male infanti-cidal and defensive behavior provides compelling evidence in favor of Hrdy's (1974) hypothesis. During attacks directed at infants, some males consistently defend particular infants, while aggressors are generally new alpha males likely to benefit reproductively from the target infant's death. The stability of this pattern opposes the view that infant fatalities are the consequence of erratic male–male aggression. Moreover, the transformation of male behavior illustrates how male mating strategies apparently involve an adaptive suite of behaviors, beginning with immigration and attainment of short-lived alpha status, leading directly to infanticide in some cases, followed by accelerated fertilization of females, then by resistance to new incoming males, an eventual fall in rank from the alpha position, and, notably, continued membership in the group and protection of infants. One of the most striking features of chacma baboon social behavior is that infanticidal alpha males later become the friends of females and the ardent defenders of young against infanticide.

Females' friends are the principal defenders against infanticide, but the responses of adult, non-friend males to attacks may also help to alleviate immediate danger to infants in some circumstances. The patterns of non-friend intervention in chacma baboons suggest that communal infant protection is fostered by one particular male's disproportionately active investment in defense. Such a mechanism may operate more generally in multimale primate groups to lower frequencies of infanticide in this social setting. Another possible deterrent arising in a multimale setting is suggested by the co-occurrence in alpha male chacma baboons of both infanticide and high mating exclusivity (see above). Where alpha males are less successful in monopolizing copulations with fertile females (e.g., Altmann *et al.* 1996), infanticide will offer them fewer advantages. In summary, preliminary data from the Moremi population suggest that the presence of multiple males in a primate group can discourage infanticide both by increasing its attendant costs and by decreasing its future benefits. If so, an important implication is that female primates can potentially alter the economics of sexually selected infanticide in their favor by manipulating male membership in groups (Altmann 1990; Packer & Pusey 1983) and their social relationships with males (Palombit *et al.* 1997). How and when they do so remain unclear.

Currently available data are remarkably consistent with the predictions of Hrdy's (1974) sexual selection hypothesis and suggest that infanticide and defense of infants are elements of an adaptive reproductive strategy of male chacma baboons.

Acknowledgments

We thank the Office of the President and the Department of Wildlife and National Parks of the Republic of Botswana for continued support of this research in the Moremi Game Reserve. We thank Mokopi Mokopi for his important contribution to the collection of field data, as well as Okavango Tours & Safaris and Gametrackers for logistical support of the research. We are especially grateful to Tim and Bryony Longden for unflagging help and hospitality. We thank Carola Borries, Sarah Hrdy, Charles Janson and Carel van Schaik for extremely helpful comments on the manuscript. Lynne Isbell provided logistical aid in the field during writing of the manuscript. The research was supported by grants from the National Science Foundation (SBR9615944), the National Institutes of Health (5 F32 MH10436-03), the Natural Sciences and Engineering Research Council of Canada, the National Geographic Society, the Wenner-Gren Foundation, the L. S. B. Leakey Foundation, the Research Foundation of the University of Pennsylvania (89–20230), and the Institute for Research in the Cognitive Sciences (University of Pennsylvania).

Appendix 6.1. *Previous observations of infanticide in chacma baboons (Papio cynocephalus ursinus)*

Population	Date	Observation Type	Infant age	Infant sex	Comments	Source
Moremi (C troop)	1 Sept. 1979	Direct	4 days	?	Newly immigrant alpha male BB seen eating infant after mother had been tranquilized	Collins *et al.* (1984)
Moremi (C troop)	8 Sept. 1979	Direct	8.1 mo	M	Infant bitten by unidentified adult male at dusk was found dead the next day with canine puncture wounds in skull	Collins *et al.* (1984)
Moremi (W troop)	6 Feb. 1980	Direct	2 days	?	Newborn infant was somehow taken from its low-ranking mother by a pregnant female; when seen again 2 days later it was dead	Collins *et al.* (1984)
Moremi (W troop)	16 Feb. 1979	Circumstantial	10.8 mo	F	Infant found with lacerations on its head and lower back; 4 weeks later, it received a second wound and then disappeared 3 days later	Collins *et al.* (1984)
Moremi (C troop)	29 June 1980	Circumstantial	1 day	M	Female discovered carrying her dead newborn infant, which had multiple lacerations and punctures (both had been seen healthy 30 min earlier)	Collins *et al.* (1984)
Moremi	Nov. 1982	Direct	6 mo	F	Natal male that became alpha in previous June attacks infant that is alone and inflicts fatal wound to head	Tarara (1987)
Drakensberg Mtns	?	Direct	2 weeks	?	Newly immigrant alpha male fatally wounds infant in the neck during an intergroup interaction	Weingrill (2000)
Drakensberg Mtns	?	Circumstantial	7 weeks	?	Mother with dried blood on her chest seen carrying a dead infant	Weingrill (2000)

Notes:
M, male; F, female; Mtns, mountains.

7

Infanticide by males and female choice in wild Thomas's langurs

Introduction

It is now widely accepted that infanticide by males has affected important features of primate social organization by selecting for various social counterstrategies of females (Hrdy 1979; Smuts & Smuts 1993; van Schaik 1996; Treves & Chapman 1996; see various chapters in this book). One of these possible female counterstrategies is female secondary transfer (Marsh 1979b; Sterck 1997; Sterck & Korstjens, Chapter 13). Female transfer decisions are expected to be strongly influenced by the identity of the male with whom a female associates. Secondary transfer is expected only where female relationships are relatively weak because it means the break-up of associations and coalitions, at least temporarily. Hence, given the predominance of female philopatry in primates, only a small number of species is expected to show this behavior. Thomas's langurs (*Presbytis thomasi*) are one of the few species known to show female secondary transfer related to infanticide avoidance. Females leave when infants have died or when they are old and independent enough to be left behind (Sterck 1997; Steenbeek 1999a). This makes the Thomas's langur an excellent choice for investigating the possible influence of infanticide on social relationships, since in species like this female choice in relation to male characteristics and strategies will be most manifest.

Species with regular female secondary transfer are of particular interest, because where the group contains only a single male, female emigration may terminate a group's lifespan. When this is the case, the lifespan of a group coincides with the tenure of its reproductive male. Thus groups go through a typical life cycle. This is also the case in Thomas's langurs.

Tenure-related changes in male and female behavior have rarely been studied, because it requires the collection of long-term field data to allow a comparison between different tenure phases. In this chapter, I give an overview of the tenure-related changes in the wild Thomas's langurs of Ketambe, which have been studied since 1987. I explore the sometimes complicated interactions between male and female strategies expected in this species. Where females use transfer to express female choice, males can in turn influence female decisions by the use of male coercion, i.e., aggression against females and infanticide (Treves & Chapman 1996; van Schaik 1996). On the other hand, resident males should protect their females against such sexual coercion.

Because the lifespan of a group depends on the tenure length of its reproductive male, I expect variation in both relative male strength and male mate competition over different phases of male tenure. Furthermore, I expect that females react to changes in infanticide risk and leave their current male when he is no longer able to defend offspring against infanticidal provocations. These expectations are examined in this chapter.

The Thomas's langurs of Ketambe

Study area and subjects

The Thomas's langurs were studied at the Ketambe Research Station (3° 41' N, 97° 39' E), Gunung Leuser National Park, situated in Northern Sumatra, Indonesia. The study area is approximately 200 ha and lies at an altitude about 350 m above sea level. It consists largely of alluvial terraces of increasing age and height above the current river level, as well as the lower slopes of a mountain range. The vegetation consists mainly of undisturbed primary rain forest as described by Rijksen (1978) and van Schaik & Mirmanto (1985). Representation of dipterocarps increases toward the higher terraces and the lower mountain slopes. The lower terraces, where most of the study groups were followed, have relatively species-poor forests with high natural turnover, remarkably high abundance of Euphorbiaceae, and relatively many Fagaceae (Djojosudharmo & van Schaik 1992).

The study subjects were wild Thomas's langurs, an anatomically specialized folivore-frugivore. Data were collected under the responsibility of C. P. van Schaik (September 1987 to August 1988), E. H. M. Sterck (November 1988 to October 1992), R.S. (November 1992 to March 1996),

A. H. Korstjens (March 1996 to March 1997), and S. A. Wich (April to December 1997), by different observers. We studied a total of 17 bisexual groups, where all individuals were recognized individually. These groups did not all exist simultaneously, because females transferred from one male to the next, but all groups had a different breeding male. Details of group size and composition and on data collection and analyses are provided elsewhere (Steenbeek 1999a).

Changes in social organization

Thomas's langur bisexual groups at Ketambe generally consist of one breeding male and one to six females with offspring. Females do not form coalitions, dominance relationships are weakly expressed (inside food patches) or not apparent at all (outside food patches), and female ranks change over time (Sterck & Steenbeek 1997). Both males and females disperse from their natal groups, and female secondary transfer to a new male is also common (Sterck 1997; Steenbeek 1999a). After their primary transfer from the natal group females change breeding males on average every 4.5 years (Steenbeek 1999a). Unlike the better-known South Asian langurs, the forest langurs of Southeast Asia show virtually no sexual dimorphism in body size, although males have substantially longer canine teeth (Plavcan *et al.* 1995; Plavcan & van Schaik 1997).

As a rule, groups go through a typical life cycle (Figure 7.1) and different tenure phases can be distinguished (see Table 7.1a). The early phase of male tenure starts when several females associate with a new breeding male and thus form a new group. When females start to reproduce, the group enters a relatively stable middle phase. During the final year of male tenure (the late phase), all females over three years of age leave the male to associate with a new male, and an all-male band remains behind, consisting of the male and his mature and immature male offspring and an occasional small juvenile female (Steenbeek 1999a; Steenbeek *et al.* 2000). Groups can start gradually, when females join over a period of time, and also end gradually when they leave the group one or more at a time. However, groups can also begin and end suddenly, namely in the case of a male takeover or the death of the breeding male (Steenbeek *et al.* 2000). At Ketambe, both forms of social change have been observed, but takeovers only accounted for around 20% of the new groups, and were relatively rare in comparison to the better-studied hanuman langurs living in disturbed areas (e.g., Sugiyama 1965; Hrdy 1977; cf. Sterck 1998).

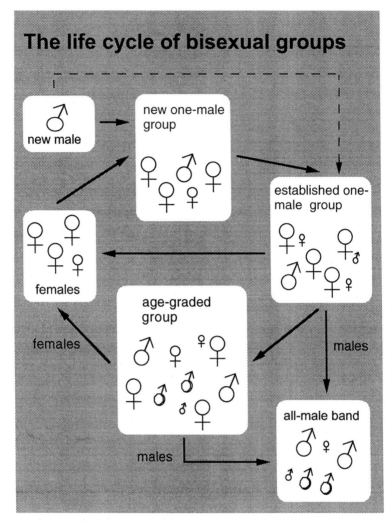

Figure 7.1. The life cycle of Thomas's langur bisexual groups (Steenbeek *et al.* 2000).

Infanticide in Thomas's langurs

Infanticide is expected to occur in two contexts: during between-group interactions and after aggressive male takeovers. In Thomas's langurs, two types of interactions between groups can be distinguished. (1) Group encounters: whole groups meet each other and, when there are aggressive interactions, they are primarily between males. (2) Male provocations: a male from another bisexual group or an all-male band silently approaches a group and suddenly attacks (Steenbeek 1999b). A male provocation could be successful only when the male was able to approach a group unnoticed because, as soon as he was spotted, the male of the target group would chase the intruder away. When a male provoked a group, individuals of that group usually fled. In 33% of the male provocations on bisexual groups involving chases, the provoking male actually chased individual females.

Infanticide attempts were observed only during male provocations, never during group encounters. Male provocations were considered to involve infanticide risk ($N = 7$) when there was contact aggression between the provoking male and a mother with infant. When a male attacked a mother with infant, the mother always defended her infant and sometimes also chased the attacking male (Steenbeek 1999b). The only time females other than the mother were observed to expel an intruding male was in the case of a male-less group (described in Steenbeek 1996). Male infanticide may have occurred after an aggressive takeover (Steenbeek 1999a). One of the two takeovers of this study concerned a group with one small infant that disappeared soon after the takeover. The cause of disappearance was unknown, but the infant was too young to survive on her own. One other takeover in 1998 took place in a group with one large male infant, which left with the ousted male (S. A. Wich, pers. comm.).

Infants that were mortally wounded in the context of infanticidal provocations by extra-group males have been observed. Sterck (1997) observed one indirect case: male–female aggression was witnessed and later a mortally wounded infant was observed. Steenbeek (1996, 1999a) and Steenbeek *et al.* (1999b) witnessed one direct attack, after which the infant died; furthermore, they saw one indirect case, in which male–female aggression was witnessed and later a mortally wounded infant was observed; they found one dead infant with canine wounds in its belly; they also witnessed several attacks where the breeding male and/or the mother could prevent harm to the infant, or where the infant recovered from wounds. Thus at least four deaths can be attributed to infanticide with reasonable certainty. We also have two additional cases in which infanticide was strongly suspected as the cause of death. These six cases out of a total of 18 dead infants led to an estimated mortality rate that might be due to infanticide of 33%. Hence, it can be concluded that Thomas's langur females face a considerable risk of infanticide. This is clearly related to a male's insufficient protective abilities or the absence of the group's male (the male-less group; Steenbeek 1996).

Infanticide and vigilance

We examined vigilance as a behavioral indicator of the importance of infanticide risk by comparing the infanticide avoidance hypothesis with the predation avoidance and mate defense hypotheses for wild Thomas's langurs. Steenbeek *et al.* (1999b) found that patterns of vigilance vary mainly in relation to predation risk and infanticide risk. Predation risk is highest at low heights, because most predators are found on or close to the ground. Infanticide risk predominates at high heights in areas that overlap with other Thomas's langur groups, because infanticidal males attack from high in the canopy, but only in overlap areas. Predation risk is reflected in patterns of vigilance of all individuals without infants: both males and females were more vigilant on or close to the ground than at other heights. Both predation risk and infanticide risk are reflected in patterns of vigilance of individuals with infants: males and females with infants were more vigilant on or close to the ground (predation risk), as well as high in the canopy in overlap areas (infanticide risk), than at other heights. Male mate competition also influences patterns of vigilance: males without infants showed higher vigilance levels in areas of home range overlap than in non-overlap areas, but only during their early tenure phase.

Vigilance can serve a variety of purposes, but infanticide avoidance has so far received scant attention in other species. However, it probably contributes to vigilance patterns observed in white-faced capuchins and vervet monkeys, and perhaps other species (Steenbeek *et al.* 1999b).

Infanticide and female transfer

Both the occurrence and timing of female transfer in Thomas's langurs were determined by infanticide avoidance (Steenbeek 1999a). Thomas's langur females transferred when coercion by extra-group males had revealed that their current breeding male was no longer able to protect the offspring. Infant mortality was twice as high for the last infant before transfer than for other infants (Figure 7.2). This higher mortality coincided with a decline in the breeding male's strength during his final tenure year. The moment of the actual transfer depended on the age of the current offspring: a female left when her infant had either died or was old enough to be left behind. Female transfer had its costs, most importantly the loss of reproduction. Natal adult females lost reproductive time by waiting for male tenure to end. Females who transferred when they could safely leave their latest offspring behind lost on average 9.2 months of

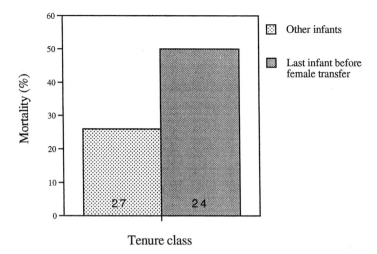

Figure 7.2. Differences in infant mortality between infants when females stayed in the group and last infants before females transferred ($P < 0.10$; Steenbeek 1999a).

reproductive time, whereas females whose last offspring born before transfer did not survive did not lose reproductive time (but had lost an infant). Females who did not wait for male tenure to end but transferred into an unfamiliar group faced other costs: not all transfer attempts were successful, and females sometimes got wounded when trying.

Tenure-related changes

Because Thomas's langur groups go through a predictable series of phases in which male strength and female vulnerability to infanticide were found to vary predictably, a systematic overview of tenure-related variation in social behavior is possible. Most comparisons are with the relatively stable middle phase. The comparisons are summarized in Tables 7.1a and 7.1b (see the footnotes for sources).

This section also includes new analyses on tenure-related changes in time spent feeding, time spent resting, time spent alone, average resting height, and wounding, for males and females separately (Tables 7.2 and 7.3). Although we did not find an influence of group size on time spent resting and feeding (Steenbeek 1999a), other factors such as male mate competition and male strength may constrain how males and females spend their time. Infanticide risk varied over male tenure: infant

Table 7.1a. *Overview of tenure-related changes that were found: males*

Males	Tenure phase		
	Early	Middle	Late
Tenure phase characteristics (4)	Females associate with male, no infants yet	Relatively stable phase	Females leave male
Duration of phase (mos (range)) (4)	8 (6.5–18)	60 (14–72)	12 (last year)
Selection pressure			
Male strength (2,3)	Same	Baseline	Lower
Proximate mechanisms			
Mate competition from extra-group males (2)	Higher	Baseline	Higher
Intensity of mate defence (2)	Higher	Baseline	Lower
Behavioral responses			
Time spent alone	Lower	Baseline	Higher
Male vigilance (1)	Overlap > Non-overlap	Overlap = Non-overlap	–
Resting height	–	Baseline	Lower
Consequences			
Time spent feeding	Lower	Baseline	Same
Time spent resting	Higher	Baseline	Same

Notes:

(1) Steenbeek *et al.* (1999b); (2) Steenbeek (1999b); (3) Steenbeek *et al.* (1999a); (4) Steenbeek (1999a); Baseline, values during the early and late tenure phase were compared to the middle phase; – = variable could not be analyzed.

Table 7.1b. *Overview of tenure-related changes that were found: females*

Females	Tenure phase		
	Early	Middle	Late
Selection pressure			
Infanticide risk: Infant mortality (4)	–	Baseline	Higher
Proximate mechanisms			
Infanticide avoidance (2,3,4)	Same	Baseline	Higher
Incite male mate competition (2,4)	Higher	Baseline	Higher
Behavioral responses			
Female immigration (4)	Many	Few	One
Time spent alone	Higher	Baseline	Lower
Resting height	Higher	Baseline	Lower
Avoidance of extra-group males (2,3)	Slightly lower	Baseline	Higher
Female interest in extra-group males (2,4)	Yes	No	Yes
Female emigration (4)	None	Few	Many
Consequences			
Female wounds (4)	Higher	Baseline	Same
Home range size (4)	Increasing for two years	Stable	Decreasing
Home range location (4)	Being in or going to middle tenure range	Middle tenure range	Some forced to leave middle tenure range
Home range overlap (4)	Higher?	Baseline	Higher
Determinants of home range size (4)	Social factors	Ecological factors	Social factors
Determinants of DJL (4)	Social and ecological factors	Ecological factors	Social factors

Notes:
For references and further explanation: see Table 7.1a. DJL, day journey length.

Table 7.2. *Statistical results for time spent feeding, resting, alone, and average resting height*

	Males			Females		
	N	Statistics	Median change	N	Statistics	Median change
Comparison early with middle phase						
Time spent feeding (min/day)	5	$T^+ = 15$ #	+135 min	14	$T^+ = 68.5$	–
Time spent resting (min/day)	5	$T^+ = 15$ #	–75 min	14	$T^+ = 61$	–
Time spent alone (% of non-feeding time)	4	–	+16.6%	12	$T^+ = 77$***	–13.5%
Average resting height (m)	3	–	–	13	$T^+ = 74$*	–2.4 m
Comparison middle with late phase						
Time spent feeding (min/day)	6	$T^+ = 14$	–	11	$T^+ = 35$	–
Time spent resting (min/day)	6	$T^+ = 14$	–	11	$T^+ = 39$	–
Time spent alone (% of non-feeding time)	6	$T^+ = 21$*	+7.8%	10	$T^+ = 45$*	–13.6%
Average resting height (m)	6	$T^+ = 21$*	–3.0 m	11	$T^+ = 60$*	–2.6 m

Notes:
Differences between tenure phases were analyzed with a Wilcoxon Matched Pairs Signed Ranks test (Siegel & Castellan 1988). For the comparison between the early and middle tenure phase, data were available for five males from five different groups and 14 female from three different groups. For the comparison between the middle and late tenure phase, data were available for six males from six different groups, and 11 females from five different groups, #, $P < 0.10$; *, $P < 0.05$; ***, $P < 0.001$

Table 7.3. *Tenure-related changes in wounding*

Age–sex class	Tenure phase	No. of individuals	Class observation days	No. of wounds (wounded individuals)	Wounds per 100 days
Adult males	Early	7	231	3 (2)	1.30
	Middle	13	529	4 (3)	0.76
	Late	6	223	2 (2)	0.90
Adult females	Early	28	1099	16 (13)	1.46
	Middle	30	1848	9 (6)	0.49
	Late	21	588	3 (3)	0.51
	Male-less group	4	108	3 (2)	2.78
Natal males	Middle	12	420	1 (1)	0.24
	Late	12	486	4 (4)	0.82
Natal females	Early	3	64	0	0.00
	Middle	12	882	0	0.00
	Late	9	287	0	0.00
Infants	Middle	46	1176	3 (3)	0.26
	Late	12	349	1 (1)	0.29
	Male-less group	3	108	3 (2)	2.78

mortality was higher for the last infant before females transferred than for infants when the mother stayed in the group (Steenbeek 1999a). Time spent alone and average resting height may both be affected by predation and infanticide risk.

Early versus middle tenure phase

Males

An overview of tenure phase characteristics and all tenure-related changes for males is presented in Table 7.1a. One new group experienced an aggressive takeover after it had existed for 4 months, which illustrates that not all males make it to the middle tenure phase. Male strength was found not to differ between the early and middle tenure phase. However, the intensity of mate competition from extra-group males did differ. During the early tenure phase, extra-group males provoked the group more often; in that period all-male band males provoked more often, and between-group encounters were more often aggressive than during the middle tenure phase. In addition, extra-group males specifically recognized early tenure males, as was shown in loud call behavior (Steenbeek *et al.* 1999a).

This increase in mate competition from extra-group males seems to be due to female behavior. During the early tenure phase, females were observed to show sexual interest in extra-group males: sometimes they copulated with extra-group males during between-group encounters or when provoked by an extra-group male, and sometimes they visited extra-group males for days (stop-overs). Thus females incited male mate competition, which shows that there must be a greater need for mate defense during the early tenure phase. This higher mate defense was expressed by a higher occurrence of herding behavior during between-group interactions (Steenbeek 1999b). The intensity of mate defense during the early tenure phase depended on the mode of group formation: males herded females more often and more aggressively after an aggressive takeover than after gradual group formation.

Early tenure males reacted to the higher competition for access to mates with extra-group males by spending less time away from the females (Table 7.2), and by spending more time vigilant, specifically in overlap areas. These behavioral responses had consequences. During the early tenure phase, males spent significantly less time feeding and significantly more time resting (Table 7.2). There are two possible explanations for the much lower feeding time and higher resting time of the male during his early tenure phase. During the middle and late tenure phases, males have to defend infants from attacks by extra-group males, which can cause a lower feeding efficiency, and consequently increased feeding time. However, because there is no increase in time spent feeding from the middle to the late tenure phase, it seems more likely that the lower feeding time during the early tenure phase is a cost of the male's extra investment in mate defense.

Male wounding (Table 7.3) could not be tested statistically because of the low expected values. The slightly higher value of adult males during the early tenure phase was related to the aggressive takeover of group B3. Only one of the adult male wounds observed during the middle tenure phase was caused within the group when the male of group K1 expelled two juvenile males; all other wounds were due to between-group aggression. Thus an increase in male mating competition did not appear to result in a higher chance of becoming wounded for early tenure males.

In summary, when females associate with a new male, they often put this male's protective abilities to the test by showing sexual interest in extra-group males, and thus eliciting their own male's defensive actions. The data suggest that, when male mate defense is successful, females will start to reproduce and the group enters the middle tenure phase.

Females

An overview of all tenure-related changes for females is presented in Table 7.1b. Important characteristics of the early tenure phase are that most females arrived, but that none left the new group. During the early tenure phase, there are no infants yet that need protection from the male, which coincides with the females' sexual interest in extra-group males. The absence of vulnerable infants is reflected in female behavior: they spent significantly more time alone (Table 7.2), rested significantly higher in the canopy (Table 7.2), and avoided extra-group males slightly less, than during the middle tenure phase, when female behavior began to reflect infanticide avoidance. In most groups, the females conceived within 2 months and the middle phase started after 8 months. However, in one group (K2), females kept showing sexual interest in extra-group males. When the male was no longer challenged by extra-group males, females finally started to reproduce after 18 months.

There are several consequences for females when they change males. The chance that a female gets injured differed significantly over the tenure phases (Table 7.3, tested as number of observation days with and without new wounds: $\chi^2 = 11.27$, df $= 2$, $P < 0.01$). Partitioned tests showed that during the early tenure phase females sustained significantly more injuries than females during the middle phase ($\chi^2 = 7.69$, $P < 0.05$, corrected for the number of partitions). Of the 16 wounds during the early tenure phase, only three were the result of between-group interactions, 13 were caused within the group (81%): 8 by the resident male (7 in groups after an aggressive takeover), and 5 after immigrations into a group with unfamiliar individuals, mostly by adult females. In contrast, during the middle tenure phase, only 44% of the wounds had a known cause within the group. These data show that male mate defense is more aggressive after a takeover; female wounds resulted from male herding behavior when females traveled away from the center of the group. This higher level of male mate defense during the early tenure phase, especially when females did not join the male voluntarily, had a cost to females: they were wounded more often.

All other differences between the early and middle tenure phase concern consequences for range use. During the first 2 years of male tenure, home range size, and, to a lesser extent day journey length (DJL), depended above all on the mode of group formation, whereas they depended mainly on group size during the middle tenure phase. When groups started suddenly, home range size increased rapidly and remained roughly constant after this. In contrast, when groups started gradually, home range size continued to increase until long after the first infant was born. Groups that formed gradually started in a small part of the range occupied by the same females with a previous male, or in an area directly adjacent to it. New groups may have started in a relatively small home range because they still shared it with the remainder of the former group (either the last few months of the late tenure phase or the first few months of the all-male band phase). New and former groups never fought over space; new groups simply seemed to wait. As soon as former groups had been reduced to all-male bands, new groups slowly moved back to the former range. Home range size increased, and reached a stable size of about four times the value at the beginning, within 2 years. This increase in home range size was not related to increases in group size. There is some indication that home range overlap was high when groups had just started, and decreased to middle tenure values during the early phase. DJL did not change with tenure, but in contrast to the middle phase, DJL during the early tenure phase was influenced by the identity of the group and the time a group had existed.

Two factors determined transfer distance, as measured by the location of the new home range in contrast to the former one: the mode of group formation and origin of the females. Groups that formed suddenly only changed the adult breeding male; females and home range location did not change. Groups that formed gradually with females which came from one and the same former group stayed in their former range or regained it within 2 years, while new groups with females that originated from two former groups utilized (or regained) both of the former ranges, roughly in proportion to the number of females that originated from each range. Females showed a strong preference to stay in a familiar area and, consequently, with familiar conspecifics. They were not philopatric in the traditional sense, but in a flexible way, because groups sometimes temporarily left the familiar habitat. The best predictor of the home range location of a group during the middle tenure phase was the home range of the former group(s) of the females.

In summary, during the early tenure phase females incite male mate competition and their behavior reflects the absence of vulnerable infants. When groups start gradually, home range size increases, and reaches a stable size of about four times the initial value within 2 years. This increase in home range size is not related to increases in group size.

Middle versus late tenure phase

Males

The most important change from the middle to the late tenure phase was the decline in the male's strength (Table 7.1a): males were no longer able to keep intruding males away from their females during aggressive group encounters; they provoked other groups less frequently or not at all, and they joined morning loud call bouts less often. Late tenure males still invested in protecting their offspring during male provocations, although increased infant mortality suggested that this defense was less effective. This decline in a male's strength could (partly) be compensated by help from adult sons in age-graded groups (Steenbeek *et al.* 2000). During the late tenure phase, females again showed sexual interest in extra-group males, but the breeding male did not increase his efforts at mate defense. The combination of a decrease in male strength and an increase of female interest in extra-group males probably caused outside pressure by extra-group males to increase again. Late tenure males reacted by spending significantly more time alone when not feeding (Table 7.2), which may reflect a lower investment in social relationships. They also rested at significantly lower heights (Table 7.2), which may reflect an attempt to avoid provocations by extra-group males. Time spent feeding and resting were not affected (Table 7.2). During the late tenure phase natal adult males tended to get wounded more often (Table 7.3). This was caused by the participation of these males in between-group aggression in age-graded groups.

In summary, during the last year of a male's tenure, he declines in strength, which probably causes the increase in mate competition. The male still invests in the protection of offspring but no longer in mate defense.

Females

The most important characteristic of the late tenure phase is that the breeding male declines in strength. As a result, females experienced two increased risks during the late tenure phase: they were more often harassed by extra-group males during between-group encounters (Steenbeek 1999b), and infant mortality tended to be twice as high for the last infant before females left the breeding male than when they stayed in the group. The increased infanticide risk makes an increase in infanticide avoidance by females necessary. Indeed, female behavior during the late tenure phase differed from the middle phase: they avoided extra-group males more by traveling away from loud calling extra-group males, spent significantly less time alone when not feeding (Table 7.2, in contrast to males), and rested significantly lower in the canopy (Table 7.2, similar to males). All these changes are best explained as adjustments to avoid provocations by extra-group males. The increased avoidance resulted in a decreased home range size, and groups sometimes even left the home range they had occupied during the middle tenure phase. The avoidance behavior of females was general: they did not seem to discriminate between extra-group males but reacted mainly to the lower strength of their late tenure male. In spite of this increase in avoidance behavior, overlap with other groups increased, and the group encounter rate remained similar. As a result, home range size and DJL were again determined by other factors than group size.

The importance of having a protector male was illustrated when in one group the male suddenly disappeared (described in Steenbeek 1996). After his disappearance, females first avoided all other groups and males. During this period neither females nor infants were wounded. However, when females showed interest in extra-group males, both females and infants sustained injuries at a higher rate (Table 7.3). Infants became wounded when attacked by extra-group males, and their mothers while defending them.

When male strength declined during late tenure, females finally left the breeding male and associated with a new one. Four of 22 secondary female emigrations took place during the middle tenure phase. One female left group M1, after her first infant in this group was wounded and, consequently died, after which she joined the male who had attacked her infant (Sterck 1997). Three females left group M2 to form M3 (group split). The two females with infants that remained with M2's male rejoined group M3 18 months later. According to the late tenure definition, the three females left during the middle phase, but during the last 18 months group M2 showed all the characteristics of the late tenure phase and seemed to have experienced a very long late tenure phase.

Females with and without infants must experience a conflict of interest during the late tenure phase: Females with infants are expected to avoid extra-group males whereas females without infants are expected to search for a new male. The data do indeed support this. The between-group encounter rate was linked to the number (and proportion) of females without infants, whereas avoidance was linked to the number of females with an infant. This does indicate that each female acted according to her own interest. Their behavior is not incompatible because infanticide did not occur during between-group encounters. During the late tenure phase, females without infants showed sexual interest in extragroup males (one copulation was observed during a between-group encounter), and females again made stop-overs: some females visited other groups and/or all-male bands before making their final transfer to a new male.

In summary, female behavior during the late tenure phase strongly reflects the increased risk due to the declined strength of the breeding male. Infanticide avoidance determines when they make their transfer to a new male. When male strength declines, females invest in keeping their last offspring alive by avoiding detection by extra-group males, which is reflected in almost all behaviors. In spite of the female's efforts, infant mortality doubles, which may explain why all three females that ended up with a new male when they had very young infants with them, in effect "sacrificed" them: the infants died soon after the transfer to a new male ($N = 2$) or soon after the takeover ($N = 1$).

Discussion

In Thomas's langurs, females are continuously exposed to sexual coercion by males and respond to this by seeking the protection of powerful males. Groups go through a typical life cycle and the lifespan of a group coincides with the tenure of its reproductive male. When a male declines in strength, female emigration terminates a group's lifespan.

Male decisions

Male primates can exhibit alternative competitive strategies. In one-male groups the first strategy concerns gaining resident male status in a group of females. Males can aggressively take over a group or lure away females from another male by using sexual coercion. The second male strategy is to try to copulate with females while not being a group member (cf., Smuts 1987a). Thomas's langur males exhibit both strategies, even though they cannot force females to copulate. In Thomas's langurs, male coercion can come from the breeding male when he herds his females away from extragroup males, and sometimes injures them when doing so. Male coercion can also come from extra-group males. Between-group interactions seem to be the context for mate acquisition, in part through sexual coercion, e.g., aggression against females and the use of infanticide. Infant mortality was twice as high for the last infant before a female transferred than for other infants. This clearly points to a higher infanticide risk and male coercive pressure during the late tenure phase. Steenbeek (1999a) concluded that male coercion is the cause of female transfer.

If male coercion causes a female to transfer, does the male benefit by having the female transfer to him? In three cases we knew which male had killed the infant. Two of the females subsequently transferred to the killer, whereas one did not. In three additional cases the killer was not known: two of the females involved stayed in their group and reproduced again with the breeding male, whereas the third one transferred. So, although we know that females transferred to the assumed killer of their infant in some cases, they did not do so in all cases. Females who stay with their infants until they can safely leave them behind lose an average 9.2 months of reproductive time, whereas females who lose the last infant before transfer do not lose reproductive time. This suggests that infanticide speeds up the transfer process.

In species with predominantly one-male–multifemale groups, male competition for the breeding position is expected to be high. In Thomas's langurs, male wounds are usually the result of fights between males, and injuries may be severe, which indicates high male mate competition. However, the behavior of males in all-male bands seems to be at odds with this conclusion. Many males over 7 years old, who were fully mature, did not start new groups when they had the opportunity; instead, they seemed to "wait" and lose reproductive time. One example is the end of group J1. Three females made a stop-over of 3 weeks in all-male band X, with at least five fully mature males. These males neither fought nor showed a very high interest in the females. After 2 weeks, a new male entered the all-male band and in a week's time, separated from the all-male band with the three females, and thus started group J2. The lack of eagerness of many mature males in acquiring females can be explained by assuming that it may sometimes be better for a male to wait than to risk injury or death from an aggressive takeover. One male who aggressively took over a group when still rather young, was wounded (and subsequently died) when another male entered still during his early tenure phase (S. A. Wich, pers. comm.). Gurmaya (1986) also reported male mortality due to mate defense in a population of Thomas's langurs living in an area of secondary forests and rubber plantations. Males of different all-male bands may be familiar with each other's strength, and delay the start of a reproductive group until the risks are minimal.

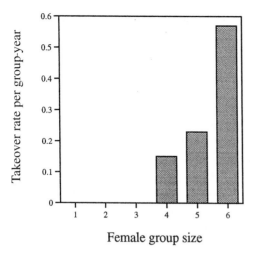

Figure 7.3. The relation between adult female group size and aggressive male takeovers (from Steenbeck 1999a).

The choice for the high-risk strategy of an aggressive takeover is expected to be affected by female group size: this strategy should become a more profitable option as there are more females to be gained. There is some preliminary evidence in favor of this suggestion. Thomas's langur groups with more adult females seem to face a larger risk of being aggressively taken over (see Fig. 7.3). This risk, in turn, may also influence group size. Steenbeck (1999a) suggested that, whereas the lower limit of Thomas's langur groups is set by predation avoidance, the upper limit is not set by feeding competition, but by infanticide avoidance (see also Crockett & Janson, Chapter 4). Females seem to maximize their fitness by staying in fairly small groups, because doing so minimizes the main risk they face – that of infanticide accompanying changes of the group male.

The male strategy of assessing the strength of competitors and waiting until his competitive abilities are relatively high is expected only when the costs of "waiting" are outweighed by the benefits of competing with minimal risks. In Thomas's langurs, males on average have only one tenure that lasts 6 years, with on average three reproducing females who each have two surviving offspring. Only once was a male observed to have a second tenure (Sterck 1997). So an average tenure male has a reproductive success of six surviving offspring. Females can reach the age of at least 18 years, and some females were observed to reproduce with a third male. If we assume that females have two surviving offspring with each of three males, the reproductive success of females is also six. The average number of adult females is somewhat higher than three but the point is that many males will have the possibility to start a reproductive group at some time.

In summary, almost all males have only one reproductive tenure and many males get a chance of starting a group at some time. Therefore, the best strategy for males is continuously to monitor the strength of competing males and start mate competition when the risks of injury are minimal. At this point, males still have two choices: lure away females through sexual coercion or aggressively take over a group. The former choice has relatively low risks because the male may be able to prevent fights. The latter is the higher risk strategy, which may be the best option for very strong males when there are no late tenure males in the area, i.e., when females cannot be induced to voluntarily leave their current breeding male. The choice between these two tactics seems to be affected by female group size: groups with more adult females seem to face a larger risk of being aggressively taken over (see also Crockett & Janson, Chapter 4).

Female decisions

Male competition for mates depends inevitably on conditions set by females and, consequently, male competition for females is inseparable from female choice (Wiley & Poston 1996). Smuts (1987a) made a distinction between direct female choice (e.g., soliciting copulations, initiating the majority of copulations, refusing sex) and indirect female choice (e.g., influencing male group membership, emigrating, or choosing a new group/male). Wiley & Poston (1996) have suggested another distinction: direct mate choice requires discrimination between attributes of individuals of the opposite sex, whereas indirect mate choice occurs when the behavior (or morphology) of one sex establishes conditions under which individuals of the opposite sex compete for access to mates. By setting the conditions for, and accepting the outcome of, competition, an individual indirectly selects a mating partner. Although some primate females may prefer to mate with unaggressive males, when they rely on males to protect them from other males, they may choose to mate with an aggressive male (Wrangham 1979). There is no firm evidence that females mate with males of superior genetic quality (Smuts 1987a), but, in species with female transfer, females may emigrate in order to avoid potentially infanticidal males (Marsh 1979b), and males may use aggression and infanticide in order to attract mates (Stewart & Harcourt 1987).

Thomas's langur females transfer when coercion by extra-group males has revealed that their current resident male is no longer able to protect the offspring. Instead of transferring to a new male, females could also wait for an aggressive takeover. Aggressive takeovers have both advantages and disadvantages when compared to gradual group ends. For all females there are two advantages. First, only strong males are expected to take the risks associated with an aggressive takeover, so females end up with a relatively strong male. Indeed, after the two takeovers in this study, females did not show any sexual interest in extra-group males. Second, females can remain in their home range. For females without infants, a takeover has additional benefits: they do not have to search for a new male and can, therefore, reproduce sooner. Females with an infant run the risk of losing it because of infanticide. There is one cost to all females: they may be wounded because of aggressive herding behavior by the new male.

So, when groups end gradually, how do Thomas's langur females choose a new male? Between-group interactions can be interpreted as a continuous test for the breeding male. During these encounters, females incite male–male competition by initiating copulations with extra-group males (Steenbeek 1999b). During the course of male tenure, females should be able, therefore, to learn about a decrease in the breeding male's strength from the outcome of aggressive group encounters and male provocations (Steenbeek 1999b), and from a male's participation in dawn-call bouts (Steenbeek et al. 1999a). That it is the females who take the initiative in leaving one male and joining another is further illustrated by the occurrence of stop-overs: when females were ready to transfer and a new male was not available, they started searching for one. In addition, in case of gradual group endings and beginnings, females did not all leave at the same time, but left when the age and survival of the last infant allowed it. Furthermore, most females were found to transfer into relatively new groups. These seem the best choice, because transfer is costly and male tenure is limited, and males of new groups can be expected to have a long tenure.

Food competition or an increase in infanticide risk may prevent females from transferring into large existing groups, but not from joining small groups or from forming new groups when these females are ready to reproduce. Transfer into a neighboring group can be costly, but if females manage to survive the first few days and stay, they lose far less reproductive time than if they had stayed in their former group, waiting for male tenure to end. However, in spite of seemingly low dispersal costs, the benefits of staying in a familiar habitat must outweigh the costs of delaying reproduction.

I concluded that Thomas's langur females do not directly choose one specific male (Steenbeek 1999a). Females merely show extra-group males that they are ready to start a new group, and wait for a male to join them. This is in contrast to gorillas, where females choose a male as much or more than they choose groups (Watts 1990). Thus, in Thomas's langurs, the choice of a new male seems to be indirect in that females incite male mate competition (according to both the definition of Smuts & Smuts (1993) and that of Wiley & Poston (1996)).

Comparisons with other taxa

Long-term studies on species with female transfer are scarce, and a comparison is only possible with some colobines, howlers and mountain gorillas. Little is published on langur species with regular female transfer, but snippets can be pieced together to suggest that similar things are happening in at least some of the other *Presbytis* species. The following langur species may resemble the Thomas's langur system, although tenure length is shorter: purple-faced langurs, *Presbytis senex*, although aggressive takeovers seem to dominate (Rudran 1973); hanuman langurs, *P. entellus*, Kanha tiger reserve population (Newton 1987); and capped langurs, *P. pileata*, although infanticide is not reported (Stanford 1991). The Tana River red colobus, *Colobus badius rufomitratus*, may also resemble the Thomas's langur in several respects (Marsh 1979b).

The same factors do not operate equally strongly in all species. Studies on howlers suggest that food competition can become a primary regulating factor for female transfer. In mantled howlers, *Alouatta palliata*, females are forced to emigrate when still juvenile. In addition, females can be very aggressive toward immigrants and female immigrants depend on the help and protection of the adult male to enter a group successfully (Glander 1992). In red howlers, the situation is even more extreme: females emigrate but do not immigrate into existing groups. Instead, they form new groups with lone males (Crockett 1984). However, Crockett & Janson (Chapter 4) present evidence that infanticide may limit group size in red howlers. Thus, by emigrating from the natal group, a female does reduce competition for the protective qualities of the male. In

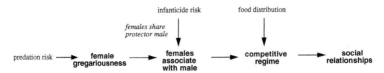

Figure 7.4. Possible pathways for the evolution of sociality in Thomas's langurs. (See also Sterck *et al.* 1997.)

Thomas's langurs, both food competition and infanticide risk may limit group size in new groups, but these effects are not strong enough for natal emigrants to leave kin.

Mountain gorillas resemble Thomas's langurs in many ways: bisexual groups are one-male or age-graded, females transfer to new males and infanticide occurs. In contrast to Thomas's langurs, mountain gorillas are mainly terrestrial, they are more folivorous, and males are much larger than females. These similarities and, more importantly, the differences with the mountain gorilla can help us to reconstruct the social evolution in the Thomas's langur. In species with male infanticide, it is thought that a female associates with a male because she needs his protective qualities. Because males differ in their protective abilities, females share a protector male and become gregarious. Both food distribution and predation risk influence female gregariousness and determine the competitive regime within the group, which in turn determines female social relationships (for a review, see Sterck *et al.* 1997). This model seems to work well for mountain gorillas where females choose a male as much or more than they choose groups (Watts 1990), and where social groups are based on male–female relationships (Harcourt 1979a), although these vary with female residence status (Watts 1992). In Thomas's langurs, by contrast, females end up with a new male through indirect female choice, and female–female relationships seem more important than male–female relationships. Thomas's langur females never start a group with only one female. Females are gregarious first and only then associate with a male. The most important difference between the two species seems to be the greater importance of predation risk for Thomas's langurs. The number of surviving offspring per female year tends to increase with group size, consistent with the benefits of gregariousness as an anti-predator strategy (Steenbeek 1999a).

Figure 7.4 illustrates the possible pathways of the evolution of Thomas's langur social groups. Predation risk may be the primary reason females stay together when they start a new group, and sets the lower limit of group size. Females then associate with a male because of infanticide risk. Food distribution determines the competitive regime that in turn determines female social relationships. Thomas's langurs experience scramble competition for food, but the upper limit of group size is not set by feeding competition but probably by infanticide avoidance.

Conclusion

When Thomas's langur females associate with a new male and form a new group, this male has to prove his protective abilities. During the early tenure phase, females elicit male mate competition and their behavior reflects that there are no infants at risk. When male mate defense is successful, females will start to reproduce and the group enters the relatively stable middle tenure phase. During the middle tenure phase, both male and female behavior reflects that there are infants at risk. During the last year of a male's tenure, his strength declines and male mate competition increases. The male still invests in the protection of offspring but no longer in mate defense. Female behavior during the late tenure phase strongly reflects the increased risks due to the declined strength of the breeding male. A female finally transfers to a new male; the timing of this move is related to infanticide avoidance. Thomas's langur females take the transfer decision when male coercion has proved that their current male is no longer able to defend the next offspring. Most females choose to stay in a familiar habitat, in spite of a loss of reproductive time, and end up with a new male through indirect mate choice.

The conclusion of this study is that the intersexual conflict, as expressed by male sexual coercion (harassment of females and infanticide) and female choice, strongly influences the Thomas's langur social system. The lower limit of Thomas's langur groups is set by predation avoidance, whereas the upper limit is not set by feeding competition, but by infanticide avoidance. The influence of infanticide risk seems as important as that of ecological factors, such as predation risk and the distribution of food. This paradigm may also hold for other species, especially for species with secondary female transfer and infanticide by males.

Acknowledgments

I thank the Indonesian Institute of Science (LIPI, Jakarta), the Indonesian Nature Conservation Service (PHPA) in Jakarta and Kutacane (Gunung Leuser National Park office) and UNAS (Universitas Nasional, Jakarta) for granting permission to use the Ketambe Research Station facilities and to conduct this field study in the Gunung Leuser National Park, Sumatra, Indonesia; and Mike Griffiths and Kathryn Monk of the Leuser Development Program for their help and friendship. I also thank Sri Suci Utami Atmoko for her help in running the project, her never-ending support and personal friendship; Professors Jan van Hooff and Carel van Schaik for their support; and Carel van Schaik, Liesbeth Sterck, Mandy Korstjens and Serge Wich for sharing data. I thank the students and assistants who helped to collect the data: Bahlias, Titi, Usman, Rahimin, Samsu, Marlan, Suprayudi, Corine Eising, Jacobine Schouten, Marleen van Buul, Mandy Korstjens, Finley Koolhoven, Ritva Meijdam, Bas duMaine, Hanneke van Ormondt, Casper Breuker, Judy Bartlett, Ruben Piek, Jeffry Oonk, Janneke Verschoor and Danielle Obbens. Finally, I thank Carel van Schaik and Maria van Noordwijk for constructive suggestions on the manuscript.

Financial support was provided by WOTRO (the Netherlands Foundation for the Advancement of Tropical Research).

8

The evolution of infanticide in rodents: a comparative analysis

Introduction

We simultaneously know more about the proximate causes of infanticide in rodents and less about its adaptive consequences and evolution than in any other taxon. Rodents are the largest mammalian order with over 440 genera and 2021 of the 4629 described species (Musser & Carleton 1993; Wilson & Reeder 1993). Infanticide by either or both males and females may be found in 46 species from 22 genera in 3 rodent groups (see below). A number of previous reviews illustrate the detailed understanding of proximate causation, development, and function gained by studying infanticide in rodents (Sherman 1981; Brooks 1984; Elwood & Ostermeyer 1984; Huck 1984; Labov 1984; Svare *et al.* 1984; vom Saal 1984; Labov *et al.* 1985; Trulio 1987; Elwood 1992; Elwood & Kennedy 1994; Ebensperger 1998a). In this review, I take a slightly different approach to study the evolution of infanticide in rodents. I first use the taxonomic distribution of infanticide by males and infanticide by females along with available phylogenetic evidence to parsimoniously reconstruct the evolution of infanticide by males, infanticide by females, and a potential female response to infanticide by males – male-induced pregnancy termination (Bruce effects; Bruce 1960). Then, using these evolutionary reconstructions, I test specific functional hypotheses about infanticide.

A thumbnail sketch of infanticide in rodents

Rodents provide some of the richest information about infanticide because studies of the proximate causation of infanticide are integrated with studies designed to document ultimate function. Studies of proximate factors influencing infanticide by male rodents have supported a number of key predictions associated with functional hypotheses – particularly the sexual selection hypothesis (Hrdy & Hausfater 1984). For instance, strain differences in infanticidal behavior in male mice (Svare *et al.* 1984; Perrigo *et al.* 1993) suggested that infanticide may be a heritable trait; a key requirement of the sexual selection hypothesis. Elegant studies examined how the timing of mating behavior (Elwood & Ostermeyer 1984; Elwood 1994), the process of ejaculating (Perrigo *et al.* 1990; Perrigo & vom Saal 1994), and familiarity with the mother (D'Amato 1993) influence male infanticidal behavior. These findings are generally consistent with another key prediction of the sexual selection hypothesis – males should not kill their own offspring. Another key prediction, verified in some rodents (e.g., vom Saal & Howard 1982), is that when a female loses a litter she becomes sexually receptive sooner than had she not lost a litter.

Functional hypotheses have been studied in the field, but field studies typically have not had the necessary control to study proximate mechanisms, although they suggest a variety of functions both for infanticide by males and for infanticide by females (Sherman 1981; Trulio 1987). Hoogland's (1995) review of his long-term studies of black-tailed prairie dogs (Latin names not in the text are given in Table 8.1) is notable. Infanticide in prairie dogs is a major source of pup mortality that accounts for as much as the partial or total loss of 39% of litters. Both males and females are infanticidal, but infanticide by females is more common. Lactating females kill and cannibalize unweaned offspring of close kin, as well as offspring of immigrant females living in their social group. Interestingly, while one lactating black-tailed prairie dog female is out killing another female's offspring, another individual may be killing the former's offspring! In contrast to Hoogland's finding, Sherman (1981) noted that ground squirrel females never killed offspring of their close relatives. Hoogland concluded that several hypotheses related to resource competition or exploitation might explain infanticide by female prairie dogs. Trulio (1987) also concluded that for female sciurid rodents, resource exploitation was the most common function of female infanticide.

Infanticide by males may be sexually selected but the link is not as clear as in some mice, carnivores or primates. Male prairie dogs who disperse into a new social group are often infanticidal. Prairie dogs – like their close relatives the marmots and *Spermophilus* ground squirrels – are monestrous and thus breed only once a year. In a study of golden marmots, another sciurid rodent, I found no strong support that females were more likely to wean young the year after male infanticide

(Blumstein 1997), but the effect size may be small. However, using a much larger dataset, Hackländer & Arnold (1999) found that female alpine marmots who terminated their pregnancy upon exposure to a novel male (see below), were in better physiological condition the next year and had higher reproductive success than females who did not terminate their pregnancies. Hoogland (1995) concluded that nourishment may also be a potential function of male infanticide. Surviving the winter is a major challenge for some sciurids and young of the year are particularly vulnerable (Arnold 1993). Some marmots hibernate socially and help to thermoregulate young. Thermoregulation is, however, costly, and alloparental care is directed at younger relatives (Arnold 1993). I found support for Sherman's (1981) hypothesis that males may kill offspring to avoid providing misdirected care. In the case of social golden marmots, males may kill unrelated juveniles to reduce costs of over-winter alloparental care (Blumstein 1997). This conclusion is supported by the pattern of infanticide reported in marmots: male alpine marmots – another social marmot that provides parental/alloparental care – are infanticidal (Coulon *et al.* 1995), while females but apparently not males of the less social yellow-bellied marmot are infanticidal (Armitage *et al.* 1979; Brody & Melcher 1985).

The evolution of infanticide in rodents

The interdisciplinary approach used to study infanticide has generated a considerable amount of knowledge since Hrdy (1974, 1979), Wilson (1975) and Sherman (1981) emphasized the potential adaptiveness of infanticide. A model for interdisciplinary research was proposed by Tinbergen (1963), who recognized four types of question that could be asked about any behavioral phenomena. Of them, evolutionary history has heretofore been a relatively neglected subject in infanticide research (but see Boonstra 1980; van Schaik & Kappeler 1997; van Schaik *et al.* 1999; van Schaik 2000). In this chapter I apply modern comparative methods (Harvey & Pagel 1991) to study the evolution of infanticide in rodents. Specifically, I reconstruct the ancestral states of infanticidal behaviors in rodents, and test hypotheses about correlated evolution between infanticide by males and male-induced pregnancy termination. In doing so, I hope to provide another line of evidence with which to evaluate the sexual selection hypothesis, as well as another way to study infanticide in rodents.

Comparative analyses may help to evaluate functional hypotheses. I make the following predictions about the evolutionary history of infanticide in rodents. First, if the sexual selection hypothesis was responsible for the evolution of infanticide in rodents, infanticide by males should be the ancestral condition in those clades where infanticide is found, and infanticide by females should evolve independently of infanticide by males. Second, if infanticide evolved as a generalized foraging strategy, or if resource competition and exploitation were responsible for the evolution of infanticide in rodents, then infanticide would be predicted to be seen in both sexes and to have evolved together in both sexes. Third, sexual selection also predicts that Bruce effects in females evolved in response to the evolution of infanticide by males. Thus there should be a significant association between the evolution and loss of infanticide by males and the evolution and loss of Bruce effects in females. In all cases, because infanticide reduces the fitness of those who lose young, we should see broad evidence of adaptations to counter infanticidal tendencies.

Methods and data

Some important caveats

All comparative analyses are fundamentally dependent on valid observations of the traits which they seek to study. It is extremely difficult to obtain an unbiased comparative dataset with which to study infanticide in rodents. Because most rodents are small, nocturnal and semi-fossorial, studying their behavior and observing infanticide poses a variety of problems. Direct observations of infanticide in many muroid rodents are lacking in nature. Fortunately for infanticide researchers, house mice and other common laboratory rodents commit infanticide. In contrast to the majority of rodents, sciurid rodents are diurnal, live in predictable locations and are generally easy to observe. This has made them an ideal group in which to study socioecological questions (Armitage 1981; Sherman 1981; Michener 1983; Blumstein & Armitage 1998, 1999), including infanticide.

Some sciurid researchers have directly observed infanticide and, as already discussed, infanticide may be a major cause of juvenile mortality. However, many reported cases of infanticide were not directly observed. This led to a discussion over the validity of inferring infanticide from indirect observations (Michener 1982; Sherman 1982). The fundamental problem is that infanticide generally occurs in underground nesting burrows and observations may, at best, be only indirect. In lieu of actually observing infanticide, a researcher may see an animal emerge with blood on its face and a litter known to be in the burrow subsequently never emerges (e.g., Hoogland 1995). Infanticide has also been inferred by seeing recently emerged young fleeing certain adults with the observer later finding carcasses with "escape wounds" (small bites on the back of the legs and back; e.g., Blumstein 1997), and by finding an animal in possession of a recently killed pup (Trulio 1996). To validate behavioral observations and inferences, Hoogland (1995) used a backhoe to immediately excavate burrows in which he and his research assistants strongly suspected infanticide had just occurred. He also caught and examined stomach contents from females assumed to be infanticidal. Excavations supported his behavioral observations: Hoogland found recently killed pups in an excavated burrow and found the remains of cannibalized young in the stomach of a suspected perpetrator. Thus, *contra* Michener (1982), it may be valid to use behavioral correlates as evidence of infanticide.

Defining infanticidal traits

To study the evolution of infanticide I first defined three traits and then "mapped" them (i.e., optimized them using parsimony) onto a series of rodent phylogenies. Because infanticide reflects an individual reproductive strategy (e.g., Sherman 1981), and because infanticide by males and females usually has different functions (e.g., Hrdy 1979), I examined the evolution of infanticide separately for males and females, based on reports both in the laboratory and the field (Table 8.1).

The first trait I defined was the presence of infanticide by males. I did not distinguish between observations of infanticide in the field or in captivity. Additionally, to minimize the number of infanticidal species excluded from analysis, I was rather liberal in my adoption of evidence for infanticide and included both direct observations, less direct inferences and anecdotes of pup cannibalism. I used both previous reviews of infanticide in rodents and current literature searches to generate the comparative dataset. Because we know relatively little about the behavior or ecology of most rodents, this review is necessarily incomplete.

I defined the second trait as the presence of female infanticide. Again, I did not distinguish between field and laboratory reports. Moreover, I did not distinguish between infanticide committed by the mother and infanticide committed by a female other than the mother (see Ebensperger 1998a) because this would introduce error into the comparative dataset (in many instances in the field the perpetrator's identity is unknown).

I defined the third trait as the presence of Bruce effects in female rodents. Bruce effects are hypothesized to have evolved to minimize the

Table 8.1. *Summary of reports of infanticide and Bruce effects in rodents*

Family	Subfamily	Species	Common name	Infanticide/ Bruce effects	Sources
Sciuridae	Sciurinae	*Paraxerus cepapi*	Tree squirrel	MN	de Villiers 1986
Sciuridae	Sciurinae	*Spermophilus beecheyi*	California ground squirrel	FN, MN	Trulio 1996
Sciuridae	Sciurinae	*Marmota caudata*	Golden marmot	MN	Blumstein 1997
Sciuridae	Sciurinae	*Marmota marmota*	Alpine marmot	B, MN	Perrin et al. 1994; Coulon et al. 1995; Hackländer & Arnold, 1999
Sciuridae	Sciurinae	*Marmota flaviventris*	Yellow-bellied marmot	FC, FN	Armitage et al. 1979; Brody & Melcher 1985
Sciuridae	Sciurinae	*Spermophilus franklinii*	Franklin's ground squirrel	MN	Sowls 1948
Sciuridae	Sciurinae	*Spermophilus tridecemlineatus*	13-lined ground squirrel	MN	Vestal 1991
Sciuridae	Sciurinae	*Spermophilus armatus*	Utah ground squirrel	FN	Burns 1968
Sciuridae	Sciurinae	*Spermophilus beldingi*	Belding's ground squirrel	FN	Sherman 1981
Sciuridae	Sciurinae	*Cynomys ludovicianus*	Black-tailed prairie dog	FN, MN	Hoogland 1995
Sciuridae	Sciurinae	*Cynomys gunnisoni*	Gunnison prairie dog	FN	Fitzgerald & Lechleitner 1974
Sciuridae	Sciurinae	*Spermophilus richardsonii*	Richardson's ground squirrel	FC, FN	Quanstrom 1968; Michener 1973
Sciuridae	Sciurinae	*Spermophilus parryii*	Arctic ground squirrel	MN	Steiner 1972; McLean 1983
Sciuridae	Sciurinae	*Spermophilus columbianus*	Columbian ground squirrel	FN, MN	Betts 1976; Balfour 1983; Dobson 1990
Sciuridae	Sciurinae	*Spermophilus townsendii*	Townsend's ground squirrel	MN	Alcorn 1940
Muridae	Arvicolinae	*Microtus pennsylvanicus*	Meadow vole	B, MC, MN	Clulow & Langford 1971; Mallory & Clulow 1977; Webster et al. 1981; Caley & Boutin 1985; Storey et al. 1994
Muridae	Arvicolinae	*Microtus montanus*	Montane vole	B	Stehn & Jannett 1981
Muridae	Arvicolinae	*Microtus townsendii*	Townsend's vole	FN	Boonstra 1978, 1980
Muridae	Arvicolinae	*Microtus canicaudus*	Gray-tailed vole	FN	Wolff & Schauber 1996
Muridae	Arvicolinae	*Microtus californicus*	California vole	B, MN	Lidicker 1979; Heske 1987
Muridae	Arvicolinae	*Microtus ochrogaster*	Prairie vole	B	Kenney et al. 1977; Heske & Nelson 1984
Muridae	Arvicolinae	*Microtus pinetorum*	Pine vole	B	Schadler 1981; Stehn & Jannett 1981

Table 8.1 (*cont.*)

Family	Subfamily	Species	Common name	Infanticide/ Bruce effects	Sources
Muridae	Arvicolinae	*Microtus agrestis*	Field vole	B, FN	Clulow & Clarke 1968; Milligan 1976; Myllymaki 1977; Agrell 1995
Muridae	Arvicolinae	*Microtus oeconomus*	Root vole	B	Jensen & Gustafsson 1984
Muridae	Arvicolinae	*Lemmiscus curtatus*	Sagebrush vole	No B	Stehn & Jannett 1981
Muridae	Arvicolinae	*Lasiopodomys brandti*	Brandt's vole	B, FC, MC	Stubbe & Janke 1994
Muridae	Arvicolinae	*Clethrionomys rutilus*	Northern red-backed vole		Koshkina & Korotkov 1975 cited in Brooks 1984
Muridae	Arvicolinae	*Clethrionomys gapperi*	Red-backed vole	B	Clulow et al. 1982
Muridae	Arvicolinae	*Clethrionomys glareolus*	Bank vole	B, FC, FN, MC, MN	Ylönen et al. 1997
Muridae	Arvicolinae	*Ondatra zibethicus*	Muskrat	FN	Errington 1963; Caley & Boutin 1985
Muridae	Arvicolinae	*Arvicola terrestris*	Water vole		Jeppson 1986
Muridae	Arvicolinae	*Dicrostonyx groenlandicus*	Collared lemming	B, FC, MC	Mallory & Brooks 1978, 1980
Muridae	Arvicolinae	*Lemmus lemmus*	Norwegian lemming	B, FC	Semb-Johansson et al. 1979; Jensen & Gustafsson 1984
Muridae	Cricetinae	*Phodopus campbelli*	Djungarian hamster	FC	Edwards et al. 1995
Muridae	Cricetinae	*Phodopus sungorus*	Siberian hamster	FC, MC	Gibber et al. 1984
Muridae	Cricetinae	*Mesocricetus auratus*	Syrian/golden hamster	No B, FC	Richards 1966; Huck 1984; Huck et al. 1988 reported pregnancy block induced by females, *not* males
Muridae	Sigmodontinae	*Neotoma lepida*	Desert woodrat	FC	Fleming 1979
Muridae	Sigmodontinae	*Peromyscus maniculatus*	Deer mice	B, FC, MC	Ayer & Whitsett 1980; Wolff 1985; Wolff & Cicirello 1989; Cicirello & Wolff 1990
Muridae	Sigmodontinae	*Peromyscus leucopus*	White-footed mouse	FC, FN, MC, MN	Wolff 1986; Wolff & Cicirello 1989; Cicirello & Wolff 1990
Muridae	Sigmodontinae	*Peromyscus californicus*	California mouse	FC, MC	Gubernick 1994; Gubernick et al. 1995

Muridae	Murinae	House mouse	*Mus musculus/domesticus*	B, FC, MC	Jakubowski & Terkel 1982; vom Saal & Howard 1982; Elwood 1986; McCarthey *et al.* 1986; Palanza *et al.* 1996
Muridae	Murinae	European wood mouse	*Apodemus sylvaticus*	FC, MC	Wilson *et al.* 1993; Montgomery *et al.* 1997
Muridae	Murinae	Norway rat	*Rattus norvegicus*	FC, MC	Jakubowski & Terkel 1985; Schultz & Lore 1993
Muridae	Murinae	Spiny mouse	*Acomys cahirinus*	MC	Makin & Porter 1984
Muridae	Gerbillinae	Mongolian gerbil	*Meriones unguiculatus*	FC, MC	Elwood 1980; Elwood & Ostermeyer 1984
Muridae	Caviinae	Guinea pig	*Cavia* sp.		Bufon 1852, cited in Labov *et al.* 1985, but see Rood 1972
Caviidae	Caviinae	Cui	*Galea musteloides*	FC	Künkele & Hoeck 1989

Notes:

B, Bruce effect reported; No B, Bruce effect studied but not reported; FC, female infanticide observed in captivity; FN, female infanticide observed in nature; MC, male infanticide observed in captivity; MN, male infanticide observed in nature.

I did not distinguish between reports of Bruce effects in captivity or the wild. For subsequent comparative analyses, I combined reports of infanticide in the field and in captivity. I scored male infanticide (or female infanticide) present if it was reported, and absent if it was not reported when female infanticide (or male infanticide) was reported. Species for which infanticide was reported but the infanticidal sex was not reported are listed, but the sex of the infanticidal animal is not noted. Patterns of presence and absence generate hypotheses about a given trait for species with no data.

cost of losing a litter to the new infanticidal male and thus reflect a female reproductive strategy to counter male infanticide (Labov 1981; Huck 1984). Alternatively, Bruce effects may be a mechanism of female choice (Schwagmeyer 1979), and evidence of pregnancy termination induced by exposure to novel females but not males (Huck *et al.* 1988) suggest other possible functions (for a review, see Labov 1981). Most experimental tests of Bruce effects found a positive effect. I also include species tested for Bruce effects where Bruce effects were not found (Stehn & Jannett 1981; Huck *et al.* 1988) and hypothesize the pattern of Bruce effects for species in which they have not been studied.

The comparative dataset

I make several assumptions when compiling observations of infanticide. First, I assume that infanticide in the laboratory is not a pathological behavior resulting from overcrowded rearing conditions (e.g., Calhoun 1962), and that as a first approximation, patterns in the laboratory may reflect patterns of infanticide in nature (e.g., Ylönen *et al.* 1997). We have come a long way in the past 20 years, and one realization is that the laboratory offers a tremendous opportunity to study apparently adaptive mechanisms of infanticide. Second, I assume no sex bias in reports of infanticide. Thus, if infanticidal behavior for a given species is reported in one sex but not the other, then it occurs in one sex and not the other. This assumption is only as good as the reports, which at times are anecdotal. Third, I also include species where infanticide was reported but the perpetrator's sex was not recorded. In including these species I illustrate lacunae in the comparative dataset that future researchers may later fill in. Moreover, by using patterns of infanticide among close relatives and assuming parsimony (Wiley *et al.* 1991), I hypothesize the sex of infanticidal perpetrators for species where the sex(es) are not specified.

Skeptics may discount many observations of infanticide in nature and Bruce effects in captivity. Nevertheless, compiling these reports generates testable hypotheses about the evolution of infanticide. Below I use a combination of observations of infanticide and Bruce effects in nature and in captivity to study the evolution of infanticide in rodents.

Comparative analyses

I conducted a series of comparative analyses. First, I used a recent phylogenetic hypothesis of higher level rodent relationships (Carleton 1984) to determine how many times infanticide (either by males or by females) evolved in the order Rodentia. Then I examined separately the evolution of infanticide within each of the two families in which infanticide was commonly reported (sciurids and muroids).

There is no published sciurid or muroid phylogeny that includes all species of interest. For these analyses, I initially inferred phylogeny from taxonomy and modified branching patterns to reflect more detailed phylogenetic information (Black 1972; Corbet 1978; Carleton 1980; Honacki *et al.* 1982; Carleton & Musser 1984; Anderson 1985; Bonhomme *et al.* 1985; Catzeflis *et al.* 1992; DeBry 1992; Musser & Carleton 1993; Engel *et al.* 1998; Steppan *et al.* 2000; C. Conroy & J. Cook unpublished data; S. Davis unpublished data; R. S. Hoffmann, pers. comm.). I do not include species in the phylogeny for which there is no information – the vast majority of muroid species. The best available microtine rodent phylogeny is not completely resolved. To permit subsequent concentrated changes tests (Maddison & Maddison 1992, and see below), I randomly resolved three polytomies in the partial phylogeny. Doing so did not change the overall parsimonious trait reconstruction found when polytomies were not randomly resolved. Available evidence suggests that the sciurid genus *Spermophilus* is polyphyletic, explaining the location of the "genus" *Cynomys*. Future investigators interested in studying specific evolutionary hypotheses of infanticide within these taxonomic groups will certainly benefit from better-supported phylogenetic hypotheses and a more complete dataset.

Given uncertain relationships between most rodents, excluding most species for which there are no reports of infanticide or Bruce effects seems an acceptable strategy. Excluding species should not change the overall parsimonious reconstruction. However, excluding species will reduce the number of branches between species. The concentrated changes test estimates the probability of two traits evolving independently by calculating a null model that changes are randomly distributed along branches (Maddison & Maddison 1992). Excluding species may therefore bias the results of the concentrated changes test, making it less likely to detect a significant association between two traits.

The evolution of infanticide

Assuming parsimony and using MacClade 3.0 (Maddison & Maddison 1992), I first optimized infanticide on the higher-level phylogeny and then optimized each of the three traits on the partial rodent phylogenies. For all optimizations, I assumed the traits were unordered and that changes were equally likely in either direction. I did not specify an outgroup and allowed MacClade to hypothesize the ancestral condition.

Did male and female infanticide evolve together?

I used Maddison and Maddison's concentrated changes test (Maddison & Maddison 1992; see also van Schaik & Kappeler 1997) to test the hypothesis that the pattern of origins and losses of infanticide by males was associated with the pattern of origins and losses of infanticide by females in both sciurid and muroid rodents. MacClade provided an exact calculation of the probability that infanticide by males and infanticide by females was independent for sciurid rodents. There were too many independent gains and losses of infanticide by males in muroid rodents to permit an exact calculation. To estimate the probability that infanticide by males and by females was independent, I selected the 1000 sample simulation option in MacClade, and calculated the probability assuming that infanticide by females was ancestral (as suggested by the parsimonious reconstruction for muroids), and that infanticide by females was either ancestral or derived.

Did Bruce effects evolve as a counterstrategy to infanticide by males?

I tested this hypothesis by using the concentrated changes test. To estimate the probability that infanticide by males and Bruce effects by females were independent, I selected the 1000 sample simulation option in MacClade, and calculated the probability assuming that Bruce effects were ancestral, and assuming that Bruce effects were either ancestral or derived.

Results and discussion

How many times did infanticide evolve in rodents?

Available evidence suggests that infanticide evolved independently in three rodent infraorders: the Sciuromorpha, the Myomorpha (muroid rodents only), and the Caviomorpha (Figure 8.1). Despite the opportunity for bias in observing infanticide, I am aware of no reports of infanticide in other rodent groups. Future workers studying the behavior and ecology of other species would be well advised to look for evidence of infanticide; reports would greatly clarify our understanding of the evolution of infanticide. Moreover, because little is known about the independent evolution of infanticide in caviomorph rodents, future study would be particularly useful and might resolve the conflicting reports of Bufon (1852 cited in Labov et al. 1985) and Rood (1972). Because

Reports of Infanticide in Rodents

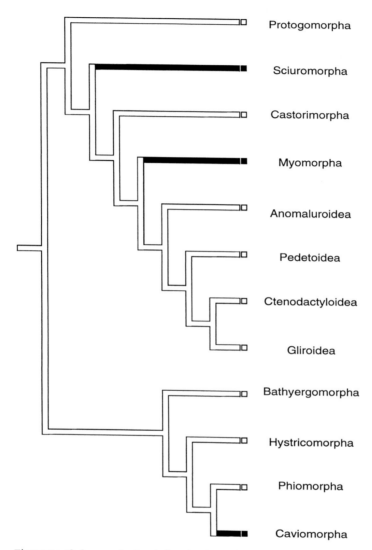

Figure 8.1. Phylogeny of rodent infraorders (-morpha suffix) and superfamilies
(-oidea suffix) illustrating independent evolutionary origins of infanticide
(phylogeny from Carleton 1984). Infanticide by both males and females is
reported in muroid rodents of the infraorder Myomorpha and the sciurid
rodents of the Sciuromorpha. There are two reports of infanticide in the
Caviomorpha. Filled bars indicate any reports of infanticide in a group, empty
bars indicate no reports of infanticide in a group.

infanticide evolved independently in sciuromorphs and myomorphs, and because there were numerous reports of infanticide in each group, I analyzed each separately in the analyses below.

How many times did infanticide and Bruce effects evolve in sciurids and muroids?

Available evidence suggests that infanticide evolved and was lost multiple times in sciurid (Figure 8.2) and muroid (Figure 8.3) rodents. Infanticide by males seems to be the ancestral condition in both sciurids and muroids. Infanticide by females is ancestral in muroids but not sciurids. Muroid reconstructions leave some ambiguity about the ancestral states for some species. Bruce effects were reported in only one sciurid – the alpine marmot – but a better dataset exists for Bruce effects in muroid rodents (Table 8.1). In muroids, patterns of extant Bruce effects suggest that the ability for females to terminate pregnancy after encountering a novel male is ancestral (Figure 8.3).

Did male and female infanticide evolve together?

In both sciurids and muroids, the pattern of origins and losses of infanticide by males and infanticide by females is non-random, suggesting that they evolved together. In sciurids, the concentrated changes test suggests that the number of gains and losses of infanticide by females evolved in unison with infanticide by males ($P = 0.024$ if infanticide by females was derived; $P = 0.024$ if either trait is ancestral). In muroids, the concentrated changes test suggests that the number of gains and losses of infanticide by females evolved in unison with infanticide by males ($P = 0.036$ if female infanticide is ancestral; $P = 0.032$ if either trait is ancestral). Because infanticide by males evolved with infanticide by females, it is likely that infanticide evolved as a generalized type of foraging strategy or as a means of competition or exploitation which later was adapted for different functions (e.g., sexual selection). Infanticide by non-parents in several mammalian taxa currently includes cannibalism (Ebensperger 1998a), suggesting that, for some species, infanticide is an ancestral foraging strategy.

Did Bruce effects evolve as a counterstrategy to infanticide by males?

The pattern of gains and losses of Bruce effects is random with respect to the pattern of gains and losses of infanticide by males in muroid rodents

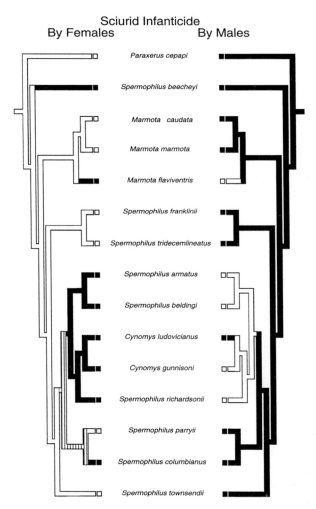

Figure 8.2. Evolutionary reconstructions of infanticide by males and by females in
sciurid rodents. For this and Figure 8.3, whether or not the square next to the
species name is filled indicates the presence or absence of a trait in the species;
no square illustrates no data. Infanticide by males (or infanticide by females)
was scored present if it was reported, and absent if it was not reported when
infanticide by females (or infanticide by males) was reported. Species for which
infanticide was reported but the infanticidal sex was not reported were scored
as no data. Patterns of presence and absence generate hypotheses about a given
trait for species with no data. A filled bar illustrates the presence of the
reconstructed trait in that lineage; an empty bar indicates the absence of the
reconstructed trait in the lineage. A striped bar illustrates ancestral lineages
that could not be reconstructed using parsimony. Traits are optimized
assuming parsimony and using MacClade 3.0 (for details, see text and
Maddison & Maddison 1992).

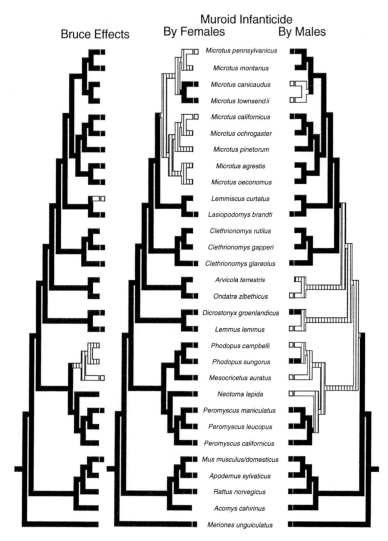

Figure 8.3. Evolutionary reconstructions of infanticide by males and by females in muroid rodents, and the evolutionary reconstruction of Bruce effects in muroid rodents. For interpretation, see legend of Figure 8.2.

($P = 0.298$ if either trait is ancestral; $P = 0.316$ if Bruce effects are considered ancestral). It is important to note that, while these data suggest that Bruce effects evolved for another unspecified purpose, they say nothing about current utility. Bruce effects may function quite well now as a strategy to minimize the costs of infanticide by males.

Bruce effects were only recently reported in alpine marmots

(Hackländer & Arnold 1999), and it is perplexing that Bruce effects are not widely reported in sciurids, particularly for those tropical species that are not monestrous. A monestrous female would have to wait at least 1 year to breed again. Rearing any young might lead to a larger pay-off to a female compared with the pay-off of resorbing or otherwise aborting her young and waiting a year to possibly mate again. It is likely, therefore, that, on average, the cost of losing a litter to a new and possibly infanticidal male is less than waiting an entire year to attempt to reproduce. The frequency of embryonic resorption in some Eurasian marmots (gray marmots (*Marmota baibacina*), Bibikov & Berendaev 1978; long-tailed/golden marmots, Davidov *et al.* 1978; Kizilov & Berendaev 1978) suggests that searching for Bruce effects and embryo abandonment more broadly might be a profitable line of research.

Other female counterstrategies

Rodent infanticide has different functions and different consequences (Labov *et al.* 1985). Agrell *et al.* (1998) argued that because females pay the greatest cost, females should evolve strategies to minimize costs associated with infanticide. Rodents illustrate several of these strategies. Multiple matings may be a strategy that a female can adopt to minimize the risk of infanticide by males by confusing paternity (Agrell *et al.* 1998) and is discussed in detail by van Noordwijk & van Schaik (Chapter 14). Mate choice may be another tactic used to counteract male infanticidal risks (Agrell *et al.* 1998). In addition to "standard" reasons given for female choice of dominant males (Andersson 1994), there are two hypothesized infanticide-avoidance benefits of mating with dominant males. First, dominant males may be more likely to kill unrelated offspring (Huck *et al.* 1988). Second, dominant males may ultimately have a longer breeding tenure than subordinate males; a female who mated with a dominant male would be less likely to have her offspring killed by a new resident male (Agrell *et al.* 1998). Female bank voles prefer mating with dominant males, a strategic decision suggested to minimize the risks of male infanticide (Horne & Ylönen 1996). While sciurid females may choose mates, there is no evidence that mate choice reduces the risks of infanticide by males.

How does infanticide influence rodent social organization?

While infanticide may be a potent selective force, there are no obvious generalizations about how infanticide by males influences rodent social organization. Perhaps this reflects the various strategies that females employ to counter infanticidal risks. More likely, this reflects the limited detailed knowledge of the socioecology of most rodents.

Some species in which males are infanticidal live in year-round complex social groups (e.g., some marmots, prairie dogs), but there is no obvious relationship between sociality and infanticide by males as reported in primates (van Schaik & Kappeler 1997). Infanticide by males is reported in some of the most socially complex (Blumstein & Armitage 1998) sciurid rodents as well as in some of the least complex sciurid rodents. For muroid rodents, infanticide by males and Bruce effects are found among both cooperative breeders with singular breeding (i.e., only a single adult female breeds in a social group) and plural breeding (i.e., more than a single adult female breeds in a social group; Table 8.1 in Solomon & Getz 1997) as well as in less social species.

Viewed more broadly as a mechanism of reproductive suppression (Wasser & Barash 1983), infanticide by females seems to be a key element of a package of traits that make marmots, and perhaps some microtines, cooperative breeders (Solomon & Getz 1997; Blumstein & Armitage 1999). Reproduction in subordinate females is often suppressed in multifemale rodent societies (Solomon & Getz 1997; Wolff 1997; Blumstein & Armitage 1999). Infanticide by females would be predicted if and when a subordinate successfully reproduces. While infanticide by females is not reported in the more social golden or alpine marmots, subdominant females in groups without new male migrants are sometimes seen with swollen nipples (indicating lactation) yet litters are never seen to emerge above ground (D.T.B., pers. obs.). Agrell *et al.* (1998) predicted that infanticide by females leads to more complex mechanisms of reproductive suppression. While this is likely, data do not currently exist to test this hypothesis.

Does the risk of infanticide by females lead to female territoriality?

Female microtine rodents are more aggressive and territorial than males, especially during the breeding season (e.g., Webster & Brooks 1981; Wolff 1985). Sherman (1981) and particularly Wolff (1985, 1993, 1997) suggested that female territoriality in altricial mammals evolved in response to the risk of infanticide by females: the pup defense hypothesis (see also Digby, Chapter 17). Females of infanticidal species must defend their altricial young; one way to do this would be by defending a home territory from other females.

If infanticide was a key factor leading to the evolution of territorial behavior, infanticide by females should be an ancestral trait in territorial species that produce altricial young. In support of Wolff's hypothesis, infanticide by females seems to be an ancestral trait in muroid rodents. However, infanticide by females is a derived trait in sciurid rodents – all of whom are territorial (suggesting that female territoriality is ancestral in this lineage) and produce altricial young. Ebensperger (1998a), citing several conflicting case studies, suggested that the pattern of rodent territorial behavior did not support the pup defense hypothesis. To what extent female territorial behavior in other taxa can be explained solely by infanticide is a question that will be better answered when we have more data on the distribution of infanticide by females.

Conclusions

The evolution of infanticide in rodents

Rodents offer the unique opportunity for richly interdisciplinary studies of infanticide. Comparative studies suggest that infanticide has evolved at least three times in rodents. Detailed studies of the muroid and sciurid rodents did not support the hypothesis that infanticide in rodents evolved as a mechanism of sexual selection. Specifically, infanticide by males and by females was the ancestral state in muroid rodents, and there was no evidence that infanticide by males was associated with the evolution of male-induced pregnancy termination.

Independent evolution within the rodents provides at least three data points for understanding the larger question of how the interplay of ecological factors, developmental constraints, and physiological constraints influence the evolution of infanticide. Comparative analyses, like these, can be applied to other taxa and will eventually provide a better understanding of the evolution of infanticide.

Unfortunately, we simply need better data for future comparative studies. I believe much can be learned by systematically applying the standardized and humane techniques developed for laboratory studies (e.g., Elwood 1991) to species not typically maintained in the laboratory. For instance, it should be possible to introduce new males systematically to females and look for evidence of both Bruce effects (e.g., Stehn & Jannett 1981) and infanticidal aggression for many of the species in which infanticide is suspected from field observations. A considerable amount of information can also be learned by conducting captive (e.g., Stubbe &

Janke 1994) or semi-natural field cage experiments (e.g., Lidicker 1979) with new species.

Infanticidal sciurid rodents offer a particularly good opportunity for studies of social strategies that females use to avoid or reduce male infanticide. Female sciurids illustrate several potential anti-infanticide strategies such as multiple mating, and female territoriality, although it remains to be seen whether Bruce effects are common. In many ways, however, sciurid sociality is much less complex (Blumstein & Armitage 1997, 1998) than that found in many primates. Thus we may not expect to find highly complex relationships evolving to reduce the likelihood of male infanticide.

Conservation implications of rodent infanticide

Understanding the distribution and function of infanticide in nature has profound implications for the conservation and management of infanticidal species. For instance, human hunts of male brown bears (*Ursus arctos*) increase the likelihood of sexually selected infanticide by increasing the frequency with which females with vulnerable young interact with novel males (Swenson *et al.* 1997). Planned hunts as a means of population control may therefore have a greater than anticipated effect on bear population size. More generally, sexually selected infanticide influences a population's effective population size, and thus affects the likelihood that a population will persist over time (Anthony & Blumstein 2000). Wildlife managers must be aware of the pattern of infanticide in nature.

The Vancouver Island marmot (*Marmota vancouverensis*) is one of the rarest mammals in the world (Bryant & Janz 1996). Current management plans include captive breeding to build up population sizes followed by the reintroduction of captively bred animals to their formally occupied range (A. Bryant, pers. comm.). Even though there is no evidence of infanticide in Vancouver Island marmots, knowledge of infanticide in other marmots should influence planned rearing patterns. For instance, novel males should not be introduced to breeding females, and breeding females should probably be separated from other breeding females. Moreover, reintroducing or translocating males into pre-existing social groups may be risky. Managers and breeders of other rodent species can modify their behavior on the basis of knowledge of the distribution of infanticide.

We have made much progress in understanding infanticide since Hrdy (1974, 1979), Wilson (1975) and Sherman (1981) emphasized adaptive socioecological hypotheses for infanticide. The development of a comparative data base will surely make the future decades as exciting as the past decades for enriching our understanding of the causes and consequences of infanticide in nature.

Acknowledgments

I thank Carel van Schaik for inviting me to write this review, and Bob Hoffmann for help in understanding rodent evolution. Bob, Carel, Leslie Digby, Luis Ebensperger and Charles Nunn made a number of very helpful comments on a previous draft. I am grateful to Jep Agrell and Walter Arnold for sending preprints and unpublished manuscripts, and I am especially grateful to Bob, Chris Conroy, Joe Cook and Scott Davis for supplying key published and unpublished rodent phylogenies. Support during manuscript preparation was provided by an Australian Research Council Postdoctoral Fellowship.

9

Infanticide by male birds

Introduction

The killing of dependent young individuals by conspecifics, what ecologists have called infanticide, has been viewed as one extreme and dramatic consequence of selection favoring those behaviors that promote the direct fitness of perpetrators (e.g., Hamilton 1964a,b). However, the evolutionary scenarios of this behavior may differ considerably depending on the identity of the perpetrators. It has been proposed that infanticide by non-kin is fundamentally different from infanticide by kin, which emphasizes the fact that the latter involves the sacrificing of shared genes for some presumed compensating benefits to the perpetrator's inclusive fitness (O'Connor 1978; Mock 1984). This infanticide by kin can be further subdivided into parental infanticide, i.e., the killing of young is committed by their own parents, and siblicide, the term used when the killing is carried out by full or half-siblings. The incidence of siblicide in taxa other than birds is, however, practically unknown (see Mock 1984).

Four functional hypotheses have been proposed by Hrdy (1979) to explain the different types of pay-offs that may accrue to infanticidal individuals. In other words, infanticide should have evolved in the following contexts: (1) the exploitation of the infant, mainly as a food source; (2) resource competition, either with the infant or with its parents; (3) parental manipulation, wherein the parents interrupt their investment in the offspring to maximize their reproductive success; and (4) sexual selection, wherein infanticide increases the success of killers in intrasexual competition for mates. While sexual selection has been emphasized as the almost exclusive explanation for the evolution of infanticide in mammals (e.g., Hrdy 1979; Sherman 1981; Packer & Pusey

1983; Brooks 1984), the study of avian infanticide has focused mostly on the other functions. In his review from 1984, Mock wrote that "there is relatively little ornithological documentation for Hrdy's (1979) sexual selection hypothesis of infanticide". New evidence accumulated in the past few years, however, indicates that sexually selected infanticide may be more frequent in birds than was believed initially.

As the focus of this book is on infanticide by males, I consider only those scenarios of infanticide by males in which the causes and consequences of the killing behavior were sexually dimorphic in appearance. According to Mock's review, avian infanticide may occur in four broad contexts that are not related to sexual selection: brood reduction systems, parental infanticide, infanticide in colonies and infanticide in cooperative and communal breeding systems. In none of these contexts is a male's advantage resulting from killing conspecifics thought to be caused by different ecological factors or to have evolutionary consequences different from those in females. In some instances males may commit infanticide with higher or lower frequency than females, but this does not imply that the ultimate causation of infanticide differs between sexes. The two following examples will aid clarification of this point.

In many species, females regularly lay more eggs than the pair can successfully rear to independence (Lack 1954). The loss of part of the brood (brood reduction) has been regularly interpreted as a density-dependent strategy to maximize parental reproductive success. In several avian taxa, the chicks themselves decisively contribute to brood reduction. The chick hatched earlier as a result of delayed onset of incubation may use physical aggression to have preferential access to the food brought to the nest, and even to kill the younger sibling. In some raptor species with obligate siblicide, sibling competition may be greater in mixed-sex clutches in which the male hatches first. Hence, this hatching sequence (male followed by female) is most likely to show nestling mortality in which sibling aggression (by the male in this case) plays an important role (Bortolotti 1986). In any case, the causes and consequences of the killing and the sex-biased frequency of occurrence result from parents manipulating their brood to achieve an optimal combination of the two sexes and to maximize fitness.

In communal and cooperative species, infanticide has been commonly reported. Competition for breeding resources is usually regarded as the main force contributing to this behavior. Selfish interests seem to have swamped the presumed effect of kin selection in this group of birds, where relatedness is usually high. Egg destruction or killing of young is in most cases committed by female members of the social group (Zahavi 1974; Vehrencamp 1977; Trail *et al.* 1981; Koenig *et al.* 1983; Mumme *et al.* 1983; Stacey & Edwards 1983). Competition is probably less intense among males, so that infanticide by males should be less frequent in this sex, as has been observed in most studies (but see Whitmore 1986), but its causes are not apparently different from those operating for females.

Taking all this into account, I hereafter equate infanticide by males with sexually selected infanticide, although this by no means implies that sexual selection cannot operate on females. In fact several authors have recently stressed the important role played by intrasexual competition in the evolution of infanticide by females in birds (Veiga 1990a; Check & Robertson 1991; Bensch & Hasselquist 1994; Hansson *et al.* 1997). However, the topic is outside the scope of this chapter.

It is worth pointing out that some cases described as male parental infanticide in the literature may be the consequence of a wrong assignation of the true genetic male parent to the offspring. For example, male white storks sometimes kill what apparently are their own chicks (e.g., Schüz, cited in Mock 1984). However, in light of recent evidence of extra-pair paternity in this and other bird species (e.g., Birkhead & Møller 1992), it is possible that males are committing prospective infanticide rather than killing their own offspring (see below). Hence, this would be a case of sexually selected infanticide, but evidence of this kind is too weak and has not been considered in the following review.

Infanticide by males: kind of evidence and frequency of occurrence

I have reviewed a total of 26 studies that give some kind of evidence of avian male (or female equivalent) infanticide. These studies refer to 13 species, while in another two species of jacanas (*Jacana jacana* and *J. spinosa*), which have sex role reversal, it is the female who destroys eggs or kills young. In these two species, males care for the eggs and chicks, and females compete among themselves for access to the limiting sex. Thus, in terms of sexual selection, this infanticide by females is functionally equivalent to infanticide by males, so that I have included these two species in the review. For only 6 of these 14 species is there more than one independent study reporting some kind of evidence of infanticidal behavior (Table 9.1). The strength of the evidence varies considerably: from definite, where the infanticidal act was witnessed, to indirect, where the infanticide was inferred from pecking marks or wounds on nestlings, to circumstantial or based on statistical approaches.

In total, infanticide by males (or functionally equivalent infanticide by females) has been demonstrated unambiguously for 14 of the 15 species considered here. Six of the studies report a single case of infanticide, while 15 include five cases or more. These 15 studies refer to nine species; in these species, infanticide can be considered to be well documented.

A high proportion of the studies are experimental: typically, the take-over of a territory is induced by removal of owners or by adding mimetic eggs to the nest to simulate intraspecific nest parasitism (Table 9.1). The experimental approach has the advantage of facilitating the demonstration of the potential for infanticide in the population (see Robertson & Stutchbury 1988), but it does not give information on the baseline frequencies of infanticide under unmanipulated conditions, thus weakening the possibility of making inferences about its functional value or evolution. All these caveats reduce to four the number of species for which there is enough information to evaluate the frequency of occurrence of infanticide by males and, in some degree, its variability between populations and between species. In the barn swallow, 16.3% of the nestling deaths were attributable to infanticide in a New York population studied by Crook & Shields (1985), while the value rose to 32.1% in a Danish population studied by Møller (1988). In a Japanese population of the little swift, infanticide occurred in 4.6% of all breeding attempts and accounted for about 18% of all offspring deaths, though probably near half of the killings were perpetrated by females (Hotta 1994). In a tropical population of the house wren, Freed (1986) concluded that male replacement occurred between 3.6% and 4.7% of the breeding attempts, depending on the nesting stage. He also stated that mortality of eggs, nestlings or fledglings occurred at 73% of these nests, and that predation could be specifically eliminated as an alternative cause of mortality. In a house sparrow population studied in Spain, infanticide was estimated to occur in 9% to 12% of all breeding attempts, depending on the kind of evidence considered, though an indeterminate portion of the killings were due to females (Veiga 1990b). Infanticide was only exceeded by predation as a cause of egg and nestling mortality in this population.

The obvious conclusion of these results is that sexually selected infanticide should not be neglected as a significant cause of mortality among

Table 9.1. *A summary of the evidence shown in the reviewed studies about infanticide by males*

Species	Type of evidence	Type of approach[a]	No. of cases[b]	Nest content	Reference
Little swift, *Apus affinis*	Definite, possible	Observational	8	Eggs, nestlings	Hotta 1994
Acorn woodpecker, *Melanerpes formicivorus*	Definite, indirect, statistical	Experimental (1)	5	Eggs	Koenig 1990
Northern jacana, *Jacana spinosa*	Possible	Observational	1	Eggs	Stephens 1982
Wattled jacana, *Jacana jacana*	Definite	Experimental	3	Chicks	Emlen *et al.* 1989
Tristram's grackle, *Onychognathus tristramii*	Definite	Observational	1	Nestlings	Hofshi *et al.* 1987
Eurasian dipper, *Cinclus cinclus*	Definite, indirect	Observational	4	Nestlings	Yoerg 1990
Eurasian dipper, *Cinclus cinclus*	Possible	Observational	1	Eggs	Wilson 1992
House wren, *Troglodytes aedon*	Definite	Experimental	19	Eggs, nestlings	Belles-Isles & Picman 1986
House wren, *Troglodytes aedon*	Indirect, statistical	Observational	22	Eggs, nestlings, fledglings	Freed 1986
House wren, *Troglodytes aedon*	Indirect, possible	Observational	12	Eggs, nestlings	Quinn & Holroyd 1989
House wren, *Troglodytes aedon*	Possible	Observational	1	Eggs	Kermott & Johnson 1990
House wren, *Troglodytes aedon*	Indirect	Experimental	10	Eggs, nestlings	Kermott *et al.* 1991
Long-billed marsh wren, *Cistothorus palustris*	Definite	Experimental	14	Eggs	Picman 1977
Barn swallow, *Hirundo rustica*	Definite, possible	Observational	9	Nestlings	Crook & Shields 1985
Barn swallow, *Hirundo rustica*	Definite, possible	Observational, experimental	18	Nestlings	Møller 1988
Barn swallow, *Hirundo rustica*	Definite	Observational	1	Nestlings	Banbura & Zielinski 1995
Tree swallow, *Tachycineta bicolor*	Definite, possible	Experimental	5	Nestlings	Robertson & Stutchbury 1988
Tree swallow, *Tachycineta bicolor*	Undescribed	Experimental	11	Nestlings	Robertson 1991
Purple martin, *Progne subis*	Definite	Observational	1	Nestlings	Allen & Nice 1952
European starling, *Sturnus vulgaris*	Possible	Experimental	6	Eggs	Smith *et al.* 1996
European starling, *Sturnus vulgaris*	Definite	Observational	2	Eggs	Merkel 1982
European starling, *Sturnus vulgaris*	Definite	Experimental (2)	30	Eggs	Pinxten *et al.* 1991

Spotless starling, *Sturnus unicolor*	Definite, possible	Observational	3	Eggs, nestlings	J. P. Veiga *et al.* unpublished data
House sparrow, *Passer domesticus*	Definite, indirect, statistical	Observational	6	Eggs, nestlings	Veiga 1990b
House sparrow, *Passer domesticus*	Definite, possible	Observational, experimental	15	Eggs, nestlings	Veiga 1993
Palestine sunbird, *Nectarinia osea*	Definite	Observational, experimental	4	Fledglings	Goldstein *et al.* 1986

Notes:

[a] Experimental evidence has usually been obtained by removing male owners, except in (1) males were temporarily detained during prelaying and laying periods, and in (2) a mimetic egg was added to the nest before or after clutch initiation.

[b] Numbers refer to nests in which infanticide was attempted or achieved.

the species in which this behavior has been relatively well documented. However, the figures given in some of these studies are probably underestimates of actual rates, given the difficulties in detecting the infanticidal behavior. This may also mean that infanticide has been overlooked in many studies of reproduction and mortality, and I can speculate that infanticide may be a general behavior in many avian species.

Patterns of infanticide by males: contrasts within and between species

A summary of mating systems, mating attributes of individuals involved in infanticidal events and the consequences of infanticide in terms of mating opportunities is shown in Table 9.2. There is considerable consistency in the patterns shown by the different studies conducted on the barn swallow and the tree swallow. Among barn swallows, infanticidal males were invariably unmated individuals, some of which had recently lost their mate. In general, infanticidal males had previously filled a territory vacated by the male owner (i.e., the female victims were "widows"), but in some cases territory takeovers and subsequent killing of nestlings occurred in the presence of the territorial male. By committing infanticide, males succeeded in many instances in mating with the victimized females and in subsequently producing clutches with them.

In the tree swallow, the mating status of infanticidal males is not well established. Due to the experimental protocol used in the two referred studies, the victimized females had in all cases lost their mates. Infanticidal males not only are able to remate with the victimized female, but in some cases they may also acquire a new female after takeover and infanticide. Interestingly, although infanticide occurred regularly after removal of territorial males, spontaneous cases of infanticide are rare in this species (Robertson & Stutchbury 1988). It is a remarkable and puzzling characteristic of the swallow group that males invariably removed nestlings but never eggs from victimized nests. Even when male replacement occurred during incubation, invading males waited until the clutch had hatched to kill the nestlings (Crook & Shields 1985; Robertson 1991). This fact that apparently works against one of the postulated functions of infanticide, namely the shortening of the interval between takeover and renesting by the female victim (see below), has not yet received a convincing explanation.

In the little swift, another aerial feeder, the pattern of infanticide did

Table 9.2. *A review of mating system, mating attributes of infanticidal males and victimized females, and consequences of the infanticide in the species in which infanticide by males has been reported*

Species	Mating system[a]	Killer's mating status	Victim's mating status[b]	Consequences
Little swift	Monogamy	Mated, unmated	Mated, widow	Remating, new female
Acorn woodpecker	Polygynandry	Mated	Mated (with killer)[c]	Remating
Northern jacana	Polyandry	Polyandrous	Polyandrous	Remating
Wattled jacana	Polyandry	Monogamous, polyandrous	Polyandrous	Remating?
Tristram's grackle	Monogamy	Unmated?	Widow?	Remating?
Eurasian dipper	Polygyny	Polygynous	Monogamous, widow	No remating or new female?
House wren	Facultative polygyny	Mated, unmated	Mated, widow?, (widow)	Remating, new female
Long-billed marsh wren	Facultative polygyny	Mated, unmated	Unknown	Unknown
Barn swallow	Monogamy	Unmated	Monogamous, widow	Remating
Tree swallow	Monogamy	Unknown	(Widow)	Remating, new female
Purple martin	Monogamy	Unknown	Unknown	Unknown
European starling	Facultative polygyny	Mated, polygynous, unknown	Mated, (widow)	Remating
Spotless starling	Polygyny	Monogamous, unmated	Polygynous	Remating, new female
House sparrow	Monogamy	Unmated	Widow, (widow)	Remating
Palestine sunbird	Monogamy	Unmated?	Mated, (widow)	Unknown

Notes:

[a] Facultative polygyny describes predominantly monogamous species in which a part of the males (usually between 5% and 15%) may have two or more mates. Polygyny (or polyandry) is used for species in which 50% or more of the territory owners have two or more mates.

[b] '(widows)' indicates that their mates were experimentally removed by researchers.

[c] The infanticidal male destroyed eggs after having been experimentally detained during prelaying or laying.

not differ much from that of the swallows. In this species, though infanticidal males were generally unmated, at least one mated male was recorded killing young. As in swallows, males tended to kill nestlings in nests in which the owner had disappeared, though takeovers when both owners were present have also been recorded. Some infanticidal males succeeded in remating with their victim, but in other cases they acquired a new female, which tended to be of higher quality than the former.

The house wren has been subject to multiple studies of infanticide. There are some differences between the patterns of infanticide in tropical and temperate populations. In the tropical house wrens studied by Freed (1986), contexts in which infanticide occurred included takeover by 1-year-old males, re-entry into the breeding population by males that had recently lost their mates, and acquisition of neighboring females by bigamous males. Almost invariably, males committed infanticide in nests of females that were mated at the time of the takeover. After infanticide, most females remained in their territories, and a high proportion started a new clutch with the replacing male following infanticide. In temperate house wrens, infanticide after the experimental removal of territory owners seemed to be committed by unmated males. Some of these males produced a clutch with the female, while the other males mated shortly after infanticide with a new female (Kermott et al. 1991). In other temperate populations, however, the benefit of infanticide was not clear because there were no further nesting attempts in nests where mortality occurred (Quinn & Holroyd 1989). House wrens seem capable of killing young in all developmental stages.

In the facultatively polygynous European starling, infanticide was committed by mated males in all reported cases. In the best documented study, victims were obligate widows after a male removal experiment (Smith et al. 1996). However, there is some observational evidence that mated females also may be targets of infanticidal polygynous males (Merkel 1982). In general, replacement males succeeded in mating with the females suffering the infanticide, although this seems to be influenced by the probability that females will produce an early replacement clutch. Additionally, males destroyed experimentally introduced parasitic eggs as a mean of protecting paternity (Pinxten et al. 1991; see below). In the closely related but highly polygynous spotless starling, infanticide seemed to be committed by both unmated and mated males, although, interestingly, polygynous males (about 90% of all territorial males) have not yet been recorded as infanticidal. After infanticide, males remated with the victim or acquired a new female (J. P. Veiga *et al.*, unpublished data). Only egg destruction, but not the killing of young, has so far been reported for both starling species.

In the house sparrow, infanticide was committed by territorial individuals that had recently lost their mates. In all cases, infanticidal males had previously taken over the nest of a disappeared male owner (either naturally or experimentally) and, in the majority of cases, they subsequently mated with the female victim and produced a clutch within the current breeding season (Veiga 1990b, 1993). House sparrows are able both to destroy eggs and to kill nestlings.

In the Eurasian dipper, two different reports of infanticide by males coincide in signaling that infanticidal males were in all instances polygynous individuals that took over neighboring pairs, or, in one case, took over the nest of a widow female. Although conclusive evidence was not obtained, in one study males apparently gained opportunities for additional matings (Wilson 1992), while in the other the victimized pair remained in its territory after the nestling death (Yoerg 1990).

In the polyandrous jacanas, infanticidal females were already mated individuals that expanded their former territory to encompass the area occupied by one ore more neighboring males (Stephens 1982; Emlen *et al.* 1989). Some of the females were even mated with more than one male when they committed the infanticide. In all cases reported, the victimized males shared their female mate with other males. In *Jacana jacana* incoming females initiated sexual behavior with the victimized males within 48 hours of killing the males' offspring, and rapid solicitation by replacement females was also mentioned in *J. spinosa* (Stephens 1982; Emlen *et al.* 1989). This strongly suggests that males regularly mated with the female killers, which laid clutches for new males within a short time interval of the takeover. Jacanas seem to be able both to destroy eggs and to kill nestlings.

In the polygynandrous acorn woodpecker, egg destruction was used by males, who had been experimentally detained during their females' fertile period, as a tactic to force renesting and improve paternity in the brood subsequently produced. However, only dominant individuals in the group destroyed the current nest (Koenig 1990).

The experimental approach used to study egg destruction behavior in the long-billed marsh wren by Picman (1977) enabled the researcher to document several aspects of this behavior, for example that both mated and unmated males destroyed eggs. However, nothing is known for this species about the outcomes of infanticidal behavior, such as the mating status of female owners of victimized nests or the mating consequences of egg destruction.

The infanticide reported in the Palestine sunbird (Goldstein *et al.* 1986) is peculiar because intruding males seemed to attack young only after they had fledged. Also, the observed cases are insufficient to establish whether males obtained some mating advantage from this behavior. Similarly, in the Tristram's grackle, the single case reported is anecdotal (Hofshi *et al.* 1987).

By examining the species listed in Tables 9.1 and 9.2, several facts are remarkable. First, with exception of the polyandrous jacanas, the poly-gynandrous acorn woodpecker and the little swift, all the species in which male sexually selected infanticide has been reported belong to the order Passeriformes. Thus some groups in which other types of infanticide occur regularly (e.g., gulls, raptors, herons) are apparently excluded from the sexually selected infanticide category. Also remarkable is the high presence of aerial feeders, and particularly members of the Hirundinidae family. Although it cannot be excluded that some common ecological attributes or phylogenetic effects are responsible for the sharing of infanticidal behavior, it is more likely that studies of infanticide have been biased toward this group because the first reports of nestling-killing by conspecific males were given for it (Allen & Nice 1952; Crook & Shields 1985).

Although information is still fragmentary, the pattern emerging from these interspecific comparisons suggests that the infanticidal tactic is present in a variety of mating systems. Under monogamy, it seems that infanticide is most frequently committed by unmated individuals, but mated individuals are also capable of taking over and killing young as a strategy to acquire costly nests (Hotta 1994). Under facultative polygyny (or polyandry), infanticide is more frequently committed by already mated males (or polyandrous females) that use takeover, followed by infanticide, as a tactic to acquire additional mates, though unmated males may also use infanticide, as under monogamy. In all systems, replacing males frequently remate with the female and start a new breeding attempt. However, females having suffered infanticide also may desert and the intruding male acquires a new female.

Takeover and infanticide: independent targets of selection

Many arguments to explain the evolution of infanticidal behavior in birds regard takeover and infanticide as intrinsically associated, i.e., they assume that infanticide is the inevitable follow-up of a nest or territory takeover. Thus any hypothesis identifying a selective advantage to territorial intrusion and mate takeovers would also apply to the infanticidal behavior. Under the restricted definition of infanticide by males used in this chapter, the main reward for males should be some kind of mating advantage in an intrasexual competitive background. It seems clear that male intrasexual competition should favor taking over a mate or a territory under a wide array of circumstances, but it is not immediately clear from this reasoning that infanticide per se provides any mating advantage or that there are no other possibilities open to invading males. There is considerable evidence for a number of species (including some of the infanticidal ones) that after a takeover males ignore or adopt the nestlings within the newly acquired territory (for a review, see Rohwer 1986), in spite of which they succeed in mating with the female owner. Hence, it seems necessary to identify separately the mating consequences of taking over a territory and killing offspring when dealing with factors that may have promoted the evolution of avian infanticide.

Despite the association between infanticide and mate acquisition, it is not obvious that a cause–effect relationship exists between the two. Table 9.2 shows that in most cases infanticide affected widows, i.e., occurred after the disappearance of the male owner. In the tree swallow, infanticide seems to be rare under unmanipulated conditions, but it is frequent after removal of owners (Robertson 1991). This seems to indicate that, in most instances, replacing males were simply seizing a vacant territory. Hence, replacing males did not need to kill young to mate and they could have developed other alternatives, such as ignoring or adopting the offspring of the recently acquired female. Only if infanticide precedes the takeover and promotes "divorce" and the subsequent mating of the killer with its victim would the usefulness of this behavior as a mating strategy be indisputable. There is, however, practically no evidence of this. Crook & Shields (1985) described a case in the barn swallow in which the victimized female deserted her nest and mate after infanticide and renested with the killer in a different nest site. This suggests that infanticide per se may be a mate usurpation tactic. In the little swift, it has been reported that, in some cases, infanticide by males occurred in the presence of both parents, so that takeover and infanticide were apparently simultaneous. Usurpers obtained not only a mate older than their former one but also a nest in a more advanced stage of construction (Hotta 1994).

Infanticide committed after a takeover may be advantageous in two ways from the point of view of mating. First, it may shorten the time elapsed between the takeover and the next fertile period of the victimized female; second, it may induce nest desertion by the female owner and the subsequent acquisition of a new female that would lay eggs earlier than the previous owner would. The first situation has been well documented in the barn swallow, the house wren and the house sparrow. In these species, females usually resumed sexual activity shortly after their young were killed and produced a new clutch with the infanticidal male earlier than if she had had to raise her offspring until independence (Crook & Shields 1985; Freed 1986; Veiga 1990b). This may be crucial because it may represent the difference between breeding or not breeding within the current season. Also, the time gained by committing infanticide may be translated into a higher probability of survival of the offspring, at least for those species in which the fledgling condition and subsequent survival strongly decrease as the season advances.

The advantage derived from the acquisition of a new female after infanticide is poorly documented. However, males commonly attract a new female to the victimized nest after infanticide and succeed in a subsequent breeding (see Table 9.2). On the other hand, in other cases, the nest remains vacant after female desertion for the rest of the breeding season. The effectiveness of this strategy probably depends on the timing of the infanticidal event (late in season there is a lower probability of attracting females) and on the breeding opportunities open to victimized females (see below). Even if females suffering infanticide desert the nest, infanticide may also result in a shortening of the time elapsed between the killing of young or eggs and the start of the male's next nesting attempt. This may occur when there is a high availability of floating females willing to mate, so that the newly acquired female immediately starts a breeding attempt with the infanticidal male.

It has been suggested that the primary benefit of infanticide may be the accelerated availability of a nest site, which can be used to attract a new mate (Robertson & Stutchbury 1988), or the acquisition of a nest in a more advanced stage of construction (Hotta 1994). However, the question is whether infanticide associated with nest acquisition has then to be classified within the sexually selected category or rather within the resource competition category. In the latter case, the determinants of the behavior in males would not differ from those of females. However, it is common in many species that the possession of a nest site is a prerequisite for males to attract a female, so that both categories of infanticide cannot be operationally separated in those species (Robertson & Stutchbury 1988).

It has been proposed also that by destroying eggs or nestlings, males reduce the breeding success of other conspecifics or induce desertion of territories and so a general decrease of intraspecific competition. Picman (1977) used this reasoning to explain the marked tendency shown by long-billed marsh wren males to destroying both nests of conspecifics and those of other species breeding in their territories. Picman designed an ingenious experiment to detect nest destruction events, which, however, did not enable him to investigate some of the consequences of infanticide most commonly argued for other species, such as rapid renesting.

In conclusion, male intrasexual competition should favor taking over a mate or a territory under a wide array of circumstances, but it is not clear that infanticide per se provides any mating advantage. Replacing males did not need to kill young to mate and they could have developed other alternatives. Only if infanticide preceded the takeover and promoted "divorce" and the subsequent mating of the killer with its victim would the usefulness of this behavior as a mating strategy be indisputable. The evidence for this is scarce. Infanticide committed after a takeover may shorten the time elapsed between the takeover and the next fertile period of the victimized female, and also may induce nest desertion by the female owner and the subsequent acquisition of a new female that would lay eggs earlier than the previous owner would have done. Accelerated availability of nesting sites, acquisition of a nest in more advanced stage of construction and the decrease of intraspecific competition have been suggested as additional benefits of infanticide.

Factors favoring the evolution of infanticide versus alternative behaviors

The classical argument to explain why sexually selected infanticide is less common in birds than in mammals is that the predominantly monogamous system of birds erodes the potential for sexually selected infanticide. In other words, a male bird is unlikely to secure an additional mate by killing a female's progeny (Mock 1984). This contention is consistent with Lack's (1968) observation that 91% of bird species are monogamous. However, after Mock's review, infanticide by males has been increasingly regarded as a sexually selected behavior in a number of species, so that it seems that infanticide rather than rare may have been rarely detected. Another interpretation of avian mating systems offers a view almost opposed to Lack's scenario, with a higher proportion of birds being polygynous (Møller 1986). Also, in more recent years, molecular techniques have demonstrated that, in many avian taxa, even socially monogamous males may develop a mixed mating strategy trying to achieve extra-pair copulation with several sexual partners (Birkhead & Møller 1992). This also creates a potential for sexually selected infanticide. It is possible that by committing infanticide an already mated male succeeds in fertilizing an additional female, but he does not subsequently invest in her offspring; in other words, males may become sexually polygynous while remaining socially monogamous. This possibility, however, has not yet been explored.

Mock's argument also does not take into account that even in monogamous species a fraction of the males in a population is often composed of floating individuals searching for a mate. This may result from skewed sex ratios or from strong limitations in the availability of nesting sites. In any case, these asymmetries in mating status promote intrasexual competition among males and, thus, the potential for evolution of takeover and the possibility of subsequent infanticide in monogamous species. In fact, examination of Table 9.2 shows that most species in which infanticide has been recorded are not polygynous. In monogamous species one would predict that mating advantages derived from infanticide should be higher for unmated than for already mated males because the acquisition of additional females is constrained by factors such as the need for caring for the offspring. However, in polygynous species, both unmated and already mated males would benefit similarly from additional matings and subsequent infanticide. The results of Table 9.2 tend to support this contention. So, underscoring the argument that mating system should not be viewed as the direct causation of infanticidal behavior, it is questionable that the potential for the evolution of sexually selected infanticide is lower in birds than in mammals.

One of the factors proposed to favor the evolution of infanticide by male birds is colonial nesting. In fact, infanticide has been frequently observed among colonial nesters, though, in most instances, the killing of young birds is not selectively committed by males, and its causes and consequences have nothing to do with sexual selection (see Fetterolf 1983; Mock 1984; Pierotti *et al.* 1988; Quinn *et al.* 1994). It has been suggested, however, that colonial breeding may provide opportunities for males to locate inadequately guarded nests (Crook & Shields 1985; Møller 1988). Again, this could in fact favor the takeover of a nest or a breeding territory, but not necessarily the subsequent infanticide, as other alternatives remain open to males. It is then necessary to identify other factors aside of coloniality responsible for the advantages of committing infanticide. An excess of unmated males in a population may also promote intrasexual competition and territory takeovers, but, as for coloniality, this factor alone does not explain why infanticide may be advantageous in relation to alternative behaviors.

As stated above, one of the most generally recognized advantages of infanticide is the shortening of the time elapsed between a takeover and the start of the next fertile period of the victimized female. This is especially important in those cases in which a delay in the start of the following nesting cycle imposes a significant reduction in reproductive success. It has been presumed that in single-brooded species, a male that commits infanticide is not able in most cases to induce the victimized female to start a new nesting attempt. On the basis of this, it has been suggested that infanticide is favored by protracted breeding seasons and by double or multiple brooding (Freed 1986; Møller 1988). Although this argument is plausible, it is also true that, even in species with a short breeding season, infanticide may still be advantageous if the killing of young occurs early enough in the breeding cycle, so that the victimized female can produce a replacement clutch. In this case, committing infanticide may represent for a male the difference between producing a successful brood or not breeding at all. So, it is not clear that infanticide pays only in species with long breeding seasons. Infanticide is frequently committed during laying and incubation (Table 9.1, and see below), when it is most likely that the victim has time to start a new clutch even in single-brooded species.

Some authors have depicted a different evolutionary scenario with the same set of ecological factors. For example, Rohwer (1986), after reviewing adoption in birds, concluded that this behavior is more common than infanticide in birds, and that it is more frequent among double-brooded species in which females remain in their nesting territory after a successful breeding attempt but desert after failure. In such cases, replacing males do better by waiting until their new mate raises her progeny before starting a new nesting attempt. On the contrary, infanticide pays more when renesting after failures occurs within breeding seasons and when new broods are rapidly produced either with the new consort or with another individual. Other authors, however, have claimed that adoption is very rare among birds (Robertson 1991). Although across species adoption and infanticide seem mutually exclusive, at present there is little information on the covariation of these traits within species.

The success of an infanticidal male in inducing a rapid remating with his victim probably depends on the mating and nesting alternatives open to females. When a female after suffering infanticide may decide to change to another nest site and/or mate, infanticide may not be the better strategy for invading males. This situation may be common in species or populations not limited by the availability of nesting sites or potential mates. In many instances, however, females have difficulty finding alternative good nesting sites or mating status. For example, after losing her former mate, a female often becomes the secondary mate of a polygynous male, which may represent a worse breeding situation (Veiga 1992; Smith *et al.* 1994). Under such limitations, a male has all the advantages to coerce his victim to remain on the nest after infanticide and produce a clutch with him. There is some evidence that females may solicit copulations from the killer shortly after infanticide (Veiga 1993), which suggests that females never attempt to shift nest or mate.

A low annual survival expectancy among breeders may discourage the postponement of breeding to a following year once a mate has been acquired, so that infanticide and rapid renesting will pay more than alternative strategies such as indifference or adoption. The concentration of infanticide by male birds in passerines (Table 9.1) and in small species in general (which have higher mortality rates than larger species) is consistent with this view. However, it remains possible that sampling biases have greatly influenced this result. Also, though indifference and adoption have been recorded for large birds, these behaviors are not infrequent among small passerines (Rohwer 1986).

In conclusion, the preponderance of monogamy among birds does not constitute a sufficient argument to explain the low frequency of infanticide in birds. Infanticide is also committed under monogamy, mainly by unmated individuals, suggesting that asymmetries in mating status create a potential for the evolution of infanticide in monogamous species as well. Protracted breeding seasons and double or multiple brooding have also been proposed as factors favoring infanticide because they enable victimized females to start a new breeding cycle. However, long breeding seasons may also favor adoption rather than infanticide if females remain in the breeding territory after a successful breeding attempt but desert after failure. In any case, the success of an infanticidal male in inducing a rapid remating with his victim probably depends on the mating and nesting alternatives open to females.

Prospective infanticide as a paternity insurance strategy

Until now, I have made reference to infanticide as a behavior enabling a replacing male to reject unrelated young in order to sire his own offspring with the victim. If a male arrives shortly before or during egg laying, he cannot be sure that at least part of the clutch has not been fertilized by the previous partner of the recently acquired female. In this case, infanticidal males may kill young hatched or eggs laid after replacement. In other words, males not only destroy nest contents at the time of territory take-over, but may also kill young hatched or eggs laid after replacement. This subset of sexually selected infanticide has been called prospective infanticide, and, to be adaptive, males must be able to evaluate paternity by killing only the offspring that appear within one fertility period (see Freed 1987).

Prospective infanticide in birds has been described in five species. In the tree swallow, it has been suggested that the timing of arrival of the replacement male is important in determining whether he adopts or kills the brood. When a male arrived during laying, so that he had some possibility of fertilizing some eggs, he adopted the nestling subsequently hatched. In contrast, males that arrived well after clutch completion or during the nestling stage, so there was no opportunity to sire any young, usually killed the nestlings. Interestingly, males that arrived shortly after clutch completion either adopted or killed the nestlings even though they should have had zero confidence of paternity (Robertson 1991). This suggests some kind of male manipulation by females (see below). In the European starling, replacement males almost invariably adopted the clutch when they replaced the previous owner before laying, but apparently always committed infanticide when replacement took place during laying, although replacement males adopted the clutch produced subsequently with their victims (Smith *et al.* 1996).

In house sparrows, replacement males invariably committed infanticide when they took over a nest after a female's fertile period had recently ended. Males that took over a nest during egg-laying destroyed eggs in about half of the cases, while those arriving before clutch initiation never committed infanticide. In all cases, males adopted the brood produced by their new mates after infanticide (Veiga 1993). In tropical house wrens, replacement males associated with clutches that were subsequently destroyed arrived closer in time to the initiation of egg-laying than did replacement males associated with intact clutches. When eggs laid after replacement survived, males provided parental care after hatching. When eggs where destroyed, females initiated additional clutches some days later, and new males provided parental care after the new clutch hatched (Freed 1987). In the cooperatively breeding acorn woodpeckers, dominant males in a group destroyed nests when they were released after having been temporarily detained during their females' fertile period, but subordinates failed to do so. This behavior apparently enabled dominant individuals to get some paternity in the clutch subsequently produced. However, it is not clear whether subordinates were denied access to the nest by dominants or all co-breeders were adjusting their behavior according to both direct and indirect fitness benefits (Koenig 1990).

Infanticide as a way of protecting paternity may be used in contexts other than prospective infanticide. For example, in the European starling, males destroyed mimetic eggs experimentally introduced before clutch initiation in 61% of cases but never after clutch initiation. This suggests that males may use some unknown cue to determine whether or not an egg appearing in the nest is their female's first egg (Pinxten *et al.* 1991).

There is a common pattern in the above studies: males almost invariably destroyed eggs and killed nestlings already present in the nest when replacement took place (Figure 9.1). However, the response varied somewhat between species when replacement occurred before or during egg-laying. In house wrens and starlings, males destroyed eggs, whereas, in the tree swallow, males always adopted the nestlings that appeared subsequently. House sparrows seems to develop an intermediate strategy, since only about half of the males arriving during laying destroyed eggs. Considering that males that copulate during laying might be successful in fertilizing at least some eggs, these interspecies contrasts could be related to the different proportion of young sired by replacing males. Unfortunately, this information is available only for the European starling. In this species, when replacement occurred between 3 or more days and 1 day before laying, the new male fathered between nearly all off-

Figure 9.1. Intruding spotless starling (*Sturnus unicolor*) male taking out an egg from the nest box of a conspecific pair (photo: Marta Arenas, Sara Sánchez and José P. Veiga).

spring and about 60%, respectively (Smith *et al.* 1996). This suggests that males were fine-tuning their behavior (infanticide versus adoption) to their probability of fathering the offspring produced after replacement and copulation with the new female. It is probable that, in the other species, in which males adopted broods when replacement occurred during laying or even later, the opportunity of the replacement male to father some young was higher. However, it is also possible that in these cases male behavior was not optimal because females might deceive males into an erroneous appreciation of their paternity expectancies. I address this aspect in the next section.

Counterstrategies of females

It has been stated for mammals, using a genetic model, that the costs of resistance to infanticide by males for females are higher than its benefits (cf. Yamamura *et al.* 1990). However, in the European starling and the tree swallow, females have been observed trying to prevent the replacing male entering the nest by behaving aggressively towards him (Robertson 1991; Smith *et al.* 1996). Nonetheless, the possibility that female birds are more efficient than female mammals in preventing infanticide by males needs further research.

Abortion of fetuses when the subsequent newborns have a high probability of suffering infanticide has been reported as an alternative female counterstrategy in mammals (Labov 1981). Stopping laying and ovulation retardation may represent an equivalent strategy of female birds. Females may use strategic ovulation retardation to assure that the replacement male has just time to fertilize the next clutch produced after replacement. In the house sparrow, females usually stop laying after male replacement and resume it after 6 to 9 days. This time interval is thought to reflect the minimum time interval between male replacement and clutch initiation that seems to prevent hostility by the incoming male (5 days), which suggests that ovulation retardation has the function of giving the male the impression of a high certainty of paternity. That delayed ovulation is adaptive is reinforced by the observation that females sometimes delayed egg-laying after male replacement even without having suffering egg losses, thereby ruling out the possibility that delayed ovulation results from energy stress on the female (Veiga 1993).

Females may also try to deceive males by making them "believe" they have some or complete paternity in a brood. For example, in tree swallows, females were seen copulating with the replacement males during incubation, apparently inducing them to adopt a brood in which they had no likelihood of paternity (Robertson 1991). Also female pied flycatchers (*Ficedula hypolenca*) may have persuaded males to adopt their broods by copulating with them during incubation (Gjershaug *et al.* 1989). It is not clear, however, that in all cases males may be deceived by this female behavior. In house sparrows, females whose previous mates were replaced during egg-laying immediately solicited copulations, but, interestingly, males initially refused to copulate (Veiga 1993). Thus studies aimed to investigate the factors involved in the evolution of the deception strategy are badly needed.

Conclusions

Infanticide by male birds, considered here as resulting from sexual selection acting differentially on this sex, has been reported for 13 species, while in another two, which have sex role reversal, infanticide was committed by the female. The amount and strength of the reported evidence varies considerably among studies. A high proportion of the reported cases of infanticide were provoked experimentally by inducing territory takeovers through the removal of male territory owners. When the evidence of infanticide was based on observations of non-manipulated populations, its frequency of occurrence seems to vary considerably among species and populations. Infanticide, however, is probably not negligible as a cause of mortality among the species in which this behavior has been relatively well documented, and it may have been overlooked in many studies of reproduction and mortality. The pattern emerging from interspecific comparisons suggests that the infanticidal tactic is more widespread among passerines than in other avian taxa, and that its presence is relatively independent of the mating system. It seems general that infanticide is committed by unmated males under monogamy, but more by already mated males under polygyny. In all systems, replacing males frequently remate with their victims and start a new breeding attempt.

Since incoming males have alternative options to committing infanticide, it seems necessary to identify separately the mating consequences of taking over a territory and of killing offspring when one is discussing the factors that may have promoted the evolution of avian infanticide. Infanticide committed after a takeover may be advantageous in two ways: (1) it may shorten the time elapsed between the takeover and the next fertile period of the victimized female, and (2) it may induce nest desertion by the female owner and the subsequent acquisition of a new female that would lay eggs earlier than the previous owner would have done. The first possibility is well documented for three species, the barn swallow, the house wren and the house sparrow. However, there is scant information on the benefits derived from the acquisition of a new mate after infanticide, though this outcome is not infrequently reported.

The preponderance of monogamy among birds, a classical argument to explain the low frequency of infanticide in this group, is not supported by the evidence currently available. Infanticide is also committed under monogamy, mainly by unmated individuals, suggesting that asymmetries in mating status create a potential for the evolution of infanticide in monogamous species. Protracted breeding seasons have also been proposed as a factor favoring infanticide because only in this circumstance is there time for a victimized female to start a new breeding cycle. Alternatively it has been argued that long breeding seasons may also favor adoption rather than infanticide if females remain in the breeding territory after a successful breeding attempt but desert after failure. In any case, the success of an infanticidal male in inducing a rapid remating with his victim probably depends on the mating and nesting alternatives open to females.

Males not only destroy nest contents at the time of territory takeover, but may also kill young hatched or destroy eggs laid after replacement. This subset of sexually selected infanticide has been termed prospective infanticide, and, to be adaptive, males must be able to evaluate paternity by killing only the offspring that appear within one fertility period. Prospective infanticide in birds has so far been described in five species. Paternity determinations in relation to replacement timing in one of these species suggest that males were fine-tuning their behavior to the probability of fathering the offspring produced after replacement and mating with the new female. Females may in turn develop counter-strategies to avoid infanticide or reduce its adverse consequences on their breeding success. It has been suggested that by copulating with replacing males, females may try to deceive them by making them believe they have paternity in the subsequently appearing brood. Females may also use ovulation retardation to assure that the replacement male has just enough time to fertilize the next clutch, giving him a high certainty of paternity.

Part III

Behavioral consequences of infanticide by males

ADRIAN TREVES

10

Prevention of infanticide: the perspective of infant primates

Introduction

Infants have the strongest incentive to avoid infanticide. Paradoxically, they are the least capable of preventing it. Primate infants are particularly vulnerable because they take a long time to develop the physical and cognitive capacity to avoid risk. Yet infants are not utterly helpless in the face of infanticidal threats. Through detection, avoidance and deterrence, infants and their caregivers prevent infanticide. If infanticide is treated as a series of steps preceding lethal injury, one can see several junctures at which animals can act to lower risk (Figure 10.1). Protectors can prevent infanticide at every step. Neonate self-protection is generally limited to the earliest steps in the process. An older infant may have more options.

In this chapter, I explore the behaviors used by primates to prevent infanticide. I begin with protectors and then turn to infant self-protection. In both sections, I search for evidence of specific adaptations to infanticide prevention. The final section specifically addresses infant transport and the coevolution of primate mothers and infants.

Protectors

Primate infants are cared for by a variety of individuals: mothers, fathers, older siblings, more distant kin, as well as some unrelated adults (Hrdy 1976; Nicolson 1987; Whitten 1987; Manson 1999; Paul 1999). Within groups, the sexes usually take different roles in infant protection. A common pattern is for females to perform direct care (e.g., transport, feeding) while males provide indirect care (e.g., guarding). Close female

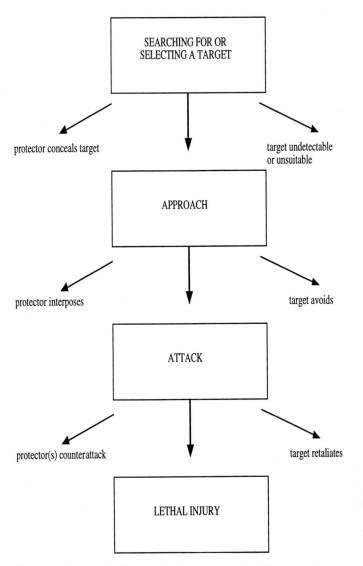

Figure 10.1. The steps leading to lethal injury to an infant. The attacker's moves are
represented by rectangles. Preventive options of caregivers and infants are
represented by arrows pointing outward at each step. Neonates are generally
restricted to preventive options at the uppermost step, while caregivers can act
at any point.

bonds may lead to protective infant handling by non-mothers (e.g., *Colobus guereza*, Oates 1977; *Semnopithecus entellus*, Borries *et al.* 1994). This form of female cooperation may be essential to infanticide prevention when resident males do not deter unrelated rivals from entering the group (e.g., *Lemur catta*, Sussman 1992; *Propithecus diadema*, Wright 1995; *Saimiri oerstedi*, Boinski 1987a; *Cercopithecus aethiops*, Henzi & Lucas 1980). But living with other females may not protect infants if female-initiated infanticide occurs (*Papio cynocephalus*, Wasser & Starling 1988; *Pan troglodytes*, Goodall *et al.* 1979; *Alouatta palliata*, Clarke & Glander 1984). In these social systems, mothers often rely on males to help prevent infanticide (e.g., *Gorilla gorilla*, Watts 1989; *Papio hamadryas*, Kummer 1968; *Alouatta seniculus*, Sekulic 1983b), or use a combination of male friends and matrilineal kin as buffers against conspecific aggression from either sex (Smuts 1985; Smuts & Smuts 1993; Treves 1998). If one can identify the relevant protectors, one can compare immature survival in groups that vary in the number of protectors.

Several researchers have related variation in immature survival to variation in group composition. For example, Koenig (1995) analyzed group composition data from wild populations of common marmosets (*Callithrix jacchus*) and found that groups with several adult males contained more infants than groups with fewer adult males (independent of group size). He attributed this to improved infant survival through caregiving by the supernumerary males (see also Sussman & Garber 1987; Snowdon 1996). The number of cooperating females may also affect immature survival or retention in the natal group. For example, in 20 populations of hanuman langurs (*Semnopithecus entellus*), groups with many females were more successful in protecting immatures than those groups with fewer females. This effect was especially noticeable in populations with high numbers of non-group males (Treves & Chapman 1996). More recently, I tested whether variation in immature numbers within groups correlated with variation in the number of adults and sex ratio within groups of howler monkeys. Groups of howler monkeys with many adult and subadult males per female contained more juveniles than expected (A. Treves, unpublished data). This result is consistent with infanticide prevention because the effect was strongest in populations with a high density of males, as expected if the pressure came from males seeking breeding positions. The approach taken in these studies derives from an early insight by van Schaik (1983). Slight revision of his pioneering methods (Treves & Chapman 1996) has produced consistent evidence that immature numbers vary as a function of group composition. However, we still have no direct evidence that infanticide is less likely to succeed in groups with many protectors.

For direct evidence of infanticide prevention by protectors, we need to examine their behavior when the risk of infanticide is elevated. Individuals can prevent infanticide if they detect and deter threats before they close to attacking distance (Figure 10.1). For example, mothers are particularly watchful when infants stray out of arm's reach (*Macaca mulatta*, Leighton-Shapiro 1986, Maestripieri 1993; *Procolobus badius*, Treves 1999; *Saimiri sciureus*, Biben *et al.* 1989). Non-mothers may also detect and deter infanticidal threats. For example, silverback gorillas (*Gorilla gorilla*) may chest-beat to warn away non-group males (Fossey 1984), or lead their groups away from solitary males (Yamagiwa 1986). Black-and-white ruffed lemur (*Varecia variegata*) fathers increase guarding in the presence of potentially infanticidal males (White *et al.,* unpublished data, cited in van Schaik & Kappeler 1997). Non-mother females also act to repel or injure threatening males. This often involves cooperative action by several females (*Lemur catta*, Pereira & Weiss 1991; *Propithecus diadema*, Wright 1995; for a review, see Smuts & Smuts 1993; Treves & Chapman 1996). For example, Sommer (1987) describes coalitions of up to seven hanuman langur females repeling infanticidal males and successfully protecting infants.

Individuals involved in direct care play a key role because they can restrain or move infants as well as detect and deter threats. Maternal movements may keep the infant out of harm's way. A possible example of this is found among wild orang-utans (*Pongo pygmaeus*), where mothers with neonates rarely associate with conspecifics, and were therefore more difficult for observers to find in one study (Rijksen 1978). Another study found that mothers slept further from fruit trees that were likely to attract conspecifics (Setiawan *et al.* 1996). In gregarious taxa, mothers with neonates may disperse when the risk of infanticide increases (Hrdy 1977). Within groups, mothers may restrain or retrieve infants to keep them safe. In an interspecific comparison, infants explored their environment later in groups with a greater number of adult males (Treves 1996), suggesting maternal restraint. More direct evidence for infant restraint under the risk of infanticide is available (Palombit, Chapter 11). In one experimental study, strange males were introduced to captive vervet monkeys (*Cercopithecus aethiops*). The mothers restrained their infants more following the appearance of a strange male (Fairbanks & McGuire

1987). Retrieval and restraint of infants is very common in situations where neither mother nor infant benefit from infant handling (Paul 1999; Silk 1999).

The protective behaviors of infants' caregivers and associates can help to prevent infanticide, as the examples given above show. However, they also occur in other dangerous situations. For example, predator threat leads mothers to restrain, move or otherwise protect infants (Hauser 1988; Cheney & Seyfarth 1990). Although female coalitions against threatening males are widespread and can prevent infanticide (Sommer 1987), they are also used to mob predators and compete with others (predators, Hrdy 1977; competitors, Wasser & Starling 1986, Smuts 1987b, van Schaik 1989). The same behavior used by male leaf-monkeys to repel rivals (van Schaik *et al.* 1992) helps in mate guarding, prevention of infanticide and protection of juveniles, who are often ousted from groups following takeovers (Reena & Ram 1991). In sum, most protective behaviors are multifunctional because they can protect against many different threats, and they can protect the actor and individuals other than infants.

Because of the multifunctional nature of protective behaviors, we currently have little evidence that protective behaviors are specifically adapted for infanticide prevention. For observational studies, differences between mothers of neonates and mothers of older infants or pregnant females are critical. We need more experimental tests to determine whether caregivers act specifically to reduce infanticide risk. For example, the presentation of unfamiliar males in captive situations, or sound playbacks of potentially infanticidal males, may elicit the defensive responses most critical for infanticide prevention.

Infant self-protection

Infant primates may be incapable of fighting off an infanticidal male or fleeing from him, but they are not utterly helpless. Avoidance of strangers is widespread among infants and may reflect an adaptive fear of aggression by associates (Rosenblum & Albert 1977). Infants may avoid detection at the outset through concealment, or through deception avoid being selected as a target.

Avoiding detection by concealment
Infanticide may be preventable if infants are never detected by unrelated males. The role of the infant in reducing its own detectability is limited,

except perhaps in taxa that leave infants in a shelter or cache them while the mother forages. Haplorhine primates rarely cache young (Kappeler 1998; but for a conspicuous example, see Fuentes & Tenaza 1995). In strepsirhines, however, most genera employ some form of caching (Table 10.1; Kappeler 1998). To date, there is only one report of infanticide in a caching genus (*Varecia*, van Schaik & Kappeler 1993, Treves 1998). Better data would come from comparisons of infanticide rates on cached and uncached young of the same population. *Avahi*, *Hapalemur* and *Varecia* may provide the most useful data because caching frequency varies over time or between populations (Mittermeier *et al.* 1994; Kappeler 1998).

Avoiding attack by deception

Preventing infanticide by concealment is more difficult if the infant and its mother are under scrutiny by unrelated males. In social systems with this characteristic, females and infants may signal deceptively. The female can mate so as to spread paternity uncertainty and the infant may confuse paternity or mimic independence. These forms of deception may work only against males who have mated with the mother. Successful deception also depends on whether an infanticidal male is targeting infants of a certain age or paternity. Three current explanations for infanticide by unrelated male primates make different predictions about the age and paternity of targeted infants.

According to the sexual selection hypothesis, males kill unrelated infants to hasten the resumption of ovulation by their mothers (Hrdy 1974, 1977, 1979). Young, unrelated infants are the primary targets as they cause the longest delay to ovulation (Leland *et al.* 1984; Sommer & Mohnot 1985; Struhsaker & Leland 1987). An alternative hypothesis for male-initiated infanticide is that infant deaths are accidents during social change, and not a deliberate strategy by males (Sugiyama 1965). This idea resurfaced as the social pathology hypothesis (e.g., Boggess 1979), which was later disproven (Newton 1986, 1988; Borries 1997). But this view may still be correct if falling and incidental wounds to infants occur during the aggression and excitement of social change without reflecting deliberate attacks by males (Sussman *et al.* 1995). In this view, young infants are victimized disproportionately because they are more susceptible and vulnerable to injury, but kin of the male may fall victim. Finally, the local resource competition hypothesis proposes that a male kills unrelated immatures to lessen competition with that male's future offspring (Dittus 1980; Silk 1983; Hiraiwa-Hasegawa & Hasegawa 1994; Agoramoorthy &

Table 10.1. *Neonate and caregiver attributes in 62 primate genera*

Suborder/ genus	Neonate transport during active period[a]	Adult female mass[b] (kg)	Neonate mass[c] (kg)	Litter size[d]	Relative litter mass[e] (%)
Haplorhini					
Allenopithecus	Mother mainly	3.440	0.242	1	7.0
Alouatta	Mother mainly	5.250	0.255	1	4.9
Aotus	Caregiver	0.936	0.091	1	9.7
Ateles	Mother mainly	8.414	0.444	1	5.3
Brachyteles	Mother mainly	8.070	–	1	–
Cacajao	Mother mainly	2.795	–	1	–
Callicebus	Caregiver	1.059	0.074	1	7.0
Callimico	Caregiver	0.525	0.054	1	10.3
Callithrix	Caregiver	0.334	0.031	2	18.4
Cebuella	Caregiver	0.122	0.016	1.8	23.1
Cebus	Mother mainly	2.468	0.227	1	9.2
Cercocebus	Mother mainly	5.655	0.557	1	9.8
Cercopithecus	Mother mainly	3.480	0.406	1	11.7
Chiropotes	?	2.677	–	1	–
Colobus	Mother mainly	7.882	0.573	1	7.3
Erythrocebus	Mother mainly	6.500	0.504	1	7.8
Gorilla	Mother mainly	80.000	2.123	1	2.7
Homo	Mother mainly	54.425	3.319	1	6.1
Hylobates	Mother mainly	6.664	0.482	1	7.2
Lagothrix	Mother mainly	7.090	0.471	1	6.6
Leontopithecus	Caregiver	0.479	0.058	1.8	21.8
Lophocebus	Mother mainly	5.890	0.6	1	10.2
Macaca	Mother mainly	6.615	0.45	1	6.8
Mandrillus	Mother mainly	12.700	0.856	1	6.7
Miopithecus	Mother mainly	1.560	0.177	1	11.4
Nasalis	Mother mainly	9.820	0.6	1	6.1
Pan	Mother mainly	38.575	1.647	1	4.3
Papio	Mother mainly	11.906	0.781	1	6.6
Pithecia	?	1.920	0.121	1	6.3
Pongo	Mother mainly	35.700	1.809	1	5.1
Presbytis	Mother mainly	6.230	–	1	–
Procolobus	Mother mainly	6.640	–	1	–
Pygathrix	?	9.300	0.428	1	4.6
Saguinus	Caregiver	0.387	0.044	1.8	20.3
Saimiri	Mother mainly	0.700	0.109	1	15.6
Semnopithecus	Mother mainly	10.533	–	1	–
Simias	Mother mainly	6.800	–	1	–
Tarsius	Cache	0.112	0.025	1	22.5
Theropithecus	Mother mainly	11.700	0.464	1	4.0
Trachypithecus	Mother mainly	6.128	0.43	1	7.0
Strepsirhini					
Allocebus	?	0.084	–	–	–
Arctocebus	Cache	0.258	0.032	1	12.4
Avahi	Mother mainly	1.049	–	1	–
Cheirogaleus	Cache	0.272	0.018	2.5	16.6

Table 10.1. (*cont.*)

Suborder/ genus	Neonate transport during active period[a]	Adult female mass[b] (kg)	Neonate mass[a] (kg)	Litter size[d]	Relative litter mass[e] (%)
Daubentonia	Cache	2.490	0.112	1	4.5
Eulemur (Petterus)	Mother mainly	1.831	0.066	1.5	5.4
Euoticus	Cache	0.261	–	1	–
Galago	Cache	0.221	0.015	1.6	11.1
Galagoides	Cache	0.109	0.012	1	11.0
Hapalemur	?	1.066	0.045	1.5	6.4
Indri	Mother mainly	6.840	–	1	–
Lemur	Mother mainly	2.210	0.066	1.2	3.6
Lepilemur	Cache	0.769	0.027	1	3.5
Loris	Cache	0.674	0.011	1.6	2.5
Microcebus	Cache	0.055	0.006	2	21.6
Mirza	Cache	0.326	0.018	2	10.7
Nycticebus	Cache	0.463	0.031	1.2	8.1
Otolemur	Cache	0.922	0.048	1.5	7.7
Perodicticus	Cache	1.023	0.039	1	3.8
Phaner	Cache	0.460	–	1	–
Propithecus	Mother mainly	4.270	0.122	1	2.9
Varecia	Cache	3.515	0.095	2.5	6.7

Notes:
[a] Caregiver, majority done by non-mothers; cache, neonates are rarely transported; ?, insufficient information for wild populations or the genus shows two or more distinct patterns (see text); mother mainly, although handlers may contribute, the mother does the majority of transport during the active period.
[b] Adult female mass from Smith & Jungers (1997).
[c] Neonatal mass from Smith & Leigh (1997).
[d] Litter size from Klopfer & Boskoff (1979), van Horn & Eaton (1979), Heltne *et al.* (1981) Haring & Wright (1989), and Kappeler (1998). Twinning rates of <5% are treated as litter sizes of 1.
[e] Relative litter mass, litter size × neonatal mass/adult female mass.

Rudran 1995). Unrelated older immatures will exhaust resources needed by females in the group, and be competitively superior to the male's own offspring when they are born. Although younger infants are easier targets, males should be equally intent on targeting older infants. In sum, the three hypotheses make different predictions (Table 10.2).

Unfortunately, we have little evidence for or against age or paternity deception among primate infants. In 1977, Hrdy proposed that infants might survive immigrations by new males by behaving so as to appear independent from their mothers. Studying captive vervet monkeys, Fairbanks & McGuire (1987) reported accelerated weaning after the

Table 10.2. *Three hypotheses for male-initiated infanticide in primates*

		Predictions about targets	
	Proposed function or motivation	Age	Relatedness to infanticidal male
Sexual selection	Hasten ovulation by mother	Young	Non-kin
Accidents during social change	Non-functional, no intent	Young	No predicted effect
Local resource competition	Eliminate competitors	Any age	Non-kin

experimental introduction of strange males. Although the mother was responsible in the latter case, such age deception would support the sexual selection hypothesis, against the accidents during social change and local resource competition hypotheses. More definitive tests are needed.

Age deception – if it exists – is probably limited to older infants. Neonates cannot behave as if they were older. However, neonates may be able to display morphological traits that conceal paternity. For example, neonatal pelage may reduce the risk of infanticide by concealing phenotypic cues to paternity (Treves 1997). Although the central prediction that neonatal coats obscure paternity has not been tested, there are other lines of evidence consistent with this view. In 74 primate taxa, neonates and adults differ in body color. Loss of the neonatal pelage typically begins with the head or dorsal surface (Treves 1997); these body parts are least open to phenotypic matching because a male cannot scrutinize his own face, head or dorsum. Neonatal coats as infanticide prevention adaptations also predict that the coat should last only as long as needed to minimize infanticide (because odd coloration makes animals more vulnerable to predators: Wolf 1985; Landeau & Terborgh 1986). The coats disappear 18–20 weeks postpartum, after the most vulnerable period for infanticide (e.g., Sommer & Mohnot 1985). Indeed, for taxa with suitable data, the loss of the neonatal coat coincided with the start of exploration away from the mother. In sum, the duration of the neonatal coat and its pattern of loss is consistent with its hypothesized function to impede phenotypic matching and its role in protecting the most vulnerable infants from infanticide. The phylogenetic pattern of neonatal coats and their relationship to mating system also supports the hypothesis.

Paternity deception may be advantageous when males use kin recognition to discriminate against infants they have not sired. This idea was anticipated in a study of birds (Davies 1992; see also Pagel 1997). To overcome uncertainty, a male may seek phenotypic markers that confirm or exclude his paternity uniquely. For example, Clarke (1983) described infant-killing by a male mantled howler (*Alouatta palliata*) with whitish spots on his feet, unique among adults. This male killed one infant in a group, while two others disappeared shortly after his takeover. He spared three infants, one of which had whitish spots on its feet. The infanticidal male may have been able to match the infant's foot color phenotype to his own (Clarke 1983).

Males benefit from kin recognition when they invest heavily in young, or when unrelated young impede the ovulation of available females. Thus paternity concealment – as proposed for the neonatal coat – may evolve when it is common for unrelated males to be present with neonates and when the mating system makes it common for two or more unrelated males to mate with a single female during her period of highest conception probability. In short, promiscuous mating systems with multiple unrelated males in residence should produce selection for neonatal coats (Treves 1997). Examples of such systems are common in the Cercopithecoidea (Melnick & Pearl 1987), less common in the Ceboidea (Strier 2000), and rare in the other primate superfamilies (Bearder 1987; Wrangham 1987a,b; van Schaik & Kappeler 1993). Consistent with the prediction, neonatal coats are most common in the Cercopithecoidea. Furthermore, testes size – a good indicator of mating systems (Harcourt *et al.* 1995) – shows an association with neonatal coat contrast.

Species with large testes (relative to body mass) tended to have discreet natal coats in which the neonate contrasted with the adult but was not striking to the eye otherwise. For example, *Papio* neonates are black while the adults range from dusky yellow to brown. On the other hand, those taxa with relatively small testes tended to have flamboyant coats, in which the neonate contrasted with the adult and was visually conspicuous as well (e.g., *Presbytis* neonates are white, gold or orange, while adults range from gray to black). Species with intermediate relative testes sizes tended to have no neonatal coats (Figure 10.2). The existence of two different types of neonatal coat was noted by Hrdy (1976) but no functional difference was suggested. I proposed that the flamboyant version was a paternity cloak with an additional warning coloration function that might deter non-group males from approaching (Treves 1997). These hypotheses for neonatal coloration in primates are consistent with available evidence but far from proven. We do not yet have the necessary direct

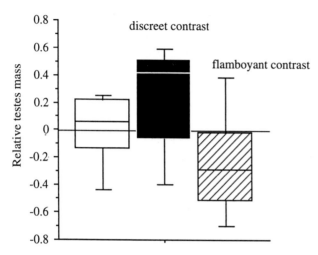

Figure 10.2. Box plots for relative testes mass split by neonatal coat contrast. Y-axis is
calculated as the residual of log(testes mass in g) regressed on log (body mass in
kg) from Harcourt *et al.* (1995). No contrast (open bar) refers to taxa in which
the color of neonate dorsal pelage is largely indistinguishable from that of
adults. Discreet contrast (black bar) refers to muted tones in which the neonate
is distinct but not eye-catching (e.g., *Papio* neonates). Flamboyant contrast
(hatched bar) refers to neonatal pelage that is eye-catching when seen against
the adult pelage (e.g., *Presbytis* neonates). The boxes span the interquartile
range, horizontal lines within the boxes designate the mean, and the vertical
bars span 1 SD. The differences are significant ($P < 0.02$, from Treves, 1997).

evidence to test whether infants are obscuring paternity or deterring the
approach of infanticidal males.

Neonatal transport

In approximately one fourth of primate genera, mothers cache neonates
during the active period (Table 10.1). In the remainder, the neonates are
transported by the mother or by helpers. Neonatal transport shows a clear
phylogenetic clustering, with all anthropoids transporting neonates
(Table 10.1). However the presence of some strepsirhine taxa that habitually transport neonates (e.g., *Propithecus, Indri*) suggests that this trait
evolved more than once in primates. Convergent evolution in the face of
major costs incurred in neonatal transport suggests a strong selective
advantage under some circumstances.

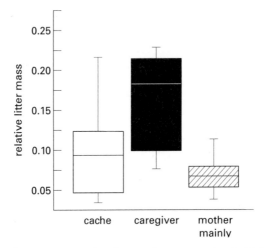

Figure 10.3. Strategies of neonatal care are associated with different patterns of
relative litter mass. Relative litter mass expressed as average neonatal mass ×
litter size/average adult female mass. Care strategies described in Table 10.1.
Box plots as in Figure 10.2. The means are significantly different ($P = 0.005$).

Animals encumbered by young rest more, capture fewer mobile prey
and fall more often than similar associates without young (Wright 1984;
Whitten 1987; Price 1992; Stanford 1992). Indeed, the heavier the neonates, the earlier caregivers begin to help mothers in those genera with
intense alloparenting (Whitten 1987). In Table 10.1, I have classified
genera as mother mainly (mother transports neonates with little help
from others), caregiver (non-mothers do the major share of neonatal
transport) or cache (neonates are mainly not transported). We get an
indication of the costs of neonatal transport and the coevolution of
mother and infant by examining relative litter mass across these three
categories of neonatal care. The genera in which mothers do the majority
of work carrying neonates have the lowest relative litter masses (Figure
10.3; Kruskal-Wallis corrected for ties, $H = 10.5$, $P = 0.005$).

This relationship is not an artifact of non-independence of related
taxa, because it holds within the strepsirhines (compare *Propithecus*,
Eulemur and *Lemur* to the caching genera *Microcebus*, *Cheirogaleus*: Table
10.1). Indeed, the same pattern emerges if we consider only those genera
in which mothers do the majority of neonatal transport and generate two
subcategories: those displaying maternal resistance to infant handling or
no infant handling in the wild and all others (data from Treves 1997; Paul
1999 excluding *Macaca* because of marked interspecific variation). Those

with some infant handling have higher relative litter masses than those with none (Mann–Whitney $Z = -2.28$, $P = 0.021$). This difference is due mainly to subcategory differences in adult female mass ($Z = -2.0$, $P = 0.045$), rather than neonatal mass itself ($Z = -0.97$, $P = 0.50$). In short, genera in which mothers do virtually all the neonatal transport are larger. This reduces the energetic and biomechanical burden of neonatal transport (for a review, see Hartwig 1996).

The caregiver genera (callitrichids plus some pitheciines) provide further clues to the evolution of neonatal transport. The callitrichids have evolved smaller adult body mass, without much change in neonatal mass, resulting in very heavy relative litter masses (Hartwig 1996). Such relatively heavy litter masses are only possible because of caregiving. Indeed, the other caregiver genera, the pitheciid *Aotus* and *Callicebus*, have adult female mass close to 1 kg and relative litter masses of 7% to 9% (Table 10.1). In these two genera, mothers transport neonates more than do caregivers in the first few weeks of life (Wright 1984; Goldizen 1987). Hence, callitrichids underwent a reduction in adult body mass that brought them into the range for cache genera (Mann–Whitney U test, $Z = -0.7$, $P = 0.50$). Cache and caregiver genera barely differ in neonate mass ($Z = -1.6$, $P = 0.10$). Finally, *Saimiri* have very large relative litter mass (15%) and tendencies toward caregiving (allonursing and early, intense allomaternal transport of young; Williams *et al.* 1994; L. Morton, unpublished data). They also show a secondary reduction in adult mass (Hartwig 1996). In sum, mothers in genera receiving help are smaller than those that transport neonates alone. In several cases, evolutionary changes in adult body mass are associated with changes in neonatal transport style.

Following body mass evolution other life history traits often change, including longer juvenile period and larger brains. However, such fundamental changes in life history traits are believed to occur only when selective pressures affect mortality schedules across a range of environments (Partridge & Harvey 1988; Kappeler 1995; Kelley 1997). On this basis, I present four non-exclusive hypotheses for the general adaptive function of neonatal transport.

1. Neonatal transport reduces the risk of infanticide on cached infants (*infanticide prevention*);
2. Neonatal transport arises when females rely less heavily on mobile prey (*sedentary feeding*);
3. Neonatal transport reduces the risk of predation in diurnal species (*predation-avoidance*)

4. Neonatal transport reduces travel costs for wide-ranging mothers (*travel economy*).

These four hypotheses each do two things. They generate testable predictions about the selective advantages of neonatal transport, while explaining the phylogenetic pattern of neonatal transport in extant primates.

Infanticide prevention

Mother–infant proximity and infant transport increase when conspecific threat is elevated. For example, in cages with conspecifics, pygmy slow loris (*Nycticebus pygmaeus*) mothers carry their neonates three times longer than do captive mothers living alone (Ehrlich & MacBride 1989). Among captive Bornean tarsiers (*Tarsius bancanus*), infants are protected by high levels of proximity to their mothers, who aggressively repel conspecifics (including the father) (Roberts 1994). Among gregarious species, mothers often actively resist infant handling because infants are at risk when far away (for review, see Paul 1999; Silk 1999). The restraint on infant exploration appears to increase in species with many resident males (Treves 1996), and within groups when new males are present (Fairbanks & McGuire 1987). The phylogenetic pattern of neonatal transport may be explained if the evolution of diurnal habits and the use of large, long-lasting resources (e.g., large crops of fruit) made mothers more conspicuous to conspecifics. In a variation on this idea, van Schaik & Kappeler (1997) proposed that gregariousness and close mother–infant association evolved together. Gregariousness of mothers might have precluded caching if mothers could not direct group movement back to cache sites.

A prediction from the infanticide prevention hypothesis is that infants will fall victim to infanticide more often when they are not in physical contact with their mothers. Also, populations that cache neonates will transport their young more frequently when the mothers are conspicuous. The latter prediction is supported by the behavior of *Otolemur crassicaudatus*. This largest of galagines transports its young more than any other and does so most when congregating at seasonal gum feeding sites used by many conspecifics (Charles-Dominique & Bearder 1979; Harcourt 1986).

Sedentary feeding

Because mothers burdened by neonates are at an energetic and biomechanical disadvantage when foraging for mobile insects (see p. 234), neonatal transport may only have appeared when a constraint was lifted. When mothers began to feed on non-mobile foods (sessile insects, plant foods), they were capable of carrying neonates without loss of foraging efficiency. This hypothesis is non-functional for it does not specify an adaptive benefit per se. However it does explain the phylogenetic pattern of neonatal transport in the plant-feeding anthropoids and strepsirhines. It yields the prediction that mothers will transport neonates less often when they are pursuing mobile prey.

Predation avoidance

If cached infants are particularly vulnerable, mothers may benefit from carrying them. I know of no evidence to support this view to date. We might expect the size of the predators and differences between diurnal and nocturnal hunting patterns to affect the cost–benefit ratios for carrying young. Hence this hypothesis can explain the phylogenetic pattern of neonatal transport if diurnality, large body size or plant food consumption increases predation risk. A prediction of the predation-avoidance hypothesis is that infants will fall prey more often when they are not in physical contact with their mothers. In addition, manipulations of predator risk should influence caching and carrying decisions.

Travel economy

Roundtrip travel to a cached infant may be more energetically expensive than transporting a neonate. The increase in daily travel requirements associated with the evolution of gregariousness, large body size and plant consumption (Charles-Dominique 1995) might have changed the energetic economy of caching. This hypothesis could be tested by comparing caching populations. For example, *Hapalemur* shows variation in neonatal transport patterns (Table 10.1). If the travel economy hypothesis is correct, the *Hapalemur* populations that transport neonates for longer periods or distances should be those that cover more distance daily, and perhaps those that feed more heavily on plants than on animals.

The evolution of neonatal transport is an intriguing problem as it is correlated with diurnality, diet and gregariousness among extant primates. Neonatal transport may be central to the evolution of anthropoids, as it seems to explain some variation in patterns of adult body mass evolution. Neonatal transport can provide several possible advantages, among them prevention of infanticide. A challenge for future research is to test these hypotheses in ways that narrow the possibilities.

Conclusion

Infanticidal attacks are the culmination of a series of moves by attackers and countermoves by infants and their allies. Long before potential attackers select a target or launch an attack, caregivers are preventing infanticide. They do this with vigilance, avoidance and deterrence displays. Infants may contribute to their own safety through concealment, and perhaps deception. Regrettably, most attention has been paid to later stages in the process when early defenses have failed. Our view of infanticide and its prevention is therefore slanted toward successful attacks. More attention to infanticide prevention and the cause of failed attack may advance our understanding greatly.

One conclusion of this review is that more information is needed. Most obviously, the entire subject of infant self-protection is speculative and preliminary. If we wish to document behaviors used in the prevention of infanticide, we need to examine individual behavior when vulnerable infants are present and when threatening males are close. We must also pay attention to all potential protectors. Infants depend on the foresight and alertness of caregivers. This leads to a second conclusion. Infants are relatively defenseless, but their caregivers can bolster their defenses considerably. Thus, infanticide intrinsically favors mother–infant association and the evolution of social defenses (van Schaik & Kappeler 1997). Indeed, it may be this social aspect of infanticide prevention that has impressed many researchers and led to suggestions that infanticide avoidance shapes primate social systems (Wrangham 1979, 1986, 1987a; Watts 1989; van Schaik & Dunbar 1990; Smuts & Smuts 1993; van Schaik & Kappeler 1993, 1997; Brereton 1995; Treves & Chapman 1996; Treves 1998). Despite all this theoretical attention, we still do not know whether variation in the number of caregivers can explain variation in the survival of infants. A productive avenue for future research will be to explore the costs and benefits of neonatal transport because the mother–infant bond is the most fundamental social relationship that may prevent infanticide.

Acknowledgments

Valuable comments were given by C. Borries, C. Chapman, C. Janson, L. Naughton, C. Snowdon, K. Washabaugh and one anonymous reviewer. Financial support was provided by NIH Grant MH 35,215 to Dr C. T. Snowdon and the University of Wisconsin-Madison.

11

Infanticide and the evolution of male–female bonds in animals

Introduction

Sexual selection theory suggests that divergent reproductive interests of the sexes impede the evolution of enduring social bonds between males and females. Males are more likely to increase reproductive success by acquiring multiple mates whereas females enhance fitness more by discriminative choice of individual mate(s) (Darwin 1871; Trivers 1972). Consequently, insofar as postcopulatory bonds limit a male's sexual access to additional fertile females and are irrelevant to antecedent female mate choice, such bonds should be relatively rare. The mostly polygynous mammals, for example, fulfill this expectation: male mating effort generally exceeds paternal effort, and affiliative interactions between the sexes center on the period of copulation or female fertility (Clutton-Brock 1989b).

And yet, males and non-fertile or anestrous females maintain stable relationships with one another in some species. The same theoretical framework that predicts the rarity of persistent heterosexual bonds also highlights a primary context for their evolution: when a male restricts his mating to a single female, a postcopulatory relationship with her is not only less costly to him, but may also offer fitness advantages to both parties. The proposed benefit to the female is the extensive parental care she receives from a male that is now certain of paternity. Thus durable male–female relationships were originally viewed as part of a coevolved suite of behaviors including monogamy, biparental care, and, in gregarious animals, "nuclear families" of parents and offspring (Morris 1967; Wittenberger & Tilson 1980; Gubernick 1994). The primarily monogamous, biparental birds have long served as vivid examples of this system (Lack 1968).

In spite of the rigor of its evolutionary precepts, this version of the "mating system" perspective fails to fully explain variation in male–female bonds. First, avian studies reveal that mating exclusivity and certainty of paternity are not natural corollaries of social monogamy (Birkhead & Møller 1992; Avise 1996), and that the impressive parental efforts of "caring" monogamous males fail to benefit their female mates as uniformly as originally assumed (Bart & Tornes 1989). Second, male–female social relationships vary independently of mating system (Kleiman & Malcolm 1981; Wickler & Seibt 1983): not only may monogamy entail weak heterosexual attachment and negligible paternal care, but strong bonds between males and anestrous females occur in polygynous mammals in which "classic" forms of direct male care, such as feeding and social thermoregulation, are absent (Palombit 1999). These data direct our attention toward alternative adaptive benefits of male–female bonds.

One of these is male protection against infanticide. A growing body of evidence (Hausfater & Hrdy 1984; Parmigiani *et al.* 1994) indicates that sexually selected infanticide often accompanies a change in breeding males, either (1) as part of a takeover strategy in which a male physically ousts another male from a breeding position; or (2) following "passive replacement" of a male that has disappeared, died, or deserted his female(s). The infanticide protection hypothesis suggests that females form bonds with males who will deter infanticide (Hrdy 1979).

Male defense and male–female social relationships

A male's protection of an infant does not *a priori* imply a social bond with its mother. In some teleost fish and amphibians, for example, males care for and defend young without the help of females, who leave males in possession of broods to seek additional mates and/or tend a second clutch (Sargent & Gross 1993). Such uniparental care by males is rare in birds, and essentially non-existent in mammals (Clutton-Brock 1991). Prolonged postnatal maternal investment in these taxa often means that a male guardian will necessarily associate at least spatially with the mother, especially when females carry their young rather than cache them in burrows or nests (van Schaik & Kappeler 1997). Propinquity may or may not, however, promote social affiliation between the sexes (Wickler 1976). If the primary cause of their association remains simply the spatiotemporal conjunction of their independent parental efforts, then males may share weak relationships with females, and be socially more attracted to their young.

There are several reasons why a cohesive social relationship between male and female may accompany infanticidal defense. First, if paternity improves male defense of an infant, then a prior association extending into prenatal periods may be selected for. Second, affiliation between mother and protector may facilitate infant protection. In mice, even modest postcopulatory cohabitation with a female markedly enhances the chances that a male will care for her offspring (Elwood 1989) rather than kill them (Soroker & Terkel 1988; Gubernick et al. 1994). Third, ongoing social relationships may be critical for maintaining male defense at maximally effective levels. In non-human primate aggression, individuals preferentially aid those with whom they have interacted affinitively in the *recent* past (Harcourt 1992). Dunbar (1980) invoked the generality of this principle in organizing primate social behavior by arguing that strong social bonds between cercopithecine mothers and daughters are the reason for the elevated frequency of coalitionary support between them. Thus, among gregarious animals maintaining dynamic, differentiated social relationships with one another, a female's ability to garner infanticide protection will often vary with her current social relationships with prospective male defenders. Females will therefore invest in social relationships with these males (Hrdy 1979; Birkhead & Møller 1992).

The economics of pair bonds and infanticide defense: benefits to females and males

I assume that male–female bonds will evolve as an anti-infanticide strategy when the benefits of the constituent behaviors exceed their costs. Limitations of space preclude consideration of costs, but attention to alternative benefits generates the competing functional hypotheses against which the infanticide protection hypothesis is tested.

Alternative potential benefits to females

The infanticide protection hypothesis identifies deterrence of infant-killing as the adaptive advantage that anestrous females gain from associating with males. Competing benefits to females derive from two sources: other forms of infant care or services provided directly to mothers (Table 11.1).

Paternal care with an immediate, physical impact on young provides examples of infant-specific benefits. These include male thermoregulatory incubation or huddling, provisioning and food sharing, carrying,

Table 11.1. *Hypothetical benefits of male–female bonds to females*

Hypotheses based on protection	
Infanticide protection hypothesis	Male protection of young from infanticide
Predation protection hypothesis	Male protection of young and/or females from predation
Female harassment hypothesis	Male protecton of young and/or females from harassment by female competitors
Sexual coercion hypothesis	Male protection of females from harassment from adult or subadult males
Other hypotheses based on infant care	
Alternative male care hypothesis	Direct, extensive male care of infant (e.g., provisioning)
Male–infant bonding hypothesis	Male–female bonds facilitate development of male–infant social bonds, which generates post-weaning benefits for juveniles

retrieving, socialization, and cleaning or grooming (Kleiman & Malcolm 1981; Woodroffe & Vincent 1994). This "alternative male care" hypothesis typically emphasizes these forms of extensive, direct male care, partly because indirect contributions (such as resource defense, territoriality, maintenance of nests or burrows, and "vigilance" behavior) may benefit mothers and the males themselves, besides infants.

A variant of this hypothesis applies to gregarious animals with a prolonged period of male–young interaction: females (and infants) initially derive no substantive benefits – defensive or otherwise – from male companions. Rather, the "male–infant bonding" hypothesis argues that a female associates with a male to promote the development of social attachment between him and her infant, which later benefits the juvenile through post-weaning male attention, e.g., intervention in competition with peers, protection from predators, etc. (Ransom & Ransom 1971; Seyfarth 1978; Nicolson & Demment 1982; Stein 1984b; Smuts 1985; Collins 1986; Strum 1987).

A second group of hypotheses posits male protection against contingencies other than infanticide (Table 11.1). Protective vigilance against predation has been advanced as the primary benefit that females derive from associating with males in some birds (Hannon 1984; Dahlgren 1990), non-human primates (van Schaik & van Noordwijk 1989; Baldellou & Henzi 1992; Rose 1994; Rose & Fedigan 1995), and other mammals (Burger & Gochfeld 1994). Alternatively, male associates may shield mothers from elevated rates of agonistic interaction with (high-ranking) female competitors (Altmann 1980) while simultaneously guarding infants from rough handling and kidnapping by these dominant females (Silk 1980; Collins *et al.* 1984; Rhine *et al.* 1988; Brain 1992; Maestripieri 1994). Finally, her bond with a male may buffer a female from harassment from adult males (Smuts & Smuts 1993). Gowaty & Buschhaus (1998) argued that pair-bonded monogamy may be an adaptive female response to the "dangerous environment" created by male aggressiveness toward them, particularly in sexual contexts. It is important to differentiate between infanticide and male sexual coercion of females, the latter of which is aggression directed explicitly at females to increase the probability of copulating with them (Clutton-Brock & Parker 1995).

Insofar as benefits apply exclusively to females, they may become crucial for mothers with dependent young for two reasons. First, females may become increasingly frequent targets of harassment or predation at this time. Second, the rate and intensity of attack may remain unaltered, but these interactions become effectively costlier to females in the context of energetically demanding lactation (Gittleman & Thompson 1988; Rogowitz 1996).

Alternative potential benefits to males

Parental or mating benefits are likely to account for male defense against infanticide (Table 11.2). The "paternal investment" hypothesis states that males increase the survival of offspring they have sired by protecting them from infanticidal attack (Palombit *et al.* 1997). The "female choice" hypothesis suggests that male care of an infant is a mating strategy that increases his chances of copulating with the mother in future reproductive cycles (Smuts & Gubernick 1992; van Schaik & Paul 1996–7; Freeman-Gallant 1997) or with additional females that observe his behavior (Gori *et al.* 1996). Females preferentially mate with these males to procure this care or because it signals a male's high genetic quality as mate (Freeman-Gallant 1996). Explanations emphasizing "parental effort" versus "mating effort" are not mutually exclusive, but the latter is obviously relevant in situations where males enhance their fitness through care of unrelated infants.

This dichotomy subsumes several other alternative hypotheses for postcopulatory association between a male and female with dependent offspring (Table 11.2). An alternative paternal investment model simply replaces infanticide defense in the above hypothesis with some other

Table 11.2. *Hypothetical benefits of male–female bonds to males*

Hypotheses based on parental effort	
Paternal investment hypothesis	Male invests in caring for and/or protecting offspring he has sired
Hypotheses based on mating effort	
Sperm contribution hypothesis	Male continues mating with female and thereby fertilizes replacement clutches in current reproductive cycle
Female choice hypothesis	Male attracts current female companion and/or additional females as mates in future reproductive cycles
Paternity guard hypothesis	Male guards his breeding and/or mated status
Pursuit of extra-pair copulation hypothesis	Male is locatable for (extra-pair) copulations with other females
Hypotheses based on non-reproductive social benefits	
Social tool hypothesis	Male exploits females and infants as social tools in agonistic interactions with other males
Immigration hypothesis	Immigrant male acquires stable residency in group through relationships with female(s)
Female allies hypothesis	Male obtains coalitionary support from female in agonistic interactions

form of paternal investment, e.g., provisioning of young, protection against predation or conspecific harassment, etc.

Likewise, males may derive several other mating benefits, besides female choice, through extended association with females. A distinctive feature of the following hypotheses is that they do not necessarily assume that females also benefit in any way from social relationships with males. Indeed, they allow for the possible *reduction* of female fitness from post-copulatory association with a male.

The "sperm contribution" hypothesis suggests that a male continues associating with a female mate in order to increase paternity of replacement clutch(es) should her first one fail. Repetitive copulation with the female achieves this goal by increasing his relative contribution to the sperm that she stores.

Avian studies reveal that the conspicuous proximity of socially monogamous males and females is frequently a paternity guard (Birkhead & Møller 1992). This hypothesis suggests that guarding by males may function genetically during fertile periods to prevent extra-pair fertilizations, or socially during both fertile and non-fertile periods to preserve mating status (Slagsvold & Lifjeld 1997).

The "pursuit of extra-pair copulation" hypothesis argues that a male maintains a postcopulatory bond with his partner because other females seeking extra-pair copulations can locate him more easily (Gowaty 1996a,b).

A final group of hypotheses focuses on potential social benefits of non-reproductive bonds with females (Table 11.2). Males may derive long-term strategic benefits in male–male competition through: (1) exploitation of female associates and/or their infants in triadic interactions; (2) coalition-ary support from females; or (3) facilitation of immigration into new groups (Smuts 1983; Strum 1983).

Empirical studies: infanticide as a cause for male–female bonds

Below I focus on research examining the predictions of the Infanticide Protection hypothesis and alternative hypotheses in a number of relatively well-studied taxa.

Insects

The mating and social system of the burying beetle (Coleoptera: Silphidae, *Necrophorus* spp.) is exceptional among insects in combining prolonged postcopulatory association of the sexes, social monogamy, extensive biparental care of young, and sexually selected infanticide (Eggert & Müller 1997). A male and female jointly prepare a small verte-brate carcass that constitutes the sole food source for developing larvae. Both sexes generally remain with the brood for the 1–4 weeks of larval development (Scott & Traniello 1990; Trumbo 1991), feeding them by regurgitation of pre-digested food, and helping them access the interior of the "carrion ball" (Eggert *et al.* 1998). Adults also actively defend the carcass, which is vulnerable to takeover by conspecifics of either sex (Scott 1990; Trumbo 1990a). If a takeover occurs, the intruding conspecific replaces the former resident of the same sex, kills larvae that are present in the nest, and pairs with the remaining adult to produce a new clutch.

Studies of the burying beetle illustrate how multiple lines of observational and experimental evidence can be marshaled to support the infanticide protection hypothesis and reject other possibilities. First, carcasses defended by a single female are significantly more vulnerable to takeover infanticide than those defended by a pair of adults (Trumbo 1990a,b, 1991; Robertson 1993) (Table 11.3). Second, although parental investment in the

Table 11.3. *Field removal experiments in species with sexually selected infanticide*

Organism	Social/mating system	Experimental results	Source
Burying beetle (*Necrophorus orbicollis*, *N. vespilloides*)	Social monogamy/ biparental care	Single females were significantly more vulnerable to takeover infanticide than heterosexual pairs; experimentally widowed females produced fewer offspring than control females due to reduced initiation of replacement clutch	Trumbo 1990a;[a] Robertson 1993;[a] Sakaluk et al. 1998
Orange-tufted sunbird (*Nectarinia osea*)	Social monogamy/ biparental care	After a single territorial male was removed from his nest, an intruder male appearing the next day interfered with female attempts to feed chicks, and attacked fledglings that left the nest; the detained territorial male was released on this day, whereupon it chased the intruder male off in an aggressive fight	Goldstein et al. 1986
Orange-tufted sunbird (*Nectarinia osea*)	Social monogamy/ biparental care	Females made compensatory increases in provisioning, but not in nest defense; nests containing three chicks suffered greater mortality due to predation, not infanticide	Markman et al. 1996
Tree swallow (*Tachycineta bicolor*)	Social monogamy	Replacement males arriving during egg-laying period generally adopted chicks, but 79% of those arriving during incubation and 71% of those arriving during nestling period killed young	Robertson 1990; Robertson & Stutchbury 1988
Starling (*Sturnis vulgaris*)	Social monogamy	Males adopted eggs if replacement occurred before egg-laying period, but destroyed eggs if replacement occurred during female's egg-laying period; 4 out of 6 females affected by infanticide produced replacement clutches with replacement male	Smith et al. 1996
Barn swallow (*Hirundo rustica*)	Social monogamy / coloniality	Males detained for less than a day at onset of nestling period (cf. 20 controls). In colonial nests, infanticide occurred among those widowed females that did not show compensatory increase in nest guarding (i.e., 40% of experimental nests). No infanticide in any control nests or following male detention among relatively more dispersed, solitarily breeding pairs. Solitary nests: no nestlings were killed (experimental or control nests); replacement males did not arrive	Møller 1988

Species	Mating system		Reference
House sparrow (*Passer domesticus*)	Facultative social monogamy/polygyny	Widows that obtained a new mate and remained socially monogamous suffered infanticide from replacement male; widows that became new mates of a polygynous male suffered infanticide from other females in the breeding unit	Veiga 1992
House wren (*Troglodytes aedon*)	Monogamy	Males removed 4–5 days post-hatching; no replacement males appeared, no infanticide	Bart & Tornes 1989
House wren (*Troglodytes aedon*)	Bigamy/monogamy	63% of replacement males in late incubation removed offspring from nests and most subsequently fledged offspring (remaining 37% failed to commit infanticide or to breed)	Kermott & Johnson 1990; Kermott *et al.* 1991
Wattled jacana (*Jacana jacana*)	Polyandry	Replacement females appeared on territories of "widower" males the day after female removal and aggressively attacked chicks; 7 of 9 of the chicks eliminated within 48 h of female removal	Emlen *et al.* 1989
Arctic ground squirrel (*Spermophilus parryii*)	Polygyny/coloniality	Following male removal at beginning of birth season, infanticide occurred when incoming intruder males were unlikely to have sired the young; overall reproductive success did not differ between experimental and control females	McLean 1983
Hanuman langur (*Presbytis entellus*)	Polygyny/one-male groups	After a single territorial male was removed, the male of adjacent group fatally attacked 4 infants, whose mothers "desert" them and did not defend; male attempted to integrate male-less group with his own group, but failed	Sugiyama 1967

Notes:
[a] These studies did not involve experimental removal of an adult from established pairs; rather, single females and heterosexual pairs were established on carrion nests and monitored for takeover infanticide.

form of larval provisioning and carcass maintenance enhances offspring survival (Eggert *et al.* 1998), the male contribution is not critical. Broods cared for by a single female survive and grow as well as do those receiving biparental care (Bartlett 1988; Scott 1989; Reinking & Müller 1990; Trumbo 1991; Müller *et al.* 1998). Third, extended residency of a male is unlikely to significantly increase his copulatory success, partly because the resident female physically interferes with his attempts to pheromonally attract additional mates (Trumbo & Eggert 1994; Eggert & Sakaluk 1995). Moreover, extended residency does not improve male reproductive success through enhanced paternity of his own mate's replacement brood should her first clutch fail: males removed before the initiation of replacement clutches had the same high paternity of replacement broods as males permitted continued mating opportunities with their mates (Sakaluk *et al.* 1998). Finally, many experimentally widowed females fail to produce a replacement clutch at all. These females may curtail reproduction because the disappearance of the male reflects his low quality or vigor as a sire, but Sakaluk *et al.* (1998) argued that it is more likely that loss of the male signals an imminent takeover, prompting the female to postpone production of offspring likely to be killed in favor of conserving the carrion ball for a replacement clutch eventually fathered by the intruder male.

There are multiple adaptive reasons for the extended residency of male *Necrophorus* on carcasses, for populations vary considerably in rates of takeover and replacement brood production, as well as in the availability of carcasses and additional female mates (Scott 1990, 1994; Sakaluk *et al.* 1998). Nevertheless, there is compelling evidence that male–female association – and the consequent social monogamy – have evolved as an anti-infanticide strategy in at least some populations.

Birds

Contemporary research has confirmed Darwin's (1871) thesis that significant intrasexual breeding competition may occur under monogamy, due to factors such as an early mating advantage, skewed sex ratios, and (though not fully appreciated by Darwin), extra-pair paternity and exclusion of "floaters" from reproduction (Andersson 1994; Palombit 1999; Veiga, Chapter 9) The incidence of sexually selected infanticide hints at a largely unstudied potential to influence the evolution of avian behavior, e.g., nest guarding, dispersal, response to extra-pair conspecifics, and social monogamy itself. Arguments that pair-bonded monogamy predominates in birds because of a conspicuous avian potential for biparental care have traditionally emphasized "conventional" forms of care, such as incubation and provisioning (Mock & Fujioka 1990). Freed (1986) was the first to propose that infanticide defense is the crucial male contribution in some populations.

The potential deterrent value of males may be suggested by observations of their nest defense and by their absence preceding successful infanticides (e.g., in 11 of 14 attacks in barn swallows (*Hirundo rustica*); Møller, 1988). Controled male removal or detention experiments measure more rigorously his effect on infanticide risk (Table 11.3). Typically, one or more intruder males (usually floaters) appear on the experimentally "widowed" female's territory within days or even hours after removal, but consequences for offspring depend upon the timing of replacement in the reproductive cycle. If replacement occurs before, and in some cases, during egg-laying, adoption of offspring is probable, but if replacement takes place during the incubation or nestling periods, infanticide becomes increasingly likely in some populations (Palombit 1999).

Study of the barn swallow underscores the importance of intervening variables – such as maternal defense and the spatial distribution of nests – in determining the value of male infanticide defense (Møller 1988). Among colonially nesting individuals, temporary detention of mated males at the onset of the nestling period precipitated immediate infanticide by unmated males, but only among those widowed females that failed to make compensatory increases in nest guarding. Among non-colonial females breeding in solitary heterosexual pairs, infanticide never followed male removal because replacement intruder males failed to settle subsequently on widowed females' territories. Thus males may help females to achieve a level of nest guarding that forestalls infanticide, without always being strictly necessary to do so.

The killing of young after experimental removal of the male strongly implicates reduced infanticidal risk as the adaptive benefit of social monogamy to females. Additional data are necessary to reject alternative hypotheses, however, as exemplified by numerous studies of the house wren (*Troglodytes aedon*). Individuals of this widely distributed passerine live in permanent socially monogamous pairs in tropical South America, and are facultatively bigamous and monogamous in northern populations. Infanticide is part of a common takeover strategy by unmated floater males (and females) in tropical latitudes (Freed 1986, 1987), and follows mate replacement in temperate populations (Kermott & Johnson 1990; Kermott *et al.* 1991; Table 11.3).

Freed (1986) argued that the need for infanticide defense by males maintains social monogamy in tropical house wrens. Two likely other benefits that males provide to females are anti-predator protection and food provisioning of nestlings. Predator defense was unimportant in Freed's study, since avian predators were absent and since subjects' nest boxes were inaccessible to terrestrial predators. Johnson & Albrecht (1993) also rejected the predator protection hypothesis for temperate house wrens by demonstrating experimentally that, although bigamous males spent much less time near the nests of their secondary females than monogamous males did near their mates' nests, both types of males detected and defended against small diurnal predators equivalently. Johnson & Albrecht (1993) warned, however, that this result does not imply that monogamous and polygynous males are equally effective in detecting and ousting *conspecific* (potentially infanticidal) floater males, since "this aspect of male care is probably less 'shareable'" than anti-predator defense. Evidence against the alternative male care hypothesis comes from removal experiments conducted late in the nestling period. These demonstrated that male food deliveries enhance offspring survival only in poor years (Bart & Tornes 1989); under average conditions, the male's participation seems to make no difference (Mock & Fujioka 1990; but see Johnson *et al.* 1992). On the other hand, the higher fledging success of primary females whose provisioning of young is aided by bigamous males would seem to support the alternative male care hypothesis. Johnson & Kermott (1993), however, attributed the lower reproductive output of secondary females to the greater need for them to stay near nests and guard young against infanticide, which severely limits food deliveries. Finally, greater extra-pair paternity among the offspring of secondary females (compared with primary females) suggests that secondary females compensate for reduced infanticidal defense from the male by pursuing an alternative strategy based on confusing paternity among potentially infanticidal males (Soukup & Thompson 1997). Although these last two hypotheses await rigorous testing, the house wren provides another case in which several pieces of evidence oppose different hypotheses in favor of the conclusion that male–female bonds have evolved as an anti-infanticide strategy.

Rodents

Infanticide is widespread in rodents (Blumstein, Chapter 8), but evidence that it has selected for female association with male defenders is meager. Even in species where heterosexual relationships and infanticide co-occur, an adaptive link between the two often remains vague. For example, deer mice (*Peromyscus maniculatus nubiterrae*) show greater pair-bond cohesion and male involvement with pups than do white-footed mice (*P. leucopus noveoboracensis*), but the two species are similar in infanticidal behavior (Wolff & Cicirello 1991) as well as in the effectiveness of maternal aggression in successfully forestalling male infanticide (Wolff 1985). In these species, the data are more consistent with the alternative male care hypothesis than the infanticide protection hypothesis.

Nevertheless, infanticide has been offered as the adaptive reason for male–female bonds in a few species such as the Malagasy giant rat (*Hypogeomys antimena*) (Sommer 1997) and, more notably, the Arctic ground squirrel (*Spermophilus parryii*). In Arctic ground squirrels, immigrant males that establish residency at the end of the mating period commit infanticide. McLean (1983) presented four results to support his claim that infanticide has selected for the extended residency of male protectors: (1) resident males were territorial primarily when pups were most vulnerable to infanticide and when food was abundant, suggesting that male territoriality has not evolved to improve female access to limited resources; (2) males' greater amount of time devoted to alert postures and exploration compared to females suggested a "lookout" role; (3) infanticide was less common when a breeding male defended a territory during the ensuing period of pup vulnerability than when he did not; (4) male removal elicited infanticide, and mean litter size at emergence among experimental colonies was significantly lower than in control areas (Table 11.3) (though the specific causes of smaller litters were unknown, and replication a year later failed to generate similar results). Evidence against the potentially competing predator protection hypothesis is unavailable.

Numerous experimental studies in captivity have measured rates of infanticide by males introduced into cages with females and pups (Blumstein, Chapter 8). Parmigiani *et al.* (1994) speculated that the frequently observed failure of maternal aggression to deter infanticide among captive rodents such as *Mus domesticus* may be due partly to the artificial absence of the stud male. In the natural setting, he might be attracted to intervene by the female's conspicuous defensive behavior. Experiments systematically varying the availability of stud males are needed to test this hypothesis.

Non-human primates

Two aspects of their biology have made primates the most frequent subjects of tests of the infanticide protection hypothesis. First, permanent association between males and anestrous females is more common than in mammals generally, characterizing almost all anthropoids and many prosimians (van Hooff & van Schaik 1992; van Schaik 1996). Second, male infanticide occurs in many polygynous species, and is best understood as a sexually selective reproductive strategy in most, but not all cases (Hiraiwa-Hasegawa & Hasegawa 1994; Hrdy *et al.* 1995).

Socioecological models have traditionally viewed heterosexual relationships in primates simply as a secondary outcome of evolutionary forces shaping female–female and male–male social interactions (Smuts 1987b; van Schaik 1996). Largely in the last decade, however, infanticide has been offered as the adaptive reason for male–female bonds in diverse taxa such as capuchin monkeys (*Cebus olivaceus*) (O'Brien 1991), Malagasy lemurs (van Schaik & Kappeler 1993), humans (*Homo sapiens*) (Smuts 1992), mountain gorillas (*Gorilla gorilla berengei*) (Wrangham 1979), gibbons (*Hylobates* spp.) (van Schaik & Dunbar 1990), and chacma baboons (*Papio cynocephalus ursinus*) (Palombit *et al.* 1997), as well as in gregarious primates universally (van Schaik & Kappeler 1997).

Mountain gorillas

Infanticide defense has long been considered the adaptive reason for group living in polygynous mountain gorillas (Wrangham 1979). Several observations suggest a protective role of the resident male in one-male groups: (1) infanticide accounts for at least 38% of infant mortality and occurs primarily after this male dies or disappears (Fossey 1984; Watts 1989); (2) strong heterosexual bonds form the core of the group (Stewart & Harcourt 1987; Watts 1990, 1996); and (3) females transfer between groups when they are near one another, thereby apparently avoiding the dangers of being alone (Harcourt 1978). Indirect data appear to reject the other two likely hypotheses for mountain gorilla sociality. First, the relative abundance and even distribution of their folivorous food resources implies an absence of feeding-related benefits of grouping to females (Wrangham 1980), and, second, immense body size should largely eliminate predation risk (Wrangham 1979). Stewart & Harcourt (1987), however, favor renewed and rigorous testing of competing hypotheses by arguing that protection from predators (leopards) is just as important as infanticide defense in promoting male–female bonds.

Chacma baboons

The chacma baboon has provided an opportunity to test more directly the infanticide protection hypothesis against other possibilities. As in savanna baboons generally (Ransom & Ransom 1971; Altmann 1980; Anderson 1983; Strum 1982; Smuts 1985; Collins 1986; Bercovitch 1991), lactating females maintain close bonds with an unrelated adult male (or two) (Seyfarth 1978). Because a single dominant male does not attract all the females of the group (as in capuchin monkeys and mountain gorillas), a chacma baboon group at a given time includes several essentially dyadic pair bonds, or "friendships" (Strum 1974), based on pronounced spatial proximity and high rates of affinitive interactions, such as grooming and infant handling. Before birth and soon after an infant's death, there is no manifestation of the "special relationship" that exists between a particular male and female during lactation (Palombit *et al.* 1997). The strict temporal conjunction of these bonds with neonate presence suggests that the benefits of friendships to females involve infant viability and survival.

An especially clear possibility in the chacma baboon population of the Moremi Game Reserve, Botswana, is that males defend the offspring of their female friends from infanticide, which accounts for at least 37% of infant mortality and appears to be a sexually selected strategy of new immigrant alpha males (Palombit *et al.*, Chapter 6). Studies of East African conspecifics, among which infanticide is exceptional (Collins *et al.* 1984), provide the alternative protective benefits of male friends (Table 11.1). They may safeguard females from postpartum agonistic interaction or semi-abusive infant "handling" from higher-ranking females (the female harassment hypothesis) or from (non-infanticidal) aggression by adult and subadult males (the sexual coercion hypothesis). A final, fourth hypothesis is that friendships do not confer any protective benefits to infants (or females) at all, but instead function as a "maternally induced" mechanism facilitating male–infant affiliation that persists into the juvenile period (the male–infant bonding hypothesis).

A crucial question has always been whether the friendship that develops between a female and a male enhances his willingness to defend her when she is attacked, as predicted by the three protection-related hypotheses. Palombit *et al.* (1997) examined this question through naturalistic observations and field playback experiments that exploited the considerable variation in male chacma baboons' orienting responses to the screams females typically give when attacked. In these experiments, we played a female's scream to the male friend and, at another time under similar conditions, to a control male who was of similar rank and also had a female friend in the group. Male friends responded significantly more strongly than control males (Figure 11.1a). Because the visual scanning elicited by a vocalization generally reflects the listener's investment in obtaining further information about the specific circumstances surrounding vocalizing (Marler *et al.* 1992), these results support the prediction that friendships predispose males to aid particular females under attack. This predisposition, however, depended upon the presence of infants. When the experiment was repeated in the period immediately following the death of a female's infant, male friends' responses were significantly weaker than control males (Figure 11.1b). A final series of experiments in which the threat calls of the simulated attacker accompanied the victim's screams suggested that male friends' responded more strongly than controls primarily when the aggressor was an infanticidal male, rather than a high-ranking female or a non-infanticidal male. In light of observational data showing that lactating females do not avoid interacting with higher-ranking, unrelated females, the combined results suggest that chacma baboon friendships confer protection-related benefits to females, and that this protection is more likely to operate in the infanticidal context rather than other harassment situations. Thus the chacma baboon provides a case where sexually selected infanticide accounts for a significant portion of infant mortality, males and females (with dependent infants) share close bonds with one another, and these friendships apparently function to reduce infanticide risk.

Gibbons

Infanticide defense has also been offered as the adaptive reason for social monogamy among gibbons. Van Schaik & Dunbar (1990) argued that individual male gibbons are capable of maintaining home ranges large enough to encompass 2–13 females, but decline to do so because a polygynous strategy would leave females vulnerable to infanticide by other males. Field studies of gibbons have generated few directly relevant data and no reports of infanticide with which to test the infanticide protection hypothesis. Comparative analysis, however, allows a preliminary test: are patterns of hylobatid behavior and ecology consistent with those observed in other primates as well as socially monogamous birds in which sexually selected infanticide and male–female bonds seem to be functionally interrelated?

Avian studies provide the comparative data base – currently lacking in

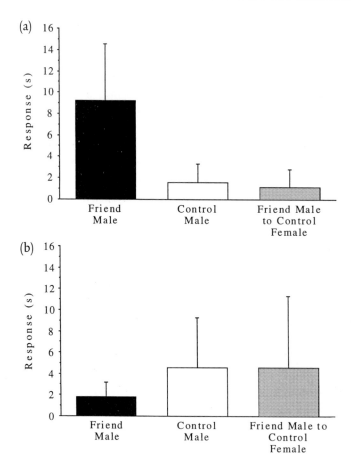

Figure 11.1. Results of playback experiments of female distress calls to males in chacma baboons. Black bars and white bars are the response of friend males and control males to playback of a particular female's scream. Gray bars indicate the friend male's response to the scream of a control (non-friend) female of similar rank and reproductive status. (a) Results during periods when female friends had infants. These indicate that male friends responded significantly more strongly than control males, and that males responded significantly more strongly to screams of female friends than to those of control females ($P < 0.01$, $N = 30$ experiments distributed equally across each condition). (b) Corresponding results but during the period immediately following the death of an infant. These indicate that male friends responded significantly less strongly than control males ($P < 0.05$), and that there was no difference in male response to the screams of female friends and control females ($N = 18$ experiments distributed equally across each condition). Response is measured as the duration of orienting toward the concealed playback speaker in the 20 s following the playback minus the duration of orienting in the same direction in the 20 s preceding the playback. Data are taken from Palombit *et al.* (1997).

mammalian research – that specifies the ecological and demographic conditions that intensify intrasexual selection among socially monogamous males and, ultimately, promote infanticide as well as anti-infanticide pair bonds (Table 11.4). Individuals of tropical populations such as house wrens are particularly relevant, for (like gibbons) they maintain permanent territories and pair bonds, breed throughout most of the year, *and* (unlike gibbons) they are known to commit infanticide.

Although aseasonal breeding and long interbirth intervals establish a clear reproductive potential for sexually selected infanticide (Palombit 1995), a fundamental demographic cause of avian infanticide appears to be lacking in gibbon populations: an abundance of unmated, fully adult "floater" males. In avian populations with sexually selected infanticide, unmated adult males, such as single floaters and/or first-time breeders, are relatively numerous (Crook & Shields 1985; Møller 1988; Veiga 1992) and account for most observed infanticides (Freed 1986; Goldstein *et al.* 1986; Møller 1988; Robertson 1990; Hotta 1994). This result is consistent with sexual selection theory. For already paired males, infanticide primarily offers an opportunity to *switch* mates. Even when bigamy is possible, as in temperate house wrens, mated males rarely use infanticide to acquire a second mate (Kermott *et al.* 1991). For unmated floaters, however, the potential value of infanticide is dramatically greater. These males have few reproductive options, not only because they lack pairmates, but also because females generally prefer mated males as extra-pair sexual partners (Birkhead *et al.* 1986). Thus infanticide as part of a takeover strategy may alter an unmated floater male's breeding status entirely. In summary, high densities of adult floaters are responsible for intensifying intrasexual mating competition and thereby generating the potential reproductive benefits of infanticide for socially monogamous male birds (Table 11.4).

These avian data provide a general biological context for evaluating the hypothesis that female gibbons are "vulnerable to infanticide by males attempting a takeover or by neighboring males attempting to expand their territories, but also by transient males" (van Schaik & Dunbar 1990). If gibbons commit infanticide, then, like birds, the cost–benefit ratio of infant-killing should be less favorable for a monogamous male that already possesses a mate and breeding territory than for a male that lacks one or both these prerequisites for reproduction. The absence of observations of infanticide in gibbons precludes direct testing of this assumption, but corroborating data come from field experiments

Table 11.4. *Conditions favoring sexually selected infanticide in socially monogamous birds, with implications for gibbons (from Palombit 1999)*

	Condition favoring infanticide in avian populations	Does condition apply to hylobatid populations?
Features of female reproduction that create potential for infanticide	Asynchronous breeding	Yes
	Long breeding season	Yes (aseasonal breeding)
Factors that reduce the costs of infanticide	Offspring often left unattended due to weak pair bond or demanding nest provisioning	No
	Coloniality/gregariousness	No
	Limited opportunity or effectiveness of maternal aggression	No
	Inexperienced ("first-time") female breeders abundant	Unknown
	High mortality among mated adults	No
	Low relatedness among (colony) neighbors	Unknown
Factors that increase the benefits of infanticide to males (by limiting breeding opportunities)	Large adult male floater population	No
	Skewed sex ratio	No
	High mortality among unmated adults	Yes (promotes "takeover" infanticide)
	Low adult mortality	Yes (promotes "takeover" infanticide)
	Low opportunity for extra-pair copulation	Unknown
	High competition for nests sites (e.g., secondary cavity nesters)	Yes: gibbons lack nest sites, but competition for territories is probably intense
	Females can not mate with males other than the infanticidal male	Unknown

in which Mitani (1984, 1985, 1987) simulated the presence of mated pairs, solitary females, and solitary males by playing back their respective songs to conspecific agile (*Hylobates agilis*) or Mueller's gibbons (*H. muelleri*). Van Schaik & Dunbar (1990) maintained that these experiments reveal a decreased willingness of mothers with dependent infants to participate in territorial interactions (as compared with non-lactating females), which they attributed to the attendant risks of male infanticide. Mitani (1984), however, emphasized the "clear and striking similarity" of female responses to the duets of mated pairs and to the songs of solitary females. Among his agile gibbon subjects, this was true for females with and without infants. If risk of infanticide motivates female response under these playback conditions, then Mitani's results suggest that mated males pose a relatively low risk of infanticide (i.e., roughly equivalent to that presented by solitary, unmated females). Also, Mitani's data suggest that female agile gibbons with infants are equally wary of interacting with unfamiliar, solitary adults of *either* sex, contrary to the predictions of the infanticide protection hypothesis in its current form. (A possible alternative explanation for both patterns is, of course, that *females* also pose an infanticidal threat.)

This difference in infanticidal behavior of unmated versus mated males is important because currently available data suggest that adult floaters of either sex are rare in most gibbon populations monitored for more than 3 years (Chivers & Raemaekers 1980; Mitani 1990; Palombit 1994a; Brockelman *et al.* 1998; but see Tilson 1981). Leighton (1987) and Mitani (1990) have consequently argued that dispersing young adults suffer high mortality. Observations that numerous breeding vacancies occurring over 6 years in hylobatid groups in Sumatra were all filled by known residents of neighboring groups, not by floaters, provide further evidence for the rarity of floaters (Palombit 1994a). Finally, a "natural experiment" transpired when two contiguous siamang territories became available for colonization after all residents died or dispersed following contraction of disease by adults (Palombit 1994a). Siamang (*Hylobates syndactylus*) did not re-occupy the total area of over 80 ha of productive lowland rainforest in the ensuing 5 years (or more), even though sympatric primates such as longtailed macaques, orang-utans and white-handed gibbons (*H. lar*) continued to live and reproduce successfully there. In most bird populations, analogous provisioning of nest boxes or removal of males rapidly exposes the presence of unmated floaters who seize the proffered reproductive opportunity. In sum, hylobatid data suggest that physically mature, unmated adult males are extremely rare, or at least rarer than in many birds with infanticide. If future demographic data confirm this pattern, then the potential threat of sexually selected infanticide to females (mated or widowed) may be lower in gibbons than it is in those socially monogamous birds in which infanticide occurs.

Most gibbon populations reveal low or "near-zero" infant mortality (Leighton 1987; Mitani 1990; Reichard & Sommer 1997: 1135), suggesting a limited role of infanticide in maintaining pair-bonded monogamy relative to other potential sources of mortality. It is unlikely that infant mortality is low in hylobatid populations simply because male protectors accompany most females. If the infanticide protection hypothesis is correct, we expect that adult male gibbons – like their catarrhine and avian counterparts – will vary in their ability and/or propensity to protect young from attack, and that this will translate directly into *variable* risk of death for young. Moreover, the argument that observations of infanticidal attacks are rare because widowhood is infrequent overlooks the fact that, if infanticide occurs at all in gibbons, it is more likely to assume a *takeover* than a passive replacement form. The high mortality of unmated males combined with low mortality among mated, territorial males should select forcefully for takeover infanticide (Table 11.4). The limited opportunity for subadult hylobatids to postpone dispersal by helping adults to raise offspring (as cooperatively breeding birds do) should intensify selection for takeover infanticide. We cannot easily predict the rate of takeover infanticide in hylobatids, but opportunities for it should be more common than for passive replacement infanticide. Nevertheless, takeover infanticide is undocumented, probably because an important causal agent suggested by avian studies is lacking, i.e., a relatively large population of fully adult, reproductively disenfranchised floater males. Young adults still residing in their (putative) natal groups are relatively abundant and might functionally constitute a floater population, but their invariable physical immaturity makes them less likely perpetrators of takeover infanticide than fully adult male floaters.

Comparative social data from primates with infanticide and "anti-infanticide" male–female bonds additionally oppose the infanticide protection hypothesis for gibbons. Two well-studied primate systems support Hrdy's (1979) and Birkhead & Møller's (1992) prediction that a female will invest substantially in a social relationship with a male who will sustain the costs of defending her offspring against infanticide.

The male–female bonds constituting chacma baboon friendships (Palombit *et al*. 1997) and one- and multimale groups of mountain gorillas (Harcourt 1979b; Sicotte 1994; Watts 1996) rely upon a greater female contribution to two pivotal mechanisms used by primates to "service" their social relationships (*sensu* Hinde (1983) and Dunbar (1988)): close proximity and allogrooming. A lactating female is more responsible than the male for maintaining the close proximity they enjoy, and she grooms him more than he grooms her. It is not the case, however, that these patterns result from a general social indifference among males; they actively maintain social relationships with *estrous* females (Sicotte 1994; R. A. Palombit, unpublished data). Similar grooming patterns incidentally characterize howler monkeys, *Alouatta* spp., where infanticide and male–female association also co-occur (Crockett & Eisenberg 1987; Chiarello 1995). Independent support for the general prediction that females will invest heavily in bonds with males bestowing a fitness-enhancing service emerges from grooming data between paired titi monkeys (*Callicebus torquatus*), a socially monogamous species in which males extensively carry infants soon after birth (Figure 11.2).

Relevant field data on hylobatid pair bonds are few, but Palombit's (1996) study of sympatric white-handed gibbon and siamang provides a starting point partly because it employed several of the same behavioral measures described above for chacma baboons and mountain gorillas. These demonstrate that male hylobatids are responsible for proximity maintenance to the same degree that female chacma baboons are with their male friends (Palombit 1999). Essentially the same is true of allogrooming between white-handed gibbon pairmates: males perform the vast majority of within-pair grooming (Figure 11.2) largely because females rarely initiate grooming sessions, routinely solicit grooming at higher rates, and ignore male solicitations for reciprocated grooming. Among siamang, grooming exchanges tend be more reciprocal, but where an asymmetry exists, it is, again, the male that contributes more.

The available data, therefore, suggest that the contributions of the sexes to maintaining pair bonds in hylobatids, particularly in white-handed gibbons, are inconsistent with the infanticide protection hypothesis and contrary to patterns exhibited in chacma baboons and mountain gorillas, where male–female association implicates infanticide defense. Further tests require data from a larger number of hylobatid pair bonds, and should also consider whether defense-inspired male–female bonds develop differently in socially monogamous versus polygynous settings.

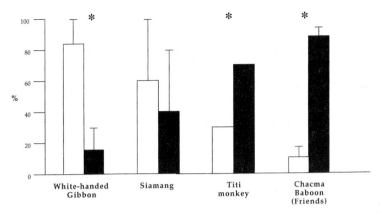

Figure 11.2. Sex differences in grooming investments of bonded adult males and females. Shown is mean (with SD where available) percentage of within-pair grooming performed by the male (white) and the female (black) for siamang and white-handed gibbons (Palombit 1996), titi monkeys (Kinzey & Wright 1982) and chacma baboons (Palombit *et al.* 1997). A star indicates a statistically significant difference ($P < 0.05$).

For example, compared with gibbons, female chacma baboons and mountain gorillas may be more vulnerable to coercion from a (much larger) male protector, whom they may attempt to mollify through a more cohesive social bond. Also, females in the two polygynous species may invest more in proximity maintenance because their male companions are frequently pursuing copulations with other females, which may motivate the behavior of male hylobatids less frequently (but see Palombit 1994a,b and Reichard 1995) or because of intense female–female competition (Palombit *et al.* 2000). These possibilities may account for some patterns of female behavior in chacma baboons and mountain gorillas, but they fail to explain why male white-handed gibbons' investment in pair bonds exceeds rather than equals that of their female mates.

Primate studies suggest that where a female benefits from some meaningful service provided by a male, her investment in their social relationships either exceeds his (e.g., infanticide protection in chacma baboons, extensive carrying of infants in titi monkeys) or equals his (limited carrying of infants in siamang) (Figure 11.2) (see also Hamilton 1984). It may be significant that in white-handed gibbons, pair bonds offer no conspicuous benefit to females, and are maintained predominantly by males. This gibbon may be an example of the "useless male" (Hrdy 1981), "unavoidable partner" (Mock & Fujioka 1990) or "pair bondage" phenomenon

(Gowaty 1996a,b): the male makes no substantive contribution to his partner's fitness, but it is simply less costly for the female to tolerate his presence than to attempt to expel him. Komers & Brotherton (1997) suggested that mammalian social monogamy generally derives from male ability to successfully monopolize a solitary female, with any accompanying advantages to females constituting evolutionary "side-effects".

The possibly negligible benefits of pair bonds to female white-handed gibbons implicate alternative selective forces operating in hylobatids. One scenario identifies mate guarding as the primary selective force behind the origin and maintenance of pair bonds in hylobatids (Palombit 1999). Where males then additionally allocate direct care to infants, as in siamang, females increase their investment in the those bonds. Ecological differences relaxing intragroup feeding competition in siamang may have facilitated the subsequent evolution of paternal care in this species but not in white-handed gibbons (Palombit 1996). A number of findings support the paternity guard hypothesis: the greater role of males in maintaining proximity to females, observations of extra-pair copulations (Palombit 1994b; Reichard 1995), frequent turnover in gibbon pair bonds (Palombit 1994a), and recent research suggesting that mate guarding may not function simply to forestall extra-pair fertilizations, but may persist *beyond* the period of female fertility to prevent loss of mate and pair-bonded status (Wagner 1991; Slagsvold & Lifjeld 1997). This argument is consistent with the view that mammalian monogamy is generally a "risk aversion" strategy of males, whereby staying with one female minimizes variance in mating success (Komers & Brotherton 1997). Thus, even if a male mammal is capable of maintaining a home range encompassing those of, say, two females, restriction of ranging to one female may still be selected for, *if* mate guarding effectively prevents loss of mate or pair-bonded status as well as extra-pair fertilizations.

In summary, four patterns taken together challenge the hypothesis that male infanticide is the selective force maintaining hylobatid pair bonds: (1) the rarity of males most likely to commit infanticide, i.e., physically as well as sexually mature unmated individuals; (2) the extremely low incidence of actual or attempted takeover infanticidal attacks under conditions apparently favoring it; (3) low infant mortality; and (4) the convergence of chacma baboons and mountain gorillas toward male–female bonds that apparently function to protect infants from infanticide, but that differ significantly from those of hylobatids.

Lemurs

A final group of primates, the gregarious lemurs, provides useful subjects for analyzing the infanticide protection hypothesis. Infant mortality is generally high, infanticide occurs (Pereira & Weiss 1991; Hood 1994; Wright 1995; Andrews 1998; Erhart & Overdorff 1998) and heterosexual pair bonds are the core of social systems in many diurnal species (Kappeler, 1993; Jolly, 1998). Preliminary support for the hypothesis derives from the correlation between permanent male–female associations and infant carrying (versus caching) in prosimians, which van Schaik & Kappeler (1997) interpreted as evidence that heterosexual bonds evolved because males protect infants transported continuously by their mothers.

The few field tests of the infanticide protection hypothesis have, however, yielded equivocal results. Contrary to a central prediction of the hypothesis, the resident male in a group of Milne–Edwards sifaka (*Propithecus diadema edwardsi*) failed to protect infants in two observed infanticidal episodes, in spite of the apparent opportunity to do so (Wright 1995). In a study of rufous lemurs (*Eulemur fulvus rufus*), Overdorff (1998) rejected the infanticide protection hypothesis because male–female bonds did not intensify during the birth season when infants were vulnerable to infanticide, females failed to limit copulation to potential male protectors, and heterosexual relations were insufficiently cohesive to constitute "pair bonds".

Additional studies should evaluate other possible causes for the co-occurrence of infant-carrying and male–female association. For example, predator avoidance through crypsis may select for both female caching and avoidance of conspicuous social relationships with males; females that rely on predator detection may carry offspring and be able associate with male protectors. Given that females in most of the species with male–female association are relatively large, diurnal, and/or intra-sexually gregarious, this anti-predation possibility merits further scrutiny.

Discussion: future directions

The female perspective

Most of the supportive evidence for the infanticide protection hypothesis addresses benefits to females of postcopulatory bonds with males. The female harassment, male coercion, and predator protection hypotheses are particularly relevant alternatives because aspects of protection may be organized in broadly similar ways to counter different threats. Discriminating among these contingencies is often difficult. For example, observational data may be insufficient to demonstrate unambiguously that "sentinel" males keep watch for infanticidal males instead of copulatory rivals, potential mates, female competitors of their companions, or predators. Experimentation is crucial, but a single manipulation may prove insufficient. For example, field observations and a male removal experiment initially suggested that male presence in the orange-tufted sunbird (*Nectarinia osea*) functions to reduce infanticide by floater males (Goldstein *et al.* 1986; Table 11.3). Building upon this work, Markman *et al.* (1996) conducted a more systematic series of male removal experiments, but with an intriguing methodological twist: they removed intruding floater males as they arrived on the widowed females' territories with the explicit intent of *preventing* infanticide. The result was that the offspring of experimental females still suffered higher morality than controls, not because of infanticide, but because of predation from jays (*Garrulus glandarius*). This finding not only suggests that male sunbirds confer multiple benefits to their mates, but raises the possibility that infanticide defense might maintain pair bonds that originated as an anti-predation strategy. In any event, the study underscores the challenges of testing functional hypotheses (Hinde 1975).

There is a critical need for mutually exclusive predictions about benefits to females under specific conditions. For example, risk of sexually selected infanticide and the need for male defense should both decline as offspring mature and approach weaning or fledging, but anti-predator protection may either increase (Andersson *et al.* 1980) or decrease (Tolonen & Korpimäki 1995) with offspring age, depending upon circumstances. Future research must also test directly the assumptions of the infanticide protection model. For example, van Schaik & Kappeler (1997) assumed that continuous proximity to the female (and her infant) is necessary for male infanticide protection in primates, but they relaxed that assumption in arguing that solitary male orang-utans (*Pongo pygmaeus*) may defend widely dispersed females from infanticide. Studies of jacanas (*Jacana* spp.) suggest that small home ranges relative to individual mobility allow infanticidal defense without requiring continuous proximity to beneficiaries (Stephens 1982; Emlen *et al.* 1989).

The male perspective

The least studied and understood aspect of the infanticide protection hypothesis concerns benefits to males. Researchers generally assume that male infanticidal defense reflects parental effort, but direct, genetic data are generally lacking (an exception is the burying beetle; for data on paternity of resident males in species with infanticide, see also Gilbert *et al.* 1991; Borries *et al.* 1999b). Copulatory patterns more often suggest the potential importance of paternity in cases of infanticide defense. In chacma baboons, for example, at least 68% of friendships established by lactating females were with males they had copulated with during conceptive cycles (Palombit *et al.* 1997).

An important goal of future research is to clarify how infanticide defense varies with probability of paternity. A suggestive preliminary result emerges again from experimental studies of chacma baboons. A male's response to playback of his female friend's scream was unassociated with his or her dominance rank at the time of the friendship (and playback), or with the strength of their social relationship, but it was positively correlated with male rank at the time his female friend conceived her infant many months earlier (Palombit *et al.* 2000). That is to say, the higher a male's rank at the time his future friend conceived her infant, the stronger was his response to playback of her distress call over 6 months later. Given the association between rank and copulatory success in male chacma baboons (Bulger 1993), this result implicates paternity as a potentially important determinant of aiding of females.

There is currently no clear evidence that males derive mating benefits by providing infanticide defense, though investigation of the mating effort function of male care has only recently begun. Smuts (1985) originally suggested that future mating preference by the female partner was an important benefit of friendships to male olive baboons, but comparable behavioral evidence was absent among chacma baboons. For example, male responses to playback experiments and male contributions to maintaining friendships fail to support the prediction of the mating effort hypothesis that males should direct greater attention to the infants of more attractive females (e.g., high-ranking females). Indeed, we might expect that male intervention in attacks on female friends would increase immediately after infants died and resumption of cycling was imminent, but males were significantly *less* responsive to their female friend's screams at this time. Nevertheless, at least one case, the Arctic ground squirrel, suggests an unspecified role of mating effort; in these

Table 11.5. *Some potential costs of bonds between females and male defenders against infanticide*

Female	Male
Ecological	
1. Feeding competition with male	1. Feeding competition with female
2. Reduced foraging efficiency	2. Reduced foraging efficiency
3. Increased predation risk	3. Increased predation risk
Social	
4. Time and energy to maintain social relationship with male protector	4. Defense of females/infants from infanticide:
5. Reduced social interaction with kin	(a) aggressive deterrence
6. Restriction of mating opportunities and mate choice:	(b) territoriality
(a) "Voluntary" by female	(c) vigilance
(b) Coerced by male partner	
7. Female–female competition for male protectors	5. Increased social interaction with higher-ranking (infanticidal) male
8. Social exploitation:	6. Restriction of mating opportunities
(a) Victim of redirected aggression	(a) "Voluntary" by male
(b) Use of infants in triadic interactions	(b) Female–female competition
	(c) Coerced by female partner

rodents males defend from infanticide young they are "unlikely to have sired" (McLean 1983).

The importance of studying costs

I have not considered costs in any detail in this chapter, but these will ultimately prove critical for understanding the evolution of male–female associations. The adaptive value of male infanticide defense will depend upon the costs as well as benefits to females of associating with him relative to the economics of alternative anti-infanticide strategies available to these females. Table 11.5 provides a preliminary outline of some potential costs.

Rodents may exemplify the importance of evaluating costs. Although infanticide is widespread in this order, the limited field data suggest that female bonds with male protectors are relatively less common than in mammals such as primates. If this is confirmed by future research, I suggest that the infrequency of this counterstrategy among rodents may be due to greater female reliance on maternal aggression to deter infanticide (Maestripieri 1992). The overall deterrent value of maternal aggression is controversial (Ebensperger 1998a), but it clearly succeeds in forestalling male infanticide in some cases (*e.g.*, Wolff 1985; Maestripieri & Alleva 1991; vom Saal *et al.* 1995). In contrast, female primates are strikingly ineffective in defense against male infanticide, even in coalitions (Hrdy 1979; Maestripieri 1992; Palombit 1999; see also Yamamura *et al.* 1990). Thus a capacity to defend her own young may select for avoiding the potentially high costs of associating with male defenders in favor of maternal aggression. It is important to note that this hypothesis does *not* predict that maternal aggression is more effective or even equally as effective in deterring infanticide as is male defense: all that is necessary is that the benefit–cost ratio for maternal defense exceeds that for male defense. Costs of associating with males are thus crucial. Consistent with this view, the currently most likely case of male–female association functioning as an anti-infanticide strategy in rodents occurs in a species in which "females could not prevent incoming males from killing" pups (McLean 1983, p. 42).

Costs to males, particularly limitations in breeding, are equally relevant. In primates, several patterns suggest that infanticide defense has evolved in systems where costs to males are relatively low (e.g., mountain gorillas, chacma baboons). First, because males in these species live in multifemale groups, restriction of breeding opportunities is less severe than it would be if infanticide protection required monogamy. Second, females appear to incur more of the time, energy and social costs of maintaining the heterosexual relationship. Third, sexual dimorphism and male social dominance over females assure priority of access to food resources for which they compete more intensively because of their association.

Infanticide defense based on social monogamy may rarely evolve precisely because of the serious breeding costs to males (even when they have paternity certainty of the protected young; Hawkes *et al.* 1995), and the ecological costs associated with co-dominance (or dominance) of females resulting from sexual monomorphism. Comparative data suggest that infanticide defense selects for social monogamy primarily when infanticidal pressure is extremely high, e.g., when relatively abundant floaters of *both* sexes pursue takeover strategies, as in house wrens and burying beetles. As with other forms of paternal care (Dunbar 1995), infanticide protection by monogamous males may be more likely to evolve *after* monogamy, rather than select for it.

Conclusion

In the near-absence of direct tests of the infanticide protection hypothesis, pertinent evidence may concern three distinct issues: (1) the effect of male presence on the risk of infanticide relative to the costs to females of bonding with males; (2) the benefits to males from providing infanticide protection relative to the costs (especially in mating); and (3) the ability to reject alternative functional hypotheses for male–female relationships.

Most studies so far have focused on only one of these questions. Current data generally oppose the hypothesis in hylobatids and are equivocal in lemurs, but support it in burying beetles, barn swallows, house wrens, Arctic squirrels, mountain gorillas, and chacma baboons. The phylogenetic diversity represented by these taxa suggests that heterosexual bonds have evolved as an anti-infanticide strategy numerous times in both polygynous and socially monogamous systems. Existing studies also focused on fitness benefits for females rather than for males.

There are clearly multiple evolutionary pathways to strong postcopulatory bonds between the sexes. Much more work is needed, including quantification of costs and variation in male–female bonds, experimental techniques such as playbacks and male removals, and explicit tests of functional possibilities against one another.

Acknowledgments

I am indebted to Charles Janson, Carel van Schaik and an anonymous reviewer for comments and suggestions on the chapter. I am extremely grateful to numerous colleagues, some of whom read previous versions of the manuscript, but all of whom provided crucially important discussion and criticism of my views on infanticide and pair bonds: Warren Brockelman, Dorothy Cheney, Sandy Harcourt, Sarah Hrdy, John Mitani, Drew Rendall, Peter Rodman, Robert Seyfarth, Pascal Sicotte, Kelly Stewart, Robert Trivers and David Watts. I thank the National Science Foundation, the National Institutes of Health, the L. S. B. Leakey Foundation, the Wenner–Gren Foundation, the University of California, the University of Pennsylvania, the California Primate Research Center, and the Institute for Research in the Cognitive Sciences for support of my work.

ANDREAS PAUL, SIGNE PREUSCHOFT, AND
CAREL P. VAN SCHAIK

12

The other side of the coin: infanticide and the evolution of affiliative male–infant interactions in Old World primates

Introduction

In most species of the animal world, fathers contribute relatively little to the well-being and provisioning of their young, often no more than their genes (for exceptions, see, e.g., Ridley 1978; Trivers 1985; Clutton-Brock 1991). This is especially true for mammals, where internal fertilization, long gestation and lactation predispose mothers to care for their offspring alone. Since only females lactate, males can contribute relatively little to rearing of young. Due to internal fertilization paternity is never certain. Moreover, during the long periods of gestation males have ample time to desert the impregnated female in order to increase their reproductive success by seeking additional fertilizations with other females. Not surprisingly then, paternal care is found in only a small minority (less than 5%) of all mammalian species (Kleiman 1977; Clutton-Brock 1991; Woodroffe & Vincent 1994). Compared to most other mammalian taxa, however, primates are characterized by a surprisingly high level of male–infant affiliation or "male care": Nearly 40% of all primate genera have been reported as exhibiting "direct male parental care" (carrying, retrieving, protecting, provisioning, grooming and/or huddling with young) – the highest percentage for any individual mammalian order (Kleiman & Malcolm 1981).

While there are reasons to remain skeptical about whether all of these observations are correctly classified as "male care" (e.g., Hrdy 1976; Packer 1980), or whether male care is really widespread among all of the species mentioned (Maestripieri 1998), this high level of male–infant affiliation is unexpected and, more importantly, does not seem to be well understood.

Could it be that the extraordinarily long dependence of primate infants on high levels of parental care makes the assistance of the father (or other males) indispensable for the successful rearing of young in this group of mammals? Probably not, because with the exception of several small-bodied New World monkeys, where male care is especially well developed and appears indeed necessary for successful (and rapid) reproduction (Koenig 1995; Garber 1997), in most primate species (including humans) males are not known for their strong involvement in routine daily care of young (e.g., Taub 1985; Whitten 1987). Moreover, there is no indication that male care is particularly well developed in those species where the dependence of the infants on the care of their mothers (and/or other individuals) is especially long. Nevertheless, the long dependence of primate infants may provide the right answer to the above enigma. In this chapter, we propose another hypothesis for the evolution of affiliative male–infant interactions in primates, which is directly related to this life history character of primates and to the topic of this book: the threat of infanticide.

Primates differ from most other mammals not only in their relatively high levels of male care but also in the relatively frequent occurrence of infanticide (see, e.g., chapters in Hausfater & Hrdy 1984; Parmigiani & vom Saal 1994; van Schaik, Chapter 3). This high incidence of infanticide is a direct consequence of the long dependence of infants on their mothers. Due to the lengthy suppression of their mothers' ovarian acitivity as a result of their suckling, primate infants create long time windows of vulnerability to infanticidal males (van Schaik & Kappeler 1997). Fathers who are able to protect their offspring would, therefore, gain rather large benefits. In fact, several observations (reviewed below) indicate that fathers are important for the successful rearing of their young and that many (though clearly not all) aspects of male behavior may be understandable in this context. Thus we argue here that these two apparently disparate phenomena, male care and infanticide, are, in fact, related to each other: they represent two sides of the same coin. Specifically, we argue that infanticide was involved in the origins of male care behavior but is probably not related to all its current functions.

The hypothesis that male care in primates may be a response to the threat of infanticide is not at all new: exactly this has been suggested as an explanation for frequent male–infant carrying in chacma baboons during the presence of potentially infanticidal newcomers (Busse & Hamilton 1981). Yet, this hypothesis remained controversial and received little systematic attention. Behavioral observations in other baboon populations indicated that males often carry apparently unrelated infants (e.g., Packer 1980; Stein 1984a; Strum 1984), and even in cases where paternity was likely, the necessary genetic tests of paternity were not available. Moreover, in those cases where such data became available such as in Barbary macaques – a species where male care is quite extensive – it turned out that fathers carried their own offspring no more often than may be expected by chance (Ménard *et al.* 1992; Paul *et al.* 1992, 1996). Consequently, other hypotheses attempting to explain affiliative male–infant interactions in primates received substantial more attention (e.g., Whitten 1987; Wright 1990; Smuts & Gubernick 1992; Paul *et al.* 1996; van Schaik & Paul 1996–7). At the same time, however, evidence was accumulating that male protection of infants from infanticide may be more important than is usually acknowledged.

When fathers matter: male care and the threat of infanticide

Apart from their relatively high levels of male care and infanticide, primates differ from most other mammalian taxa in another important way. In the majority of mammals adult males and females associate only briefly during short periods of female receptivity and year-round male–female association is found in at most 34% of species (median of six orders is 27.5%). Quite in contrast, in the majority of primates (76% of the species) males remain associated with females throughout the year (van Schaik 2000). Comparative tests provide strong support for the idea that the risk of infanticide is ultimately responsible for the evolution of permanent male–female association in primates (van Schaik & Kappeler 1997). Obviously, this hypothesis also predicts that males should protect their putative offspring from infanticidal males. Observational and experimental evidence suggests that this is indeed the case. We restrict this review to Old World primates.

Hanuman langurs

Apart from infanticidal attacks, adult hanuman langur males (*Presbytis entellus*) living in one-male groups rarely interact with infants at all (Hrdy 1977; Vogel 1984; Whitten 1987). Given that defeated leaders are usually evicted from their troop, these males no longer have a chance to protect their offspring against infanticidal males. In multimale groups of hanuman langurs, however, half of all defeated alpha males stayed within their group (Launhardt 1998). These males were frequently observed in the vicinity of infants if newly immigrant males were present, and they frequently defended their putative offspring against attacks by high-ranking newcomers. Long-term observations and DNA analyses revealed that only genetic fathers or males who had been residents when the respective infant was conceived protected infants. Newcomers, on the other hand, were never observed to defend infants (Borries 1997; Borries *et al.* 1999b; Borries & Koenig, Chapter 5).

Vervet monkeys

Vervet monkeys (*Cercopithecus aethiops*) represent another species where "male care" is clearly not pronounced (Whitten 1987). Nevertheless, play-back experiments revealed that likely fathers responded more strongly than other males to an infant's distress calls (Hauser 1986). While this might be interpreted as the male's willingness to protect his offspring against predators such as leopards or martial eagles, Hauser (1986: 70) noted that "for most of these predators, an adult male vervet would have little to no chance in defending his potential offspring from an attack". Thus protection against infanticide, which has been observed in the respective population, appears to be at least as parsimonious an explanation.

Geladas

Observations on geladas (*Theropithecus gelada*) suggest that many interactions between males and infants may serve to protect the male's offspring against potentially infanticidal males. Harem leaders frequently interact with infants in the presence of intruders attempting to take over the harem, and defeated harem leaders who stayed in their group as old followers have been described as extremely solicitous of the unit's youngest infants during the months immediately following a takeover (Dunbar 1984a,b). In captivity, infanticide by new harem leaders has been observed in the absence of old followers (Angst & Thommen 1977; Moos *et al.* 1985; cf. Mori *et al.* 1997).

Savanna baboons

Savanna baboons (*Papio cynocephalus* spp.) belong to those species where males are commonly observed to carry or hold infants (Whitten 1987), often during tense interactions with other males. Several hypotheses have been proposed to explain both dyadic and triadic male–infant–male interactions in baboons. For convenience, and because most researchers agree that more than one hypothesis may apply (e.g., Smuts 1985; Smith & Whitten 1988), we restrict our presentation here to the question of whether the available data are consistent with the protection against infanticide hypothesis (for other explanations, see the Discussion section).

The strongest support for the hypothesis that male savanna baboons affiliate with putative offspring in order to protect them from infanticidal males comes from studies on wild chacma baboons (*Papio c. ursinus*) in Moremi, Botswana. As noted above, Busse & Hamilton (1981) were the first to suggest that in this population probable fathers carry their unweaned offspring in order to protect them from potentially infanticidal immigrant males. In this study, almost all carriers were likely or probable fathers (for observations in another chacma baboon population, see also Anderson 1992), while nearly all opponents were recent and high-ranking immigrants who could not have fathered the infant and would benefit from killing it. Other observations revealed that, in groups in which the alpha male was a long-term resident, adult males rarely interacted with infants. After the immigration of two additional males, however, the resident males began to carry infants frequently in the presence of these newcomers (Busse 1984b).

Male–infant affiliations in baboons are typically accompanied by special relationships or "friendships" between the male and the infant's mother (Strum 1984; Smuts 1985). Recent work on the Moremi baboons indicated that these friendships are typically established as a result of prior sexual activity and that they are intensified following the birth of an infant, but terminated after its death (Palombit *et al.* 1997). All these findings strongly support the hypothesis that protection of putative offspring against infanticide (which is a major source of infant mortality in this population) may be the primary selective force for the establishment of affiliative male–female relationships in this population of savanna baboons. Experimental research provided further evidence for this hypothesis: Males responded strongly to playbacks of the distress calls of their female friends, but they virtually ignored screams of other females, and within days after the death of their female friend's infant, they also showed significantly weaker responses to the screams of their female friends (Palombit *et al.* 1997).

Observations from other baboon populations (*P. c. anubis* and *P. c.*

cynocephalus) are less intriguing, but in several ways in general agreement with the Moremi data. First, almost all researchers agree that the male's relationship with the infant's mother is the single most important variable for the establishment of an affiliative relationship between a male and an infant (Strum 1984; Smuts 1985), but they disagree to what extent these males are also likely fathers of the infants they carry. For example, while Packer (1980) noted in a population of olive baboons that there was no tendency that males affiliated preferentially with infants of females with whom they had consorted (see also Packer & Pusey 1985), Smuts (1985) reported that there was a "general trend that females' friends were also likely fathers or possible fathers" (Smuts 1985: 161). Although several other researchers (Altmann 1980; Stein 1984a; Strum 1984; Noë & Sluijter 1990; Bercovitch 1991) reached similar conclusions to Smuts (whose own data were rather ambiguous), the problem is as yet unresolved since DNA fingerprint data are not available.

Second, encounters with other males are a common context where male baboons carry infants, and often, but by no means always, the opponents are recent immigrants. Collins (1986), for example, observed that 39% of triadic male–infant carrying in a population of yellow baboons was by residents interacting with newcomers, but the reverse was never seen (see also Packer 1980, for olive baboons). Third, as in the study by Busse & Hamilton (1981), males generally carry young, unweaned infants, which are much more vulnerable to infanticide than older infants, far more often than expected by chance (e.g., Collins 1986). Fourth, in one-male groups where no other males are present, resident males appear to be uninterested in interacting with infants (Stein & Stacey 1981). Finally, one anecdotal observation of an (unsuccessful) attempt by a male baboon to rescue his 24-month-old affiliate from a fatal attack by another male (Altmann 1980) provides further support for the protection hypothesis.

Mangabeys

As in geladas and baboons, male–infant interactions in mangabeys appear to occur predominantly in the presence of newly dominant males. In crested mangabeys (*Cercocebus galeritus*) a rank reversal between the two adult males of the group was followed by infant carrying of the deposed alpha in the presence of the new alpha male. Infant carrying by adult males had not been observed in this group prior to the dominance rank reversal (Gust 1994a). Similarly, in a captive group of sooty mangabeys

(*Cercocebus atys*, now *C. torquatus atys*) male–infant interactions occurred frequently after a rank reversal between the two highest ranking males of the group (Busse & Gordon 1984). All interactions involved the former alpha male carrying infants while in proximity of the new alpha male, and 40% of the episodes occurred after the new alpha male had actually threatened the infant. Affiliative interactions between males and infants of the genus *Lophocebus* also appear to occur often in the vicinity of other males (Chalmers 1968).

Macaques

Although along with baboons, several macaque species have been intensively studied and although several of them are characterized by more or less intensive male–infant affiliation (for reviews, see Whitten 1987; Maestripieri 1998), strong evidence that males act as to protect their putative offspring from potentially infanticidal males appears to be rather restricted. In fact, as is discussed further below, many male–infant interactions in macaques are not easily explained within the framework of the infanticide avoidance hypothesis. The currently best available evidence in support of the hypothesis comes from long-term research on wild long-tailed macaques (*Macaca fascicularis*) living at the Ketambe Research Area in northern Sumatra.

Longtailed macaques are one of the species usually characterized by male indifference toward small infants (Whitten 1987; Maestripieri 1998). After being deposed from the alpha position, however, most males not only remain in the group where they had sired most of the infants (de Ruiter *et al.* 1994) until these offspring are weaned but they are also found significantly more often near their putative offspring than are other males (van Noordwijk & van Schaik 1988). Several strongly suspected (though not eye-witnessed) cases of infanticide in the population occurred during takeovers (de Ruiter *et al.* 1994), when the infants were not associating with the former alpha male. During these periods, infant mortality is far above average levels (M. A. van Noordwijk & C. P. van Schaik, unpublished data).

Indirect support for the hypothesis also comes from wild rhesus monkeys (*M. mulatta*), where the only observed case of infanticide occurred after the death of the former alpha male (Camperio Ciani 1984). As with male longtailed macaques, rhesus males rarely interact with infants at all (e.g., Lindburg 1971), but occasional instances of male protection of infants against other males (Hill 1986) as well as cases of "adoption" of orphaned infants by high-ranking males (e.g., Taylor *et al.* 1978; Berman 1982) have been reported.

Observations of males who consistently avoid young infants with natal coat color may also be significant in this context. Interestingly, stumptailed macaques (*M. arctoides*), for whom this behavior has been most often reported (Hendy-Neely & Rhine 1977; Bruce *et al.* 1988; Bauers & Hearn 1994; Maestripieri 1994; for bonnet macaques, see also Simonds 1974), appear to be characterized by considerable variation in the degree of adult male affiliation with infants. For example, while Gouzoules (1975: 413) reported that "adult males showed interest in, and interacted with infants almost as much as females did" (see also Rhine & Hendy-Neely 1978), Maestripieri (1994: 107) noted that "adult males showed very few clear instances of caregiving behavior such as embracing, carrying or grooming the infant" (see also Estrada 1984). Apparently, at least some males are much more involved with infants than others. For example, Bauers & Hearn (1994) reported that both in a wild and in a captive group only alpha males carried young infants and "baby-sat" them while the mother was foraging (see also Hendy-Neely & Rhine 1977). The most plausible explanation for the avoidance behavior of the other males seems to be that these animals anticipate that approaching an infant would be regarded as a serious threat to that infant by its relatives (see also Treves, Chapter 10).

Mountain gorillas

Male mountain gorillas (*Gorilla gorilla beringei*) are known to be tolerant and protective toward infants (Fossey 1983), but they are not known for heavy involvement in daily care of infants. But the quality or quantity of a gorilla father's interactions with his offspring may be much less important for their development than his mere presence. If a gorilla father dies, his unweaned offspring have little chance of survival – they are likely to be killed by his successor (Watts 1989).

Conclusion

These observations suggest a major protective role for likely sires. First, genetic paternity analyses carried out in conjunction with behavioral observations during the past decade provide strong evidence that paternity certainty for most if not all of the males in the above cited cases is sufficiently high (hanuman langurs, Borries *et al.*, 1999b; savanna baboons, Altmann *et al.* 1996; mangabeys, Gust *et al.* 1998; longtailed macaques, de Ruiter *et al.* 1994; stumptailed macaques, Bauers & Hearn 1994). At the same time, the strong correlation between male dominance rank and reproductive success found in these studies ensures that infanticide is an adaptive reproductive strategy for males who manage to suddenly rise from very low to top rank. Taken together, these results make it quite plausible that affiliative male–infant interactions in Old World primates initially "may have functioned exclusively to protect the infant against infanticide", as Packer & Pusey (1985: 180) assumed.

Beyond protection: triadic male–infant–male interactions in baboons and macaques

Observations

We already noted that by no means all male–infant interactions found in primates can easily be explained within the framework of the infanticide protection hypothesis. In fact, primatologists have long been aware that not everything that looks like paternal care truly is paternal care: males may also exploit infants for their own (ultimately reproductive) interests (e.g., Hrdy 1976; Packer 1980). Triadic male–infant–male interactions in macaques and, to some extent probably also in baboons, provide a classic example. Yet, we argue that even here the infanticide protection hypothesis may be helpful in understanding the evolution, though not necessarily the current function, of these behaviors.

Barbary macaques are one well-studied example (Figure 12.1) where "intensive caretaking" (Whitten 1987) of males is not related to paternity and the risk of infanticide (though the latter has been observed; see Angst & Thommen 1977). Barbary macaque males frequently carry infants to other males or approach males that are holding infants, whereupon both males sit together, "displaying a series of exaggerated and stereotypic behaviors to each other and to the infant" (Taub 1980: 189; for more detailed descriptions, see Deag 1974, 1980; Taub 1978). Not rarely (11% to 23% of the interactions; Paul, unpublished data; Taub 1980) the infant is transferred from one male to the other.

Field studies reported that triadic interaction occurred about once per hour, or once every 8 to 11 hours, if the rates are adjusted for the number of males in the groups (Taub 1985). During the mating season triadic interactions are relatively infrequent (A. Paul, unpublished), but at the end of the birth season, when many black infants are available, males may initiate triadic interactions as often as once per hour (Paul *et al.* 1996). Triadic

Figure 12.1. Triadic male–infant–male interaction in free-ranging Barbary maquaques (photo Signe Preuschoft).

interactions are usually initiated by the lower-ranking male, although this rule was found to be valid only for interactions that were intiated by a male carrying an infant (Deag 1980; Taub 1980; Paul 1984; Paul *et al.* 1996). Moreover, triadic interactions occur much more frequently between males with a small rank distance than between males with a large rank distance (Paul *et al.* 1996; see also Stein's (1984a) reanalysis of Taub's (1980) data). Usually, the males separate immediately after termination of the interaction, which may last from a few seconds to several minutes, but in a substantial number of cases both males remain in close proximity to each other (Deag 1980; Taub 1980; Paul *et al.* 1996). Among adult males, friendly approaches in the absence of an infant are rare, and grooming outside the context of triadic interactions is virtually absent (Deag 1980; Paul *et al.* 1996). Typically, triadic interactions are not triggered by overt agonistic behavior (Deag 1980; Taub 1980; Paul *et al.* 1996), but they occur more frequently during tense situations than in completely relaxed situations (Paul *et al.* 1996).

All studies agree that the males preferentially carry and use certain infants and tend to ignore all others. Kinship appears to be unimportant, however: In Paul *et al.*'s (1996) study only in a small minority of the episodes (15 out of 180) either the carrier or his opponent was a maternal or paternal relative of the infant, and in no case were both males related to the infant (for data on dyadic male–infant interactions and paternity, see also Ménard *et al.* 1992). However, in most cases at least one of the males had a special relationship with the infant, and during triadic interactions males preferentially used those infants that had a special relationship with their opponent.

Within the genus *Macaca* triadic male–infant–male interactions similar to those observed in Barbary macaques have been observed in stumptailed macaques (Gouzoules 1975; Hendy-Neely & Rhine 1977; Estrada 1984), longtailed macaques (de Waal *et al.* 1976), and virtually all species of the *sinica* group of macaques (Maestripieri 1998), although the only other species in which triadic male–infant–male interactions occur at a rate, intensity and degree of ritualization similar to Barbary macaques appear to be Tibetan macaques (Ogawa 1995a; Zhao 1996).

Triadic male–infant–male interactions in macaques resemble those observed in baboons and mangabeys insofar as they are usually initiated by the subordinate male (*M. sylvanus*, see above; *M. radiata*, Silk & Samuels 1984; *M. arctoides*, Gouzoules 1975; *M. thibetana*, Ogawa 1995a; Zhao 1996). However, they differ from those observed in other papionins in two important aspects. First, while among savanna baboons, geladas and mangabeys infants are predominantly carried by potential fathers in the presence of newcomers or newly dominant males, this is usually not the case in triadic interactions observed in macaques. Second, during triadic interactions male baboons, geladas and mangabeys simply "sit in relative proximity, staring at the baby or glancing at each other" (Strum 1984: 151). In sharp contrast, male macaques carrying infants frequently establish close body contact with each other and the infant, holding it simultaneously while exchanging friendly signals. Since the infant is used as a physical bridge between the two males, these interactions have also been termed "bridging" by various authors (e.g., Estrada & Sandoval 1977; Ogawa 1995a). Moreover, as noted above, at least in Barbary macaques infants are often transferred from one male to the other. All these behaviors would be unlikely if potentially infanticidal males were involved.

Interpretation: the agonistic buffering hypothesis

As noticed by Hrdy (1976: 108), male–infant interactions do not necessarily benefit the infant. Indeed, while there are reasons to suggest that infant-carrying by males often serves to protect the infant from potentially infanticidal males, a considerable body of observations suggests that males also use infants to protect *themselves* against attacks from higher-ranking males (hamadryas baboons, Kummer 1967; savanna baboons, e.g., Packer 1980, for a review, see Smith & Whitten 1988; geladas, Dunbar 1984a; macaques, Gouzoules 1975, Deag 1980, Silk & Samuels 1984, Ogawa 1995a, Paul *et al.* 1996). In their classic study on male–infant interactions in Barbary macaques, Deag & Crook (1971) labeled these interactions "agonistic buffering", suggesting that "an animal deliberately using a baby as a 'buffer' in a situation where an approach without that 'buffer' would lead to an increased likelihood of an aggressive response by the higher-ranking animal" (Deag & Crook 1971: 196).

While this hypothesis was readily accepted by many authors (e.g., Hrdy 1976), it remained unclear why infants should function as "buffers". Obviously, the understanding of why the use of infants should inhibit aggression would be fundamental for the understanding of the evolutionary origin of this complex behavioral phenomenon. The notion of an innate releasing mechanism – Lorenz's "Kindchenschema" (Lorenz 1943) – previously so readily embraced, may have obscured this problem. However, in view of accumulating reports of infanticide in primates the idea of the Kindchenschema as an innate, and largely unfailing appeasement mechanism becomes more and more confusing and loses much of its appeal. We cannot ignore that (1) primate infants are not per se protected against lethal aggression from conspecific males, (2) such aggression is by no means exceptional or abnormal, but rather systematic (see van Schaik, Chapter 2), and (3) that specific infantile features such as natal coats do not invariably evoke friendly behavior (Treves 1997; Chapter 10). Once again, the solution to the puzzle may have to do with the risk of infanticide.

The hypothesis that the risk of infanticide might provide an indirect explanation for why infants function as "buffers" comes in different versions. The first one assumes that males carry unrelated infants to more dominant males who are likely sires of the infants. By using infants related to their opponents males would reduce the likelihood of an escalated fight because any severe aggression performed by the dominant male would put his own putative offspring at a serious risk (Popp 1978; Popp & DeVore 1979). Although this hypothesis received little support from data on triadic interactions in savanna baboons – the species for which it was originally developed (for a review, see Smith & Whitten 1988), it may neatly explain "tripartite" male–infant–male interactions among hamadryas baboons (Kummer 1967) and geladas (Dunbar 1984a), where young followers threatened by the harem leader often grasp an infant and carry it in front of the adult male. Dunbar (1988) also suggested that this hypothesis may explain triadic male–infant interactions in macaques, where subordinate males (i.e., unlikely fathers) carry infants to dominant males (i.e., more likely fathers). Analyses of triadic interactions in Barbary macaques, where genetic paternity data were available, did not provide strong support for that hypothesis (Paul *et al.* 1996; for Tibetan macaques, see also Ogawa 1995a). Nevertheless, it should be noted that, although in the case of Barbary macaques recipients of triadic interactions were rarely actual fathers, they often were likely fathers of the infants involved. Moreover, the use of infants familiar with the recipient (Ogawa 1995b; Paul *et al.* 1996; see also Taub 1980) appears to function as a proximate mechanism inhibiting aggression even if there is no genetic relationship between the infant and the recipient.

The second version argues that by carrying his own putative offspring during a tense encounter with a rival, a male is signaling his readiness to escalate maximally in order to protect his offspring. Since the pay-off for escalation would be usually much smaller for his opponent, the rival is expected to respect the signal and refrain from aggression (Packer & Pusey 1985). Here, again, the evidence in support of the hypothesis appears to be ambiguous. Incomplete information about paternities notwithstanding, it is clear that in many cases the carrier is an unlikely or even impossible father (e.g., Stein 1984a). Moreover, as noted above, the pattern of triadic interactions in macaques, where close body contact between both males and the infant is established and both males exchange friendly signals, appears to be inconsistent with this hypothesis. Nevertheless, the idea may shed light on the evolutionary origin of the behavior: Packer & Pusey (1985) suggested that, once widespread, honest signals of readiness to escalate were corrupted by subordinates who carried infants despite not being the infants' actual father at all.

The last version is a mixture of both of the above, but does not make the genetic relationship between male and infant a necessary precondition. Instead, it is argued that males attacking infants are often the subject of massive counteraggression by other group members: the infant's mother, female relatives and often one or more males (see, e.g., Altmann 1980; Tarara 1987; Valderrama *et al.* 1990; Agoramoorthy & Rudran 1995; Borries 1997). An obvious hypothesis, therefore, is that males use infants as an implicit threat that any escalation by the opponent will evoke massive retaliation (Busse 1984b; Dunbar 1984a; Stein 1984a; Collins 1986).

Note that, in all three versions of the hypothesis, males who succeed in preventing escalated aggression may simultaneously reduce the infant's risk of infanticide. If so, the infants benefited from such behavior and may be selected to cooperate in triadic male–infant–male interactions. Indeed, in many cases, males who are obviously not the infants' sires carry the infants into these interactions. However, this observation can be reconciled with the infanticide protection hypothesis if we follow Packer & Pusey's (1985) argument that, even though the behavior started with fathers' carrying their own offspring, this was no evolutionarily stable strategy (ESS) and subordinates soon began to fake protection and motivation to fight back. Thus, subsequent to its initial usage, other males have begun to use familiar infants in the same way as formerly only fathers did, provided they had established a special relationship with the infant. As a result, these interactions no longer serve to reduce infanticide risk to the infants, although they are historically based on infant protection behavior. Ironically, they may well lead to the contrary: the exposure of infants to some risk (Packer 1980), although the actual risk of being severely wounded during these interactions appears to be minimal (Stein 1984a).

Obviously, these ideas are not necessarily mutually exclusive. Depending on social circumstances, males may use different tactics in order to gain a competitive advantage. The behavior of geladas provides a well-known example. Harem holders, young followers and old followers may all use infants for different purposes (Dunbar 1984a, 1988). Most noteworthy is, however, that protecting infants from the potentially lethal aggression by other males lies at the heart of all hypotheses explaining the evolution of triadic male–infant–male interactions in primates.

Infants as mediators of cooperative male–male relationships

Although the "agonistic buffering" hypothesis received substantial empirical support for both baboons (e.g., Stein 1984; Strum 1984; Collins 1986) and macaques (Deag 1980; Silk & Samuels 1984; Ogawa 1995a; Paul *et al.* 1996), it soon became clear that agonistic buffering, in its strict sense, "is not the whole story" (Deag 1980: 78; see also Taub 1980). The "agonistic buffering" hypothesis explains neither "why males initiate male–baby–male interactions with males subordinate to themselves" (Deag 1980: 78; see also Taub 1980; Paul *et al.* 1996), nor why males rarely pick up infants as a response to an actual threat from a higher-ranking male. Indeed, at least in macaques few triadic interactions are preceded by agonistic interactions between the participants (Deag 1980; Taub 1980; Silk & Samuels 1984; Ogawa 1995a; Paul *et al.* 1996). Why, then, do males use infants to initiate interactions with other males?

In all multimale groups, adult males find themselves in proximity to other males, most of whom are not close relatives. Their reproductive interests rarely coincide and even if they do and cooperative defense of a group of females would be indicated, this system can always be invaded by free-riders. Hence, stable male–male alliances are unlikely to become an ESS among unrelated males, and, consequently, such alliances are found in only a handful of primates where males are the philopatric sex (Plavcan *et al.* 1995; van Schaik 1996). Moreover, since males compete primarily for non-shareable resources, i.e., fertilizations, tolerance and affiliative bonding are expected to occur less easily among male than among female primates (van Hooff & van Schaik 1992, 1994). Consequently, within-group coalitions and alliances among males are uncommon in most species of non-human primates (Chapais 1995; for New World monkeys, see also Janson 1997; for gregarious lemurs see also Kappeler 1999). Nevertheless, recent studies have shown that there is considerable variation in the relationships between adult males (Hill & van Hooff 1994; van Hooff & van Schaik 1994), and in at least some species of Old World monkeys within-group coalitions among males appear to be relatively common (savanna baboons, e.g., Hall & DeVore 1965; Packer 1977; Bercovitch 1988; Noë & Sluijter 1990; bonnet macaques, Silk 1992a,b; Barbary macaques, Witt *et al.* 1981; Kuester & Paul 1992; J. Kuester & S. Preuschoft, unpublished data). Yet, even in these species, social relationships between adult males are essentially tense (Smuts & Watanabe 1990; Kuester & Paul 1992; Silk 1994; Preuschoft *et al.* 1998; Preuschoft & Paul 2000). Moreover, the males have dangerous weapons, and where they do form short-lasting and opportunistic coalitions, they must find a way to establish and maintain their social relationships in a situation where potentially lethal damaging fights could flare up at any minute. How can males control this risk of escalation?

To minimize the disruptive effects of competition and to establish and maintain cooperative relationships with other males, effective conflict management is essential (de Waal 1989; Aureli & de Waal 2000). Highly ritualized interactions appear to be evolutionarily designed to enhance and reinforce such bonds (Smuts & Watanabe 1990; see also Colmenares *et al.* 2000; Preuschoft & van Schaik 2000). In savanna baboons, Smuts and Watanabe (1990: 166) suggested that males use "greetings", i.e., ritualized sexual acts such as presenting, mounting and genital touching, "to negotiate and constitute aspects of their relationships, including cooperation". Notably, in this species "greetings constitute the only context in which adult males routinely exchange friendly behaviors" (*ibid.*: 148).

Virtually the same is true for triadic male–infant–male interactions in male Barbary macaques. Outside the context of triadic interactions, adult males almost never exhibit affiliative behaviors toward one another, such as grooming or huddling (Deag 1980; Paul *et al.* 1996; J. Kuester & S. Preuschoft, unpublished data). Not surprisingly, therefore, several authors have suggested that triadic interactions with infants not only serve as an effective form of conflict management (Kuester & Paul 2000), but also to enhance and reinforce cooperative bonds (Paul *et al.* 1996; see also Deag & Crook 1971; Kurland 1977; Stein 1984). Infants are, in this sense, not only used as a physical bridge between two males, they also constitute a social bridge or, as Itani (1959) put it, are used as a "passport". Since primates cooperate mainly by supporting each other in agonistic conflicts (Harcourt & de Waal 1992), males succeeding in building an alliance may gain substantial benefits for themselves, while at the same time reducing the risk of infanticide to the infant they use.

Support for the contention that triadic male–infant interactions in Barbary macaques facilitate coalitionary bonds between males is still circumstantial (see also Smuts & Watanabe 1990). First, these interactions appear to enhance affinitive relationships between males; second, their use of infants familiar with the opponent (Paul *et al.* 1996; for Tibetan macaques, see also Ogawa 1995b) can be regarded as an especially sophisticated tactic to reach this goal; and third, they interact predominantly with males relatively close in rank, i.e., the most likely coalition partners (cf. Noë & Sluijter 1995). Some observations on other species are also suggestive. In wild longtailed macaques, for instance, triadic interactions occurred primarily among allies during the rare periods when resident males formed defensive coalitions against challengers for the top-dominant position (M. A. van Noordwijk & C. P. van Schaik, unpublished results). These male encounters always were tense; the males sat facing each other, holding each other's lower legs, and generally embraced while lip-smacking. This limited context for triadic encounters is consistent with the coalition hypothesis. However, longitudinal studies, which are necessary to evaluate the hypothesis more fully, are not yet available.

Interspecific patterns

A second way to test this hypothesis is to compare species. The incidence of triadic male–infant–male interactions and of male alliances should be positively associated across species. However, comparative analysis is difficult. Quantitative data on triadic male–infant–male interactions and male–male coalitions, even if available, are often not directly comparable. How often coalitions occur, for instance, depends on a variety of factors such as the number and quality of agonistic interactions and the number of available partners. Any specification of whether male–male coalitions in a specific group or species are "common", "frequent" or "rare" should therefore be treated with great caution. Hence, the following comparison is preliminary.

The frequency of male–infant interactions varies considerably across species. For example, in several well-studied macaque species, such as rhesus and Japanese macaques, affiliative male–infant interactions in general, and triadic interactions in particular, are virtually absent (Maestripieri 1998). The same species are characterized by very low rates of male–male coalitions (e.g., Watanabe 1979; Chapais 1983).

Barbary macaques appear to represent the other extreme: as noted above, triadic male–infant–male interactions are very common in this species, and the same appears to be true for male–male coalitions. J. Kuester & S. Preuschoft (unpublished data), for example, found that nearly half of all agonistic encounters between adult males and three quarters of the undecided ones (see Preuschoft *et al.* 1998) ended with support interventions by other males. Tibetan macaques resemble Barbary macaques in their rate of triadic interactions (Ogawa 1995a; Zhao 1996); male–male alliances in this less well studied species are known to occur (Deng & Zhao 1987), though no quantitative data on coalitionary behavior are available yet. However, Ogawa (1995a) noted that triadic interactions rarely occurred in agonistic contexts, and that these interactions also appeared to enhance affiliative relationships between the interacting males.

In the remaining species triadic male–infant interactions occur at a much lower rate. For example, Silk & Samuels (1984) recorded 68 events during 1125 hours of observation (0.06 or one per 16.5 hour) in a large captive group of bonnet macaques. Adjusted for the number of males in the group, there was about one episode every 500 observation hours per male actor. Similar rates have been reported from several populations of savanna baboons, where about one episode per male actor occurs every 100 (Moremi) to 1250 observation hours (Gombe; for a review, see Taub 1985).

A comparative test

In order to assess whether triadic male–infant–male interactions and male–male alliances show correlated evolution, we need to force the available evidence into simple categories, remembering the caveats we made earlier. Table 12.1 provides a preliminary categorization. We tested for correlated evolution in the cercopithecine radiation using Maddison's (1990) concentrated changes test, implemented in the MacClade program (Maddison & Maddison 1992). We make the following assumptions: (1) the presence of triadic interactions is the evolutionary precondition for intensive male coalitions in this taxonomic group, and not the other way around; (2) the rare occurrence of triadic interactions in M. *fascicularis* is considered presence; and (3) species in which coalitions are scored as rare are considered as having no coalitions. The phylogeny (Figure 12.2) was based largely on the data of Purvis (1995), with changes in generic relations from Fleagle (1999) and in *Macaca* (except the position of M. *arctoides*) from Morales & Melnick (1998).

Under these assumptions and with this phylogeny, the hypothesis that male alliances arose independently of triadic male–infant interactions can be rejected at the $P = 0.01$ level (the four reconstructed gains were all on branches with reconstructed presence of triadic interactions). Changing the position of M. *arctoides* does not affect this conclusion. However, if we relax assumption (3) and instead consider all species with rare male–male coalitions to have male alliances, there is no relationship left. Thus it seems that triadic male–infant–male interactions were only a precondition for the evolution of intensive male–male alliances. Obviously, this conclusion remains tentative due to the uncertainties associated with binary classifications and the phylogeny used.

Discussion

As mammals go, primate males show a striking level of involvement with infants (Kleiman & Malcolm 1981). But since males are rarely heavily involved in routine daily care and provisioning of young it seems rather

Table 12.1. *Male–infant and male–male relationships in relation to demographic parameters in Old World primates living in multimale groups*

Species	Number of adult males in natural groups[a]	Breeding season duration (days)[a]	Intensity of male–infant handling[b]	Triadic male–infant–male interactions	Frequency of male–male coalitions[c]	Refs[c]
Colobus badius	3.5	244	0	No	variable	1
Presbytis entellus	2.5	180	0	No	No	2
Cercopithecus aethiops	3	92	0–1	No	No	3
Miopithecus talapoin	13	59	0	No		4
Macaca sylvanus	9	76	3	Yes	Yes (common)	5
Macaca nemestrina	3	365	0	No	Yes (rare)	6
Macaca silenus	1.75	365	0	No		7
Macaca tonkeana		365	0	No		8
Macaca maurus	4.25	365	0	No	No	9
Macaca nigra	5.8	365	0	No		10
Macaca fascicularis	4	123	1	Yes (rare)	Yes (rare)	11
Macaca mulatta	2.5	82	0–1	No	Yes (rare)	12
Macaca fuscata	3	46	0–1	No	Yes (rare)	13
Macaca cyclopis	3.3	Seasonal				14
Macaca assamensis		Seasonal	2	Yes	Yes	15
Macaca thibetana	4.5	89	3	Yes	Yes	16
Macaca radiata	7	92	1–2	Yes	Yes (common)	17
Macaca sinica	5	66	2	Yes		18
Macaca arctoides		365	1–2	Yes	Yes (rare)	19
Cercocebus torquatus			1	Yes	Yes	20
Cercocebus galeritus	2	59	0	No	No	21
Mandrillus sphinx		Seasonal	0	No		22
Lophocebus albigena	3	212	1–2	No		23
Papio anubis	14	365	2	Yes	Yes (common)	24

Table 12.1. (cont.)

Species	Number of adult males in natural groups[a]	Breeding season duration (days)[a]	Intensity of male–infant handling[b]	Triadic male–infant–male interactions	Frequency of male–male coalitions	Refs[c]
Papio cynocephalus	8	365	2	Yes	Yes (common)	25
Papio ursinus	7	365	2	Yes	Yes (rare)	26
Pan troglodytes	10	365	1	No	Yes (common)	27
Pan paniscus	8	365	1	No	Rare	28

Notes:

[a] After Mitani *et al.* 1996a, Table II, except for the following: *P. entellus*, Borries 1997; *M. cyclopis*, Kawamura *et al.* 1991 (small groups appear to be predominantly 1-male groups: Wu & Lin 1992); *M. maurus*: Matsumura 1991; *M. nigra*, O'Brien & Kinnaird 1997.

[b] After Whitten 1987: 0, no affiliation; 1, occasional affiliation; 2, affiliation; 3, intensive caregiving.

[c] References: (1) Struhsaker 1975; Starin 1994; (2) Borries 1997; Borries *et al.* 1999b; C. Borries (pers. comm.); (3) Fairbanks 1990; Lancaster 1971; Whitten 1987; (4) Rowell 1974; (5) Deag 1980; Deag & Crook 1971; Kuester & Paul 1992; J. Kuester & S. Preuschoft, unpublished data; Paul *et al.* 1996; Taub 1980, 1984; Whitten 1987; Witt *et al.* 1981; (6) Maestripieri 1994; Whitten 1987; Oi 1990; (7) Kumar & Kurup 1981; (8) S. Preuschoft, pers. obs.; (9) S. Matsumura, pers. comm., Matsumura 1997; Watanabe & Matsumura 1996; (10) Hadidian 1979, cited by Maestripieri 1998; (11) Auerbach & Taub 1979; Cichy 1996; van Noordwijk & van Schaik 1988, and unpubl. obs.; (12) Chapais 1983; Hill 1986; Meikle & Vessey 1981; Vessey & Meikle 1984; (13) Gouzoules 1984; Hiraiwa 1981; Watanabe 1979; (14) no data; (15) Bernstein & Cooper 1998; (16) Deng & Zhao 1987; Ogawa 1995a; Zhao 1996; (17) Simonds 1974; Silk 1992a,b, 1994, 1999; Silk & Samuels 1984; Sugiyama 1971; (18) Baker-Dittus 1985 cited by Maestripieri 1998; Dittus 1975; (19) Bauers & Hearn 1994; Bernstein 1980; Estrada 1984; Gouzoules 1975; Hendy-Neely & Rhine 1977; Maestripieri 1994; Smith & Pfeffer-Smith 1984; (20) Bernstein 1976; Bernstein *et al.* 1983; Busse & Gordon 1984; Gust 1994a; (21) Gust 1994a; (22) A. T. C. Feistner, pers. comm.; (23) Chalmers 1968; (24) Bercovitch 1988; Packer 1977, 1980; Ransom & Ransom 1971; Smuts 1985; Strum 1984, 1987; (25) Altmann 1980; Collins 1986; Noë & Sluijter 1990; Stein 1984a; (26) Bulger 1993; Busse 1984b; Busse & Hamilton 1981; Hall & DeVore 1965; Palombit *et al.* 1997; Saayman 1971; (27) Goodall 1986; Nishida 1983a,b; Nishida & Hiraiwa-Hasegawa 1987; de Waal 1982, 1992; (28) Furuichi & Ihobe 1994; Ihobe 1992; Kuroda 1992.

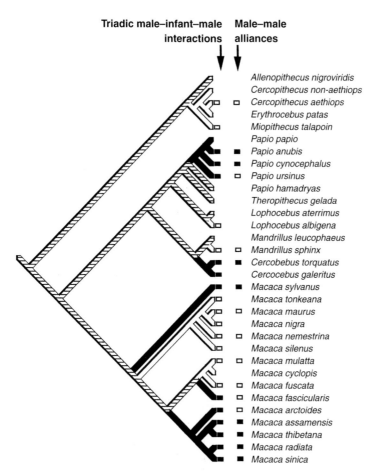

Figure 12.2. The evolution of triadic male–infant–male interactions in cercopithecines and its relationship to that of intensive male–male alliances. The branches of the cladogram reflect reconstructed character states for triadic interactions: open, absent; closed, present; shaded, equivocal. Male–male alliances are depicted at the tips: open, absent; closed, present.

unlikely that the unusually high level of male–infant affiliation in primates is related to the high costs of reproduction in this group of mammals. Instead, we have suggested that the threat of infanticide provides a better explanation: male care and infanticide represent two sides of the same coin.

At its face, male care does not appear to be related to the threat of infanticide. Male caregiving behaviors are not particularly common in species where infanticide is a significant source of infant mortality (hanuman langurs: Sommer 1994; Borries 1997; Borries & Koenig, Chapter 5; howler monkeys, Clarke & Glander 1984; mountain gorillas, Watts 1989; but for chacma baboons, see Palombit *et al.* 1997). However, as most clearly shown by hanuman langurs and mountain gorillas, the threat of infanticide may nonetheless be an important selective force regulating affiliative male–infant relationships. Moreover, as Smuts (1985: 114) suggested: "most of the observed infanticidal attacks succeeded precisely because there were no males present willing to protect the infant, and that such attacks would be much more common if male protection did not occur."

This conclusion may extend to humans, where fathers, although tolerant and protective, are not known to be heavily involved in routine daily care and provisioning of their offspring (see, e.g., Betzig 1992; Hewlett 1991, for traditional; and Lamb *et al.* 1987, for modern human societies). Yet, the mere presence of the father may be essential for the survival of their offspring, who are much more likely to be killed by stepfathers or other males than children living with both (genetic) parents (see, e.g., Daly & Wilson 1988; Hurtado & Hill 1992). Moreover, the infanticide protection hypothesis may provide a better explanation for the observation that among humans divorce rates peak after 4 years of marriage – i.e., at weaning age! – than the idea that mothers need their mates' help in caring for and provisioning their offspring (Fisher 1992).

The infanticide protection hypothesis may also explain why male care is more common in multimale groups than in one-male groups (Snowdon & Suomi 1982). While this observation has been ascribed more often to the fact that females in multimale groups can, at least principally, choose the male with the best paternal qualities (Smuts & Gubernick 1992), an alternative interpretation seems also possible: there are always males around who may gain reproductive benefits from killing unrelated infants. In fact, there is increasing evidence that infanticide is not only a source of considerable infant mortality in one-male groups but also in multimale groups (Collins *et al.* 1984; Izawa & Lozano 1991; Borries 1997; Palombit *et al.* 1997; van Schaik, Chapter 2; Borries & Koenig, Chapter 5).

An alternative interpretation for male–infant affiliation is that it represents male mating effort (Smuts & Gubernick 1992; Hawkes *et al.* 1995), an idea for which there is considerable empirical support (for a review, see van Schaik & Paul 1996–7; but see Tardif & Bales 1997). However, this hypothesis would not predict that male baboons react to the death of their female friends' infants by a sudden decline in interest (Palombit *et al.* 1997). Since their female friends are now able to conceive again, exactly the contrary would be expected. Moreover, both triadic male–infant–male interactions and adoptions of motherless infants by males (for a review, see Thierry & Anderson 1986) are obviously difficult to explain by the mating effort hypothesis. Since males do not lactate (and unweaned infants therefore rarely survive that adventure, see Taylor *et al.* 1978; Hasegawa & Hiraiwa 1980), the primary selective force leading males to adopt motherless infants seems to be protection of the infant against aggression by conspecifics. Although infanticidal males would obviously not benefit from killing motherless infants (unless resource competition is involved), such cases have been observed (Tarara 1987). (They do not necessarily contradict the sexual selection hypothesis; see van Schaik, Chapter 2). Protection against infanticide by likely sires is therefore an important function of male–infant affiliation, but male–infant relationships are too diverse to claim that infanticide avoidance is the major, let alone the sole, function.

We also suggested that triadic male–infant–male interactions, in which infants are used as agonistic buffers, are historically based on the protective responses given to infants attacked by adult males. However, their current function is thought to relate to the need for male–male coalitions, which are found mainly in various baboons and macaques. Preliminary tests are not unfavorable for this hypothesis. The absence of infant carrying by males and triadic male–infant–male interactions in chimpanzees, a species with elaborate male alliances, is not inconsistent with the hypothesis, because male chimpanzees affiliate frequently without infants and, being philopatric and thus variously related (cf. Morin *et al.* 1994), are expected to be more tolerant.

Savanna baboons are also a special case. Although they generally fit neatly into the present framework, male–male coalitions appear to be rare in chacma baboons – despite the fact that in this subspecies triadic male–infant–male interactions appear to be even more common than among their East African counterparts (Taub 1985). The most likely explanation seems to be that, as suggested by Smuts & Watanabe (1990), "greetings" may be better suited to making friends in baboons than triadic male–infant–male interactions because in contrast to macaques baboon "greetings" constitute the only context in which close body contact is established and friendly signals are exchanged (perhaps not coincidentally "genital touching" – a common behavior during baboon "greetings" also occurs regularly during triadic male–infant interactions in Barbary macaques, A. P. and S. P., pers. obs.). Thus triadic male–infant–male interactions in baboons may be more closely related to "classic agonistic buffering" or, as it seems to be the case in the Moremi baboons, protection against infanticide.

In any case, the threat of infanticide appears not only to have shaped the lives of female primates but also those of males in many, and sometimes quite unexpected ways.

Acknowledgments

We very grateful to Carola Borries, Anna Feistner and Shuichi Matsumura for providing unpublished information.

13

Female dispersal and infanticide avoidance in primates

Introduction

Sex-biased dispersal

Dispersal is a common feature of bird and mammal societies. Philopatry, however, is generally assumed to be the most advantageous situation, because it assures knowledge of the home range and offers better opportunities for cooperation with relatives. Several benefits of foregoing these advantages have been proposed (better mating opportunities, including inbreeding avoidance; reduction of within-group competition; coercion) (Greenwood 1980; Pusey & Packer 1987a; Moore 1993). The balance between the costs and benefits of dispersal obviously differs between the sexes, because dispersal is often highly sex biased. Most bird species show a female bias in dispersal, whereas most mammal species show a male bias (Greenwood 1980; Dobson & Jones 1985; Liberg & Schantz 1985). The best explanation proposed so far for such a strong sex bias is inbreeding avoidance.

In mammals it is generally assumed that the costs of dispersal are higher for females than for males, because loss of a range or allies affects female reproductive success more than male reproductive success, explaining the prevalence of male-biased dispersal (Waser *et al.*, 1986; Pusey & Packer 1987a; Clutton-Brock 1989a). Female dispersal, however, is found in some mammal species and a considerable number of primate species (Moore 1984; Clutton-Brock 1989a; Strier 1994).

Costs and benefits of dispersal

Different costs and benefits of female philopatry and dispersal can be distinguished. Females in gregarious animals such as diurnal primates can remain in their natal range or in their natal group. Accordingly, dispersal can be separated into locational dispersal (site desertion) and social dispersal or group desertion (Isbell & van Vuren 1996; Sterck 1998). The possible benefits of female philopatry are many. Females that remain in their natal area have better knowledge of their range (Waser & Jones 1983). This may enhance predator avoidance (e.g., Isbell *et al.* 1990) and may reduce the search costs for food (e.g., Pope 1989). Females that remain with their natal group retain their allies in within- or between-group conflicts (Wrangham 1980; van Schaik 1989).

There are also possible costs of female philopatry. Females in a population with male philopatry or with a long tenure of breeding males may run the risk of inbreeding (Clutton-Brock 1989a). The costs of sharing a range will be high when food sources cannot be shared (Waser & Jones 1983). Moreover, large groups will experience strong within-group competition (for a review, see Pusey & Packer 1987a). In addition, females in large groups may experience an increased infanticide risk (Crockett & Janson, Chapter 4).

Dispersal has several benefits. Locational dispersal can lead to a range in a higher-quality habitat. Dispersing to another group can yield new, more or better mating partners (i.e., not related, with better genes or better able to protect females and offspring against predators or infanticidal males). The costs of dispersal are related to the loss of a familiar range or group (see above). In addition, when the dispersing individual is solitary for some time, the risk of predation may be much higher than in a group. The act of dispersal may also have costs. Gaining entrance in a new group may be costly due to resistance of group members (cf. Glander 1992). Moreover, dispersing females may postpone reproduction (cf. Sterck 1997). The costs of dispersal may be low when food is relatively easily found or when ranges of the group of origin and destination overlap (Isbell & van Vuren 1996).

Female dispersal and infanticide risk

In this chapter, we address a possible connection in primates between female dispersal and male infanticide. Infanticide is usually conducted by a male unrelated to the infant that has just gained the dominant position in the group (Struhsaker & Leland 1987; van Schaik, Chapter 2). This can be an immigrant or resident male. Alternatively, males may kill infants during between-group encounters (Watts 1989; Sterck 1997, 1998). Resident males, especially potential fathers, are often effective in protecting their infants against infanticide (e.g., Sugiyama 1966; Paul *et al.*, Chapter 12).

There are three hypotheses that relate female dispersal to an ultimate reduction in the risk of infanticide. The first hypothesis, "remaining with the father", concerns protection of the current infant. A female increases the survival chances of her current infant when she disperses with its ousted father after a new male has taken over the breeding position in her group (Rudran 1973; Hrdy 1977). The second hypothesis, "choosing the best male", concerns protection of future offspring, because a female chooses the male that is a better protector against infanticide than the current male (Marsh 1979b). The third hypothesis, "reducing the chance of male takeovers", concerns protection of both current and future offspring. Resident females reduce the risk of infanticide in their current group by evicting additional breeding females (Crockett & Janson, Chapter 4). Through this action, they reduce the rate of male takeovers in their group and, thereby, reduce the risk of infanticide. In the first two hypotheses the dispersing females reduce their risk of infanticide, whereas in the last hypothesis it is not the dispersing females but the philopatric ones that benefit.

The ultimate benefits of dispersal, whether they concern food competition, predation risk, inbreeding avoidance, mating opportunities or infanticide risk, have to be measured by a comparison of the new situation and the situation as it would have been without dispersal. Only when the circumstances are better in the new situation can an ultimate advantage be proven. Proximate mechanisms that are related to any of the above measures may give an indication of the ultimate importance of a particular phenomenon, but do not yield decisive evidence.

Ultimate benefits, however, can be achieved only through proximate mechanisms, although the relationship between proximate mechanism and ultimate benefits does not have to be a close one. For example, the eviction of new breeding females in red howler monkeys was first thought to be related to food competition (Crockett 1984). An analysis of the data, however, revealed that not so much food competition but infanticide risk, was the main correlate of eviction, although no direct link between female dispersal and infanticide was evident (Crockett & Janson, Chapter 4). Thus dispersal events related to infanticide or to situations with a high infanticide risk may, but do not necessarily, lead to a reduced rate of infanticide in the new situation.

Proximate mechanisms related to infanticide avoidance

Notwithstanding the fact that the relationship between proximate mechanisms and ultimate explanations is not necessarily close, we further investigate proximate mechanisms that may be related to infanticide avoidance. Alternative explanations for these patterns of behavior are sought. The first hypothesis, i.e., "remaining with the father", has a direct link with a proximate mechanism. After the eviction of a resident male, females with infants should disperse with him, whereas females without infants should readily accept the new male. Such observations are not expected under any other ultimate hypothesis concerning female dispersal. This hypothesis requires that the risk of infanticide is high after a new male takes over a group. In addition, only dispersal of parous females is expected. However, other situations are possible in which the costs of dispersal are outweighed by the benefits for parous females only.

A prediction from the second hypothesis, i.e., "choosing the best male", concerns the reproductive status of dispersing females. A female may disperse and choose a new male when future infants will be at risk in her current group. This applies to both nulliparous and parous females. The current infants of parous females are at risk when infanticide can take place during between-group encounters. A dispersing parous female will join a new male not related to her previous offspring. He may behave infanticidally toward a previous infant. Therefore, parous females that disperse are expected to do so without infants. However, alternative explanations for such a dispersal pattern are possible, because unrelated males are probably not the only hazard that dispersing infants face. During dispersal or after immigration infants may experience reduced survival chances due to a higher predation risk, aggressive opposition to its mother's immigration by females of the new group or reduced milk intake due to a lower food intake of the dispersing female. For all these reasons, females may also preferentially disperse without infants. In addition, when females disperse to a new male, a version of the "remaining with the father hypothesis" predicts that females with an infant are expected to remain with the father in the former group, thus leading to a bias for non-lactating females to disperse.

According to the third hypothesis, "reducing the chance of male takeovers", the risk of infanticide increases with an increasing number of breeding females. Infanticide should take place after male takeovers. Females are expected to evict maturing, nulliparous females. Alternatively, maturing females may be evicted because they compete for limited resources such as food or male parental care. Maturing females may also disperse to avoid inbreeding (Clutton-Brock 1989a). Such females, however, are expected to disperse voluntarily.

In reviews of female dispersal, the distinction between pre-reproductive and reproductive dispersal is often not made explicitly (Moore 1984; Pusey & Packer 1987a; Strier 1994; Isbell & van Vuren 1996) or accounts are limited to nulliparous females (Clutton-Brock 1989a). The three hypotheses relating infanticide risk to female dispersal, however, make explicit predictions regarding the female reproductive stage.

To test the predictions of the three hypotheses and the proposed proximate mechanism, data on dispersal of female primates were collected from the literature. Proximate patterns relating to female dispersal are described. Measures of the occurrence of female dispersal are related to possible proximate factors. The evidence for an ultimate relationship between infanticide avoidance and female dispersal is reviewed.

Methods

Data extracted from the literature

Literature was searched for reports of female dispersal in diurnal primates. For each species, each site at which female dispersal was observed is recorded separately. Cases might have been missed, as female dispersal is often mentioned in passing. In total, we found reports of female dispersal for 37 primate species and 48 populations.

Females were considered to disperse when they were known to immigrate into a group or were observed to wander alone after emigration. We recorded whether the female was nulliparous (pre-reproductive dispersal) or parous (reproductive dispersal) during dispersal. The incidence of pre-reproductive dispersal was considered to be rare when 50% or fewer of all immatures dispersed before they reached adulthood and it was considered common when more than 50% dispersed before becoming adult. For parous females it was noted whether they dispersed together with an infant or without an infant. Pregnant females were considered to disperse without an infant. Some females dispersed with a young, but weaned, offspring. These were counted as females that dispersed without an infant. We calculated the percentage of all dispersing females that were nulliparous and the percentage of parous females that dispersed without infants of all dispersing parous females.

The rate of female dispersal (females per female-year) was calculated for all females, for immature females and for adult females. The female dispersal rate was the total number of dispersing females divided by the number of total female-years in the population. The number of female-years was the number of groups times the average number of females in a group times the duration of the study in years. The female-years of immature females included infants, juveniles and subadult females, because in a number of species dispersing immature females could be infants, juveniles or subadults. When the sex of immatures was not reported, we assumed that half of the immatures were female. The female-years of adult females included only adult females.

Male infanticide was recorded as present when it was observed in the species (van Schaik, Chapter 2) and as a potential risk when females did not show postpartum estrus (van Schaik, Chapter 3). Males could commit infanticide after obtaining the alpha position or during between-group encounters. When infanticide was observed both during between-group encounters and after takeovers, it was counted as occurring during between-group encounters.

Behavioral patterns that could be related to female dispersal were recorded: eviction of females by females; aggression from resident females against immigrant females; male philopatry; and the length of male tenure relative to female age at menarche.

Statistics
The data were analyzed with Pearson correlations or non-parametric statistics and analyzed in SPSS for Macintosh. The test were two-tailed unless mentioned and the results were considered significant at $\alpha < 0.05$.

Results

General patterns in female dispersal
The first two ultimate hypotheses predict that parous females will disperse in relation to infanticide, whereas the third hypothesis predicts that mainly nulliparous females disperse. The proportion of pre-reproductive versus reproductive dispersal differs enormously between species and populations (Table 13.1, Figure 13.1). Exclusive dispersal of only one female class can be expected in small samples. When populations with four or fewer cases of female dispersal were omitted, many of the extreme cases disappeared. Still, exclusive pre-reproductive dispersal was found in three populations, i.e., the muriqui (*Brachyteles arachnoides*), the red

howler monkey (*Alouatta seniculus*) and the chimpanzees (*Pan troglodytes*) of Taï. Exclusively reproductive dispersal was present in four populations, i.e., the hanuman langur (*Presbytis entellus*) of Kanha, the capped langur (*Presbytis pileata*), the mountain baboon (*Papio ursinus*) and the yellow baboon (*Papio cynocephalus*). In the mountain baboon, only adult females were marked and recognized, so the result may be a consequence of the research methods. Most species and populations with a reasonable number of dispersing females, however, showed both pre-reproductive and reproductive dispersal.

In 8 of the 15 populations with both types of dispersal and five or more dispersal events, pre-reproductive dispersal was more common than reproductive dispersal. This implies that, after the first dispersal event, most females remain in their new group for the remainder of their life. This applied to the ring-tailed lemur (*Lemur catta*), the mantled howler monkey (*Alouatta palliata*), the chimpanzees of Gombe and Mahale, the red colobus monkey (*Colobus badius*) of Abuko, the saddleback tamarin (*Saguinus fuscicollis*), the longtailed macaque (*Macaca fascicularis*) and the hamadryas baboon (*Papio hamadryas*).

In the other seven populations, reproductive dispersal was equally common as or more common than pre-reproductive dispersal. This implies that in species where most or all females disperse, in principle all females disperse several times during their lifetime. This applies to (possibly) the mountain gorilla (*Gorilla gorilla berengei*), the white sifaka (*Propithecus verrauxi*), the Thomas's langur (*P. thomasi*), the red colobus monkey of Tana, the silver langur (*P. cristata*) and the common marmoset (*Callitrix jacchus*) of Dois Irmaos Forest. The seventh case concerns the hanuman langur of Abu, a population where dispersal is rare.

Male infanticide and female dispersal

Male infanticide is thought to be a common phenomenon in primates (Hrdy *et al.* 1995). The risk of infanticide is present in 34 of the 37 species with female dispersal (cf. van Schaik, Chapter 2). Infanticide has been observed or was strongly suspected to take place in 16 of the 37 species with female dispersal, whereas it was known to be absent in three species (Table 13.1). The presence of male infanticide in a population did not significantly affect either the percentage of dispersing nulliparous females (Mann–Whitney U-test: $N_1 = 6$, $N_2 = 22$, $U = 51$, $Z_{(ties)} = -0.846$, $P = 0.40$) or the percentage of parous females dispersing without infant of all dispersing parous females (Mann–Whitney U-test: $N_1 = 1$, $N_2 = 18$, $U =$

Table 13.1. *Reproductive status of dispersing females in primates, population features and the*

Ref.	Species	Site	Incidence of pre-reproductive dispersal (amount)	Nulliparous female (N)	Total parous females (N)	Parous without infant (N)	Parous with infant (N)	% parous females without offspring of all dispersing parous females	% of dispersing females that are nulliparous	Female dispersal rate (all females)
1	Eulemur rubiventer	Ranomafana	?	1[d]	0	–	–	–	100	–
2	Lemur catta[b]	Duke University Primate Center	Common (many)	8	1	1	0	100	89	–
3	Propithecus verreauxi	Beza Mahafaly	?	2	3	–	–	–	40	–
4	Propithecus tattersalli	–	?	0	1	0	1	0	0	–
5	Saguinus nigricollis	River Caquetá	?	1	0	–	–	–	100	–
6	Callitrix jacchus	Tapacurá	?	1	1	–	–	–	50	0.17
7	Callitrix jacchus	Dois Irmaos Forest	?	1	9[c]	–	–	–	10	1.15[d]
8	Callitrix jacchus	EFLEX-IBAMA	?	2	1	1	0	100	67	0.17
9	Saguinus fuscicollis	Cocha Cashu, Manu	Common (most)	9	2	–	–	–	82	0.07
10	Alouatta palliata (immigrants)	Hacienda La Pacifica	Common (most)	29[e]	4[c]	4	0	100	88	–
11	Alouatta seniculus	Hato Masaguaral – Mata	Common (most)	14	0	–	–	–	100	–
12	Alouatta caraya	Río Riachuelo	?	2	1	1	0	100	67	0.03
13	Ateles paniscus chamek	Cocha Cashu, Manu	Common (most)	2	0	–	–	–	100	0.01
14	Brachyteles arachnoides	Fazenda Montes Claros	Common (most)	11	0	–	–	–	100	0.05
15	Saimiri oerstedi	P.N. Corcovado	Common (all)	Several	7	6[f]	1	86	–	–
16	Papio hamadryas	Erer	Common (all)	24	19	17	2	89	56	0.19
17	Papio cynocephalus	Mikumi	?	0	12[g]	9	3	75	0	–

ccurrence of female eviction (12 September 1999)

Female dispersal rate immature subadult males	Female dispersal rate (adult females)	Male infanticide present	Potential risk male infanticide	Timing male infanticide	Female eviction	Female aggression against immigrants	Male philopatry	Male tenure (mo) + female menarche (mo)
	–	?	Yes	?	Yes	?	No	?
	–	Yes	Yes	?	Yes	No	No	?
	–	?	Yes	?	Yes	?	No	?
	–	?	Yes	?	Yes	?	No	?
	–	No	No	?	?	?	?	?
15	0.18	No	No	?	?	?	No	?
33[d]	1.58[d]	No	No	?	?	?	No	?
6	0.08	No	No	?	?	?	?	?
13	0.02	No	No	?	?	?	Yes (occasionally)	22 > [13 mos first conception]
	–	Yes	Yes	a	Yes	Yes	Yes (occasionally)	46 > 36
	–	Yes	Yes	a	Yes	Yes	No	60–90 > 48
05	0.02	Yes	Yes	?	?	?	No	
04	0	?	Yes	?	?	?	Yes	?
10	0	?	Yes	?	No	Yes (some?)	Yes (usually)	?
	–	?	Yes	?	No	No	Yes (usually)	?
20	0.17	Yes	Yes	?	No	No	Yes	36–84 = 52–67
	–	Yes	Yes	?	No	No	No	?

Table 13.1. (cont.)

Ref.	Species	Site	Incidence of pre-reproductive dispersal (amount)	Nulli-parous female (N)	Total parous females (N)	Parous without infant (N)	Parous with infant (N)	% parous females without offspring of all dispersing parous females	% of dispersing females that are nulliparous	Female dispersal rate (all females)
18	*Papio ursinus*	Suikerbos-rand	?	0	9	6	2	75	0	–
19	*Macaca fascicularis*	Ketambe	Rare (few)	3	2	2	0	100	60	–
20	*Colobus badius*	Abuko N. R.	Common (most)	23	5	4[h]	1[i]	80	82	0.14
21	*Colobus badius*	Kibale	Common (all)	Several	1	–	–	–	–	–
22	*Colobus badius*	Tana[j]	Common (most?)	3	10	10	0	100	23	0.33
23	*Colobus guereza*	Bole	Rare (none)	0	1	1	0	100	0	–
24	*Colobus polykomos*	Tiwai	?	0	1	1	0	100	0	–
24	*Colobus satanus*	Lopé	?	0	3	2	1	67	0	–
24	*Colobus verus*	Tiwai	?	0	1	1	0	100	0	–
25	*Nasalis larvatus*	Samunsam	?	?	3	1	2	33	–	–
26	*Presbytis cristata*	Kuala Selangor	?	1	6	6	0	100	14	0.02
27	*Presbytis entellus*	Kanha	?	0	6	5	1	83	0	–
28	*Presbytis entellus*	Jodhpur	rare (few)	1	0	–	–	–	100	
29	*Presbytis entellus*	Dharwar	rare (some)	1 (some)	some	–	–	–	–	–
30	*Presbytis entellus*	Ramnagar	rare (few)	2	0	–	–	–	100	0.06[k]
31	*Presbytis entellus*	Abu[l]	rare (few)	3	3	0	31	0	50	0.03
32	*Presbytis femoralis*	Perawang	?	0	4[m]	4[m]	0	100	0	0.10
33	*Presbytis johnii*	Ootacamund	?	0	4[n]	3	1[n]	75	0	0.16
34	*Presbytis melalophos*	Kuala Lompat	?	0	1	0	1	0	0	0.02
35	*Presbytis pileata*	Madhupur	?	0	6	6	0	100	0	0.05
36	*Presbytis rubicunda*	Sepilok	?	2[o]	0	–	–	–	100	–

male dispersal rate immature subadult males)	Female dispersal rate (adult females)	Male infanticide present	Potential risk male infanticide	Timing male infanticide	Female eviction	Female aggression against immigrants	Male philopatry	Male tenure (mo) + female menarche (mo)
	0.07	Yes	Yes	?	No	No	No	?
	–	Yes	Yes	?	?	?	No	?
30	0.04	Yes	Yes	?	No	Mild	Yes (virtually all)	Male philo-patry >34
	–	Yes	Yes	a	No	No	Yes (virtually all)	?
27	0.35	Yes	Yes	a	?	?	No	?
	–	Yes	Yes	?	?	?	?	?
	–	?	Yes	?	?	?	?	?
	–	?	Yes	?	?	?	?	?
	–	?	Yes	?	?	?	?	?
	–	?	Yes	?	?	?	No	?
01	0.04	Yes	Yes	a	?	?	No	?
	–	Yes	Yes	a, d	?	?	No	?
		Yes	Yes	a	No	?	No	27<29
	–	Yes	Yes	a	?	?	No	?
	–	Yes	Yes	a	?	(No)	No	?
04	0.03	Yes	Yes	a	?	No	No	?
	0.30	?	Yes	?	?	Mild	?	?
	0.33	?	Yes	?	?	No	No	?
	0.03	?	Yes	?	?	No	?	?
	0.10	?	Yes	?	?	No	?	?
	–	?	Yes	?	?	?	?	?

Table 13.1. (cont.)

Ref.	Species	Site	Incidence of pre-reproductive dispersal (amount)	Nulli-parous female (N)	Total parous females (N)	Parous without infant (N)	Parous with infant (N)	% parous females without offspring of all dispersing parous females	% of dispersing females that are nulliparous	Female dispersal rate (all females)
37	Presbytis senex	Polonnaruwa	?	1	2	1	1[p]	50	33	0.01
38	Presbytis thomasi	Ketambe	Common (all)	7	23	21[q]	2[q]	91	23	0.06
39	Hylobates lar	Mo Singto	Common (most/all)	1	0	–	–	–	100	0.01
40	Hylobates lar	Ketambe	Common (most/all)	2	0	–	–	–	100	0.07
40	Hylobates syndactilus	Ketambe	Common (most/all)	1	1	1	0	100	50	0.14
41	Gorilla gorilla berengei	Karisoke	Common (most)	8	9	9	0	100	47	0.01[s]
42	Pan paniscus	Wamba	Common (all)	3	0	–	–	–	100	–
43	Pan troglodytes	Gombe	Rare (13–50%)	10	2[t]	1	1	50	83	0.02
44	Pan troglodytes	Mahale	Common (most)	18	9	9	0	100	67	0.05
45	Pan troglodytes	Taï	common (almost all)	13	0	–	–	–	100	0.06

Notes:

Ref: reference number (1) Overdorff 1991; (2) Sussman 1991; Vick & Pereira 1989; (3) Richard et al. 1993; (4) Meyers 1993, cited in Kappeler 1999; (5) Izawa 1978; (6) Scanlon et al. 1988; (7) Mendes Pontes & da Cruz 1995; Roda & Mendes Pontes 1998; (8) Digby & Baretto 1993; Digby 1995b; (9) Goldizen & Terborgh 1989; Goldizen et al. 1996; (10) Clarke 1983; Glander 1992; Clarke & Zucker 1994; (11) Crockett 1984; Crockett & Sekulic 1984; Crockett & Pope 1993; (12) Rumiz 1990; (13) Symington 1988; Strier 1994; (14) Strier 1991; Strier et al. 1993; (15) Boinski 1987a,b; Boinski & Mitchell 1994; (16) Abegglen 1976; Sigg et al. 1982; (17) Rasmussen 1981; (18) Anderson 1987; (19) van Noordwijk 1985; van Noordwijk & van Schaik 1985, 1999; van Schaik & van Noordwijk 1988; C. P. van Schaik pers. comm.; (20) Starin 1991, 1994; (21) Struhsaker & Leland 1979; 1985; Strusaker & Pope 1991; (22) Marsh 1979a,b; (23) Dunbar & Dunbar 1974; Dunbar 1987; (24) Oates et al. 1994; (25) Bennett & Sebastian 1988; (26) Wolf 1984; Wolf & Fleagle 1977; (27) Newton 1986, 1987; (28) Sommer 1987; Sommer & Rajpurohit 1989; Sommer et al. 1992; (29) Y. Sugiyama pers. comm.; 1965; Yoshiba 1968; (30) Borries 1997; Koenig et al. 1998; (31) Hrdy 1977; (32) Megantara 1989; (33) Poirier 1970; (34) Bennett 1983; (35) Stanford 1991; (36) Davies 1987; (37) Rudran 1973; (38) Sterck 1997; Steenbeek 1999a; Steenbeek et al. 2000; (39) Brockelman et al. 1998; (40) Palombit 1994a; (41) Harcourt et al. 1976; Watts 1989, 1991b, 1994; (42) Idani 1991; Boesch & Boesch 2000; (43) Pusey 1979; Goodall 1986; Boesch & Boesch 2000; (44) Nishida et al. 1985, 1990; Uehara et al. 1994; (45) Boesch & Boesch 2000.

Column 4: Incidence of pre-reproductive dispersal (amount of dispersing nulliparous females). Column 14: Male infanticide present: cf. Table 2.2. Column 15: Potential risk male infanticide: cf. Table 2.2. Column 16: Timing male

Female dispersal rate immature subadult (males)	Female dispersal rate (adult females)	Male infanticide present	Potential risk male infanticide	Timing male infanticide	Female eviction	Female aggression against immigrants	Male philopatry	Male tenure (mo) + female menarche (mo)
01	0.01	Yes	Yes	a, d	?	?	No	?
05	0.07^f	Yes	Yes	d	No	mild	No	59 > 52
11	0	?	Yes	?	?	?	No	?
15	0	?	Yes	?	No	No	No	?
14	0.15	?	Yes	?	No	?	No	?
	–	Yes	Yes	d	No	Yes	No	96 > 89 (–120)
	–	?	Yes	?	?	little	Yes	Male tenure > 170
04	0.01	Yes	Yes	d	No	Yes	Yes	Male tenure > 120–138
07	0.03	Yes	Yes	d	No	?	Yes	Male residency > 120–132
12	0	Yes	Yes	?	No		Yes	Male residency > 164

infanticide: a, after takeovers; d, during between-group encounters.

^{a} Possibly two maturing females dispersed (Overdorff 1991).

^{b} Data from a captive, but semi-free-ranging population (Vick & Pereira 1989).

^{c} A number of these females returned to their original group after a few months.

^{d} These dispersal rates may be unnaturally high due to repeated capturing.

^{e} Twenty-two immigrants and 7 females that moved through and whose group of immigration was not known; a female that moved through one of the known groups and later immigrated elsewhere was counted once; four females of groups that lost their male fused with neighbours (Glander 1992: 428), and this was counted as immigration; one of these females later transferred to an other group, and this was not counted separately.

^{f} One of the immigrating adult females was pregnant.

^{g} Two of these females emigrated temporarily; Rasmussen (1981: 238) speculates that the 10 immigrant females may have left the group during the initial phases of habituation.

^{h} Two parous females transferred after their infants were killed during infanticidal attacks (Starin 1991: 181).

^{i} This female joined permanently one group after a 5 month period during which she and her infant had "dual group membership" of this and another group.

^{j} The number of dispersing females is derived from immigrating females.

^{k} Taken from Koenig et al. (1998).

Notes: to table 13.1 (*cont.*)

[l] Temporary splitting of the group. These three females first remained with the ousted male and subsequently joined their former group in the presence of infants. Two of these infants (both 13 months old) were partially weaned in the group with the ousted male (Hrdy 1977: 266). All three (aged 6, 13 and 13 months) were immediately weaned after their mothers had returned to their old group that contained a new dominant male. One of the infants (13 months old) was attacked and wounded by the new resident male on the day of immigration.

[m] One female really transferred; three females only stayed for a short time with the new group.

[n] Home range of the group was being logged; no data on the survival of the infant.

[o] The dispersal of these females was likely; after they disappeared two habituated females of similar size were found in the presence of a new and relatively shy male in an other part of the research area.

[p] Female with infant migrated with ousted male.

[q] Of the five immatures that transferred with their mother three were weaned and two were not. These two latter infants died soon after their mother entered the new group.

[r] Based on calculations of Chapter 4, Table 4.1 and data in Chapter 8 of Steenbeek 1999a; we use these data, because we want to compare them with the total and immature dispersal rate; Steenbeek's own estimate of adult female dispersal was 0.14 females per year.

[s] The female dispersal rate is a rough estimate, as the number of young immature females are not included.

[t] Both were primiparous and possibly peripheral females.

4.5, $Z = 0.889$, $P = 0.38$). A similar result was obtained when the potential risk of male infanticide was used.

The female dispersal rate is expected to be higher in populations where male infanticide is present. The adult female dispersal rate, however, was not affected by infanticide (Mann–Whitney U-test: $N_1 = 4$, $N_2 = 12$, $U = 11.5$, $Z = 1.522$, $P = 0.13$) whereas the immature dispersal rate (Mann–Whitney U-test: $N_1 = 4$, $N_2 = 11$, $U = 6$, $Z = 2.095$, $P = 0.04$) was actually significantly lower in species with infanticide. In addition, the incidence of pre-reproductive dispersal was not related to the presence of infanticide (Fisher Exact test: $FI = 0.593$, $P = 1.00$). A similar result is obtained when the potential risk of male infanticide was used.

Comparing the measures of female dispersal

The occurrence of female dispersal was measured in three different ways: as the dispersal rate of immature females, adult females and all females; as the percentage nulliparous females of all dispersing females; and as the incidence of pre-reproductive dispersal. These measures were not known for all populations. The relationships between the different measures were investigated. The dispersal rate of immature and adult females were not correlated (Figure 13.2; Correlation coefficient: $N = 23$, $r = 0.085$, $P > 0.05$). A high dispersal rate of adult females does not necessarily imply a high rate of nulliparous dispersal. This indicates that the adult and immature dispersal rates are independent.

The incidence of pre-reproductive dispersal, categorized for each

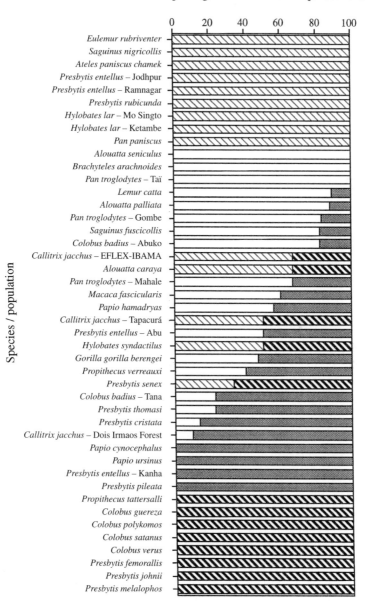

Figure 13.1. Parity of dispersing females. ind., individual. □ % nulliparous females > ind. ▨ % parous females > ind. ▨ % nulliparous females ≤ 4 ind. ▨ % parous female ≤ 4 ind.

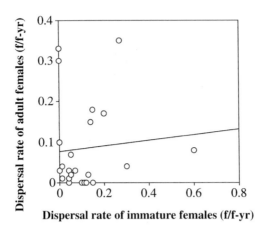

Figure 13.2. The relationship between the dispersal rates (females per female-year) of immature and adult females. The data of the *Callitrix jacchus* at Dois Irmaos Forest were probably unrealistically high and were therefore omitted from the figure.

species as rare or common, corresponds well with the immature dispersal rate (Mann–Whitney U-test: $N_1=3$, $N_2=12$, $U=1$, $Z=2.463$, $P=0.01$), as it should. However, it was not reflected in the percentage of nulliparous females among the dispersers (Mann–Whitney U-test: $N_1=6$, $N_2=17$, $U=43.5$, $Z_{(ties)}=-0.542$, $P=0.59$). Neither was it reflected in the adult female dispersal rate (Mann Whitney U-test: $N_1=3$, $N_2=12$, $U=17$, $z_{(ties)}=0.147$, $P=0.88$). This finding confirms that the dispersal of immature and adult females are independent phenomena.

The percentage of nulliparous females among dispersing females, and therefore also its complement, the percentage of parous females, was not related to the dispersal rate of immature females ($N=23$, $r=0.317$, NS) or all females ($N=23$, $r=0.211$, NS). The percentage of nulliparous females decreases, and thus the percentage of parous females increases with an increasing dispersal rate of adult females ($N=23$, $r=-0.610$, $P<0.01$). In other words, whereas the percentage of dispersing females that is nulliparous does not depend upon the absolute dispersal rate of immature females, this correlation was found for parous females. Thus, when a higher percentage of the dispersing females is parous, in general the incidence of dispersal of parous females is also high, suggesting that each female disperses more than once in her lifetime.

Reproductive stage of females during reproductive dispersal
In most species and populations with female dispersal, at least some parous females were known to disperse. When all species and populations are considered, the reproductive stage of 129 dispersing parous females was known (Table 13.1). Of these females, 18 (14%) dispersed with infant, whereas 111 (86%) dispersed without dependent offspring. Therefore, most parous females dispersed without dependent offspring.

Testing the hypotheses

"Remaining with the father" hypothesis
Parous females may disperse with their infant in order to remain with the ousted father of their offspring. Of the females that dispersed with their infant, only one female (a purple-faced langur, (*Presbytis senex*); Rudran 1973) dispersed in the presence of the ousted resident male of her former group.

Three hanuman langur females with infants from Abu were found in the presence of the father of their offspring in their former range (Table 13.1 and 13.2), while the other females and a new male used the same range. These three females may have remained with the ousted male, but the actual splitting of this group was not observed. After a short period, these females with infants joined the group with the new male. This event is presented here as female dispersal, but it is possible that these females actually never split from the main body of the group, because all events probably happened within a month's time. In the latter case, it is an example of females remaining temporarily with the ousted male. All other females with infants joined groups with unfamiliar males. Therefore, the majority of these females did not aim to protect their current infant by dispersing.

For this hypothesis, infanticide was expected after a male takeover. However, the percentage of dispersing parous females without infants of all dispersing parous females was not affected by the timing of infanticide (Mann–Whitney U-test: $N_1 = 6$, $N_2 = 4$, $U = 9$, $Z_{(ties)} = 0.685$, $P = 0.49$).

"Choosing the best male" hypothesis
Many studies report that dispersing parous females do not have an infant with them. Some of these females lost their infants due to infanticidal attacks (gorillas, Watts 1989; red colobus monkey, Starin 1991; Thomas's langurs, Sterck 1997) or to unknown causes (chimpanzees of Mahale, Nishida *et al.* 1985; mountain baboons, Anderson 1987). In addition, some

Table 13.2. *The reproductive stage of dispersing parous females and population interbirth interval and weaning age*

Ref.	Species	Inter-birth interval (mos)	Weaning age (mos)	Expected distribution (without vs. with)	Female without vs. with infant	Binomial test (P)
15	*Saimiri oerstedi*	–	–	–	7:1	–
16	*Papio hamadryas*	24	–	–	17:2	–
17	*Papio cynocephalus*	–	–	–	9:3	–
18	*Papio ursinus*	–	–	–	6:2	–
20	*Colobus badius* – A.	29	–	–	4:1	–
22	*Colobus badius* – T.	25	–	–	10:0	–
26	*Presbytis cristata*	23	–	–	6:0	–
27	*Presbytis entellus* – K.	27	13–20[a]	10:17	5:1	0.0369 (0.40)
35	*Presbytis pileata*	–	–	–	6:0	–
38	*Presbytis thomasi*	24	17	7:17	21:2	0.0000 (0.30)
41	*Gorilla gorilla berengei*	48	21[b]	27:21	9:0	0.0008 (0.55)
44	*Pan troglodytes* – M.	72	54[c]	18:54	9:0	0.0000 (0.25)

Notes:

Ref., see Table 13.2.

[a] The weaning age from Abu (Hrdy 1977) was taken.

[b] We took 21 months, because Fossey (1983) wrote: weaning becomes most traumatic for the infant around the middle to the end of the second year.

[c] The oldest infant to have nipple contact was 4 years and 11 mos old. The average age at weaning was not mentioned. However, "the five year period before weaning . . ." indicates that 5 years is some upper limit and nipple contact was recorded up to and including an age of 4 years (Hiraiwa-Hasegawa 1990). Therefore, we took the average weaning time at the middle of the fifth year of life, i.e., at 54 months.

females leave their infants behind (Thomas's langur, Sterck 1997) or wean it before or immediately after entering the new group (hanuman langur of Abu, Hrdy 1977; hamadryas baboons, Sigg *et al.* 1982; chimpanzees of Mahale, Nishida *et al.* 1985).

That many females leave without an infant does not imply that this event occurs more often than expected. This pattern was tested for the 12 species and populations with five or more dispersing parous females. The expected values were based on the weaning age, i.e., expected proportion of females with infant, and the interbirth interval minus the weaning age, i.e., expected proportion of females without infant (Table 13.2) (cf. Sterck 1997). In the four populations where the expected distribution could be calculated, females without infants dispersed significantly more often than females with infants (Table 13.2). A similar pattern was found in the other populations and it is obvious that in all populations with a substantial number of observed cases of dispersal by parous females, these females usually were not accompanied by an infant. Therefore, parous females seemed to reduce the potential costs of dispersal by timing their dispersal when they are not lactating.

Although, in general, parous females leave without offspring, in some species or populations the pattern was the reverse. More or equal amounts of dispersal of females with infants than those without infants was found in 6 out of 36 species and populations with dispersal of parous females. These were the white sifaka, the proboscis monkey (*Nasalis larvatus*), the hanuman langur of Abu, the banded langur (*Presbytis melalophos*), the purple-faced langur (*P. senex*) of Polonnaruwa and the chimpanzee of Gombe. In all six populations, only a small number of dispersal events by parous females, namely one, three, three, one, two and two, were documented. Thus the populations with the exceptional pattern are also populations with only a few observations of dispersal by parous females. Moreover, these populations were not overrepresented in the dataset of populations with three or fewer dispersing parous females (Binomial test: $N = 14$; $k = 6$; $p = q = 0.5$, $P = 0.395$, one-tailed). In addition, there may actually be only four of the six cases in which dispersal of parous females with infant is more or equally common to dispersal of females without infant. The dispersal of the female hanuman langurs could be considered temporary group splitting (see above) and in the chimpanzees the immigration of these females was not definitely confirmed, as they could also have been peripheral females. Altogether, the importance of these observations is limited.

It can be concluded that females with infants were not eager to disperse. Patterns in group-splitting indicate that the females joining the new male are often those females that do not have a dependent infant, whereas those with a dependent infant remain with the resident male (Table 13.3). This result is consistent with the remaining with the father hypothesis for the females that stay with the resident male and the choosing the best male hypothesis for the females that join the new male.

Alternative explanations for preferential dispersal of females without infants include increased predation risk. We could not test for this possibility. Females may also refrain from taking an infant because resident females may be aggressive toward immigrating females. However, the occurrence of female aggression against immigrants was not related to the proportion of parous females dispersing without infants (Mann–Whitney U-test: $N_1 = 13$, $N_2 = 3$, $U = 14$, $Z_{(ties)} = -0.761$, $P = 0.45$). If immigrant females did not have sufficient milk for their infant, these should gradually languish after the female dispersed. Infants of the Thomas's langur died soon after their mother transferred to a new male (Steenbeek 1999a). Whether this was due to their mother having not sufficient milk was unclear. Countering such an interpretation is the observation that a female Thomas's langur that had transferred to a new group and had lost her infant (most likely due to infanticide: Ol in Sterck 1997) still produced milk the first days in the new group (Jacqueline van Oijen, pers. comm.). An alternative explanation is maternal neglect of an infant that is doomed to die through male infanticide (Steenbeek 1999a). Thus most other explanations that can be tested do not explain the patterns very well.

"Reducing the probability of takeover" hypothesis

Females that evict female group members may benefit through a decreased risk of infanticide. Female eviction is found in lemurs: the red-bellied lemur (*Eulemur rubriventer*), the ring-tailed lemur, the white sifaka, and the golden-crowned sifaka (*Propithecus tattersalli*). It is also found in howler monkeys: the mantled howler and the red howler monkey. In three of the six species, male infanticide exists and, in the howlers, it occurs after males gain the alpha position. The eviction of nulliparous females may lead to a relatively high incidence of dispersal of these females. The percentage of nulliparous females in all dispersing females, however, was not different in species with and without female eviction (Mann–Whitney U-test: $N_1 = 13$, $N_2 = 6$, $U = 34$, $Z_{(ties)} = 0.446$, $P = 0.66$) nor

Table 13.3. Reproductive stage of dispersing females and female group members that remained with the resident male

Ref.	Species	Location	Groups (left – entered)	Female remains in group		Dispersing female		Fisher Exact test
				Without infant	With infant	Without infant	With infant	(significance)
15	*Saimiri oerstedi*	Corcovado	Groups not mentioned	?[a]	5–8[a]	3	1	–
18	*Papio ursinus*	Suikerbosrand	Several groups	1	14	5	2	FI = 9.152 (0.003)
22	*Colobus badius*	Tana	Groups not mentioned	1	7	6	0	FI = 10.439 (0.001)
31	*Presbytis entellus*	Abu	Toad Rock: splitear-toad	0	3	5	0	FI = 7.035 (0.008)
46	*Presbytis entellus*	Kanha	Groups not mentioned	5–8[a]	3	3–6[a]	0	–
32	*Presbytis femoralis*	Perawang	TS24	3	2	3	0	FI = 1.267 (0.26)
36	*Presbytis rubicunda*	Sepilok	Groups without name	0	2	2	0	FI = 3.312 (0.08)
38	*Presbytis thomasi*	Ketambe	Merah-jingga	0	3	2	0	FI = 4.012 (0.02)
38	*Presbytis thomasi*	Ketambe	Biru-hijau	0	2	2	0	FI = 3.312 (0.08)
44	*Pan troglodytes*	Mahale	K–M	0	2	10	0	FI = 8.005 (0.005)

Notes:
Ref., see Table 13.1; (46) Newton 1986.
[a] The exact number of females in this category was not mentioned in the original article.

did the incidence of pre-reproductive dispersal differ (Fisher Exact test: FI = 0.346, P = 1.00). No variance in female eviction was found for the populations where the immature dispersal rate was known. In addition, the occurrence of infanticide after takeover or during between-group encounters did not affect the percentage of dispersing nulliparous females (Mann–Whitney U-test: N_1 = 6, N_2 = 7, U = 11.5, $Z_{(ties)}$ = 1.367, P = 0.17), the immature dispersal rate (Mann–Whitney U-test: N_1 = 4, N_2 = 3, U = 6, $Z_{(ties)}$ = 0.001, P = 1.00) and the incidence of pre-reproductive dispersal (Fisher Exact test: FI = 0.675, P = 0.58).

Alternatively, nulliparous females may disperse to avoid inbreeding with their father. These females are expected to disperse voluntarily. Indeed, when the inbreeding risk was measured as the combination of either male philopatry or a longer male tenure than female age at menarche, in populations without eviction the percentage of dispersing females that were nulliparous was higher in populations with than in populations without a risk of inbreeding (Mann–Whitney U-test: N_1 = 17, N_2 = 11, U = 50, Z = 2.075, P = 0.04) and the incidence of pre-reproductive dispersal tended to be higher (Fisher Exact test: FI = 4.695, P = 0.06), although it was not related to the immature dispersal rate (Mann–Whitney U-test: N_1 = 9, N_2 = 11, U = 46.5, Z = 0.229, P = 0.82). Similar or weaker results were obtained when using separately male philopatry or a longer male tenure than female age at menarche.

Discussion

Female dispersal can be pre-reproductive and reproductive. A distinction between these two classes of dispersal is often not explicitly made. Their causes, however, may differ. The data on primates compiled here reveal that, in most populations and species with a reasonable number of observed dispersal events, dispersal of both nulliparous and parous females was observed, but their rates were independent. Because this finding suggests that different causes are involved in the dispersal of nulliparous and parous females, a distinction between pre-reproductive and reproductive dispersal should be made. Inbreeding avoidance is often cited as the main ultimate cause of female dispersal, yet it cannot account for dispersal of parous females. One alternative explanation is infanticide avoidance. Below, we discuss whether infanticide avoidance is an important factor in female dispersal in primates. A direct effect of infanticide occurrence on female dispersal rates and patterns was not found. The hypotheses linking infanticide to female dispersal, however, do predict different correlations and were therefore viewed separately.

"Remaining with the father" hypothesis

Proximate mechanisms

A minority of the dispersing parous females was accompanied by an infant. These females may protect their infants against infanticide when they disperse in the presence of the ousted male of their group. Although this is an often-cited cause of female dispersal, it was actually very uncommon (Table 13.4). Only in the purple-faced langurs was one definite case found.

The majority of the females dispersing with infants joined a new, and therefore potentially infanticidal, male. Whether these females dispersed with infants was not related to the general pattern of timing of infanticide in the populations concerned. The survival chances of these infants differ. In the Thomas's langur, infants succumbed soon after they transfer with their mother (Steenbeek 1999a). In other cases, females immediately wean them (e.g., hamadryas baboon, Sigg *et al.* 1982), the actual immigration of the female is doubted (chimpanzees at Gombe), the fate of the infant is not known (e.g., Nilgiri langur (*Presbytis johnni*); Poirier 1970) or it (obviously) survived (e.g., red-backed squirrel monkey (*Saimiri oerstedi*), Boinski 1987b; yellow baboon, Rasmussen 1981). This indicates that ranging with unknown males does not always lead to death of an infant.

Similarly, there are reported short visits of females with a small infant to neighboring groups that did not result in male infanticide for the black spider monkey (*Ateles paniscus*; van Roosmalen 1985), the proboscis monkey (Moore 1993) and bonobos (*Pan paniscus*; Furuichi 1989). This could perhaps be due to a low risk of infanticide. Indeed, infanticide has not been observed in these species (van Schaik, Chapter 2). Alternatively, in populations with infanticidal males, not all males are necessarily infanticidal (Sommer 1994). If females can assess the probability that a particular male will become infanticidal, they may adjust their behavior accordingly.

This does not imply that remaining with the father of an infant is not an effective means of avoiding infanticide. When a considerable part of the females emigrated from a group, females without infants joined the new male, whereas females with infant remained with the resident male (Table 13.3). Such a pattern may be attributed to infanticide avoidance.

Table 13.4. *A descriptive overview of the occurrence of female dispersal in relation to infanticide avoidance and alternative explanations*

Female parity	Hypothesis	Proximate observations	Species (no.)	Species (%)	Populations (no.)	Populations (%)
Parous		*Females with infants*	14		15	
	Hyp 1: remain with father	Stay with father	1 of 14	7	1 of 15	7
		All parous females	29		36	
	Hyp 2: choosing best male	Females without infants	25 of 29	86	27 of 36	75
		Females without infants and infanticide during between-group encounter	5 of 29	17	6 of 36	17
		Females without infants	25		27	
	Hyp 2: choosing best male	Infanticide during between-group encounter	5 of 25	20	6 of 27	22
	Alternative for hyp 2: between-female competition	Female aggression against female immigrants	3 of 25	12	3 of 27	12
Nulliparous		*Nulliparous females*	25		35	
	Hyp 3: reducing the chance of male takeover	Eviction	5 of 25	20	5 of 35	14
		Eviction and infanticide after male takeover	2 of 25	8	2 of 35	6
	Alternative for hyp 3: inbreeding avoidance	Male philopatry or long male residence	12 of 25	48	15 of 35	43
		Voluntary female dispersal and male philopatry or long male residence	10 of 25	40	13 of 35	37

Notes:
Hyp., hypothesis.

However, it may also be caused by other hazards (see below) that infants face during dispersal.

In conclusion, reduction of the current infant's infanticide risk is potentially an explanation for female dispersal. Its importance as a general explanation, however, is limited, because few parous females disperse at the required reproductive stage.

Ultimate cause

This hypothesis predicts that the risk of infanticide is lower for infants of females that remain with the father than for those that associate with a new male. Evidence for this hypothesis is very limited, but yields the expected pattern. In the purple-faced langur, the infant of the dispersing female survived until at least an age of 10 months, whereas the infant of the original group went missing (Rudran 1973: 170–2). Obviously, more data need to be collected for a more definitive conclusion.

"Choosing the best male" hypothesis

Proximate mechanisms

Parous females that joined an unknown male were much more often without than with infants. In all cases with sufficient data, this distribution was significantly different from the expected pattern. Therefore, in general, parous females time their dispersal such that they disperse without infant. This may be linked to a lowering of the rate of infanticide. A connection between the percentage of parous females dispersing with infants and the timing of infanticide, however, was not found.

In species where female dispersal is the result of female choice for protective males, the females are expected to disperse soon after their infant dies through infanticide. Indeed, some females dispersed shortly after their infant had been killed by an extra-group male during between-group encounters (mountain gorilla, red colobus monkey and Thomas's langur). Some of these females subsequently joined the infanticidal male (Watts, 1989; Sterck 1997). For at least these cases, infanticide seems to stimulate female dispersal directly. Similarly, females with infants in a group that had lost the resident male(s) avoided strange males that are potentially infanticidal (mountain gorilla, Harcourt *et al.* 1976; Watts 1989; Thomas's langur, Steenbeek 1996). These observations indicate that a connection of female dispersal with male infanticide is possible. However, these observations are not conclusive and do not prove than infanticide avoidance is an ultimate cause of female dispersal.

Alternative explanations, i.e., increased predation risk, aggression from resident females toward immigrating females and problems with finding sufficient food to produce milk for an infant, may also explain this pattern. No data on predation risk were included in our analyses. Female transfer, however, is more often observed in populations with a low than a high predation risk (Anderson 1986). Female aggression against immigrants does not explain the percentage of parous females that disperses without infants. Dispersing females are expected to be less well fed than philopatric females, at least when they range alone (Crockett & Pope 1993). No such data are available on females that immediately join a new group. Female transfer, however, is more likely when ranges overlap (Isbell & van Vuren 1996) and the food intake of these females may not be hampered. This may indicate that obtaining sufficient food is often not a problem for dispersing females. Altogether, no proximate cause obtains definitive support, although the data indicate that choosing the best male may explain more cases than female aggression against immigrants (Table 13.4). Dispersal of parous females without infants may well result from infanticide avoidance (cf. Thomas's langurs, Sterck 1997), but detailed analyses of the alternative hypotheses for a population are clearly required.

Ultimate cause

Females may exert mate choice through female dispersal. The "choosing the best male" hypothesis implies that these females choose males that are better protectors against male infanticide than the male with which they currently associate. The first to suggest such an effect was Marsh (1979b). His data on the red colobus monkey of Tana, however, do not make such a comparison possible. Watts (1990) suggested, but did not test, the same effect for mountain gorillas. The data on the Thomas's langurs are more conclusive. The death rate of infants born after a female joined a new male is only half that of infants born before a female left a male. This difference tended toward significance (Steenbeek 1999a). That the male was less able to defend the infants in his group shortly before females left him was also reflected in the higher incidence of surprise attacks of non-resident males on the group that could result in infanticide (Steenbeek 1999a; Chapter 7).

A reduction of the infanticide risk may also account for the dispersal of parous female chimpanzees at Mahale. These females left their community (K) when only two adult males remained and most joined a community (M) with a larger number of adult males. Their dispersal may have resulted in increased protection against predators (Nishida *et al.* 1985) or against males from other groups (Uehara *et al.* 1994). These possibilities, however, were not tested.

If females choose males with better protective abilities, they may also choose in other respects for a "better" male. A dispersing female would get better genes for her offspring, because the more successful male is chosen. This "good gene" hypothesis, however, does not falsify the infanticide avoidance hypothesis, as these two hypotheses are not mutually exclusive. The good genes hypothesis requires that the offspring of the new male will be of higher quality than that of the previous male, whereas the infanticide avoidance hypothesis requires that the survival chances of infants are higher with the new than with the old male.

"Reducing the chance of male takeovers" hypothesis

Proximate mechanism

Crockett & Janson (Chapter 4) argued that red howler monkey females evict females that are approaching reproductive age to reduce the risk of male takeovers. Consequently, these females reduce the risk of infanticide. Female eviction has also been observed in some other primate species, but is generally rare. Where it is observed, a link with pre-reproductive dispersal was not evident.

Pre-reproductive dispersal is often explained by inbreeding avoidance. Indeed, pre-reproductive dispersal is much more common in species and populations with male philopatry than in those with male dispersal. Therefore, dispersal of pre-reproductive females is probably in general related to inbreeding avoidance. In line with this hypothesis, female dispersal in most species and populations is voluntary (Table 13.4).

Ultimate cause

The effect of female eviction on infanticide risk has been established for the red howler monkeys (Crockett & Janson, Chapter 4). The females that remain in the group reduce this risk. The data collected for the current chapter would indicate that this is rather rare, as eviction of females is found in a minority of the species. Eviction, however, may not be a necessary feature of this system.

In the red howler monkey, emigrating females pay a high price. They range solitarily for a considerable time (Crockett & Pope 1993) during which time, the risk of predation is probably high. In addition, the quality of their diet is low (Pope 1989) and it is very difficult for them to gain a breeding status elsewhere (Crockett & Pope 1993). The evictions in this population, however, may not so much be due to the fact that only (or mainly) the resident females gain from a reduction in group size, but may be caused by the high costs of dispersal. Where the costs of female dispersal are relatively low females may voluntarily distribute themselves evenly over groups, thereby reducing the risk both for females that remain in the group and for females that disperse.

Main conclusion

In general, there are not sufficient data to test whether the evolution of female dispersal is related to infanticide risk. Only in the red howler monkeys are the data conclusive (Crockett & Janson, Chapter 4) and can female eviction be explained by the reducing the chance of male takeovers hypothesis. Data on Thomas's langurs, chimpanzees of Mahale and gorillas indicate that the choosing the best male hypothesis may account for female dispersal in these species, but more data or analyses of the data in the light of this hypothesis are needed. Evidence for the remaining with the father hypothesis is even more scant, but may account for a few cases of female dispersal.

Proximate mechanisms that may relate to infanticide risk were found. These indicate that females may time their dispersal in order to reduce the risk of infanticide. Alternative explanations, however, are possible and could not be excluded with the current data. When infanticide accounts for the pattern in proximate data, this suggest that the choosing the best male hypothesis may account for many cases of dispersal by parous females. Less often will the remaining with the father hypothesis account for their dispersal. Dispersal of nulliparous females will usually be due to inbreeding avoidance. Only in a limited number of cases will it be due to the reducing the chance of male takeovers hypothesis, unless eviction of females is not necessary.

Potentially, infanticide avoidance may affect female dispersal patterns in several ways. The current state of knowledge, however, allows only for some tentative conclusions of its explanatory value. At the same time, this can also be said for all other explanations of female dispersal with only one exception: inbreeding avoidance is a likely explanation for the emigration of nulliparous females from their natal group (Clutton-Brock 1989a; this chapter). However, inbreeding avoidance cannot account for the dispersal of parous females. Therefore much more research on female dispersal in primates is needed and the hypothesis that dispersal functions to reduce the risk of infanticide should be one of the possible hypotheses considered.

Acknowledgments

We thank Carel van Schaik and Charles Janson for the invitation to write this chapter. Their constructive comments greatly improved it. E.H.M.S. thanks Filippo Aureli for his encouraging reaction on an early version of the paper. A.H.K. was financed during the writing of the paper by the DFG (Deutsche Forschungs Gemeinschaft) and MPIV (Max-Planck Institut für Verhaltensphysiologie) in Germany.

14

Reproductive patterns in eutherian mammals: adaptations against infanticide?

Introduction

Although all female eutherian mammals share the same basic reproductive physiology, they show great variation in their sexual behavior (van Tienhoven 1983; Short 1984; Flowerdew 1987). In some families, especially among primates, females can mate over a period of weeks and may actively pursue polyandrous mating, whereas in others, for example some bovids, females restrict sexual behavior to a single mating with a single male per conception. Likewise, females of some species regularly mate while pregnant, whereas in most others mating stops upon fertilization. Details of the regulation of ovarian cycles also vary greatly. For example, in the ovarian cycles of many carnivores, extensive stimulation by mating is required for ovulation to occur (known as induced ovulation), whereas in many rodents ovulation is spontaneous but mating induces the formation of the corpus luteum; in primates, in contrast, the whole cycle proceeds spontaneously regardless of whether any mating occurs.

Traditionally, such variation in ovarian cycles and sexual behavior has been regarded as a feature that reflected merely evolutionary history rather than current function. Reproduction is critical to fitness, however, and, increasingly, adaptive explanations are sought for these variations in sexual behavior, its regulation and other features of reproduction (e.g., Eberhard 1996). Ever since Hrdy's (1977, 1979) research, it has been argued that some features of sexual behavior are best considered an adaptive counterstrategy to the threat of infanticide by males. In general, males who have mated with a sexually attractive female are less likely to attack her offspring and more likely to protect them (Huck *et al.* 1982; Labov *et al.*

1985; van Schaik, Chapter 2). By mating polyandrously, females of vulnerable species may solve the problem of concentrating paternity in the male most likely to be available to provide protection, but simultaneously confusing enough so that the other males will refrain from infanticide (cf. van Schaik *et al.* 1999). There is some evidence that polyandrous mating indeed reduces the risk of infanticide (e.g., Dunbar 1984b; Robbins 1995; Feh 1999; for more detailed discussion, see van Schaik *et al.* 1999, Chapter 15). However, once physiological adaptations have evolved these may constrain the feasibility of behavioral adaptations to current circumstances.

At the time of writing, tests of the hypothesis that female sexual behavior serves in part to reduce the risk of infanticide by males have focused almost entirely on primates (Hrdy & Whitten 1987; Small 1993; van Schaik *et al.* 1999; Chapter 15, but see Eaton 1978; Agrell *et al.* 1998). In this chapter, we survey the variability in female sexual behavior in relation to vulnerability to infanticide among eutherian mammals in general. In addition to detailed information on female sexual behavior we need a reasonable assessment of a species' vulnerability to infanticide. Since information on most taxa is very incomplete we cannot, at this stage, evaluate the anti-infanticide function of reproductive behavior against all other possible functions. However, we do discuss, below, the most common explanations for features of reproductive behavior that could also affect infanticide risk. It is useful to examine possible anti-infanticide functions of reproductive behavior at two different stages: preconception and postconception. Before conception, a female may achieve paternity confusion by mating polyandrously, resulting in a real, but varying, chance of paternity for all mating partners (see also van Schaik *et al.* 1999, Chapter 15).

By mating after conception, females may manipulate a male's assessment of paternity if males cannot recognize (early) pregnancy and base their assessment on aspects of their mating history with the female (for further elaboration, see van Schaik *et al.*, Chapter 15; van Schaik, Chapter 2). This form of manipulation may be used when one or more new males appear during pregnancy and it may be especially useful in one-male situations when the previous male is effectively eliminated. It may also be an extension of preconception paternity confusion, whenever there is potential for polyandrous mating; this gives the female more opportunity to concentrate (but not limit) paternity chances in the dominant male and at the same time confuse paternity assessment by the other available males.

Alternatively, females confronted with a likely infanticidal male after they have conceived may cut their losses by ceasing to invest in their unborn offspring. After a major upheaval in the male dominance hierarchy or when a new male appears who is able to dominate or even oust a female's previous mating partner(s), females may refrain from facilitating implantation or may resorb or abort embryos ("Bruce effects,", Huck, 1984, Huck *et al.*, 1985, Blumstein, Chapter 8; induced abortion, e.g., Pereira 1983, Berger 1986, Agoramoorthy *et al.* 1988). Since, in most circumstances, it will be cheaper for the female to reduce the risk of infanticide through postconception sexual behavior rather than abandonment of the reproductive investment, we assume that (unborn) infant abandonment occurs when postconception matings cannot be used effectively. This could be because it is impossible to hide hormonal or other signals of pregnancy or because males remain infanticidal for a clearly circumscribed period postmating, as in some rodents (Elwood & Kennedy 1994; Perrigo & vom Saal 1994). (We do not discuss the female's options after birth here; for an overview of these options, see Agrell *et al.* 1998.)

Predictions to be tested

We can now delineate which female reproductive countertactics are expected in species vulnerable to infanticide by males relative to related non-vulnerable species:

1. *Preconception mating (paternity confusion)*: we expect females to mate polyandrously in species vulnerable to infanticide, at least where the demographical context allows this. Females may also have longer mating periods to achieve this polyandry.
2. *Postconception mating (paternity assessment manipulation)*: we expect females to engage in postconception sexual behavior in species vulnerable to infanticide, especially when new males appear postconception.
3. *Bruce effects*: Embryo abandonment in response to new males is expected in vulnerable species.

These counterstrategies are not mutually exclusive and we expect them to interact depending on the social conditions and their costs to the female:

Polyandrous mating and postconception mating serve overlapping purposes, and may thus be positively correlated across species.

Postconception mating and Bruce effects are expected to be largely mutually exclusive.

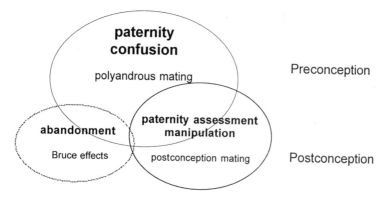

Figure 14.1. Predicted female sexual strategies against infanticide by males, and their mutual relationships.

> Sexual counterstrategies, both pre- and postconception, being less costly to the female, are expected to be more common than offspring abandonment.

Figure 14.1 illustrates these predictions and their mutual relationships. Provided we can develop an independent measure of the risk of infanticide by males, these predictions can be put to the test in a comparative analysis of the reproductive behavior of females in species of mammals vulnerable to infanticide by males relative to that of females in species that are not. However, given the quality and extent of our knowledge, the tests presented here can be no more than illustrations and explorations into the possible evidence for sexual strategies molded by the risk of infanticide by males. In addition, in the light of the critical role of reproductive behavior in an individual's fitness, most aspects of sexual behavior may represent adaptations to various pressures, and sometimes may reflect compromises between opposing pressures. Hence, we should also examine the presence of additional or alternative functions that may confound the predictions made above, as well as the possible costs of polyandrous mating.

Other factors affecting sexual behavior

Polyandrous mating: other possible benefits   Recently, many studies have examined polyandry and its proposed benefits (Table 14.1, and references therein), although most of them do not consider a reduced risk of infanticide through confused paternity as a possible benefit. The assumed benefits to multiple mating can be direct (benefiting the female

Table 14.1. *Suggested advantages of polyandrous mating, and their applicability to primates and other mammals*

	Relevant to spp. with single young?	Could it explain: Long mating period	Could it explain: Post-conception mating	Evidence for primates	Ref.
Direct benefits					
Replenish sperm supply	Yes	No	No	No	1, 2
Insurance against male sterility	Yes	No	No	Possible	1, 3
Female–female competition through sperm depletion	Yes	Yes	Yes	No	4, 5
Nutrients in seminal fluid	Yes	Yes	Yes	No	1, 2
Courtship feeding	Yes	Yes	Yes	No	1
Consortship ecological advantage	Yes	Yes	Yes	No	6
Protection against predation	Yes	Yes	Yes	No	2
Reduced harassment "convenience polyandry"	Yes	No	No	?	3
Genetic benefits					
Genetic diversity in litter	No			No	1
Larger litter	No			No	3
"Good genes" through sperm competition	Yes	Yes	No	Possible	1, 2
Delayed benefits					
Paternal care by more males	Yes	Yes	Yes	No	3, 7, 8
Reduced risk of infanticide	Yes	Yes	Yes	Yes	5, 9, 10

References:

(1) Keller & Reeve 1995; (2) Yasui 1998; (3) Hoogland 1998;
(4) Eaton 1978; (5) Small 1990; (6) Rasmussen 1985; (7) Smuts & Gubernick 1992;
(8) van Schaik & Paul 1996–7; (9) Hrdy 1979; (10) van Schaik *et al.* 1999.

while she is sexually active), genetic (increasing the quality of a female's offspring) or delayed (increasing the survival chances of the offspring through modification of male behavior). Here, we include only those alternative benefits that could be expected to apply to mammals.

DIRECT BENEFITS

Fertilization insurance   When a certain proportion of males is sterile, polyandrous mating would maximize the chances of fertilization. However, one expects strong selection on ejaculate size, sperm motility, longevity, ability to penetrate the ovum (Keller & Reeve 1995), and other traits reducing the risk of male sterility. Indeed, routine monandrous mating is found in a variety of mammals (e.g., ungulates, Berger & Cunningham 1994, Byers 1997; primates, Brockman & Whitten 1996). Nonetheless, temporary depletion could occur in species with a concentrated breeding season, such as, for example, elephant seals (LeBoeuf 1972), gray seals (Amos *et al.* 1993) and prairie dogs (Hoogland 1998).

Female competition for sperm   In social groups with a female-biased sex ratio, females may engage in multiple matings with the same male and thus reduce the amount of sperm available for other females to the point that their fertilization chances are compromised (Small 1990). This scenario is unlikely to be common, since males would benefit from maintaining fertilization potential under a wide range of conditions. Extreme interspecific variation in the size of testes (e.g., Eberhard 1996) and shape of sperm (Gomendio & Roldan 1993) shows that natural selection can bring about enhanced fertilization ability where needed. Some primate studies appear to support the depletion hypothesis (gelada baboons, Dunbar 1980; lion-tailed macaques, Kumar 1987; hanuman langurs, Sommer *et al.* 1992; captive hamadryas baboons, Zinner *et al.* 1994). However, these studies refer to large one-male groups with extremely female-biased adult sex ratios, relative to the species' range, and in all cases more mundane explanations are available, such as rank-dependent access to food for females and hence likelihood of ovulation.

Ecological or social benefits   Females could receive access to food or social protection just before or after mating. For instance, chimpanzee females may obtain more meat when they have sexual swellings (Stanford *et al.* 1994) or transfer into another community (Wallis & Lemon 1986), and sexually attractive female callitrichids may elicit more care for current infants (Sánchez *et al.* 1999). In general, however, dramatic courtship benefits do not accrue in most species. On the contrary, in some species, male mate guarding clearly compromises the female's energy budget (e.g., savanna baboons, Rasmussen 1985). Moreover, in most species, males do not provide any personal services to receptive females. Thus direct benefits accompanying sexual attractivity are far from universal in primates and virtually not mentioned in the scarce data on other orders.

Convenience polyandry   Males may attack females when they mate with other males or in order to prevent them from leaving a mating association. Females may opt for acceptance of polyandrous mating rather than putting up a fight and risk injury where they cannot easily prevent matings under harassment (Smuts & Smuts, 1993; Mesnick 1997). Thus females benefit from polyandry by avoiding higher costs. Behavioral observations make it possible to distinguish this convenience polyandry from the active polyandry proposed by the infanticide-avoidance hypothesis.

GENETIC BENEFITS

Possible genetic benefits arising from polyandrous matings are genetic diversity among offspring in the same litter and bet-hedging (i.e., avoidance of low fitness combinations in future generations; Yasui 1998). However, these benefits accrue only in species with large litters. Among these, mixed paternities are indeed common (e.g., rodents, Topping & Millar 1998, Goossens *et al.* 1998, Keil & Sachser 1998; carnivores, Schenk & Kovacs 1995, Girman *et al.* 1997).

DELAYED BENEFITS

Paternal care by males   It has been suggested (Hrdy 1981, Hrdy & Whitten 1987) that polyandrous matings would incite more males to show paternal care toward the infants. However, there is little evidence for this: among primates at least, male care for infants (carrying and provisioning) is very poorly correlated with potential paternity (Smuts & Gubernick 1992; but see Anderson 1992; Marlowe 1999), although protection is (Anderson 1992; Borries *et al.* 1999a). Among rodents, male care seems to be limited to offspring of monandrously mating females (Ribble 1991; Gajda & Brooks 1993; Cantoni & Brown 1997) The species most likely to fit this prediction are the social carnivores with unrelated adult helpers (Derix & van Hooff 1995). More generally, polyandry may tend to make more males tolerant toward the offspring but this benefit can be subsumed under the infanticide-avoidance function.

Polyandrous mating: possible costs   Obviously there are potential costs to polyandrous matings as well. First, the transfer of sexually transmitted diseases is likely to be enhanced (for a review, see Lockhart *et al.*

1996). Second, although allowing matings may reduce the risk of injury (see above), polyandry may also lead to increased harassment by males who respond to a female's (potential) mating with another male (Smuts & Smuts 1993; cf. Matsumota-Oda, 1998). Third, polyandry will dilute paternity and thus may reduce male parental care for infants, including protection against infanticidal males (see above). Finally, possible genetic benefits from mate choice might be reduced when females mate with males in a seemingly indiscriminate fashion. However, polyandrous mating does not necessarily eliminate postmating female choice (Eberhard 1996).

In conclusion, despite the costs, various alternative or additional benefits of polyandrous matings may exist. The most widespread of these may be convenience polyandry and (where females produce litters) increased genetic diversity or bet-hedging. Nevertheless, the initial prediction that polyandrous mating serves to reduce the risk of infanticide by males may hold in a wide variety of circumstances, sometimes in addition to other benefits.

Prolonged mating periods   When females have incomplete control over their mating partners, they may need longer mating periods in order to overcome monopolization attempts by dominant males to achieve the actively sought amount of polyandry (cf. van Schaik *et al.*, Chapter 15). This leads to a correlation between polyandrous matings and long mating periods in such species (as in catarrhine primates; van Schaik *et al.* 1999). Here, the confounding effects noted for polyandrous mating should also apply to duration of the mating period. However, when polyandrous matings are an unintended by-product of male harassment ("convenience polyandry"), we expect short mating periods. Hence, when prolonged mating periods are found in species with single births, in which polyandry cannot increase genetic diversity in the litter, these extended mating periods may primarily reflect infanticide avoidance. This prediction is made more plausible because long mating periods also have additional costs. First, they take time and energy away from other profitable activities. Second, conspicuous activities such as male mating competition, courtship and scent marking may increase a female's predation risk (e.g., Koskela *et al.* 1996). Third, longer mating almost inevitably means more harassment (see above).

Postconception mating   For species in which males assess their paternity through sexual behavior the paternity assessment manipulation strategy predicts that females in vulnerable circumstances should engage in postconception matings. However, matings, and especially courtship, can potentially bring ecological or social benefits to the female even when this happens during pregnancy. If such benefits were substantial, natural selection would act to produce attractive periods during pregnancy, regardless of changes in the social environment. However, among mammals such courtship benefits seem to be rare (see above). Hence, the prediction that mating during pregnancy serves to reduce infanticide risk by manipulating the male's paternity assessment stands.

Bruce effects   The longer after conception that embryo abandonment takes place, the higher is its cost. Thus we expect that infant abandonment is confined to situations where continuing the pregnancy would threaten the life of the female or the offspring. The female's life could be threatened when there is a dramatic drop in food supply or the female is victim to a serious disease. The offspring's life could be threatened by the presence of a dominant and potentially infanticidal conspecific (male or female, see Blumstein, Chapter 8; Digby, Chapter 17). When a healthy female abandons her unborn offspring when exposed to a new male, it is therefore difficult to avoid the interpretation of an adaptive response to the threat of male infanticide.

This discussion shows that tests of the predictions of the infanticide-avoidance hypothesis face two major problems. First, the predictions may not be unique because reproductive behavior may serve a variety of other functions. This is especially likely for polyandrous mating and the duration of the mating period, although the predictions may still apply to more limited sets of species, in particular those that produce single offspring and are not subject to male harassment. However, embryo abandonment, from implantation failure to abortion, in response to exposure to a novel dominant male, is uniquely predicted by the infanticide-avoidance hypothesis, which is also the most consistent with postconception mating in response to novel males or in a multimale demographic setting.

The second problem is that natural history, social organization and population density are expected to affect opportunities for infanticide by males in species that are vulnerable to it, and likewise are likely to affect aspects of sexual behavior. Because mammals differ greatly in these respects, there are many confounding variables not controled for in these comparisons.

Methods

Reproductive behavior

We did a broad literature survey focusing on reviews of reproduction in eutherian mammal families (but excluding the huge order Chiroptera for which our knowledge is too limited). The major sources were: Hayssen *et al.* (1993), Estes (1991), supplemented by van Schaik *et al.* (1999) for primates; Alderton (1996), Blumstein (Chapter 8) for rodents; Eaton (1978), Gittleman (1986), Alderton (1993, 1994), Moehlman (1989), and Mead (1994) for carnivores; Reeves *et al.* (1992) for pinnipeds; and Clapham (1997) for whales; see Table 14.2). We thus obtained information on both vulnerability to infanticide by males (see below) and sexual behavior for 211 species (admittedly only a small fraction of the roughly 3000 eligible species).

Data on sexual behavior recorded were the number of males typically mating with a female for a given pregnancy, the duration of the actual mating period, and the incidence of postconception mating, embryo abandonment or Bruce effects, and of postpartum mating and conception. The mating period was defined as the average interval between first and last mating within one cycle. The number of mating partners was taken both from direct observations and from data on mixed paternities; the number of mating partners was classified as *multiple* if a single male accounted for fewer than 50% of observed matings or paternities in mixed litters; as *single male plus* if one male accounted for the majority of matings, but other male(s) were reported to mate as well during >10% of the cycles; and *single* if females were reported to mate with a single male almost exclusively, with "extra-pair-matings" in fewer than 10% of the cycles. For mating period we used mainly the objective compilation of Hayssen *et al.* (1993), who used the term estrus. This refers either to mating period (as in classic estrus) or to the period between first and last mating under natural conditions, if available.

Assessing vulnerability

Although infanticide by males is now known for many mammal species (see van Schaik, Chapter 2), the list is certainly far from complete. To test the predictions presented above, it is important to be able to distinguish in the dataset between absence of infanticide due to lack of observations and absence due to low vulnerability. We assume that life history, in particular the relative length of gestation and lactation, affects vulnerability.

Table 14.2. *Reproductive features, ecology and vulnerability to infanticide in mammals*

Family	Common name	Species	Male-infX observed	Vuln. estimate	Lac/ Ges	PPE	Season- ality	Births /year	Obs no. males	Mating period	Post. conc. mating	Induced ovulation	Delayed implant	Addi- tional refs.
Artiodactyla – Bovidae	Pronghorn	Antilocapra americana	0	0	∨		2	1	s	1–2 hrs	N			1
Artiodactyla – Bovidae	Bison	Bison bison	0	0	∨		2	1	s	<1h	N			2
Artiodactyla – Bovidae	Bontebok	Damaliscus dorcas	0	1	∧		1	1	s+	1		Yes?	No	3
Artiodactyla – Bovidae	Kirk's dikdik	Madoqua kirkii	0	0	∨	Y	1	∧	s	2	N			4
Artiodactyla – Bovidae	Bighorn sheep	Ovis canadensis	0	0	∨		2		s+	3	N			5
Artiodactyla – Giraffidae	Giraffe	Giraffa	0	0	∨	Y	0	∨	s+	1	N			6
Artiodactyla – Hippopotamidae	Common hippopotamus	Hippopotamus amphibius	Obs	1	∧		1	∨	s+	6				7
Artiodactyla – Suidae	Warthog	Phacochoerus aethiopicus	0	0	∨		1	1		3				
Carnivora – Canidae	Arctic fox	Alopex lagopus	0	1	∧		2	1	s	5				
Carnivora – Canidae	Side-striped jackal	Canis adustus	0	0	∨		2	1	s					8;9
Carnivora – Canidae	Golden jackal	Canis aureus	0	1	1		2	1	s	4				9
Carnivora – Canidae	Coyote	Canis latrans	0	1	≪		2	1	s	10				8
Carnivora – Canidae	Wolf	Canis lupus	0	0	∨		1	∧	s+	11				9
Carnivora – Canidae	Black-backed jackal	Canis mesomelas	0	1	≪		1	∧	s	7				9
Carnivora – Canidae	Simian wolf	Canis simensis	0	1	∧				M					10
Carnivora – Canidae	Crab-eating zorro	Cerdocyon thous	0	1	∧		0	∧	s					9
Carnivora – Canidae	Maned wolf	Chrysocyon brachyurus	0	1	∧		2	1	s	5				9
Carnivora – Canidae	Dhole	Cuon alpinus	0	1	≪		1	1	M					9;11
Carnivora – Canidae	Wild dog	Lyaon pictus	0	1	≪	N	1	1	s+		N			9;12
Carnivora – Canidae	Racoon dog	Nyctereutes procyonoides	0	1	∧		2	1	s	4				9
Carnivora – Canidae	Bat-eared fox	Otocyon megalotis	0	1	∧		2	∧	s	7				9;13
Carnivora – Canidae	Bush dog	Speothos venaticus	0	1	∧			∧		4				8
Carnivora – Canidae	Grey fox	Urocyon cinereoargenteus	0	0	∨	N			s					8
Carnivora – Canidae	Kit fox	Vulpes velox	0	1	≪	N	2		s					8

Order – Family	Common name	Species													
Carnivora – Canidae	Red fox	*Vulpes vulpes*	o	1	∧∧	N	2	∧	1	s+	3				8;14
Carnivora – Canidae	Fennec	*Vulpes zerda*	o	1	∧	If lost	2	∧	1(2)	s	2			Oblig.	8
Carnivora – Felidae	Cheetah	*Acinonyx jubatus*	o	1	∧	N	0	∨		M	4			No	8
Carnivora – Felidae	Caracal	*Caracal caracal*	o	1	∧	N	0	1		M	3				15
Carnivora – Felidae	Feral cat	*Felis catus*	o	1	∧		1	∧		M		Y			16
Carnivora – Felidae	Swamp cat or jungle cat	*Felis chaus*	o	1	∧			∧		M	5				
Carnivora – Felidae	Puma	*Felis concolor*	Obs	1	1	If lost	0	∨		s+	9	Y			15;16;17
Carnivora – Felidae	Black-footed cat	*Felis nigripes*	o	0	∨		2	1		s+	1				
Carnivora – Felidae	Serval	*Felis serval*	o	1	∧		0	∧		s+	2				
Carnivora – Felidae	Europ. wild cat	*Felis sylvestris*	o	1	∧	N	2	∧		s+	4	Y			16
Carnivora – Felidae	Ocelot	*Leopardus pardalis*	Obs	1	∨		2	∨		s+	5				
Carnivora – Felidae	Lynx	*Lynx lynx*	o	1	∧		2	1		s+			Some		
Carnivora – Felidae	Bobcat	*Lynx rufus*	Obs	1	∧		2	1		s+	2		Some		
Carnivora – Felidae	Lion	*Panthera leo*	Obs	1	∧∧	N				M	7	Y	Some		16;18
Carnivora – Felidae	Jaguar	*Panthera onca*	o	1	∧			∨		s+	12				
Carnivora – Felidae	Leopard	*Panthera pardus*	Obs	1	∧			∨		s+	7				15;8
Carnivora – Felidae	Tiger	*Panthera tigris*	Obs	1	∧	N		1		s+	7	Y	Yes		16
Carnivora – Felidae	Leopard cat	*Prionailurus bengalensis*	o	0	∨		0	1		M	7				15
Carnivora – Felidae	Fishing cat	*Prionailurus viverrina*	o	0	∨		0			M	7				15
Carnivora – Hyaenidae	Spotted hyaena	*Crocuta crocuta*	Obs	1	∧	N		∨		s+	1		No		19
Carnivora – Hyaenidae	Brown hyaena	*Hyaena brunnea*	o	1	∧	N	0	∨		M	10	Y			8
Carnivora – Hyaenidae	Striped hyaena	*Hyaena hyaena*	o	0	∧		0			s+	1				8
Carnivora – Hyaenidae	Aardwolf	*Proteles protoles/cristatus*	o	1	∧					s+	2				20;21
Carnivora – Mustelidae	Tayra	*Eira barbara*	o	1	∧					M	12				22;23
Carnivora – Mustelidae	Sea otter	*Enhydra lutris*	Obs	1	1	Y	0	1		M				Facult.	24
Carnivora – Mustelidae	Wolverine	*Gulo gulo*	o	1	∨		2	∨		s+				Oblig.	24
Carnivora – Mustelidae	Zorilla	*Ictonyx striatus*	o	1	∧	If lost	2	1(2)		s+				No	8
Carnivora – Mustelidae	River otter	*Lutra canadensis*	o	0	∨	Y	2	1		s+/M				Oblig.	24;25
Carnivora – Mustelidae	European otter	*Lutra lutra*	o	1	∧	N	1	∨		s+/M	14		No		24;26

Table 14.2. (*cont.*)

Family	Common name	Species	Male-infX observed	Vuln. estimate	Lac/ Ges	PPE	Season-ality	Births /year	Obs no. males	Mating period	Post. conc. mating	Induced ovulation	Delayed implant	Addi-tional refs.
Carnivora – Mustelidae	Smooth-coated otter	*Lutrogale perspicillata*	o	1	∧				s	14			No	24
Carnivora – Mustelidae	European pine marten	*Martes martes*	o	0	∨	N	2	1	M	1		Prob	Oblig.	22
Carnivora – Mustelidae	Sable	*Martes zibellina*	o	0	∨		2	1		2			Oblig.	22
Carnivora – Mustelidae	Das/ badger	*Meles meles*	o	0	∨	Y	2	1	M	5			Oblig.	27;28
Carnivora – Mustelidae	Striped skunk	*Mephitis mephitis*	Obs	1	∧		2	1		7			Facult.	29
Carnivora – Mustelidae	Short-tailed weasel/ ermine/stoat	*Mustela erminea*	o	0	∨		2	1	M			Yes	Oblig.	30
Carnivora – Mustelidae	Long-tailed weasel	*Mustela frenata*	Obs	1	∧	If lost	2	1(2)	M			Yes	No	30
Carnivora – Mustelidae	European mink	*Mustela lutreola*	o	1	∧		2						Oblig.	22
Carnivora – Mustelidae	Least weasel	*Mustela nivalis*	o	0	∨		0	3	M	5		Yes	No	22;30
Carnivora – Mustelidae	African striped weasel	*Poecilogale albinucha*	o	1	∧	N	1		s+				No	8
Carnivora – Mustelidae	Giant otter	*Pteronura brasiliensis*	o	1	∧	If lost	1		s				No	
Carnivora – Procyonidae	Red panda	*Ailurus fulgens*	o	1	1		2	1	s+	1			Suspected	31
Carnivora – Procyonidae	Ringtailed cat	*Bassariscus astutus*	o	1	∧					2				
Carnivora – Procyonidae	Coati	*Nasua nasua*	Obs	1	∧		2		M	14	N	Some		32
Carnivora – Procyonidae	Kinkajou	*Potos flavus*	o	1	⫽					18				
Carnivora – Procyonidae	Raccoon	*Procyon lotor*	o	1	∧	If lost	2	1	s+	3			Oblig.	33;34
Carnivora – Ursidae	Giant panda	*Ailuropoda melanoleuca*	o	1	∧	N	2	√	M	3	N	Maybe	Suspected	35
Carnivora – Ursidae	Sunbear	*Helarctos malayanus*	o	1	⫽		0		s+	2		Yes	Suspected	35
Carnivora – Ursidae	Black bear	*Ursus americanus*	Obs	1	∧	N	2	√	M	2		Yes	Oblig.	36
Carnivora – Ursidae	Brown bear	*Ursus arctos*	Obs	1	≫	N	2	√	M			Yes	Oblig.	
Carnivora – Ursidae	Polar bear	*Ursus maritimus*	Obs	1	∧		2	√	M	7		Yes	Suspected	
Carnivora – Ursidae	Himalayan black bear	*Ursus thibetanus*	o	0	∨		2	1	s+			Yes	Oblig.	

Order – Family	Common name	Species												Ref
Carnivora – Viverridae	African civet	*Civettictis civetta*	0	∧	1	1	∧					No		8
Carnivora – Viverridae	Yellow mongoose	*Cynictis penicillata*	0	∨	N	2	1	s+	4					8
Carnivora – Viverridae	Striped civet	*Fossa fossana*	0	∨		2	1	s						37
Carnivora – Viverridae	Ring-tailed galidia	*Galidia elegans*	0	∨		2	1	s						37
Carnivora – Viverridae	Genet	*Genetta genetta*	1	∧	1	1	∧	s+				No		8
Carnivora – Viverridae	Ichneumon	*Herpestes ichneumon*	0	1	N	1	∧	s+				No		8
Carnivora – Viverridae	Slender mongoose	*Herpestes sanguineus*	0	∨		1	∧					No		8
Carnivora – Viverridae	Banded mongoose	*Mungos mungo*	0	∨	Y	1	∧	M	7	N				8
Carnivora – Viverridae	Meerkat	*Suricata suricatta*	0	∨	Y	1	∧	M	6	N				38
Cetacea – Mysticeti – Balaenidae	Right whale	*Eubalaena australis*	1	1	N	1	∨	M		(y)				39
Cetacea – Mysticeti – Balaenidae	Bowhead whale	*Eubalaena mysticetus*	0	∨	N	1	∨	M		(y)				39
Cetaces – Mysticeti – Eschrichtiidae	Gray whale	*Eschichtus robustus*	0	∨	N		∨	M						39
Cetacea – Odontoceti – Delphinidae	Bottle-nosed dolphin	*Tursiops* spp.	0	∧	If lost		∨	M	8	(y)				40
Insectivora – Soricidae	Musk shrew	*Suncus murinus*	0	∨	Y				14	Y	Yes	Some		41
Macroscelidea – Macroscelididae	Golden-rumped elephant shrew	*Rhynchocyon chrysopygus*	0	∨	Y		∧	s		N		Some		37; 42
Perissodactyla – Equidae	Common zebra	*Equus burchelli*	Obs	≫	Y			s	1	N				8
Perissodactyla – Equidae	Horse	*Equus caballus*	Obs	≪	Y			s/s+	3	N				43; 44
Perissodactyla – Rhinocerotidae	White rhinoceros	*Ceratotherium simum*	0	∨		0		s	1	N				8
Perissodactyla – Rhinocerotidae	Black rhinoceros	*Diceros bicornis*	0	∧		0		M	3	N				8
Pinnipedia – Odobenidae	Walrus	*Odobenus rosmarus*	0	∧	N	2	∨	s+				Oblig.		45
Pinnipedia – Otariidae	Galapagos fur seal	*Arctocephalus galapagoensis*	0	∧	Y	2	∨	s	1			Oblig.		45; 46
Pinnipedia – Otariidae	Antarctic fur seal	*Arctocephalus gazella*	0	∨	Y	2	1	s	1			Oblig.		45
Pinnipedia – Otariidae	Southern fur seal	*Arctocephalus pusillus*	0	1	Y	2	1	s	1			Oblig.		45

Table 14.2. (cont.)

Family	Common name	Species	Male-infX observed	Vuln. estimate	Lac/ Ges	PPE	Season-ality	Births /year	Obs no. males	Mating period	Post. conc. mating	Induced ovulation	Delayed implant	Addi-tional refs.
Pinnipedia – Otariidae	Subantarctic fur seal	*Arctocephalus tropicalis*	o	o	∨	Y	2	1	s	1			Oblig.	45
Pinnipedia – Otariidae	Northern fur seal	*Callorhinus ursinus*	o	o	∨	Y	2	1	s	1			Oblig.	45
Pinnipedia – Otariidae	Steller sea lion	*Eumetopias jubatus*	o	1	1	Y	2	1	s	1			Oblig.	45; 47
Pinnipedia – Otariidae	Australian sea lion	*Neophoca cinerea*	o	1	1	Y	2	∨	s	1			Oblig.	45;47
Pinnipedia – Otariidae	South American sea lion	*Otaria byronia*	o	1	1	Y	2	∨	s	<1		Maybe	Oblig.	45; 47
Pinnipedia – Otariidae	New Zealand sea lion	*Phocartos hookeri*	o	o	∨	Y	2	1	s	1			Oblig.	45
Pinnipedia – Otariidae	Sea lion	*Zalophus californicus*	o	o	∨	Y	2	1	M	1			Oblig.	45; 47
Pinnipedia – Phocidae	Hooded seal	*Cystophora cristata*	o	o	∨	Y	2	1	s+	2			Oblig.	45
Pinnipedia – Phocidae	Gray seal	*Halichoerus grypus*	o	o	∨	Y	2	1	s+			Maybe	Oblig.	45; 48; 49
Pinnipedia – Phocidae	Weddell seal	*Leptonychotes weddelli*	o	o	∨		2	1	s+				Oblig.	45
Pinnipedia – Phocidae	Crabeater seal	*Lobodon carcinophagus*	o	o	∨	Y	2	1	s	1			Oblig.	45
Pinnipedia – Phocidae	Northern elephant seal	*Mirounga angustirostris*	o	o	∨		2	1	M	4			Oblig.	45; 50
Pinnipedia – Phocidae	Southern elephant seal	*Mirounga leonina*	o	o	∨		2	1	M	4			Oblig.	45
Pinnipedia – Phocidae	Harp seal	*Phoca groenlandica*	o	o	∨	Y	2	1	s+					45
Pinnipedia – Phocidae	Spotted seal	*Phoca largha*	o	o	∨		2	1	s	1				45
Pinnipedia – Phocidae	Baikal seal	*Phoca sibirica*	o	o	∨		2	1	M				Oblig.	45
Pinnipedia – Phocidae	Harbor seal	*Phoca vitulina*	o	o	∨		2	1	M	2			Oblig.	45
Proboscidae – Elephantidae	Elephant	*Loxodonta africana*	o	1	∧	N			s+	2	N			8; 51
Rodentia – Cavidae	Guinea pig	*Cavia sp.*	o	o	∨	Y			s+					52
Rodentia – Cavidae	Mara	*Dolichotis patagonium*	o	1	1				s+					52
Rodentia – Cavidae	Cui	*Galea musteloides*	o	o	∨	Y			M	1+		Yes	Suspected	53

Taxon	Common name	Species										Ref.
Rodentia – Muridae, Sigmodontinae	Bushy-tailed woodrat	*Neotoma cinerea*	o	o	∨	Y		s	2		Yes?	54
Rodentia – Muridae, Sigmodontinae	California mouse	*Peromyscus californicus*	Obs C	1	‖		∧	s				55
Rodentia – Muridae, Sigmodontinae	Deer mouse	*Peromyscus maniculatus*	Obs C	1	‖			s+				54
Rodentia – Muridae, Sigmodontinae	White-footed mouse	*Peromyscus polionotus*	o	1	‖		∧	s				
Rodentia – Muridae – Arvicolinae	Water vole	*Arvicola terrestris*	o	o	∨	Y		s+		Y	Yes	56
Rodentia – Muridae – Arvicolinae	Collared lemming	*Dicrostonyx richardsoni*	Obs	1	∨	Y		s+			Facult.	57
Rodentia – Muridae – Arvicolinae	Prairie vole	*Microtus ochrogaster*	o	1	‖	Y		s+			Yes	58;59
Rodentia – Muridae – Cricetinae	Djungarian hamster	*Phodopus sungorus campbelli*	Obs C	1	∧	Y		s	5 hrs	N	Yes?	60
Rodentia – Muridae – Gerbillinae	Mongolian gerbil	*Meriones unguiculatus*	Obs C	1	∧			s+				58
Rodentia – Muridae – Murinae	Wood mouse	*Apodemus sylvaticus*	o	o	∨	Y			1			61
Rodentia – Muridae – Murinae	Norway rat	*Rattus norvegicus*	Obs	1	∧	Y	∧	M	10 hrs	N		41
Rodentia – Sciuridae	Gunnison's prairie dog	*Cynomys gunnisoni*	o	1	∧			M	3 hrs	N		62
Rodentia – Sciuridae	Black-tailed prairie dog	*Cynomys ludovicianus*	Obs	1	∧		1	s+		N		63
Rodentia – Sciuridae	Golden marmot	*Marmota caudata aurea*	Obs	1	∧			s+				64;65
Rodentia – Sciuridae	Alpine marmot	*Marmota marmota*	Obs	1	∧			M	3 hrs			66;67
Rodentia – Sciuridae	Belding's ground squirrel	*Spermophilus beldingi*	o	1	‖			M		N	Yes	54
Rodentia – Sciuridae	European ground squirrels	*Spermophilus citellus*	o	1	‖		2		1		Yes	68

Table 14.2. (cont.)

Family	Common name	Species	Male-infX observed	Vuln. estimate	Lac/ Ges	PPE	Season-ality	Births /year	Obs no. males	Mating period	Post. conc. mating	Induced ovulation	Delayed implant	Addi-tional refs.
Rodentia – Sciuridae	Thirteenlined ground squirrel	*Spermophilus tridecemlineatus*	Obs	1	≥			∧	M	1–6 h	N	Yes		69
Rodentia – Sciuridae	Eastern chipmunk	*Tamias striatus*	o	1	∧	N		∧	s+	1	N	No?		70
Rodentia – Sciuridae	Cape ground squirrel	*Xerus inauris*	o	1	∧	N		∨	M	3 h	N			71
Sirenia – Trichechidae	West Indian manatee	*Trichechus manatus*	o	1	∧	N			M	14				45
Primates														
Lemuriformes – Cheirogaleidae	Eastern/Greater dwarf lemur	*Cheirogaleus major*	o	0	∨		2			3				
Lemuriformes – Cheirogaleidae	Fat-tailed dwarf lemur	*Cheirogaleus medius*	o	0	∨	Y	2		M	3	N			
Lemuriformes – Cheirogaleidae	Gray mouse lemur	*Microcebus murinus*	o	0	∨	Y	2	∧		1				
Lemuriformes – Indriidae	Verreaux's sifaka	*Propithecus verreauxi*	Obs	1	∧	N	2	1	M	2	Y			
Lemuriformes – Lemuridae	Brown lemur	*Eulemur fulvus*	Obs	1	∧	N	2			2				
Lemuriformes – Lemuridae	Black lemur	*Eulemur macaco*	Obs	1	∧	N	2							
Lemuriformes – Lemuridae	Ring-tailed lemur	*Lemur catta*	Obs	1	∨	N	2	∨	M	1	N			
Lemuriformes – Lemuridae	Ruffed or variegated lemur	*Varecia variegata*	Obs	0	∨	N	2	1	s+	2				
Lemuriformes – Lorisidae	Angwantibo	*Arctocebus calabarensis*	o	0	∨	Y	0			1				

Order – Family	Common name	Scientific name									
Lemuriformes – Lorisidae	Lesser bushbaby	*Galago senegalensis*	o	o	<	Y	1	>		2	N
Lemuriformes – Lorisidae	Slender loris	*Loris tardigradus*	o	1	≪	Y	1			2	N
Lemuriformes – Lorisidae	Slow loris	*Nycticebus coucang*	o	o	<	Y	o			1	
Lemuriformes – Lorisidae	Potto	*Perodicticus potto*	o	o	<	Y	1			2	N
Lemuriformes – Tarsiidae	Western or Bornean tarsier	*Tarsius bancanus*	o	o	<	Y	1	>	s+	1	
Ceboidea – Callimiconidae	Goeldi's marmoset	*Callimico goeldii*	o	o	<	Y	1	>	s	7	Y
Ceboidea – Callitrichidae	Common marmoset	*Callithrix jacchus*	o	o	<	Y	o	>	s	28	Y
Ceboidea – Callitrichidae	Pygmy marmoset	*Cebuella pygmaea*	o	o	<	Y	o	>	s+	6	Y
Ceboidea – Callitrichidae	Lion tamarin	*Leontopithecus rosalia*	o	o	<	Y	1	>	s+	18	Y
Ceboidea – Callitrichidae	Saddleback tamarin	*Saguinus fuscicollis*	o	o	<	N	1	>	s+	10	Y
Ceboidea – Callitrichidae	Cottontop tamarin	*Saguinus oedipus*	o	o	<	Y	1	>	s+	11	Y
Ceboidea – Cebidae	Black howler	*Alouatta caraya*	Obs	1		N	o	<	s+	3	Y
Ceboidea – Cebidae	Mantled howler	*Alouatta palliata*	Obs	1	>	N	o	<	M	3	Y
Ceboidea – Cebidae	Red howler	*Alouatta seniculus*	Obs	1	>	N	o	<	s+	3	Y
Ceboidea – Cebidae	Night or Owl monkey	*Aotus trivirgatus*	o	o	<	N	1		s		Y
Ceboidea – Cebidae	Black spider monkey	*Ateles paniscus*	o	1	>	N	1	<	M	6	N
Ceboidea – Cebidae	Woolly spider monkey (Muriqui)	*Brachyteles arachnoides*	o	1	>	N	1	<	M	2	
Ceboidea – Cebidae	White-fronted capuchin	*Cebus albifrons*	o	1	>	N	1		M	5	N
Ceboidea – Cebidae	Brown capuchin	*Cebus apella*	Obs	1	>	N	2	<	M	5	N
Ceboidea – Cebidae	White-throated capuchin	*Cebus capucinus*	Obs	1	>	N	1	<	M		Y
Ceboidea – Cebidae	Wedge-capped capuchin	*Cebus olivaceus*	Obs	1	<	N	2	<	M		N

Table 14.2. (*cont.*)

Family	Common name	Species	Male-infX observed	Vuln. estimate	Lac/ Ges	PPE	Season-ality	Births /year	Obs no. males	Mating period	Post. conc. mating	Induced ovulation	Delayed implant	Addi-tional refs.
Ceboidea – Cebidae	Woolly monkey	*Lagothrix lagotricha*	0	1	∧	N	1	∨	M	8	N			
Ceboidea – Cebidae	Saki	*Pithecia* spp.	0	0	∨	N	2	1	s+		N			
Ceboidea – Cebidae	Red-backed squirrel monkey	*Saimiri oerstedi*	0	0		N	2	1	s+	2	N			
Ceboidea – Cebidae	Common squirrel monkey	*Saimiri sciureus*	0	1	≫	N	2	∨	s+	2	N			
Cercopithecoidea – Cercopithecidae	Gray-cheeked mangabey	*Cercocebus albigena*	0	1	∧	N	0	∨	M	4	N			
Cercopithecoidea – Cercopithecidae	Agile mangabey	*Cercocebus galeritus*	Obs	1		N		∨	M		Y			
Cercopithecoidea – Cercopithecidae	White collared mangabey	*Cercocebus torquatus*	Obs	1		N	1	∨	M		Y			
Cercopithecoidea – Cercopithecidae	Vervet monkey	*Cercopithecus aethiops*	Obs	1	∧	N	2	∨		33	Y			
Cercopithecoidea – Cercopithecidae	Redtailed monkey	*Cercopithecus ascanius*	Obs	1	∧	N	2	∨	s+	3 [++]	Y			
Cercopithecoidea – Cercopithecidae	Blue or Sykes's monkey	*Cercopithecus mitis*	Obs	1		N	1	∨	s+	2 [++]	Y			
Cercopithecoidea – Cercopithecidae	Patas monkey	*Erythrocebus patas*	0	1		N	2	∨	s+	1 [++]	Y			
Cercopithecoidea – Cercopithecidae	Stump-tailed macaque	*Macaca arctoides*	0	1	∧	N	1	∨	M	30	Y			
Cercopithecoidea – Cercopithecidae	Longtailed macaque	*Macaca fascicularis*	Obs	1	∧	N	1	∨	M	15	Y			
Cercopithecoidea – Cercopithecidae	Japanese macaque	*Macaca fuscata*	0	1	≫	N	2	∨	M	11	Y			

Cercopithecoidea – Cercopithecidae	Common name	Scientific name								
Cercopithecoidea – Cercopithecidae	Moor macaque	*Macaca maurus*	o	1	<	N	1	<	M	15	N
Cercopithecoidea – Cercopithecidae	Rhesus macaque	*Macaca mulatta*	Obs	1	<	N	2	<	M	9	Y
Cercopithecoidea – Cercopithecidae	Pig-tailed macaque	*Macaca nemestrina*	Obs	1	<	N	1	<	M	13	Y
Cercopithecoidea – Cercopithecidae	Celebes or Crested black macaque	*Macaca nigra*	o	1	<	N	1	<	M	9	N
Cercopithecoidea – Cercopithecidae	Bonnet macaque	*Macaca radiata*	o	1	<	N	2	<	M	5	Y
Cercopithecoidea – Cercopithecidae	Lion-tailed macaque	*Macaca silenus*	Obs	1	<	N	1	<	s+	18	
Cercopithecoidea – Cercopithecidae	Toque macaque	*Macaca sinica*	o	1	<	N	2	<		14	
Cercopithecoidea – Cercopithecidae	Barbary macaque	*Macaca sylvanus*	Obs	1	<	N	2	<	M	14	Y
Cercopithecoidea – Cercopithecidae	Tibetan macaque	*Macaca thibetana*	o	1	<	N	2	<	M		
Cercopithecoidea – Cercopithecidae	Tonkean macaque	*Macaca tonkeana*	o	1	<	N	1	<	M	10	Y
Cercopithecoidea – Cercopithecidae	Drill	*Mandrillus leucophaeus*	Obs	1	<	N	1	<			Y
Cercopithecoidea – Cercopithecidae	Talapoin	*Miopithecus talapoin*	o	1	<	N	2	<		11	N
Cercopithecoidea – Cercopithecidae	Proboscis monkey	*Nasalis larvatus*	o	1	<	N	1	<	s		Y
Cercopithecoidea – Cercopithecidae	Olive or Anubis baboon	*Papio anubis*	Obs	1	<	N	0	<	M	6	N
Cercopithecoidea – Cercopithecidae	Yellow baboon	*Papio cynocephalus*	Obs	1	<	N	0	<	M	9	N

Table 14.2. (*cont.*)

Family	Common name	Species	Male-infX observed	Vuln. estimate	Lac/ Ges	PPE	Season-ality	Births /year	Obs no. males	Mating period	Post. conc. mating	Induced ovulation	Delayed implant	Addi- tional refs.
Cercopithecoidea – Cercopithecidae	Hamadryas baboon	*Papio hamadryas*	Obs	1	>	N	1	√	s+	5	N			
Cercopithecoidea – Cercopithecidae	Chacma baboon	*Papio ursinus*	Obs	1	>	N	o	√	M	9	N			
Cercopithecoidea – Cercopithecidae	Silver leaf monkey	*Presbytis cristata*	Obs	1	>	N	1	√	s						
Cercopithecoidea – Cercopithecidae	Gray or hanuman langur	*Presbytis entellus*	Obs	1	>	N	1	√	s+/M	6	Y			
Cercopithecoidea – Cercopithecidae	Dusky leaf monkey	*Presbytis obscura*	o	1	>	N	1	√	s			N			
Cercopithecoidea – Cercopithecidae	Purple-faced langur	*Presbytis senex*	Obs	1	1	N	1	√	s			Y			
Cercopithecoidea – Cercopithecidae	Gelada baboon	*Theropithecus gelada*	Obs	1	>	N	1	√	s+	9	N			
Hominoidea	Gorilla	*Gorilla gorilla*	Obs	1	>	N	o	√	s+	2	Y			
Hominoidea	Gibbon	*Hylobates spp*	o	1	>	N	o	√	s[+]	4	Y			
Hominoidea	Bonobo or Pygmy chimpanzee	*Pan paniscus*	o	1	>	N	o	√	M	15	Y			
Hominoidea	Chimpanzee	*Pan troglodytes*	Obs	1	>	N	o	√	M	14	Y			
Hominoidea	Orang-utan	*Pongo pygmaeus*	o	1	>	N	o	√	s+	13	Y			

Notes:

Columns:

4. Observed infanticide (inf X) by males from van Schaik (Chapter 2): Obs C, infanticide only known from captivity; Obs, infanticide observed; o, infanticide not known to occur.

5. Vulnerability estimate as explained in text.

6. Lac/Ges: ratio of durations of lactation and gestation. '>' signifies lac/ges ratio of >1; '<' means lac/ges ratio of <1.

7. PPE: postpartum estrus with possibility of postpartum conception resulting in largely overlapping lactation and subsequent gestation. N, no; Y, yes.

8. Seasonality: 2, >67% births in peak 3 months season; 1, birth peak with 33–67% of births within 3 months period; o, no birth seasonality.

9. Births/year: number of litters per year with surviving offspring. '>' stands for multiple births per year; '<' is mean interbirth intervals exceeding 1 year.

10. Observed number of males mating per gestation: s, >90% copulations per conception with single male; s+, 50–90% copulations with single male; M, <50 % copulations with single male.

11. Mating period in days, unless otherwise specified: median for species taken from Hayssen et al. (1993), unless otherwise indicated.

12. Postconception mating: Y, frequently observed; N, not commonly observed.

13. Induced ovulation: ovulation induced by mating.

14. Delayed implantation: Oblig. (obligatory), always present ususally many months; Facult. (facultative), variable in duration and depending on maternal condition often related to presence of lactating offspring.

Sources: Hayssen et al. 1993; Nowak 1999; for primates van Schaik et al. 1999.
Additional references by number:

(1) Byers 1997; (2) Berger & Cunningham 1994; (3) David 1975; (4) Brotherton & Manser 1997; (5) Hogg & Forbes 1997; (6) Dagg & Foster 1976; (7) Lewison 1998; (8) Estes 1991; (9) Alderton 1994; (10) Sillero-Zubiri et al. 1996; (11)Venkataraman 1998; (12) Girman et al. 1997; (13) Moehlman 1989; (14) Cavallini 1998; (15) Alderton 1993; (16) Eaton 1978; (17) Busch 1996; (18) Pusey & Packer 1994; (19) Kruuk 1972; (20) Sliwa & Richardson 1998; (21) Richardson 1991; (22) Mead 1994; (23) Poglayen-Neuwall et al. 1989; (24) Estes 1989; (25) Wright 1963; (26) Chanin 1985; (27) Da Silva et al. 1993; (28) Neal 1986; (29) Mead 1989; (30) King 1989; (31) Roberts & Kessler 1979; (32) Gompper et al. 1997; (33) Gehrt & Fritzell 1996; (34) Gehrt & Fritzell 1999; (35) Schaller et al. 1985; (36) Schenk & Kovacs 1995; (37) Haltenorth et al. 1979; (38) Doolan & MacDonald 1997a; (39) Clapham 1997; (40) Connor et al. 1996; (41) Rissman 1995; (42) FitzGibbon 1997; (43) Berger 1986; (44) Feh 1999; (45) Reeves et al. 1992; (46) Trillmich 1986; (47) Campagna & LeBoeuf 1988; (48) Boness & James 1979; (49) Amos et al. 1993; (50) LeBoeuf 1972; (51) Moss 1983; (52) Alderton 1996; (53) Keil & Sachser 1998; (54) Topping & Millar 1998; (55) Ribble 1991; (56) Jeppson 1986; (57) Gajda & Brooks 1993; (58) Solomon & Getz 1997; (59) Getz & Carter 1998; (60)Wynne-Edwards & Lisk 1984; (61) Stopka & MacDonald 1998; (62) Hoogland 1998; (63) Hoogland 1995; (64) Blumstein & Arnold 1998; (65) Blumstein 1997; (66) Goossens et al. 1998; (67) King & Allainé 1998; (68) Millesi et al. 1998; (69) Schwagmeyer 1984; (70) Wishner 1982; (71) Waterman 1998

Van Schaik (Chapter 2) used the lactation/gestation length ratio as a crude index of vulnerability. In this chapter, we use a smaller number of species for which we have additional information (see Table 14.2), including the presence of reproductively dormant periods or the incidence of facultative delayed implantation to classify most species as vulnerable or not, depending on whether a male potentially derives a reproductive advantage from killing offspring.

Thus we classify species as vulnerable if lactation lasts longer than gestation and where reproduction is not strictly annual and seasonal. They are also considered to be vulnerable when there is evidence of either facultatively delayed implantation (implying that gestation could be shortened if current offspring are lost) or a shortened interbirth interval after a lost litter. In contrast, we classify species as non-vulnerable if there is postpartum conception resulting in largely overlapping lactation and subsequent gestation (in some cases in a dormant stage). They are also classified as non-vulnerable if gestation lasts clearly longer than lactation in the absence of evidence for dormant periods.

Validation of vulnerability criteria

If a female can conceive again immediately after the birth of her offspring, a male is unlikely to benefit from committing infanticide. Postpartum conception can be viable only in species where lactational dependence of the current offspring is shorter than the duration of gestation (cf. van Schaik 2000; in some species this may include a delayed implantation phase). Among primates postpartum estrus is reported for 13 of 17 non-vulnerable species and for none of the 32 species with documented infanticide or the 21 additional species that are considered vulnerable (but without recorded infanticide) based on the lactation/gestation ratio and seasonality ($G_{adj} = 45.4$; $df = 2$; $P < 0.001$). Among non-primates, postpartum estrus is recorded for 31 of 63 ($= 49\%$) non-vulnerable species, but only in 4 of 26 ($= 15\%$) species with confirmed infanticide and in 6 of 51 ($= 12\%$) additional vulnerable species ($G_{adj} = 22.31$; $df = 2$; $P; < 0.001$). The majority (21 of 31) of non-primate species with postpartum estrus also have delayed implantation. Among these species, those with facultative delayed implantation can still be vulnerable if the survival of current offspring delays the birth of the next – this is the case in at least 7 of the 10 species with postpartum estrus that are vulnerable or have a confirmed infanticide risk. Thus absence of postpartum estrus is a good, but not perfect, indicator of a species' vulnerability to infanticide by males.

We tested whether our estimate of vulnerability correctly predicts infanticide by males. The adjusted estimate yielded a dataset of 71 primate and 140 other mammal species of which 51 and 77, respectively, were considered to be vulnerable. Among the vulnerable species, infanticide by males has been documented for 31 primates and 26 other mammals whereas among the non-vulnerable species, infanticide was recorded only in one primate species (*Varecia*, see below). Thus, of the 58 species with reported infanticide, 57 were classified as vulnerable.

It should be noted that these adjusted estimates are still likely to underestimate vulnerability. Vulnerability could be increased in species with large litters, where the survival of one litter may affect the size and survival of the next (Stearns 1992). Similarly, in some species, especially hunting carnivores, offspring dependence on the mother lasts much longer than lactation; so an offspring's survival to ecological independence is likely to postpone the mother's subsequent reproduction, rendering her offspring vulnerable to infanticide by males. At the time of writing, the available information on the actual effects of these factors is too sketchy to be included in the estimate for all species.

Since our data base is heavily biased toward primates, this order will be treated separately from the very incomplete and taxonomically unbalanced information on other orders. For primates, we can provide an even more detailed assessment of vulnerability (Table 14.3, based largely on van Schaik *et al.* 1999). Most primate mothers carry their (usually single) newborns with them wherever they go, but many strepsirrhines leave their offspring hidden in the vegetation or a treehole while foraging ("infant caching"), whereas among callittrichids group members other than the mother carry the offspring (often twins) from a very early age. These three different modes of infant care/carrying are correlated with vulnerability to infanticide. In contrast to the species in which the mother is the main carrier, both infant cachers and communal infant carriers have a short lactation period in relation to gestation, rendering them not very vulnerable to sexually selected infanticide, and making postpartum conception a viable option (see Table 14.3). Accordingly, infanticide has been recorded only for primates in which the burden of carrying rests solely on the mother, with one exception among the cachers (ruffed lemur, *Varecia variegata*)[1].

---

1. The observed infanticide in the ruffed lemur occurred under the expected circumstances (F. J. White *et al.*, unpublished data). It was classified as non-vulnerable on the basis of its seasonality in combination with an interbirth interval of 1 (Table 14.2) however, field data may show a longer interbirth interval for surviving litters.

Table 14.3. *Summary of female sexual behavior in primates by mode of infant care: infant is "cached" while the mother forages, infant is carried by the mother and infants (mostly twins) are carried by group members other than the mother (based on data presented by van Schaik et al. 1999)*

	Mother caches infant	Mother carries infant	Group members carry infants
Lactation > gestation	0%	94%	0%
	N = 10	N = 34	N = 7
Infanticide by male observed	7 %	50%	0%
	N = 14	N = 66	N = 8
Postpartum estrus	82%	0%	63%
	N = 11	N = 65	N = 8
Postconception mating	0%	58%	100%
	N = 6	N = 48	N = 6
Average mating period (in days)	1.8 ± 0.8	8.1 ± 7	13.3 ± 8.3
	N = 12	N = 45	N = 6
No. of mating males			
Single	0	5	4
Single plus	3	14	4
Multiple	1	27	0
Exaggerated sex skin	0%	38%	0%
	N = 13	N = 64	N = 8
Estrous calls	33%	18%	0%
Copulation calls	22%	51%	0%
	N = 9	N = 45	N = 8

Results

Polyandrous mating

Mammalian females tend toward polyandrous mating: in only 29% (of 126) of non-primate and 14% (of 58) of primate species do females typically mate with a single male only (Figure 14.2). To separate the genetic benefits due to mixed paternity in a litter from other benefits, we can compare species with single offspring to those that produce litters. Among primates, monandry is equally rare in both groups of species (13% of 48 species versus 20% of 10 species, respectively). Among non-primate mammals with single offspring, females in 42% (of 38) of the species mate monandrously, whereas they do so in 24% of 88 species with larger litters ($G_{adj} = 4.04, P < 0.05$, one-tailed). Thus, while monandry is more common in non-primate mammals with single offspring, there are likely to be

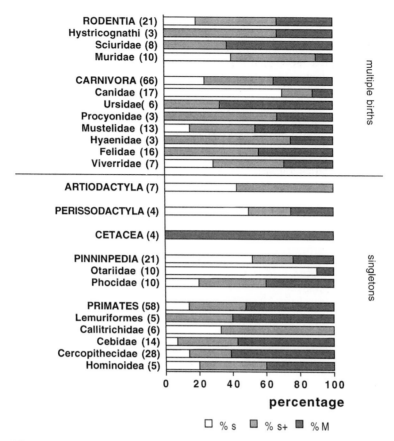

RODENTIA (21)
Hystricognathi (3)
Sciuridae (8)
Muridae (10)

CARNIVORA (66)
Canidae (17)
Ursidae(6)
Procyonidae (3)
Mustelidae (13)
Hyaenidae (3)
Felidae (16)
Viverridae (7)

ARTIODACTYLA (7)

PERISSODACTYLA (4)

CETACEA (4)

PINNINPEDIA (21)
Otariidae (10)
Phocidae (10)

PRIMATES (58)
Lemuriformes (5)
Callitrichidae (6)
Cebidae (14)
Cercopithecidae (28)
Hominoidea (5)

multiple births

singletons

0   20   40   60   80   100
percentage

□ % S   ▨ % S+   ■ % M

Figure 14.2. Percentage of species per taxonomic group in which females typically mate with "s": a single male ($<$10% extra-pair copulations); "s+": a "single male plus" (50% to 90% of copulations with the same male); or "M": multiple males ($>$50% of cycles with more than 1 male). The number of species for each family/order contributing to the data is indicated in parentheses. (Note: orders with only one species in Table 14.2 are not included in this figure.)

additional benefits to polyandrous mating beyond the genetic ones in primates and non-primates alike.

Primate females vulnerable to infanticide mate polyandrously significantly more often (62% of 47 species) than females of non-vulnerable species (9% of 11; see Figure 14.3, Table 14.3). The primate species in which females typically mate with a single male are those that live in social groups with only one male. However, females of these species occasionally mate with extra-group males. Moreover, when the

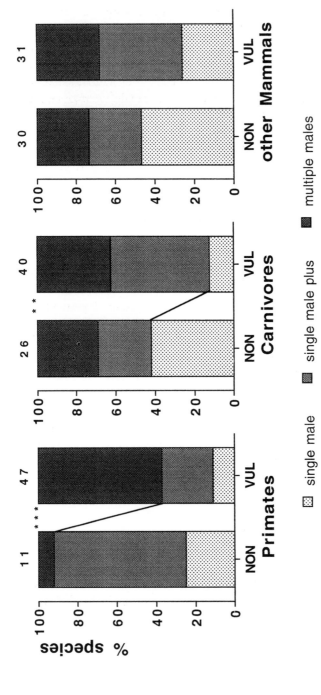

Figure 14.3. Tendency toward polyandrous mating in relation to vulnerability to infanticide by males. *Primates:* females of vulnerable species mate more often with multiple males (comparison of number of mating partners per vulnerability (VUL) class: $G_{adj} = 9.69$, df = 2; $P < 0.01$; proportion of multiple mates per class: $G_{adj} = 10.57$, df = 1; $P < 0.005$). *Carnivores:* females in non-vulnerable (NON) species mate more often with a single male (total comparison $G_{adj} = 7.66$, df = 2, $P < 0.025$ and for proportion of single mates $G_{adj} = 7.28$, df = 1, $P < 0.01$ (indicated by the connecting line and stars)). The difference between vulnerability classes was not significant for 61 other mammals representing 7 orders (with 1 to 21 species; $G_{adj} = 2.95$, df = 2, NS).

male is ousted or dies, these females engage in postconception mating (see below).

Among 66 carnivore species, females in 42% of 26 non-vulnerable species mate with a single male, compared with only 13% of 40 vulnerable species (Figure 14.3).In many vulnerable carnivores, females face considerable difficulties in mating polyandrously. In canids, for instance, extrapair matings might endanger male care; nonetheless, polyandry occurs (wolf, Derix & van Hooff 1995; dhole, Venkataraman 1998) or even with nomadic solitary males (red fox, Cavallini 1998). In solitary felids, male territories often overlap those of several females. Nonetheless, females often also mate with males of adjacent territories and occasionally nomadic males (Eaton 1978; Alderton 1993; Busch 1996). In the only gregarious felid, lions, polyandrous mating is routine.

Unfortunately, details of the number of mating partners under natural conditions are scarce for most other mammals. Except for the Sciuridae, most knowledge of rodent reproduction comes from captivity, where secondary predictions of female preference for certain males are more often tested than their actual tendency toward polyandry (cf. Ebensperger 1998b). Polyandry often varies with the circumstances, for example encounter frequency of potential mates (see, e.g., Agrell et al. (1998) on *Microtus agrestis*). Given the opportunity, most rodent females seem to mate with more than one male, except for some Muridae (in our sample, 4 of 11 mated with a single male). Among 31 vulnerable "other" mammal species, 26% of females mate with a single male only, compared with 47% (of 30) in non-vulnerable species – a non-significant difference (Figure 14.3). This dataset is a mix of species from seven different orders with different life styles.[2]

In sum, the prediction that females of species vulnerable to infanticide will mate with more than one male is mostly supported, especially in primates and carnivores.

---

2. Among the pinnipeds 5 of 22 species were considered vulnerable to infanticide, one of which, the southern fur seal (*Arctocephalus pusillus*), only marginally so. Like other Otariidae, the Galapagos fur seal (*Arctocephalus galapagoensis*) and Australian sea lion (*Neophoca cinerea*) mate with a single male postpartum and have delayed implantation. Unlike other pinnipeds, they often continue to nurse an older offspring after the birth of the next, but the youngest one often does not survive (Reeves *et al.* 1992; Nowak 1999). Although clearly vulnerable to sexually selected infanticide (the postpartum mating may result in a viable offspring only if the current pup does not survive) we have not found records of infanticide for these species.

Means of achieving polyandry

In many taxa, females actively pursue additional matings, achieving polyandry despite vehement male monopolization attempts. In most social groups with two males, females manage to mate with the second male as well; and even females typically living in one-male groups may occasionally mate "secretly" with extra-group males (gibbon, Palombit 1994b, Reichard 1995, bontebok, David 1975; horse, Feh 1999).

To mate polyandrously, females often alert males as to their receptive status. The universal mammalian way of communicating physiological state is through olfactory signals. Females of several mammals, especially in carnivores and primates, also give estrous calls that attract males. In northern elephant seals, females give loud vocalizations when mounted when they are not receptive or mounted by a young or subordinate male. These calls often induce higher-ranking males to interfere and stop the attempted mating (Cox & LeBoeuf 1977). However, if females call during or after mating and sperm has been transferred, these calls are likely to serve to attract additional males and thus promote polyandrous mating (van Noordwijk 1985; O'Connell & Cowlishaw 1994; van Schaik *et al.* 1999). These calls are known for about 30% of primate species. Females of all primate species with distinct copulation calls routinely mate with more than one male if available, although not all polyandrously mating species do call. Data on calls during the mating period for other taxonomic groups are very sketchy, and hence their possible role in achieving polyandry cannot be assessed.

Modest vulval swellings serve as short-range visual advertisements of sexual receptivity among various mammals (e.g., in several martens and a weasel, King 1989, Mead 1994; and the giant panda, Schaller *et al.* 1985). Conspicuous, exaggerated, swellings are limited to Old World monkeys and apes living in social settings where polyandrous mating is possible (van Schaik *et al.* 1999). Males are clearly attracted to these exaggerated swellings (e.g., Bielert & Girolami 1986), and they are therefore likely to promote polyandrous matings (for more detailed discussion, see Nunn 1999c). Interestingly, among catarrhine and hominoid primates, all 17 species with exaggerated visual advertisement also have copulation calls, whereas species without vocalizations have no $(N=7)$ or small visual signals at most $(N=4)$ (data in appendix of van Schaik *et al.* 1999), suggesting complementarity of these signals.

In conclusion, polyandrous mating often does not qualify as convenience polyandry.

Extended mating period

If the only function of mating were to achieve conception, a mammalian mating period (measured as the average interval between first and last mating within one cycle) would not need to last longer than a day, at most, and only a single mating or a rapid series of matings would be needed. Indeed, in a few mammal species a single mating per pregnancy is the norm, for example in the bison (Berger & Cunningham 1994) and the Galapagos fur seal (Trillmich 1986).

Figure 14.4 shows that the average length of the mating period varies both between and within families. Although rodents, which tend to be smaller than members of other orders, tend to have short mating periods, there is no general body size effect (as illustrated in Figure 14.4). In the majority of species, mating period exceeds the minimum time required to achieve conception, especially in the fissiped carnivores and primates: females of 90% (of 49) of fissiped species mated on more than 1 day per cycle, and 92% (of 59) of primate species do. This strongly suggests additional functions for mating beyond conception. Indeed, prolonged mating periods of more than 1 week were found in 41% (of 59) of primates and 13% (of 91) of non-primates, mainly carnivores.

Among primates, the effect of vulnerability on mating period is affected by the non-vulnerable callitrichids, some of which have very long mating periods (Table 14.3). We speculate that their sexuality is derived from that of other infant carriers and has acquired the derived function of exchange of male infant care for mating access (cf. Sánchez *et al.* 1999). The other infant carriers have far longer mating periods than the non-vulnerable infant cachers (Table 14.3; cf. van Schaik *et al.*, Chapter 15).

Among the 11 mammalian families with data on mating periods for at least one species in both vulnerability categories, mating periods are longer for vulnerable species in 6, shorter in 3 and equal in 2. Thus mating period does not show a strong effect of vulnerability to infanticide by males.

Postconception mating

Mating after conception is common in several primates species, sometimes even with a cyclical recurrence (for a review, see van Schaik *et al.* 1999). In the current sample, females of 59% (of 61) of primate species are known to have at least some postconception mating. We expected post conception mating to occur in vulnerable species, especially after dramatic postconception changes in the presence of males. Among primates,

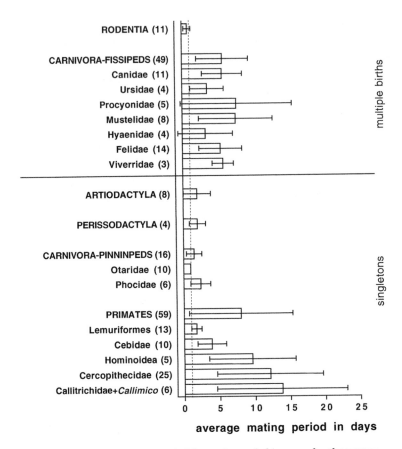

Figure 14.4. Average duration (± SD) of the mating period (measured as the average interval between first and last mating within one cycle) for several mammal families. The number of species for each family contributing to the data is indicated in parentheses. Indicated in the upper panel are species that mostly give birth to litters; in lower panel, species with mostly single offspring per birth. A mating period of 1 day is indicated by the dashed line. (Note: orders with only 1 species in Table 14.2 are not included in this figure.)

60% (of 10) non-vulnerable species and 61% (of 44) vulnerable species have postconception mating at least occasionally. However, among the non-vulnerable species, postconception mating is absent in the infant cachers and only seen among the callitrichids (Table 14.3).

Among other mammals, postconception mating has been reported for several felids, especially upon exposure to new males (Eaton 1978). Although little is known about the mating behavior of most cetaceans, postconception mating has been suggested for bottle-nosed dolphins

(Connor *et al.* 1996) and the right whale (Clapham 1997). All these species are thought to be vulnerable to infanticide by males. Among rodents, postconception mating has been documented for the water vole (Jeppson 1986) after a pregnant female moves to a nesthole in the territory of a different male. Numerous manipulative studies on rodents have also shown that postconception mating can reduce the chances that a male will commit infanticide (Huck *et al.* 1982; Huck 1984; Ebensperger 1998a). The musk shrew (*Suncus murinus*, Rissman 1995), an insectivore without periodic estrus, which ovulates some 18 hours after mating, continues to mate during pregnancy, at least in captivity. Both the vole and the shrew are not considered vulnerable to infanticide by males. However, in both species postconception mating is reported mainly after exposure to a new male.

It is still unclear whether lack of field evidence of postconception mating in other orders is due to lack of observations or occurrence. However, despite this lack of data, the common occurrence among the vulnerable species in two of the most vulnerable orders, primates and car-nivores, is suggestive of an anti-infanticide function.

Abandonment of unborn offspring

The "Bruce effect", the failure of a zygote to implant after exposure to a new (dominant) male (or female!), was first described for rodents. It is rarely reported in other mammals. There are several possible reasons for this (see Ebensperger 1998a). First, because unambiguous documentation is difficult it may be underreported in most taxa or regarded as failure to conceive. Second, it may truly be rare because the kind of social change precipitating the phenomenon, exposure to a new male shortly after conception, is less common in non-rodents. Third, the apparent high incidence of male-induced implantation failure in rodents is an artifact of the ease with which they can be manipulated in captivity. Thus a recent study on gray-tailed voles (*Microtus canicaudatus*) by de la Maza *et al.* (1999) reported that females in large enclosures showed no Bruce effect follow-ing the replacement of males. Similarly, other studies reported that females with sexual exposure to multiple males were less likely to abandon embryos (Huck 1984). At this stage, it is difficult to decide among these possibilities, although it is unlikely that the Bruce effect is simply an artifact. However, it is clear that quantitative evaluation of the possible function of the abandonment of unborn offspring is as yet impossible.

In a broader sense, embryo abandonment in the form of abortion or resorption of implanted embryos (often included as Bruce effects) shortly following the appearance of a new (dominant) male has been reported for numerous rodents (Huck 1984; Ebensperger 1998a), horses during the first half of gestation (Berger 1986) and primates. In primates, despite decades of studies of individually known wild primates, relatively few examples exist of abortions linked to the appearance of a new dominant male (savanna baboons, Pereira 1983; gelada baboons, Mori & Dunbar 1985; hanuman langurs, Agoramoorthy *et al.* 1988; white-handed gibbon (following death of mate), Palombit 1995). Although it is possible that the phenomenon is seriously underreported, in most of the documented cases the new males were considered to be unusually aggressive. Thus Agoramoorthy *et al.* (1988) suggested that males would attack weaned offspring to cause stress in recently pregnant mothers as a low-risk male strategy to induce abortions. In their study after takeovers that were followed by at least one female aborting her fetus, 70% of females who succeeded in giving birth suffered infant loss due to subsequent infanticide.

Lionesses have a low rate of conception for several cycles after their pride has been taken over by new males. The most likely explanation for this temporarily low fecundity is lack of ovulation (which is induced by mating to some extent) or conception instead of early abortions or Bruce effects, since the phenomenon lasts for several cycles after the appearance of the new males, apparently until these new males are well established and unlikely to be ousted (Pusey & Packer 1994; see also Steenbeek 1999a for a similar situation in Thomas's langurs).

In conclusion, embryo abandonment is observed in a range of species in circumstances where subsequent infanticide is highly likely and the response is thus likely to be adaptive. However, problems of documenting it in the wild make this conclusion only preliminary.

Interactions among the various tactics

Since polyandrous mating and postconception mating both serve to confuse paternity or its assessment, we expect a positive correlation between these two tactics across species. This prediction is upheld among non-primate mammals, where all 10 species with known post-conception mating are also known to mate polyandrously before conception. However, in primates the correlation is poor (Tables 14.2 and 14.3). More detailed knowledge of these species may explain this: first, there may be selection against postconception mating in some polyandrous species (e.g., baboons). Second, females in monandrous species cannot mate polyandrously but may benefit from paternity assessment manipulation when their mate is replaced after conception and thus selection may also favor some postconception mating under stable conditions.

We had predicted a negative correlation between embryo abandonment and postconception mating, and also that embryo abandonment should be less common than mating tactics serving to reduce infanticide risk. It is difficult to test these predictions because reliable data on postconception mating are scarce in rodents, whereas reliable data on embryo abandonment is scarce for other mammals. Nonetheless, if embryo abandonment in rodents is not an artifact (cf. de la Maza *et al.* 1999), the rarity of postconception mating in this taxon is consistent with the prediction. We already noted that abortion is occasionally observed in other mammals after immigration by aggressive males. Among these species, some do not have postconception matings (horses, savanna baboons), but others do (gibbons, hanuman langurs). Among non-rodents, the scarcity of data on abortion and decreased fecundity after the appearance of a new male suggests that male-induced abortion is a rare event restricted to situations where alternative tactics are unlikely to prevent infanticide. However, this conclusion is very tentative.

Discussion

This chapter asked whether mammalian reproductive behavior during the prenatal period may serve an anti-infanticide function by manipulating paternity or the males' assessment of paternity. In particular, we hypothesized that polyandrous mating, long mating periods, matings during pregnancy, or abandonment of unborn offspring could all function to reduce the risk of infanticidal attacks in vulnerable species. Vulnerability to infanticide by males is largely concentrated in three or four orders: primates, carnivores, (sciurognath) rodents and probably cetaceans (cf. van Schaik, Chapter 2).

Support for the predictions was mixed but promising. Among carnivores and primates, polyandrous mating was far more common in vulnerable species than in non-vulnerable species, but in the mixed collection of other mammals the difference was not significant. Only in primates was there some support for the prediction of longer mating periods in vulnerable species. Postconception mating, the pattern most likely to be unique to the anti-infanticide hypothesis, is indeed concentrated in species vulnerable to infanticide in carnivores, cetaceans and primates, although in the latter order a group of non-vulnerable species also shows extensive postconception mating. The distribution of implantation failure is too poorly known to evaluate the prediction, although its occurrence in rodents seems most common in the expected context (exposure to new dominant male). Embryo abandonment in other taxa seems more common in vulnerable species without postconception mating. Obviously, for this first round of testing, critical information on many species is lacking and firm conclusions cannot yet be drawn.

The main reason for the imperfect fit with predictions was known at the outset: the difficulty of generating competing exclusive predictions. First, reproductive behavior is subject to many different selective forces, and infanticide risk is only one among several survival issues for the adults (e.g., genetic benefits of mate choice, or constraints on mating behavior due to predation risk). Second, details of the species' natural history, such as the local sex ratio, degree of territoriality, potential for harassment or the risk of takeover by non-sire males, affect the need for countertactics in vulnerable species. We discuss these two possibilities below.

Alternative pressures on mammalian reproductive behavior

Several authors have proposed genetic benefits to polyandrous mating that may apply only to species producing larger litters or clutches (Table 14.1). However, the effect of litter size on polyandrous mating is modest, and polyandry is common among species producing single young. Indeed, Yasui (1998) recently concluded that genetic benefits do not seem to furnish a major selective advantage to polyandry in mammals.

One cost of polyandrous mating may be the loss of male care for young (though not in primates; Smuts & Gubernick 1992). Paternal care can be substantial in some monogamously mating Muridae, shortening the female's interbirth interval by some 30% (California mouse, Cantoni & Brown 1997). In Djungarian hamsters and bank voles (*Clethrionomys glareolus*) exposure to more males does not lead to polyandry, but rather to delayed conception (especially when their dominance order is not clear; Wynne Edwards & Lisk 1984; Horne & Ylönen 1996). In at least three of the four rodents in our dataset in which females mate with a single male only, these males provide paternal care (California mouse, white-footed mouse and Djungarian hamster; Table 14.2). Likewise, the five vulnerable carnivore species (three canids and two mustelids) that mate with a single male all live in long-term pairs with male care. Thus, where females can raise more young when the sire provides care, natural selection is likely to favor monandry rather than polyandry (provided that takeover risk is sufficiently low).

Although the length of the mating period was not found to be related to infanticide risk, one of the other factors affecting its duration is probably direct benefits. In many socially monogamous species, females may derive more infant care from prolonged mating periods (primates, Stribley *et al.* 1987, Converse *et al.* 1995). Alternatively, in some orders, especially carnivores, long mating periods may reflect the presence of induced ovulation (canids, Estes 1991, Derix & van Hooff 1995). In addition, mating outside the periovulatory period may function to establish and maintain helping and protection bonds between males and females.

Effects of social and biological constraints
Social organization may affect the feasibility for certain tactics or the actual risk of infanticide despite *a priori* vulnerability. Thus, in a species living in pairs where males are strictly territorial and widely spaced, the potential for polyandry may be limited (though securing paternal care may also contribute to the lack of polyandry; see above). On the other hand, if it is found that vulnerable species are significantly more likely to have the demographic potential for polyandry this in itself is likely to be meaningful. For instance, female groups are far more likely to contain multiple males in carnivores and primates, both being taxa with many vulnerable species, than in ungulates, in which vulnerability is rare (van Schaik 2000). Female reproductive behavior may therefore modify the demographic context to one in which reproductive countertactics can be more effective (see also Nunn & van Schaik, Chapter 16).

Female mammals may vary systematically in their susceptibility to male harassment. A systematic effect of vulnerability to infanticide on the length of the mating period may be absent in non-primates because the potential for sexual harassment is strong enough only in primates to force females into longer mating periods in species with polyandrous mating (cf. Table 14.3; see also van Schaik *et al.*, Chapter 15). In most other orders, females may have enough control over their mating behavior not to require long mating periods to achieve the intended degree of polyandry. Thus females in some species achieve matings with many males in only a few hours (e.g., up to 8 males in 3 hours in the Cape ground squirrel; Waterman 1998). The most important predictor of infanticide in vulnerable species is the rate of sudden replacement of the dominant male. A variety of factors may affect this rate, many of them determined by social organization, life style and life history. Because patterns in these factors tend to follow taxonomic boundaries, one way to eliminate their confounding effect is to conduct more detailed tests within orders or families. Let us give one example of this approach. Many canids live in social groups with long-lasting pair bonds. They tend to mate predominantly with a single partner who subsequently contributes to the protection and provisioning of the young, sometimes with additional helpers. These canids may actually face a very low risk of infanticide, because rates of violent male takeover are very low (Moehlman 1989). Among wild dogs, for example, most individuals are related to at least some of the pups, and many individuals migrate to groups adjacent to close relatives (Girman *et al.* 1997). In solitary carnivores (cats, bears and mustelids), in contrast, the turnover rate of males tends to be high even within a single mating season (Sandell 1989), leading to a much higher risk of infanticide. This difference may explain why highly polyandrous or postconception mating is less common among canids than among other carnivores with a life history that makes them vulnerable to infanticide by males.

Additional countertactics?

It is important to highlight our ignorance of many fundamental aspects of reproductive physiology. For instance, the degree to which ovulation or corpus luteum formation are induced by mating varies from taxon to taxon. These features may affect the female's options for reproductive countertactics against infanticide by males. On the other hand, these same features may themselves be ancient adaptations, though their functions are unknown (for speculation, see Conaway 1971). Here we ask whether some other peculiar features of mammalian reproductive behavior should be interpreted as serving to reduce the risk of infanticide.

Mammals vary considerably in the number of cycles before conception. Connor *et al.* (1996) suggested that having multiple receptive periods before conception enhanced bottlenose dolphin females' ability to have mating relationships with many males, and thus confuse paternity assessment by those males. In the case of male harassment it could even enable a female to endure matings by non-preferred males in preconception receptive periods and then conceive during another cycle with a preferred mate. These preconception matings could serve the same function as postconception matings in manipulating male paternity assessment and we would expect the two to be correlated. But our data base is too anecdotal for a test: we need more data on the distribution of multiple cycles/receptive periods, in relation to seasonality, correlates of limited female power against harassment (such as swellings and vocalizations in primates), number of males available, continuation of receptive periods postconception and the risk of infanticide by males.

Another potential countertactic is delayed implantation, which is found in at least 12 different mammal families, spread over eight orders (for reviews, see Daniel 1970; Mead 1989). In species with a long dormant phase (usually before implantation of the blastocyst), a female can mate shortly after giving birth irrespective of the survival of her newborns. In most of these species, reproduction is annual, with a restricted mating season and separation of the sexes during most of the year (Mead 1989; Estes 1989; exception Ursidae). Thus males are unlikely to benefit from infanticide.

Once this form of delayed implantation is in place, females could also show Bruce effects without additional loss of time if a new (dominant) male appears after mating. Indeed, a second mating period after arrival of a new male has been reported for the American mink (*Mustela vison*; for a review, see Mead 1994). However, because the second mating results in superfetation, i.e., without the complete loss of the older blastocysts, Bruce effects probably do not occur.

In some rodents (Daniel 1970; Elwood & Ostermeyer 1984; Huck 1984) and fissiped carnivores (see Table 14.2), implantation delay is brief, but variable and induced by nursing. In this situation, males may benefit considerably from killing current dependent offspring, although this gain may be limited in seasonal breeders. Thus, although delayed implantation as a rule cannot be considered to have evolved as a counter-tactic to infanticide, it may have secondarily acquired this function in a few species with short and variable delays. However, there is no evidence for this hypothesis.

Conclusion

These preliminary analyses strongly suggest that at least the number of mates a female has and whether she engages in postconception mating reflect her vulnerability to infanticide. However, perhaps the most important function of this chapter is to call for more descriptive data on reproductive behavior for wild mammals, both on actual mating patterns from the female's perspective and on the effect of surviving offspring on the timing (and quality/number) of the next as an indication of potential benefits for a non-sire of eliminating current offspring.

Acknowledgments

We thank Sarah Hrdy for continuing inspirational support and Charles Janson and Charles Nunn for comments on the manuscript.

CAREL P. VAN SCHAIK, J. KEITH HODGES, AND CHARLES L.
NUNN

15

Paternity confusion and the ovarian cycles of female primates

Introduction

It has often been noted that female primates tend to have extended mating periods in their ovarian cycles, tend to mate polyandrously and also tend to mate during pregnancy (cf. Hrdy 1979; Hrdy & Whitten 1987). Since females in species vulnerable to infanticide show these features to a greater extent, this behavior was interpreted as serving to confuse paternity (cf. van Schaik *et al.* 1999; van Noordwijk & van Schaik, Chapter 14). The extent to which such mating tactics succeed in confusing paternity depends on the outcome of an "arms race" between males and females concerning the amount of information on paternity available to males. In order to examine more closely the claim that sexual behavior in primates serves at least in part to reduce infanticide risk, we must examine the physiological basis for paternity confusion, as well as for its complement, paternity concentration. Since ovarian cycles vary considerably in detail among taxa (e.g., Short 1984), we limit this examination to primates, the best-known order in this respect. We ask therefore how the ovarian cycles of female primates are organized in relation to the need for strategies to reduce infanticide risk. Two features are examined in particular which we will argue are related to the benefits to females of unpredictability in the timing of ovulation: (1) the large variance in the length of the pre-ovulatory or follicular phase both within and between individuals, and (2) interspecific variation in the mean length of the follicular phase.

Before reviewing the data on primate ovarian cycles, we must first put these cycles in the context of the observed sexual countertactics and the constraints of primate physiology (cf. Nunn, 1999c; van Schaik *et al.* 1999). A male's decision to protect an infant against attacks, or conversely to direct attacks at it, must be based on his assessment of paternity, yet male mammals cannot recognize their infants (Elwood & Kennedy 1994; see also van Schaik, Chapter 2; Treves, Chapter 10). A male's estimate of paternity is therefore probabilistic, depending on his mating history with the infant's mother. The question is therefore if and how females can manipulate the actual probability of paternity for the various male players involved. We therefore address the female dilemma of concentrating versus confusing paternity (van Schaik *et al.* 1999), and we consider the extent to which females are able to conceal ovulatory events.

The female dilemma

As discussed in several chapters in this volume, females of various taxa have evolved social counterstrategies to reduce the risk of infanticide by males. A prominent counterstrategy is association with the likely sire, who turns into a protector (van Schaik & Kappeler 1997). Successful infanticide is confined largely to situations where this protector male is incapacitated or has disappeared. A male will protect an infant, and be closely associated with it if, on average, the likelihood of paternity is high enough to outweigh the costs, especially those of the risks of injury while defending the offspring. Hence, in order to gain this protection against attacks by outside males, a female needs to provide the male with a high enough probability of paternity to make it selectively advantageous for the male to engage in protective behavior. Other, more deceptive strategies that would produce a low probability of paternity to the potentially protective male would not be an evolutionarily stable strategy (ESS), because males foregoing protection and instead pursuing additional matings would outreproduce protective males.

The female should therefore *concentrate* paternity into the male most capable of protecting her offspring against infanticidal attacks, wherever these attacks are likely to occur. The easiest way to achieve protection is by having a well-circumscribed period of attractivity and proceptivity, culminating in a short period of receptivity, in other words estrus. There is abundant evidence that sexually attractive primate females are guarded by dominant males (e.g., Hausfater 1975; Tutin 1979; Glander 1980; Harcourt *et al.* 1980; Schürmann 1982; van Noordwijk 1985; Noë & Sluijter 1990; Aujard *et al.* 1998). Especially if the receptive period is brief, these guarding males are the most likely sires. They are also the most likely to be present and capable of protecting the infant when it is born some time later.

Alternatively, the female may live in a social unit in which she is surrounded by several males during her receptive period, either because she lives in a multimale social unit, or because a new male invades at some other time, for instance during gestation. Regardless of anatomical advertising, dominant male primates mate preferentially with females of maximum attractivity and proceptivity, and often guard them to monopolize mating during this period. If the receptive period is long enough, males lower in the hierarchy will also mate at times when the female is somewhat less sexually attractive; they can do so because the dominant males are either pursuing other, more attractive females, or face ecological costs to prolonged mate-guarding (cf. Rasmussen 1985; Alberts *et al.* 1996; Nunn 1999c). The female would benefit from mating polyandrously if copulations during reduced attractivity led to paternity with a low, but non-zero probability; in other words when she *confuses* paternity. In this way, the other males in the upper regions of the dominance hierarchy, who may take over top dominance if the alpha male dies or is injured, would benefit less from committing infanticide, possibly to the point of making the behavior disadvantageous. Obviously, the very lowest-ranking males may only be allotted vanishingly small chances of paternity in this way, and if for some reason these males rapidly rise in rank and acquire top dominance, the female still faces a serious infanticide risk (e.g., Hamai *et al.* 1992; de Ruiter *et al.* 1994). In addition to confusing paternity, females may also manipulate male paternity assessments when they can attract males to mate while they are pregnant (see van Schaik, Chapter 2). (This possibility is discussed below.)

This hypothesis of female paternity confusion and manipulation through sexual behavior faces a serious challenge. It must somehow explain how females can simultaneously concentrate and confuse paternity. The hypothesis must assume, therefore, that females can assign each male chances of paternity that are optimal from her perspective. This assumption is examined in a later section.

It should be noted that the female dilemma arises only when demographic conditions provide an opportunity for polyandrous mating and immigration by powerful males or rapid turnover of rank among within-group males occurs at times when females cannot be sexually active. Thus, if females are never surrounded by more than one male during their receptive periods, no elaborate sexual countertactics would ever be feasible, and complete reliance on male protectors is to be expected. Likewise, if intrusions by strange males never occur, females should have no reason to concentrate paternity in the dominant male, since no protector is required, and the only benefit from concentrating paternity in particular males might be genetic.

The concept of concealed ovulation

A female can solve the female dilemma if she can conceal ovulation and mate with all the males who are likely to be infanticidal. A male's assessment of his chances of fertilization must be derived from his estimate of the female's proximity to ovulation based on anatomical, chemical and behavioral cues. All he can determine is the female's attractivity, which is tightly linked to estrogen levels (Robinson & Goy 1986).

Ovulation itself cannot possibly be observed directly. There are no data to suggest that ovulation can be inferred indirectly by signals emanating from the preovulatory rises in luteinizing hormone (e.g., Hodges *et al.* 1979) or other hormonal substances, though cyclic changes in unidentified urinary components (possibly pheromones) have been described and their potential involvement in communication of reproductive status proposed, at least in the cottontop tamarin (*Saguinus oedipus*; Ziegler *et al.* 1993).

The conclusion that males cannot accurately determine the time of ovulation is underscored by the phenomenon of pregnancy matings when males find females sexually attractive enough to mate, at a time when fertile ovulation is obviously impossible. Females tend to be less attractive during pregnancy, and in established groups attract mainly lower-ranking and young males (e.g., gorilla, Watts 1991a; sooty mangabey, Gust 1994b). Nonetheless, dominant adult males may still attack mating pairs (e.g., hanuman langurs, Sommer 1989a). Newly immigrated males lack the temporal comparison to judge the female's relative attractivity because females vary in maximum attractivity, and may therefore be more likely to mate than long-term residents. Pregnant females may also be more proceptive (e.g., patas, Loy 1981) or receptive (e.g., chimpanzee, Wallis & Lemon 1986) than cycling females. Field studies of primates with one-male groups suggest that when a newly immigrated male mates with a pregnant female, he is less likely to attack her infant (see van Schaik, Chapter 2; but see also Borries & Koenig, Chapter 5), though the reduction of infanticidal tendencies disappears the closer matings are to term.

Assuming that ovulation is fundamentally hidden to males, it is conceivable that a female mating strategy has evolved that provides a high enough chance of paternity to dominants to protect infants where this is needed, and yet also allots enough paternity chances to other males in the group to forestall infanticidal attacks when these males become dominant. In order to evaluate this possibility we need to review patterns of primate sexual behavior in relation to ovarian state.

Estrogen levels undergo a predictable rise during the second half of the follicular phase, followed by an abrupt decline immediately preceding ovulation, at a time when progesterone levels are rapidly increasing. Estrogens affect female behavior, change vaginal histology and secretory activity, cause skin tightening or even edema, and, in species with sex skins, increase the size of the swelling or the intensity of coloration (Robinson & Goy 1986). Progesterone opposes many of these effects of estrogen, at least in primates (Campbell & Turek 1981). Thus males are expected to use a variety of cues that indicate that ovulation is approaching. In many species, females visually advertise the coming of ovulation with a morphological change around the vulva. In most of those species, these changes are relatively inconspicuous, but they are predictable and reliable signals of impending ovulation. In others, these swellings or coloration changes are exaggerated, and visible from a long distance, clearly signaling something to distant observers. Other potential forms of advertisement include olfactory cues from the female's urine or vaginal and/or scent gland secretions as documented for certain callitrichid species (e.g., Epple 1986). Even in catarrhines, which lack a functioning vomeronasal organ (Dixson 1983), males often inspect and sniff at females' perineal regions, and there is evidence that this information can be used to assess female reproductive state (Keverne 1976). Specialized vocalizations and female proceptive behavior may also provide information on reproductive status. Thus a variety of morphological, physiological and behavioral cues indicate that the female is close to ovulation, although direct evidence that females are able to signal (and males detect) the *precise* time at which ovulation occurs remains lacking.

The existence of such a variety of cues suggests that in species advertising reproductive state, matings should at least be concentrated around a particular stage of the cycle (i.e., during the days preceding and directly following ovulation). By and large, this is what is found. However, some studies have shown that mating rates do not vary dramatically over the cycle (for overviews, see Dixson 1983; Nadler 1992), not even in species with advertisement of impending ovulation, such as rhesus macaques or chimpanzees. Most of these studies used pair-tests: a male and a female are placed together for a brief period of time (30 minutes or 1 hour) but are housed alone the rest of the time. Closer inspection reveals that the lack of effect of cycle stage on mating is an artifact of the pair-test paradigm, probably because it involves repeated separation and reunion of mates, who then engage in (male-initiated) sex upon each reunion. In apes, peaks in the periovulatory period appear when females are given complete control over access to males (Nadler 1992). Likewise, species showing no differentiation in pair-tests usually show far more pronounced peaks in mating around ovulation when living in permanent social groups (Wallen 1995). Hence females in advertising species do not conceal the fact that they are close to ovulation.

It is often thought that, where females do not advertise the periovulatory period through morphological or olfactory signals, they conceal it (e.g., Sillén-Tulberg & Møller 1993). This view assumes that, if a female advertises her ovarian state through morphological signals, the signal is honest, but that, if the female does not advertise, this allows her to manipulate the male's assessment of her proximity to ovulation. At first sight, data on non-advertising species support this view. For instance, in groups of stumptailed macaques, females receive matings at an approximately constant rate throughout their ovarian cycles. Even here, however, closer inspection reveals that the alpha male, who is by far the most frequent mate of the females, mates throughout the cycle, whereas other males show a very distinct mating peak near ovulation (Nieuwenhuijsen *et al.* 1986; Steklis & Fox 1988). In callithrichids housed in permanent pairs, mating takes place throughout the cycle. Nonetheless, male behavior indicates that they are quite capable of perceiving the periovulatory period (cottontop tamarin, Ziegler *et al.* 1993; pygmy marmoset, Converse *et al.* 1995). In situations where pair-tests show at least some mating throughout the cycle, females still show pronounced peaks around ovulation in proceptivity (e.g., orang-utan, Nadler 1992), often using special ritualized signals such as head flagging or shaking and tongue flicking (e.g., common marmoset, Kendrick and Dixson 1983). Hence, in all these species, a variety of cues and signals are available to males to build up an estimate of the female's ovarian state. The only remaining possible exception is the vervet, in which Andelman (1987) recorded mating throughout the cycle. Although her data suggest a peak around the estimated time of conception, the extremely low rate of mating makes this case the hardest to dismiss (cf. Burt 1992). In general, therefore, males can clearly perceive whether the female is near ovulation in the follicular phase, and when she is postovulatory in the luteal phase.

In conclusion, *concealed* ovulation, in the sense that females withhold all information to males about their proximity to ovulation, is not found in primates (see also Burt 1992). Instead, all mammalian females have concealed ovulation in the sense that its exact timing is fundamentally hidden from males. In many primates, the mating peak around the approximate time of ovulation covers a relatively broad window even in species advertising their attractivity, suggesting that ovulation is not signaled in a very precise way, and may in fact be *unpredictable*. This window is so broad that it encompasses times at which fertilization is either very unlikely or impossible if ovulation were predictably related to attractivity. Hence unpredictable ovulation should lead to less paternity certainty and may thus constitute a tactic used by primate females to confuse paternity. In the next section, we examine this proposal.

Unpredictable ovulation and follicular phase variance

Primate endocrinologists have long been aware of variation in the length of the two phases of the primate ovarian cycle and in the interval between the timing of external signals and of ovulation (e.g., Dukelow 1977), but most have not entertained the possibility that this variation is adaptive. However, if ovulation is unpredictable, males may only have an approximate assessment of paternity, especially because primate sperm remains viable for only a few days at most (Gomendio & Roldan 1993). Unpredictable ovulation from the female's perspective can be achieved by adding some random error to the control system, or by having it directly controlled by the female's internal state, for example depending on the quality of the stimulation by the current mate. However, some randomness is probably required to produce the ESS of optimal paternity biasing.

Unpredictable timing of ovulation means that there is large margin of error in when the follicular phase ends, resulting in a high variance in its duration. (This higher variance will also require a longer mean length of the follicular phase, which we examine in the next section.) Evaluation of this proposition requires data on the timing of ovulation, independent of external signs or female behavior. This independent evidence can derive from timed matings, vaginal histology, or (most commonly) endocrinological profiles, which can now be obtained non-invasively from

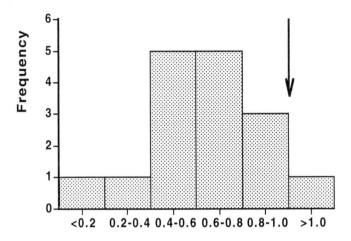

Figure 15.1. The relative standard deviation of mean length of the luteal phase relative to that in the follicular phase, in a sample of 16 primate species (multiple studies of any given species were averaged), to illustrate the relative lack of variability in the luteal phase. Data taken from Hayssen *et al.* (1993), with more recent studies added for *Pan paniscus* (Heistermann *et al.* 1996), *Macaca tonkeana* (Thierry *et al.* 1996), *M. silenus* (M. Heistermann, unpublished data) and *Presbytis entellus* (Heistermann *et al.* 1995).

urine or feces (Hodges *et al.* 1979; Lasley and Kirkpatrick 1991; Hodges 1992; Heistermann *et al.* 1996).

The most basic prediction is that the follicular phase shows high variance in duration. Because no extra noise is expected for the luteal phase, it is easiest to test this prediction by comparing variance in the length of the follicular phase with that in luteal phase length. This simple prediction is already difficult to test because many studies have censored their data by considering cycles exceeding some given length from the sample as somehow in error. This, of course, will bias estimates of variance. Nonetheless, a compilation of those studies of primate ovarian cycles that report on the length of follicular and luteal phases and their variability (Figure 15.1; see legend for sources) show a clear tendency for the standard deviation of the length of the luteal phase to be less than that of the follicular phase (one-group *t*-test: $t = 5.17$, df $= 15$, $P < 0.001$). Since, in most biological phenomena, variance and mean are correlated, and since the luteal phase is somewhat longer than the follicular phase in most species in the sample, this conclusion is conservative.

Another way to examine variability is to compare ranges of follicular phase length. Obviously, the observed range depends strongly on sample size, but since most studies report equal numbers of follicular and luteal phases, one can compare the ranges of the two phases within studies. Although in a few cases the absolute range was similar, relative ranges (range divided by mean length of the phase) were always less for the luteal phase ($N = 14$ studies of 11 species; same sources as listed in legend of Figure 15.1). In fact, in 8 of the 14 studies, relative range for the luteal phase was less than half of that for the follicular phase. To illustrate this difference, compare the range in the length of the follicular phase observed in various chimpanzee studies (10 to 32.5 days) to that observed in the luteal phase (12 to 16 days in the same dataset; Hayssen *et al.* 1993). Thus the first prediction is abundantly supported by empirical data on primates.

The second, more specific, prediction is that there will be high variance in the interval between ovulation time and the timing of easily noted signals, such as peak proceptive behavior or maximum size or coloration of a sex skin. Nunn (1999c) compiled studies that provided information on the time course of the signal relative to independently determined ovulation. From his compilation, we can determine how easy it is for a male to predict the timing of ovulation, and thus to derive the optimum mate-guarding strategy so as to maximize his chances of paternity, as well as to derive the best estimate of his chances of paternity so as to optimize the decision to protect, be neutral, or attack the infant. The data are all from catarrhines, and from species with swellings or coloration (though not all exaggerated).

To predict ovulation, we must look at the stability of the interval between some easily identified landmark, the onset of maximum swelling, and ovulation. It appears there is considerable error in this interval: from a 1 day range in the gorilla, which lacks an exaggerated swelling, to a 21 day range in the bonobo (Figure 15.2). Thus once a female is maximally swollen she will ovulate, but the timing of this ovulation is difficult to predict. Data from the bonobo study (Heistermann *et al.* 1996) illustrate this variability. Over nine cycles, maximum swelling lasted from 4 to 26 days, and the onset of maximum swelling in relation to the estrogen peak ranged from 2 to 21 days. The two extreme intervals (2 and 21 days) were both in conception cycles, underscoring that this is natural variability in normal cycles.

In theory, males could also estimate their paternity post hoc, because

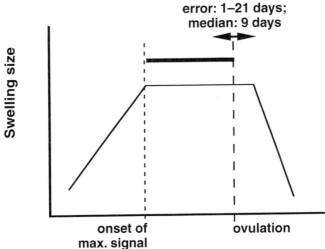

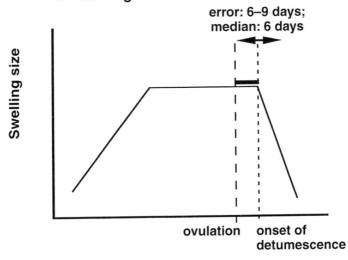

Figure 15.2. Error in the interval between the onset of maximum advertising and the subsequent timing of ovulation (top) and between the timing of ovulation and the subsequent decline in advertising (bottom). Summarized from studies compiled by Nunn (1999c).

matings after detumescence usually take place well after ovulation and mating even half a day after ovulation is unlikely to be fertile. Thus the timing of detumescence could provide an accurate retrodictive estimate of the timing of ovulation provided males can accurately remember the timing of their last consort period. However, as shown in the second panel of Figure 15.2, the relatively brief interval between detumescence or color breakdown and ovulation is still remarkably variable. Most studies show a range of about 6 days of possible ovulation (which explains why some mating activity is still observed after detumescence in many species). Hence, natural selection cannot successfully hone male counting ability, however indirect (cf. Perrigo & vom Saal 1994): regardless of whether males count forward or backward, their estimates of paternity remain highly probabilistic, unless one male monopolizes the female over her entire cycle.

In species lacking morphological advertising, it is more difficult for researchers to assess the extent of this noise. Data collected by T. Ziegler (unpublished results) for hanuman langurs also show extreme variability in the timing of behavioral estrus (i.e., frequent mating activity) relative to the hormonally determined timing of ovulation, suggesting that ovulation can take place more or less any time during the proceptive period (Figure 15.3). Data on humans compiled by Martin (1992) show an exceptionally broad probability distribution of ovulation, suggesting highly variable duration of the follicular phase. Thus species without morphological advertising may be at least as unpredictable in their ovulation, and perhaps more so.

In conclusion, although male primates in polyandrous mating situations have endocrine and behavioral indications that ovulation is impending or has occurred, they are probably unable to predict or retrodict it with accuracy.

The most detailed prediction is that the relative magnitude of this variability depends on the balance of the need for concentration and confusion (assuming that protection against infanticide is the only benefit from concentrating paternity, rather than other services such as infant carrying or protection against predators). The data are currently too scarce to perform a systematic test of this prediction, but it is perhaps significant that the error in bonobos is the largest known (Heistermann *et al.* 1996). In bonobos, no male immigration is known to occur; hence females would benefit less from paternity concentration.

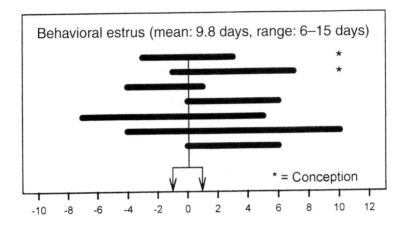

Figure 15.3. Duration of estrus (mating activity) in individual female hanuman langurs in Ramnagar in relation to the time of ovulation as determined by fecal progestin measurements (T. Ziegler, unpublished data).

Testing the consequences: paternity distribution

Uncertainty in ovulation timing provides females with the means to bias paternity toward one male who will act as protector for the infant, while still confusing it to the extent that some low-ranking males occasionally sire offspring. If this uncertainty is indeed found to the same extent in nature, this means that top dominant males in multimale groups, despite their tenacious attempts at mate guarding the female around the time of likely conception, should not obtain all paternities. Recent paternity studies using DNA, especially those conducted in natural demographic conditions, support this prediction (Paul 1997). Some of the patterns apparent in the paternity distribution are also consistent with variation in the need for paternity concentration in protectors. First, most of the other males with paternity were second- or third-ranking males, those most likely to take over the group if the current dominant male disappears. Second, in very large groups with clearly seasonal breeding, concentration of paternity in the top-ranking males is much less (Oi 1996; Paul 1997). The need for protectors in these conditions may be less because the top dominant male cannot predictably sire offspring when several females are receptive simultaneously and many male competitors are active (Nunn 1999b).

An interesting exception is the red howler monkey where second-ranking males are not observed to mate and never sire any offspring (Pope 1990), and females only occasionally mate during pregnancy (Sekulic 1983a; cf. Glander 1980; Jones 1997). It is possible, therefore, that howler females are unable to confuse paternity or its assessment. Indeed, if the alpha male dies or is injured, the second-ranking male is likely to kill infants (Izawa & Lozano 1991). (See also Crockett & Janson, Chapter 4.)

Conclusion

The studies reviewed here show that female primates, whether or not they advertise their approximate reproductive state, can potentially manipulate paternity by concentrating it largely but not entirely in top dominant males. Females can do this by adding error to the timing of ovulation relative to peak attractivity and proceptivity.

Duration of the follicular phase

Despite considerable variation in follicular phase duration within species, most species show clear tendencies toward repeatable mean values. Indeed, primate taxa show systematic variation in the mean length of the ovarian cycles and in the mean length of the two components (Table 15.1; based on data compiled by van Schaik *et al.* 1999; J. K. Hodges *et al.*, unpublished data). The variation in total length is determined by independent taxonomic patterns in the length of the follicular and luteal phases. If we recognize three main radiations (prosimians, platyrrhines, catarrhines; compressing prosimians into one category owing to poor data), strong patterns emerge. Total ovarian cycle length is longer in prosimians and catarrhines than in platyrrhines, but this pattern is caused by two independent taxonomic patterns in follicular and luteal length. Follicular phase length is about twice as long in catarrhines as in prosimians and platyrrhines, whereas luteal phase length is almost twice as long in prosimians as in the two anthropoid radiations. Nested analysis of variance indicates a very different pattern of taxonomic distribution for the two phases (J. K. Hodges *et al.*, unpublished data), with most of the variation in the length of the luteal phase being found at the level of suborders, whereas that in follicular phase is found mainly at the superfamily and species level.

These two separate patterns probably have separate explanations. As

Table 15.1. *Mean length of ovarian cycles and their components in the main primate radiations (data from van Schaik* et al. *1999 and J. K. Hodges* et al., *unpublished data). The* post hoc *comparisons are based on the Bonferoni–Dunn procedure*

Radiation	Total			Follicular			Luteal		
	N	x	SD	N	x	SD	N	x	SD
Prosimians	19	33.77	7.91	3	7.33	4.04	4	26.38	1.78
Platyrrhines	20	20.03	5.84	8	6.90	2.15	8	13.54	4.83
Catarrhines	38	31.01	3.67	22	14.99	3.37	21	14.94	1.58
ANOVA:	$F_{[2,74]} = 35.71^{***}$			$F_{[2,30]} = 23.08^{***}$			$F_{[2,30]} = 33.81^{***}$		
Post-hoc:	Pla < Pro = Cat			Pro = Pla < Cat			Pro > Pla = Cat		

Notes:
N, number of observations; x, mean length of ovarian cycle; SD, standard deviation. Pro, prosimians; Pla, platyrrhines; Cat, catarrhines. *** $P < 0.001$.

to luteal phase variation, J. K. Hodges *et al.* (unpublished data) suggest that the length of the luteal phase may be related to the nature and timing of the signals involved in the maternal recognition of pregnancy. Here we concentrate on explanations for the pronounced interspecific variation in the length of the follicular phase.

The basic prediction is that mating periods, and thus follicular phases, should be long enough for the female to attract matings by all the males most likely to commit infanticide if they have no paternity. The length of time required to achieve this should depend on the control females have over mate selection and on the demographic situation. Where females have complete control over the identity of their mates and the timing of their matings with them, high variance in ovulation timing is not needed, and so follicular phase length can be brief and may or may not depend on demographic context. Where females need to reduce the monopolization and coercion potential of one or more dominant males in multimale groups, we expect a longer mean follicular phase (along with more female-induced polyandrous mating and an increased margin of error in the timing of ovulation).

We perform three sets of analyses that examine slightly different aspects of the general hypothesis that the length of the follicular phase reflects female mating tactics. First, we examine a hypothesized (but as yet untested) major taxonomic difference and its consequences for follicular phase variation: catarrhine females seem to experience more coercion than platyrrhine females. We generate a number of proxies for female control to examine this hypothesized contrast. Second, we use the same proxy measures to test hypotheses related to the duration of the follicular phase in the catarrhines. By focusing on one clade where sexual behaviors and their endocrinological controls are more constant, we eliminate the possible effects of (partly unknown) confounding variables. Finally, we examine several demographic predictions across all species and within only the catarrhines. In particular, we test whether follicular phase length is longer when females mate with more than one male, or at least when there is more than one potential mate (i.e., multimale groups).

Taxonomic contrast in female control

Table 15.1 indicates that catarrhine females have longer follicular phases than females in other primate radiations. We tested this possibility by using the method of independent contrasts (Felsenstein 1985; Harvey & Pagel 1991), which tests whether evolutionary changes in one trait correlate with evolutionary changes in another trait. We can therefore determine whether the change ("contrast") in follicular phase length over the branch of the phylogeny (Purvis 1995) that connects platyrrhines with catarrhines is larger than contrasts within these clades (e.g., Garland et al. 1993). In this case, only 6 out of 27 contrasts in follicular phase length are larger in absolute magnitude than the contrast that corresponds to the transition of interest that connects catarrhines with platyrrhines, but this difference is not quite significant ($t_s = 1.61$; $P = 0.12$).

Since all infant-carrying species are vulnerable to infanticide by males (apart from the communally breeding platyrrhines: van Schaik et al. 1999), the systematic differences in follicular phase between radiations indicates that either demography or female behavioral freedom differs between platyrrhines and catarrhines. Demography does not show major differences (e.g., Smuts et al. 1987; van Schaik & Kappeler 1996), suggesting that some aspect of female control is the causal factor in explaining species differences in follicular phase length.

Behavioral observations support this hypothesis. Table 15.2 compiles reports of significant male harassment for four primate radiations (prosimians are lumped because virtually nothing is known about the behavior of nocturnal lemurs, lorisids or tarsiers in the wild). Although the catarrhines tend to be better studied, the lack of reports

Table 15.2. *Evidence for male harassment and female behavioral freedom in four primate radiations. Based on Smuts (1987a), Smuts & Smuts (1993), and sources mentioned in the text*

	Prosimians[a]	Platyrrhines	Cercopithecoids	Hominoids
Male harassment				
Males attacking estrous females/ consorting females	–	–	Macaques Savanna baboons	Chimpanzee Gorilla Orang-utan
Males seriously injuring females	–	–	Macaques Baboons	Chimpanzee Gorilla
Forced copulations	–	–	–	Orang-utan
Female control				
Female dominance in male–female dyads	Many lemurs	Capuchins Squirrel monkey	3 Macaque spp. Patas Talapoin Guenons	Gibbon
Female aggression to male in mating context	Ring-tailed lemur Ruffed lemur	?	–	–
Lack of male aggression in mating context	?	Spider monkey Muriqui Capuchins	–	–
Cooperative harassment by males needed	–	Spider monkey Squirrel monkey	–	–

Notes:
[a] Only infant-carrying species considered.

of harassment in platyrrhines or prosimians is striking. Likewise, female control over mating decisions is also clearly greater in these two radiations.

A plethora of observations on sociosexual behavior supports this generalization. Catarrhine males may mate-guard a female for days or weeks (see Introduction), and males may direct aggression to a fertile female (see Table 15.2). Females, nonetheless, actively pursue polyandry in many species (Small 1993; van Schaik *et al.* 1999). Prosimian females, in contrast, tend to have estrus-like behavior: females control sexual access by courting males through aggression except during brief receptive periods (e.g., Foerg 1982a,b; Pereira & Weiss 1991; Richard 1992; Morland 1993; Winn 1994). In addition, they tend to take the initiative or are selective in mating (Foerg 1982a; Richard 1992; Pereira & Weiss 1991; Morland 1993). Even in their short mating periods, females can actively achieve polyandry (e.g., Brockman & Whitten 1996). Much the same holds for platyrrhines. Although mate-guarding does occur (e.g., black howlers, Glander 1980), female-initiated mating is the rule (capuchins, Janson 1984; squirrel monkeys, Boinski 1987b) and females can easily refuse matings (marmosets, Kendrick & Dixson 1983). Platyrrhine females, too, can achieve polyandry in brief mating periods. In the woolly spider monkey, for example, a female attracts (through a twittering call) up to nine males around her and then mates with up to five of them in a single day (Milton 1985; see also Strier & Ziegler 1997).

The incidence of coercion is difficult to measure. We therefore also assessed the degree of female control by using several variables that should proxy a female's risk of coercion (see also Nunn & van Schaik, Chapter 16). Where males are larger or have more dangerous weapons, the cost of resistance is higher; thus sexual dimorphism in body and canine size is expected to reduce female control (Smuts & Smuts 1993). Likewise, coercion is probably less common in arboreal settings (cf., Gowaty 1997b), as females have more escape routes in the trees, immobilization of females by males is more difficult on wobbly branches, and the risks of falling are far more serious.

Thus, if the information in Table 15.2 is correct, then catarrhines should have greater dimorphism in body and canine size, and they should more often be terrestrial. Because lemurs lack dimorphism in body and canine size (Kappeler 1991) and only one species is terrestrial, it is more conservative to test this prediction in anthropoids only, where at least some of these traits show more overlap.

Substrate use

No living platyrrhine is terrestrial (out of 53 species that can be characterized as one or the other), whereas 33 out of 91 catarrhine species are terrestrial (involving 12 different genera in each category). Evolutionary patterns indicate that this trait is primitive for some catarrhine clades (e.g., the cercopithecines). Yet terrestriality is also evolutionarily flexible within catarrhines, with five evolutionary gains and six losses reconstructed on the Purvis (1995) phylogeny. Even if this variability indicates some arbitrariness in how species are classified into arboreal and terrestrial categories rather than evolutionary flexibility, it nevertheless indicates that catarrhines are generally closer to the cut-off between arboreal and terrestrial, while no extant platyrrhines come close to that boundary.

Body size dimorphism

Size dimorphism in platyrrhines shows some overlap with catarrhines: some howlers and capuchins are fairly dimorphic, whereas pair-living and some arboreal polygynous catarrhines (e.g., smaller colobines) are virtually monomorphic (Plavcan & van Schaik 1997; Smith & Jungers 1997). Nevertheless, body size dimorphism is greater in the catarrhines than in the platyrrhines, suggesting that catarrhine females are at greater risk of coercion (Figure 15.4a). Using species for which information is available for both sexes, male catarrhines are on average 45% heavier than females ($N = 82$ species), but male platyrrhines are only 14% heavier than females ($N = 44$). As above, we tested whether there is a significant difference between the two clades. Although the contrast between catarrhines and platyrrhines is in the predicted direction, it is not significantly larger than the other observed contrasts in the tree ($t_s = 0.69; N = 106; P = 0.50$).

This difference is not a by-product of the difference between the two clades in terrestriality (see Figure 15.4a). Catarrhines exhibit greater dimorphism even when terrestrial taxa within this group are excluded (catarrhine males are 32% heavier than females, $N = 53$; the orang-utan was also excluded because its quadrumanal locomotion may provide males with greater substrate contact and thus more opportunities for coercion). Hence, both factors independently contribute to the risk of coercion: dimorphism increases the potential for coercion through increased physical size, while terrestriality allows the male to more effectively wield his greater physical threat (see above).

Canine dimorphism

All polygynous catarrhines show clear-cut canine dimorphism (Plavcan *et al.* 1995). However, platyrrhine males have much smaller canines than catarrhine males (average residual from an isometric line: platyrrhine average $= -0.022$, $N = 17$; catarrhine average $= +0.33$, $N = 48$, data from Plavcan *et al.* 1995). Here, however, the issue is the ratio of male canines to female canines: when males have larger canines than females, then female control is reduced, and so sexual counterstrategies are expected. We therefore used the General Canine Dimorphism Index (GCDI; Plavcan & van

(a)

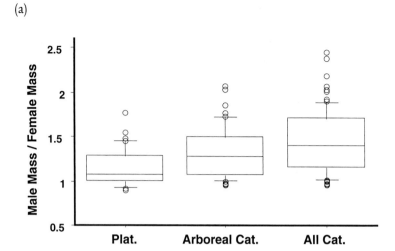

(b)

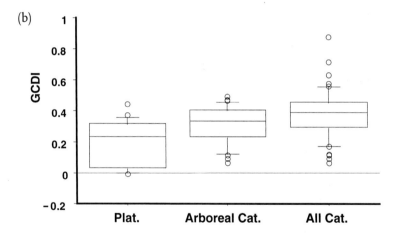

Figure 15.4. Coercion measures in platyrrhines and catarrhines. Box plots showing
medians and interquartile and 90 percentile ranges. Plat., platyrrhines;
Arboreal Cats., arboreal catarrhines; All Cats., all catarrhines. (a) Catarrhines
have greater body size dimorphism than platyrrhines, indicating greater
susceptibility to male coercion in catarrhines. Substrate use also differs
between these two radiations and may contribute to this pattern. However,
terrestriality probably also increases the risk of coercion independently of its
effects on dimorphism (see text). (b) Catarrhines also have greater canine
dimorphism. This dimorphism puts females at even greater risk of coercion
and further reduces female control over mating decisions. GCDI, General
Canine Dimorphism Index.

Schaik 1992) to capture in a single measure three aspects of canine dimorphism: their width in two dimensions and their height. Platyrrhines have a much smaller GCDI (average = 0.19, $N = 21$) than catarrhines (average = 0.37, $N = 52$), and this difference is highly significant (Figure 15.4b; $t_{71} = 4.73$; $P < 0.0001$). The transition to catarrhines is accompanied by a positive increase in canine dimorphism; however, this change is not significantly greater than the other contrasts ($t_s = 0.73$, $N = 18$, $P = 0.47$). Terrestriality does affect GCDI as well (Figure 15.4b), but, as was the case with body size dimorphism, it only accentuates the systematic difference between platyrrhines and catarrhines.

On the whole, then, patterns of substrate use and dimorphism support the hypothesized difference in female behavioral freedom between the clades (Table 15.2). This difference in the risk of coercion may therefore underlie the short follicular phases of prosimians and platyrrhines compared with those of cercopithecoids and hominoids.

Female control and follicular phase length in catarrhines

The same predictions can be tested more directly by relating the same proxies for sexual coercion to the length of the follicular phase. Since data for platyrrhines are few, we test this relationship only in catarrhines, where the range of coercion is greater and the sample is larger (including the platyrrhines, however, does not affect the results). Even for catarrhines low statistical power is expected, given the relative crudeness of our proxies and the small sample sizes. Moreover, one-male and arboreal species are underrepresented in the sample.

Among catarrhines, terrestrial species have slightly longer follicular phases than arboreal ones (15.6 versus 14, $N = 16$, 5), but this difference is not statistically significant ($F = 0.65$, $N = 21$, $P = 0.43$; nearly identical results are obtained in phylogenetic analysis). Species with greater sexual dimorphism tend to have longer follicular phases in non-phylogenetic ($b = 0.54$, $N = 21$, $F = 3.70$, $P = 0.07$) and in phylogenetic analyses ($b = 0.50$, $N = 20$, $F = 2.90$, $P = 0.11$). When the non-phylogenetic analysis is restricted to arboreal taxa, which reduces the sample to five taxa, the difference remains nearly significant ($b = 0.46$, $N = 5$, $F = 8.48$, $P = 0.06$). Finally, canine dimorphism is positively correlated with follicular phase length, although again not significantly ($b = 0.25$, $N = 15$, $F = 1.16$, $P = 0.30$). Surprisingly, however, this difference is statistically significant after phylogeny has been controlled for ($b = 0.56$, $N = 14$, $F = 6.04$, $P =$

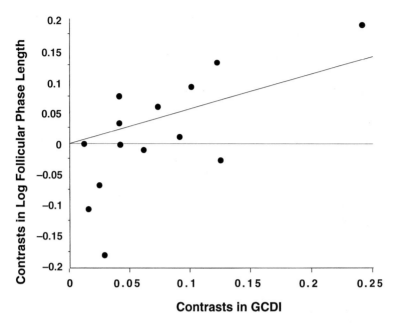

Figure 15.5. Relationship between follicular phase length and canine dimorphism in catarrhine primates. The graph shows contasts, which represent evolutionary change, and the line is forced through the origin (see Garland *et al.* 1992), though the pattern is statistically significant even when the line is not forced. Thus evolutionary increases in canine dimorphism are associated with evolutionary increases in follicular phase length. GCDI, General Canine Dimorphism Index.

0.029), indicating that evolutionary increases in canine dimorphism are correlated with evolutionary increases in follicular phase length (Figure 15.5). In sum, while only one of the proxies used here shows a statistically significant correlation, trends are generally in the predicted direction and many of these are nearly significant.

Demographic context and the potential for polyandrous matings

Another set of predictions deals more directly with the demographic context. Where females have behavioral control, we do not expect a correlation between the potential for polyandrous mating and the length of the follicular phase. However, where females are subject to harassment and long mate guarding by males, they may need longer follicular phases (and therefore also more variable ones) to bring about polyandrous mating. Hence, in catarrhines, variation in follicular phase length among species should reflect the demographic setting, with longer follicular phases expected where females can potentially mate with multiple males (i.e., multimale groups, multilevel societies, or one-male groups subject to regular multimale influxes). We can examine these predictions by examining the actual pattern of female matings, or the number of potential matings, as proxied using the number of males in the group.

Number of mating partners

To illustrate this effect of the number of mates consider the apes, in which follicular phase lengths are known for all species. Female gibbons mate with a single male (extra-pair matings, though observed, are rare; Palombit 1994b; Reichard 1995), and display a mean follicular phase length of approximately 8.4 days (all data on follicular phase length taken from J. K. Hodges *et al.*, unpublished data). In the orang-utan, females mate preferentially with a single adult male (Schürmann 1982; Fox 1998), although other males mate as well, but mainly outside the periovulatory period (Fox 1998); mean follicular phase length is 12.9 days. In the gorilla, two-male groups occur alongside the more common one-male groups, and some polyandrous mating takes place (Robbins 1995); follicular phase length is 18.4 days. Finally, the two thoroughly promiscuous chimpanzee species both have extended follicular phases with mean values approximating to 20 days.

These general patterns are supported when one looks across all species (categories taken from van Schaik *et al.* 1999). When females tend to mate with only one male, follicular phases are relatively short (mean of 8.4 days, $N = 4$). However, when some promiscuous mating occurs, the follicular phase is slightly longer (11.5 days, $N = 8$), and follicular phases increase again when females regularly mate with more than one male (15.2 days, $N = 15$). These differences are statistically significant in non-phylogenetic tests ($F = 4.25$, $N = 27$, $P = 0.026$), but fail to reach statistical significance in phylogenetic tests, even though 5 of the 7 contrasts in number of mating partners are in the predicted direction (mean contrast $= 0.028$, $t_s = 0.98$, not significantly different from 0 in a one-tailed test ($P = 0.18$)). Statistical significance is not improved after removal of the two contrasts in platyrrhines (mean $= 0.022$, $t_s = 1.01$, $P = 0.18$).

This analysis assumes that species that require a longer period of mating will have longer follicular phases. Mating period is difficult to measure, but patterns in anthropoids indicate at least a positive relationship between these variables ($b = 0.11$, $F = 1.10$, $N = 27$; $P = 0.31$). The pattern improves but remains non-significant when species without circumscribed mating periods are excluded (*Callithrix jacchus*, *Macaca arctoides* and *Cercopithecus aethiops*, $b=0.18$, $F=1.95$, $N=24$, $P=0.18$), but restricting the analysis to catarrhines does not improve statistical significance ($b=0.03$, $N=17$, $F=0.08$, $P=0.78$). Similar results obtain in phylogenetic analyses. They imply that mating periods and follicular phases show only a weak relationship in practice. Thus it is better to say that longer follicular phases make possible longer mating periods, when these are needed (which may not be often).

Number of males in group

We predict a positive relationship between the number of males in the group and follicular phase duration: when there are more males, then females will require more time to mate with them. In non-phylogenetic analysis, there is a slight but non-significant relationship between the number of males and follicular phase length (all species: $b=0.12$, $N=27$, $F=0.88$, $P=0.36$; catarrhines only: $b=0.046$, $N=17$, $F=0.33$, $P=0.57$). However, in a phylogenetic analysis restricted to catarrhines, the slope is actually negative, although not significantly so ($b=-0.06$, $N=16$, $F=1.76$, $P=0.20$). The result is the same if the phylogenetic analysis is expanded to look across all primates ($b=-0.07$, $F=1.52$, $N=26$, $P=0.23$). Hence, there is no support for this demographic prediction.

Unfortunately, there is not enough variation in the number of potentially mating males among the species with known follicular phase length, in that most of our information comes from multimale primate taxa. In addition, this analysis is probably confounded by other factors. For example, some multimale species are seasonal breeders, where male control will be relaxed owing to greater female overlap of receptive periods (see van Schaik *et al.* 1999; Nunn 1999b). Thus the number of males may not be the ideal test of this hypothesis at the species level.

A more sensitive test of this prediction would be to perform comparisons within species where the number of males in a group is variable. Such a comparison is possible in hanuman langurs. Data on female mating periods and cycles are available from both Jodhpur, where the population is composed of one-male groups (Sommer *et al.* 1992) and from Ramnagar population, which contains mainly multimale groups (Borries & Koenig, Chapter 5). The mating period is 5.5 days longer in the multimale groups of Ramnagar (9.5 (± 2.9) days, $N=15$, T. Ziegler, unpublished data) than in the one-male groups of Jodhpur (4.0 (± 2.2) days; $N=206$: Sommer *et al.* 1992). The follicular phase is 12.8 (± 3.8) days ($N=6$) in the multimale population at Ramnagar, some 2.6 days longer than the 10.2 ($\pm$ 2.9) days in Jodhpur ($N = 68$; reconstructed from Sommer *et al.* 1992), confirming the prediction at this more immediate level.

Exaggerated female sexual swellings

Exaggerated swellings have been hypothesized to function to reduce a male's monopolization potential (Nunn 1999c). Furthermore, these swellings are limited to some (but not all) multimale taxa, and only in the catarrhines. We can therefore use this information to test whether species with exaggerated swellings also have longer follicular phases.

Among catarrhines, those with exaggerated swellings have longer follicular phases than those without swellings (17.5 versus 12.7 days, $N = 11$ and 10, respectively). This difference is highly significant in non-phylogenetic analysis ($F = 15.61$, $P < 0.001$). In phylogenetic tests, the mean of the four contrasts available is significantly greater than 0 ($t_s = 4.06$, $N = 4$, $P = 0.014$, one-tailed test). An alternative explanation for this pattern is that species with exaggerated swellings may require a longer follicular phase for the larger signal to reach maximum size. This alternative cannot easily be tested.

Discussion

Because male mammals cannot unequivocally recognize their offspring, female mammals can potentially manipulate male paternity. Where females are at risk of infanticide by males, they would benefit from concentrating paternity into the locally dominant male if there is an appreciable risk of intrusion of new dominant males. They would also benefit from confusing paternity when the local neighborhood contains multiple males that have a reasonable probability of having displaced the male who is currently the local dominant by the time the females have produced their offspring. In many primate species, females would benefit from achieving both. We argued here that females can manipulate paternity chances by making ovulation unpredictable (see also van Schaik *et al.* 1999; Nunn 1999c). Examination of endocrinological studies indicates that many primate females show remarkably unpredictable ovulation.

It is difficult to imagine that unpredictable ovulation could have selective benefits other than paternity manipulation. However, one could entertain the null hypothesis that the unpredictability truly reflects physiological constraint: the organism simply cannot make the physiological links more rigid. However, this idea is not very plausible in light of the brief mating periods tightly associated with clear signaling in other mammals (see van Noordwijk & van Schaik, Chapter 14). A critical prediction of the adaptive hypothesis (as opposed to this by-product hypothesis) is that primate species not expected to require paternity confusion should have more predictable timing of ovulation (i.e., a less variable time interval between measurable signal and ovulation) than those that confuse paternity. Such data are currently unavailable, but can be collected.

At the same time, species differ reliably in the mean duration of the follicular phase. Follicular phase length showed a striking taxonomic pattern, with brief lengths of about 7 days in prosimians and platyrrhines and about twice that length in catarrhines. We argued that this variation should be related primarily to the female's need for paternity manipulation. Where females have total control over their mating behavior, we expected the follicular phase to be short. By contrast, where they lack control and are surrounded by several males, we expected the follicular phase to be long. We showed that sexual coercion is clearly much less in prosimians and platyrrhines than in catarrhines, corresponding with a large difference in the length of the follicular phase. We found that catarrhines are significantly more terrestrial and more dimorphic, especially with respect to canine dimensions than platyrrhines (and prosimians), supporting this hypothesis. However, there may be various other potentially causal factors that show a similar taxonomic distribution (e.g., the extent to which animals rely on olfaction). More direct tests of the effects of proxies for sexual coercion on follicular phase were therefore conducted as well. Follicular phase length in catarrhines is predicted reasonably well by the physical potential for male harassment, but less strongly by the demographic component. The latter may be because selection has molded follicular phase length to deal with extreme demographic situations, whereas the field data refer to a variety of demographic settings. Alternatively, the number of species for which mean follicular phase length is available is simply too small to produce convincing tests. Interestingly, the predicted relationship between the mean number of mates per cycle and length of follicular phase predicted among catarrhines was found to hold within hanuman langurs. The time is ripe for experimental studies of the effect of the number of males on female ovarian cycles in primates.

We would be more justified in accepting the paternity confusion hypothesis as the best explanation for interspecific patterns in the length of the follicular phase if it could be demonstrated that alternative hypotheses do not account for the same patterns. Unfortunately, we are not aware of any well-developed alternative hypotheses. It is often thought that the length of the follicular phase is simply an allometric function of body size. On the face of it, this hypothesis is supported: across all species, female body mass is strongly related to follicular phase length ($b = 0.25, F = 39.5, N = 33, P < 0.0001$). However, this is an artifact of simultaneous grade shifts in body size and follicular phase length. If the analysis is repeated using independent contrasts, the relationship disappears ($b = 0.061, F = 0.27, N = 32, P = 0.61$). Thus, for now, the infanticide avoidance hypothesis provides the best explanation for the patterns in follicular phase length across primates, but it is clear that further testing and explicit evaluation against alternative hypotheses is needed before we can accept this hypothesis with confidence.

Implications for humans

We would like to end with a note on humans. Human females are widely regarded as the quintessential concealed ovulators: there is no advertising and mating is thought to be equally common throughout the cycle. Indeed, many ideas for the evolution of concealed ovulation incorrectly assumed that it was a unique human adaptation (i.e., an autapomorphy; Andelman 1987). However, the situation in human pairs is similar to that of other primate species in pair-tests, with partners being in and out of contact on a regular basis. As we noted earlier, such periodic reunions always produced a much weaker relationship between female ovarian state and mating behavior in the primate species tested, even in species with clear advertising. In apes, such effects largely disappeared if females were given control over their mating behavior (Nadler 1992). Accordingly, female proceptivity in humans peaks around midcycle in lesbian pairs (Matteo & Rissman 1984; but see also Steklis & Whiteman 1989). Thus humans, too, do not seem to conceal ovulation in the classic sense. As we noted, the studies on timing of ovulation compiled by Martin (1992) indicate that human females have highly unpredictable ovulation. If we assume that no special selective processes have operated during hominid evolution, we should conclude that human females have historically been concerned with paternity confusion to avoid infanticide within groups, but hardly with paternity concentration. This suggests that the threat of infanticide by newly immigrated strange males was minimal, and also that male care for infants, if it was important, represented mating effort rather than parenting effort (cf. Smuts & Gubernick 1992; Paul *et al.*, Chapter 12). The predicted lack of male intrusion is consistent with the strong patrilocal and territorial tendencies of many pre-settlement societies (Rodseth *et al.* 1991). This interpretation also assumes that hominid societies showed polyandrous mating, and that pair bonds were far from exclusive. This suggestion, too, is consistent with the ethnography of hunter–gatherer societies (Hawkes 1993). Obviously, if novel selective pressures began to operate during hominid evolution, this conclusion is not warranted.

Acknowledgments

We thank Maria van Noordwijk and Dietmar Zinner for discussions, and Thomas Ziegler for making unpublished observations available. The work reported here was started while C.v.S. was at the Deutsches Primatenzentrum, supported by a Forschungspreis of the Alexander von Humboldt Foundation. C.L.N. was supported by an NSF graduate student fellowship.

16

Social evolution in primates: the relative roles of ecology and intersexual conflict

Introduction

Infanticide by males is costly to females and so should select for female counterstrategies (Hrdy 1979; Smuts & Smuts 1993; van Schaik 1996; Ebensperger 1998a). Some of these hypothesized female counter-strategies are social, such as special relationships with "protector" males (Palombit *et al.* 1997). Female counterstrategies to infanticide may there-fore explain important variation in primate social systems, including monogamy (van Schaik & Dunbar 1990), patterns of male–female associa-tion (van Schaik & Kappeler 1997) and female coalitions (Treves & Chapman 1996).

Female primates also face ecological pressures that may account for cross-species variation in social systems. For example, predation pressure may select for larger female groups, which then leads to greater within-group competition for food resources (van Schaik 1989; Sterck *et al.* 1997). Additionally, predation pressure may select for an increase in the number of males in the group (Hall & DeVore 1965; van Schaik & Hörstermann 1994; Hill & Lee 1998), especially when males are more vigilant than females (van Schaik & van Noordwijk 1989) and when they deliver more effective communal defense (Stanford 1998).

The goal of this chapter is to examine the relative roles of ecology and intersexual conflict in primate social evolution. To achieve this goal we need independent estimates of the relevant ecological and intersexual pressures, but three issues complicate this effort. First, quantification of the relevant factors is difficult. For example, infanticide and predation are rare events, and when they do occur they often happen quickly and out of direct sight. Second, observed rates of infanticide or predation do not represent their intrinsic risks because these observations occur after the counterstrategies are in place (e.g., Hill & Dunbar 1998; Janson 1998). Finally, the goal of identifying the relative roles of ecology and intersexual conflict is hindered by similar adaptive responses. Returning to the example from above, females may reduce the risk of predation by increasing the number of males in the group, but they may also reduce infanticide risk with the same strategy: females may encourage the formation of multimale groups (perhaps by synchronizing their timing of fertility; Altmann 1990; Nunn 1999b) and then mate polyandrously to confuse paternity (Hrdy 1979; van Schaik et al. 1999). Hence, some social system parameters, such as the number of males in the group, cannot always be attributed solely to ecological or intersexual pressures.

In this chapter, we deal with the first two complications by using measures of ecological and intersexual factors that proxy the *risks* associated with each selective factor rather than the observed *rates*. This important distinction between risks and rates is rarely considered in studies of primate social evolution (Hill & Dunbar 1998; Janson 1998) and so we return to it in the Discussion, below. We deal with the complication of similar adaptive responses by controlling for confounding variables and by examining potential interactions among the ecological and intersexual factors. In most cases we do not test competing predictions because many social system parameters are likely to be influenced by multiple selective pressures, which renders a strong inference approach less useful and perhaps even misleading (see Quinn & Dunham 1983).

While some of the questions addressed here have been examined in previous comparative research, especially in studies of ecological factors, our analyses use phylogenetic comparative methods to test for correlated evolutionary change in the relevant variables. Our analyses are conducted across species (rather than within species) because the goal is to determine whether ecological and intersexual factors explain species differences in primate social systems. However, observational and comparative studies within species are reviewed when such studies inform our cross-species approach. To set the stage for these analyses, we begin by reviewing the major hypotheses for variation in primate social systems.

Ecological and intersexual pressures in primate social evolution

In the standard socioecological model, environmental risks and resources determine the spatiotemporal distribution of females. This female distribution then determines male strategies to monopolize fertile matings

(Emlen & Oring 1977; Wrangham 1979). Overlaying this basic model are selective factors involving intersexual conflict (Figure 16.1), especially sexual coercion (Smuts & Smuts 1993; Clutton-Brock & Parker 1995). Sexual coercion includes aggressive behaviors that coerce a female to mate (e.g., aggression and forced copulations), infanticide to shorten the time until she returns to fertility, and herding behavior to keep the female away from other males.

The factors relevant to primate social evolution can therefore be categorized as ecological or intersexual, although these factors probably interact and modify one another to varying degrees (Figure 16.1). The primary ecological factors thought to affect social arrangements are predation risk and the competitive regime. Group living is a common adaptive response to predation risk among mobile animals (Alexander 1974; Bertram 1978). In primates, predation represents a significant selective force (e.g., Anderson 1986; Hill & Lee 1998), and primate groups are indeed larger under increased predation, as suggested by comparisons among populations that vary in the array of predators while attempting to hold other factors constant (Kummer *et al.* 1985; van Schaik and van Noordwijk 1985; Struhsaker 2000). As noted above, an increase in predation risk is also thought to select for additional males in the group (van Schaik & Hörstermann 1994; Hill & Lee 1998).

Once females live in groups to reduce predation, the competitive regime among females becomes a factor that influences the social system (van Schaik 1989; Sterck *et al.* 1997). Competition depends on the distribution of the size and quality of food patches relative to female spatial clumping (Janson 1988). Where the potential for within-group contest competition is high, females are thought to form female-bonded groups: they form decided dominance relationships to assist their relatives in competition for food, resulting in alliances with relatives and female philopatry (van Schaik 1989; Isbell 1991; Sterck *et al.* 1997). Here, we use Sterck *et al.*'s (1997) terminology for describing female social relationships, as these categories provide explicit information on dispersal tendencies and the nature of the dominance relationships among females. Thus resident-nepotistic corresponds to female-bonded and is contrasted with resident-nepotistic-tolerant and resident-egalitarian, where dominance ranks are less despotic or non-existent, respectively. Dispersal-egalitarian corresponds to non-female-bonded (see also Wrangham 1980; van Schaik 1989).

The evidence from detailed behavioral studies again supports the

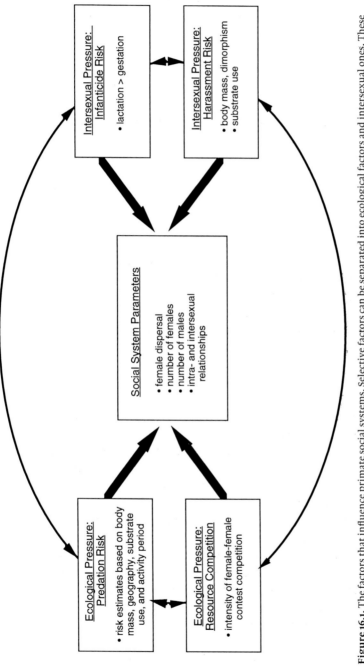

Figure 16.1. The factors that influence primate social systems. Selective factors can be separated into ecological factors and intersexual ones. These factors modify social system parameters. The factors also modify one another to various degrees, as indicated by the arrows on the outside of the figure.

hypothesized effects of within-group competition on female dominance relations, philopatry and alliances (e.g., Mitchell *et al.* 1991; Barton *et al.* 1996; Isbell & Pruetz 1998; Koenig *et al.* 1998). Between-group competition may also be important when population densities are high, but it probably does not represent a primary selective force that affects female social relationships (Cheney 1992; Cowlishaw 1995; Sterck *et al.* 1997; Matsumura 1999).

The ecological pressures of predation and female intrasexual competition are intertwined: an increase in predation risk selects for larger or more spatially cohesive female groups, which may lead to greater within-group feeding competition (Janson 1988). Differences exist across species in how a unit increase in group size affects competition levels, but living in a large group is clearly an important factor that increases competition (Janson & Goldsmith 1995; Sterck *et al.* 1997).

The other major socioecological factors are intersexual rather than ecological (Figure 16.1). These selective factors usually involve forms of male sexual coercion, which can be separated into infanticide and harassment (Smuts & Smuts 1993). Infanticide by males is thoroughly discussed in other chapters and may have selected for a variety of social counter-strategies, including year-round male–female association (van Schaik & Kappeler 1997), female breeding dispersal (Watts 1989; Sterck 1997; Sterck & Korstjens, Chapter 13), male bonding when females are solitary (van Schaik 1996), the presence of multiple males in female groups (Altmann 1990; van Schaik 1996), and associations of females that form defensive coalitions against infanticidal males (Marsh 1979a; Sommer 1987; see also Treves 1998).

Many of these examples of intersexual conflict are based on observational and comparative studies within species. Few cross-species analyses have been performed (but see van Schaik & Kappeler 1997), and so the generality of these mechanisms as an explanation for cross-species patterns in primate social systems is unknown. However, the effects of infanticide at this level are difficult to test because clear predictions are not always possible. For example, infanticide risk has been hypothesized to lead to multimale social systems with promiscuous mating in some taxa (Altmann 1990; van Schaik 1996), as suggested by intraspecific patterns in rates of infanticide (e.g., Newton 1986; Robbins 1995). However, other intraspecific studies have suggested that infanticide risk leads to smaller female groups with one male because such groups are less attractive take-over targets and infanticide risk does not seem to be reduced in multimale groups (Crockett & Janson, Chapter 4). It therefore seems probable that ecological factors modify the solutions used by females. Consequently, a cross-species perspective is needed to identify which ecological constraints are most salient. One goal of this chapter is to take the first steps toward identifying the ecological factors that probably modify female counterstrategies to male coercion.

Harassment includes the other male coercive behaviors mentioned above that result in immediate mating opportunities (rather than the delayed benefits of infanticide; see also Clutton-Brock & Parker 1995). Examples of harassment include active aggression against females that resist mating and "herding" behavior to keep females away from other males. Female sexual behavior is a prominent counterstrategy to both harassment and infanticide: females can manipulate males by changing the costs and benefits of mating and mate-guarding, for example through the sexual signaling system (Nunn 1999c; van Schaik *et al.* 1999, Chapter 15).

Harassment by males may also affect the social system, though very little is known about its effect both within and across species. A number of authors have provided hypotheses for social counterstrategies in particular cases (i.e., male protectors as "hired guns"; Wrangham 1979; Wrangham & Rubenstein 1986; Smuts & Smuts 1993). More recently, Brereton (1995) proposed that harassment explains variation in important socioecological parameters across species. This "coercion–defense" hypothesis posits that females respond to harassment by forming kin-based alliances against coercive males, thus providing an alternative explanation for female-bonded (i.e., resident-nepotistic) groups. Female alliances against coercion lead to an intersexual "arms race" under this hypothesis, including male counterstrategies involving increased dimorphism, and subsequently the evolution of either female dispersal, "hired-gun" protector males, or male brotherhoods that prevent the entry of harassing males from outside the group or community. The coercion–defense hypothesis has been critically evaluated elsewhere (Sterck *et al.* 1997; Treves 1998). Here we examine the testability of the hypothesis and provide some additional assessments of its validity at the interspecific level.

Predictions and methods

The hypothesized selective factors are ideally tested using variables that estimate the risks faced by individuals before counterstrategies are in

Table 16.1. *Categorization of predation levels in diurnal species*

	Predation risk categories		
	High	Medium	Low
Terrestrial			
Open habitat	All species (7)		
Wooded habitat	<5 kg (7)	5–20 kg (12)	>20 kg (3)
Arboreal			
Africa, Neotropics	<3 kg (9, 33)	3–8 kg (12, 9)	>8 kg (1, 4)
Asia, Madagascar	<2 kg (0, 6)	2–6 kg (10, 5)	>6 kg (21, 2)
Islands without large predators		<5 kg (1)	>5 kg (3)

Notes:
Sample sizes are in parentheses, separated by region for arboreal species.

place. We examined four sets of predictions aimed at testing each of the four selective factors depicted in Figure 16.1. Tests of these predictions required estimates (or at least indirect "proxies") for the relevant ecological and intersexual risks. In some cases, we used subsets of the available data to control for similar adaptive responses and possible interactions among the hypothesized selective factors.

Predation

When predation risk increases, female group size is predicted to increase (van Schaik 1983). The number of males should also increase if males help to defend females and their offspring from predators (Hall & DeVore 1965; van Schaik & Hörstermann 1994; Hill & Lee 1998).

As a measure of predation risk, we grouped species into three risk categories (see Appendix 16.1 and Table 16.1) that relate several socioecological and geographic factors to body size, which is probably the primary modifier of predation risk (see Cheney & Wrangham 1987; Janson & Goldsmith 1995). Among diurnal primates, arboreal species are subject to a smaller array of predators than terrestrial ones, and terrestrial species that live in wooded areas can flee up trees and so experience lower predation risk than species in open areas far from tree cover (Crook & Gartlan 1966). Thus substrate use (arboreal versus terrestrial) and habitat use (open versus wooded) played major roles in our categorizations (Table

16.1). Geography is also an important factor: the absence of large aerial predators in Madagascar and Asia places these primates at lower risk than species in the New World and mainland Africa (e.g., Struhsaker 1981). Nocturnal species that face serious predation risk are thought to follow a cryptic anti-predation strategy (Bearder 1987), and so group size is not expected to be modified by predation risk among these species. We therefore excluded nocturnal species from our tests. Some small-bodied arboreal species also follow a cryptic anti-predation strategy (Terborgh 1983; Janson & Goldsmith 1995), and so we re-examined the comparative patterns after excluding these probably cryptic diurnal species (as based on the criteria in Table 16.1).

These species classifications are perhaps less precise than those based on field data involving rates and outcomes of predator encounters (Hill & Lee 1998) or studies that model these probabilities explicitly based on habitat and predator characteristics (Cowlishaw 1997). However, our classification is more general and so can be applied to many more species. Furthermore, the classification used here is highly congruent with one independently derived by Hill & Lee (1998) for cercopithecoids ($r_s = 0.53$; N $= 38$; $P < 0.01$, using their highest values for each species).

As an estimate of body mass we used mean female values taken from Smith & Jungers (1997). Our dependent variables were the number of females and males in the foraging group (e.g., Clutton-Brock & Harvey 1977) because this is the unit that is expected to respond to predation risk. Data on foraging group size came from an unpublished data base that has been used in previous and ongoing comparative studies (e.g., Nunn & van Schaik 2000; C. L. Nunn & R. A. Barton, unpublished data).

Female intrasexual competition

When intragroup resource competition increases, we expect female residence, decided dominance, and nepotism (Sterck et al. 1997).

As a measure of resource competition we used the intensity of female–female competition compiled by Plavcan et al. (1995). Although Plavcan et al. (1995) found that female canine size is best predicted by measures of aggression among females, they also found that canine size is modified by the presence of coalitions among females, and so we preferred their behavioral measure of competition in our tests. We limited our analysis to species that contain, on average, over two females per social group (i.e., gregarious females), as this is where intragroup competition and nepotistic rank acquisition become relevant.

A measure of female social relationships was the dependent variable.

We used character states provided by Sterck *et al.* (1997) except that *Lemur catta* and *Propithecus verreauxi* were categorized as resident-egalitarian rather than dispersal-egalitarian. We also combined the resident-nepotistic and resident-nepotistic-tolerant categories because too little information is available in the Sterck *et al.* (1997) dataset (only *Macaca nigra* is listed explicitly as tolerant).

Infanticide risk

We examined hypotheses for the effect of infanticide on the number of males, female dispersal and female sexual behavior. To estimate the risk of infanticide (the independent variable), we used the ratio of lactation to gestation. Our rationale was based on van Schaik (2000, Chapter 3): when lactation is longer than gestation, then postpartum fertilization is not possible and males will benefit from killing a female's infant (see also van Noordwijk & van Schaik, Chapter 14).

Two predictions are available regarding the effect of infanticide risk on the number of males in the group. On the one hand, infanticide risk should lead to more females and more males when paternity confusion is possible and ecological constraints allow larger groups (Altmann 1990; van Schaik 1996). On the other hand, infanticide risk should lead to smaller female groups and fewer males when paternity confusion is not possible (Crockett & Janson, Chapter 4; Steenbeek 1999a). Unfortunately, estimates of "paternity confusion potential" and the relevant ecological constraints are not available. We therefore used the patterns to generate hypotheses for future research.

Similarly ambiguous predictions arise for the effects of infanticide on female breeding dispersal. Dispersal has been hypothesized to provide females with a means to escape infanticidal males (Marsh 1979a; Watts 1989; Sterck 1997; Crockett & Janson, Chapter 4; Sterck & Korstjens, Chapter 13). In contrast, philopatry may allow females to share the costs of infanticide defense with close kin (e.g., the "conspecific threat" hypothesis; Treves 1998). Thus, infanticide risk may select either for or against philopatry, with the outcome a function of ecological pressures and constraints (including those discussed above involving predation and competition).

Finally, infanticide risk is expected to impact female sexual behavior, as pointed out by other chapters in this volume (e.g., van Noordwijk & van Schaik, Chapter 14; van Schaik *et al.*, Chapter 15). In particular, infanticide risk should lead to more "elaborate" sexual behaviors that enable females to mate with multiple males to confuse paternity. We therefore predicted a longer duration of receptivity and postconception matings when infanticide risk increases.

Data on the ratio of lactation to gestation were taken from van Schaik (unpublished data, cf. Chapter 3). Data on male number were taken from the unpublished dataset mentioned above. In this set of tests, however, we used population group size rather than foraging group size (see Appendix 16.1), as paternity confusion would be needed among all males in the group, not just the more flexible foraging group. Information on secondary breeding dispersal is limited (Sterck *et al.* 1997). However, natal dispersal is a likely correlate of secondary dispersal, and so we assumed that when females exhibit natal dispersal, secondary dispersal is also possible (cf. Sterck & Korstjens, Chapter 13). Information on dispersal was taken from the primary literature (see Appendix 16.1) and information on sexual behaviors was taken from van Schaik *et al.* (1999).

Harassment

Clear predictions are again not available for testing the effects of harassment. We discuss these problems in more detail below. However, we did test whether some patterns are correlated with body mass (from Smith & Jungers 1997), a likely estimate of male harassment and coercion in general, and we assessed whether male "brotherhoods" function to reduce harassment committed by males outside the group or community.

Statistics and phylogenetic comparative methods

A number of studies have demonstrated the importance of incorporating phylogenetic information in cross-species comparative tests (Harvey & Pagel 1991; Martins & Garland 1991; Gittleman & Luh 1992). When phylogenetic history is not included, Type I error rates increase because the degrees of freedom are not properly partitioned (Pagel 1993). At a more fundamental level, comparative methods allow tests of whether evolutionary changes in one trait are correlated with evolutionary changes in another trait, and so they are more direct tests of hypotheses that predict correlated evolutionary change. Phylogenetic tests also allow the assessment and statistical control of confounding variables, including "grade shifts" (e.g., Price 1997; Purvis & Webster 1999; C. L. Nunn & R. A. Barton, unpublished data; cf. Martin 1996).

For all tests we used the most complete hypothesis of primate phylogeny that also provides times of divergence (Purvis 1995; Purvis & Webster 1999). Our analyses involved continuous characters, discrete characters, or a combination of continuous and discrete characters. The method of independent contrasts applies to continuous data by calculating independent evolutionary change in two or more characters (Felsenstein 1985). We used the computer program CAIC to calculate standardized contrasts (Purvis & Rambaut 1995). The concentrated changes test applies to discrete data and tests whether evolutionary gains in a dependent character are concentrated on portions of the tree characterized by a particular character state in an independent character (Maddison 1990). We used the computer package MacClade to calculate statistical significance levels (Maddison & Maddison 1992). The BRUNCH algorithm applies to a combination of continuous and discrete data and is based on the logic of independent contrasts. This method examines whether evolutionary transitions in a discrete character are associated with consistent directional change in a continuous character (Purvis & Rambaut 1995). We used the computer program CAIC to calculate contrasts in discrete and continuous characters (Purvis & Rambaut 1995).

In some cases, we also reconstructed ancestral character states using MacClade (Maddison & Maddison 1992). To deal with possible error in reconstructing ancestral states (see Omland 1997), sensitivity tests were conducted using the ACCTRAN and DELTRAN options. ACCTRAN forces ambiguous reconstructions to occur closer to the root and therefore reduces the number of transitions, while DELTRAN delays transitions and so increases the number of reconstructed evolutionary changes (see Maddison & Maddison 1992). Polytomous nodes in the Purvis (1995) phylogeny were randomly resolved to run tests in MacClade that require a fully dichotomous tree (i.e., the concentrated changes test and ACCTRAN and DELTRAN reconstructions).

For hypotheses that make explicit directional predictions we performed one-tailed tests with the significance criterion set at $\alpha = 0.05$. In some cases, variables were transformed to meet the assumptions of the statistical and phylogenetic tests. In particular, the data were often $\log_{10}$ transformed to correct problems of unequal variances in non-phylogenetic analyses and to meet the assumptions of independent contrasts in phylogenetic tests (e.g., Garland *et al.* 1992).

Results: "bivariate" tests

Predation
Among diurnal species, those categorized as having medium or high predation risk have more females in the group (mean 3.7 females in low pre-

dation, versus 6.5 and 5.6 in medium and high predation). However, the difference among these categories was not statistically significant in a non-phylogenetic test $(F_{[2,99]} = 2.14, P = 0.12)$. Some small-bodied species follow a cryptic anti-predation strategy (Terborgh 1983), and so we re-examined the patterns after removing these species (see Methods). Results with this more restricted dataset were highly significant $(F_{[2,70]} = 5.94, P = 0.004)$ and in the predicted direction (in high predation risk categories the number of females increases to 9.4; Figure 16.2a). We also examined whether evolutionary transitions ("contrasts") in predation risk are consistently accompanied by an increase in the number of females (the BRUNCH algorithm; Purvis & Rambaut 1995). This type of phylogenetic analysis is easier to interpret when there are only two character states, and so we combined the medium and high predation risk categories (based on the above results). Ten of the 13 contrasts available for this test are in the predicted positive direction, with a mean increase of 2.7 females over evolutionary transitions to higher predation risk $(t_s = 3.04, P = 0.005$, one-tailed; Figure 16.2a).

The number of males also differs by predation risk categories in diurnal species $(F_{[2,100]} = 3.11, P = 0.049)$, with a larger number of males found in medium- and high-risk categories (1.7 males versus 2.8 males in both medium- and high-risk categories). Results become very highly significant when the probably cryptic diurnal species are excluded $(F_{[2,71]} = 5.04, P = 0.009$; Figure 16.2b). This result was also upheld in phylogenetic tests: the number of males increased in 10 of 13 transitions to higher predation risk (mean increase = 1.0, $t_s = 1.92, P = 0.04$, one-tailed; Figure 16.2b).

It may be that an increase in the number of females accounts for the increase in the number of males because a single male can no longer monopolize exclusive access to a larger female group (Andelman 1986; Altmann 1990; Mitani et al. 1996a; Nunn 1999b). We therefore also examined the residual variation from a regression of contrasts in log-transformed male number on contrasts in log-transformed female number. If higher predation leads to more males independently of female number, then shifts to higher predation should be associated with positive residuals, indicating an increase in the number males that is greater than is expected on the basis of the number of females (see Garland et al. 1993). This is not the case, however: seven residuals are positive and six are negative. We therefore cannot demonstrate an independent effect of predation on the number of males at this broad interspecific level, although

(a)

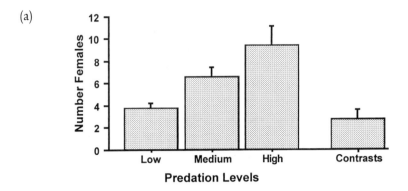

(b)

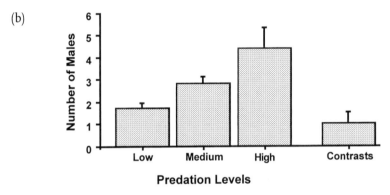

Figure 16.2. The influence of predation risk on the number of females and males in primate groups. Bars indicate means, with 1 SE indicated above. Small-bodied, probably cryptic species are removed. (a) The number of females under different levels of predation risk, and contrasts in the number of females from low predation to medium or high predation. (b) The number of males under different levels of predation risk, and contrasts in the number of males from low predation to medium or high predation. Contrasts represent evolutionary change, and so these results show that female number increases with increasing predation risk. The effect of predation risk on male number is less easy to assess because the number of males may simply reflect the number of females (see text).

both van Schaik & Hörstermann (1994) and Hill & Lee (1998) demonstrated an independent effect of predation in different analyses that controlled for female number.

Female intrasexual competition

An increase in female intrasexual competition is expected to lead to female philopatry and nepotistic rank inheritance (van Schaik 1989;

Sterck *et al.* 1997). We used information on competition levels from Plavcan *et al.* (1995), as these categories are based on information compiled independently from the pattern investigated here. Ten of the 20 gregarious species characterized by high female intrasexual competition are resident-nepotistic, while none of the 13 gregarious species characterized as having low competition is resident-nepotistic. This difference is highly significant in non-phylogenetic analysis ($\chi^2 = 9.33$, df = 1, $P = 0.002$).

To control for phylogeny we used the concentrated changes test (Maddison 1990). Because some ambiguity exists in the actual number of gains that have taken place (Figures 16.3 and 16.4), we used a range of estimated evolutionary gains in the relevant traits based on ACCTRAN and DELTRAN reconstructions (Maddison & Maddison 1992). These two reconstruction methods gave four to five gains of high female competition (the independent variable), and three to five gains of the resident-nepotistic social system (the dependent variable). When the maximum of five gains in resident-nepotism was used, the observed pattern is statistically significant, regardless of how many gains in high competition are assumed (with five transitions to high competition, $P = 0.03$; with four transitions, $P = 0.05$). However, when fewer gains were assumed in the dependent variable, results were no longer significant. This decline in significance is expected because the statistical power decreases when there are fewer reconstructed evolutionary changes in this variable. Thus, only when gregarious females experience contest competition are they likely to be philopatric and to form nepotistic dominance hierarchies.

Infanticide

Infanticide risk is expected to lead to a variety of social and sexual counterstrategies by females. As noted in the Methods, however, predictions for social strategies are not always straightforward. In these cases, where no clear predictions are possible, we treated the analyses as exploratory and so used two-tailed statistical tests.

We first examined the effect of infanticide risk on the number of males in the social group, which has been predicted to increase or decrease with increasing infanticide risk. The relationship between lactation/gestation ratios and the number of males was positive and therefore supports the hypothesis that the number of males increases ($b = 0.41$). However, this slope estimate was not statistically different from 0 ($F_{[1,41]} = 1.67$, $P = 0.20$, two-tailed). Further ambiguity was indicated in phylogenetic analysis based on independent contrasts, which gave a negative slope ($b = -0.04$),

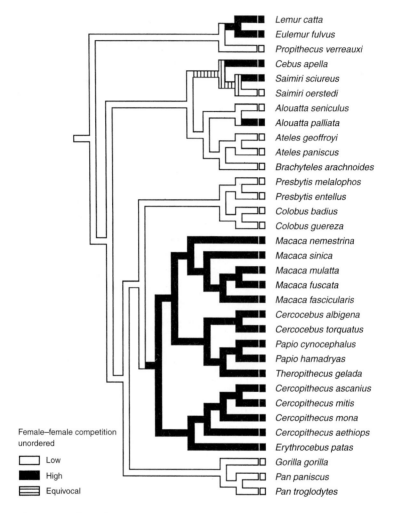

Figure 16.3. The evolution of female intrasexual competition. The competition levels
from Plavcan *et al.* (1995) are mapped onto Purvis' (1995) phylogeny. Some
ambiguity exists in the New World primates (indicated by striped bars), and so
ACCTRAN and DELTRAN reconstructions were used. ACCTRAN
reconstructions "accelerate" transitions toward the root of the tree, while
DELTRAN reconstructions "delay" transitions so that homoplasy is assumed
to be more common (for more information, see Maddison & Maddison 1992).

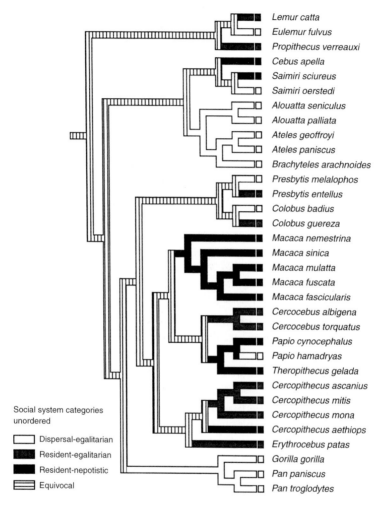

Figure 16.4. The evolution of female social relationships. The social system categories provided by Sterck *et al.* (1997) are mapped onto Purvis' (1995) phylogeny. These reconstructions are even more equivocal than those in Figure 16.3 (indicated by striped bars). We therefore used ACCTRAN and DELTRAN reconstructions (Maddison & Maddison 1992) to identify a range of values for statistical tests (see text).

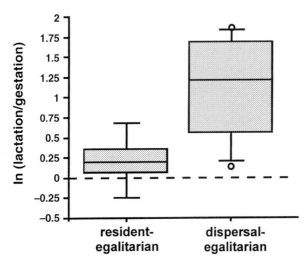

Figure. 16.5. Female dispersal patterns and the risk of infanticide. Among egalitarian species, those characterized by female dispersal have higher lactation/gestation ratios (ln-transformed), suggesting that they are at higher risk of infanticide. The analysis is limited to egalitarian species to control for ecological influences on dispersal.

although this slope was also not statistically different from 0 ($F_{[1,40]} = 0.22$, $P = 0.64$, two-tailed). With this proxy for risk, then, there is no evidence that infanticide risk consistently influences the number of males in primate groups.

We next examined the effect of infanticide risk on female dispersal, which has also been proposed to increase or decrease with increasing risk of infanticide (Treves 1998; Crockett & Janson, Chapter 4). However, lactation/gestation ratios were nearly indistinguishable in female-dispersal and female-resident species ($F_{[1,46]} = 0.002$, $P = 0.96$, two-tailed; only five of eight transitions to female dispersal have positive lactation/gestation contrasts, $t_s = 0.02$, NS). Confounding ecological variables may limit our resolution, and so we also examined patterns in subsets of the data that included only folivores (Crockett & Janson, Chapter 4), but this left only two female-resident species (the gelada (*Theropithecus gelada*) and the hanuman langur (*Presbytis entellus*)). Folivory probably involves reduced competition, which may allow for female dispersal, and so we also looked only among the species in the Sterck *et al.* (1997) dataset that are egalitarian (Figure 16.5). This gave a statistically significant result ($F_{[1,10]} = 7.24$, $P = 0.02$, two-tailed), with higher lactation/gestation ratios among species that disperse; however, the pattern was not significant in another analysis using the low competition species from Plavcan *et al.* (1995; $F_{[1,14]} = 2.05$, $P = 0.17$, two-tailed). To examine patterns phylogenetically, we removed contrasts to female dispersal that had corresponding transitions in diet or competition, but results remained non-significant. To summarize, a pattern emerged in one non-phylogenetic test when female intrasexual competition levels were controlled statistically. Although not conclusive, this result suggests that ecology structures the options available for females to reduce infanticide risk. In particular, female dispersal may be less of an option when alliances among kin are strategies for feeding competition (cf. Treves 1998).

Our final set of predictions involves female sexual behaviors (see also van Schaik *et al.* 1999, Chapter 15). Species with higher lactation/gestation ratios have longer mating durations ($b = 0.40$, $F_{[1,39]} = 3.32$, $P = 0.038$), consistent with the role of sexual behavior in modifying the risk of infanticide (see also van Schaik *et al.* 1999). Thus, for negative ratios (indicating less risk of infanticide), the median mating period is 2 days ($N = 13$), while for positive ratios the median is 7 days ($N = 28$). However, in phylogenetic analysis, the expected positive relationship between contrasts in lactation/gestation ratios and contrasts in mating periods was not found ($b = -0.01$, $F_{[1,39]} = 0.007$, $P > 0.50$, one-tailed). We also examined the pattern after removing the nocturnal species and the communally breeding callitrichids (where lactation/gestation ratios tend to be negative), but this had no effect on the general conclusion that no relationship exists once phylogeny is taken into account ($b = 0.06$, $F_{[1,28]} = 0.15$, $P = 0.35$, one-tailed).

Postconception matings may also serve to confuse paternity (Hrdy 1979), and so we expect their incidence to increase with lactation/gestation ratios. Lactation/gestation ratios were indeed higher in species where females have been observed to mate during pregnancy (median of 0.51 versus 0.17), but this difference was not statistically significant ($t_s = 0.92$, $N = 35$, $P = 0.18$, one-tailed). In analyses that controled for phylogeny, five out of the seven cladistic reconstructions of postconception matings were accompanied by higher lactation/gestation ratios, and the mean increase in lactation relative to gestation was nearly significant ($t_s = 1.64$, $P = 0.08$, one-tailed, after removal of one contrast involving the communally breeding callitrichids). We consider some reasons for these ambiguous results in the Discussion, including the possibility that our proxy for infanticide risk was unable to detect the subtle patterns examined here.

Harassment

Brereton (1995) proposed that primate social systems are a female response to male harassment. However, this "coercion–defense hypothesis" makes remarkably few predictions. For example, should we expect to see dimorphism increase before the evolution of female-bonded groups, when coercion is hypothesized to first arise, or after the evolution of female-bonded groups, as a male response to counteract female alliances against coercion? Is dimorphism then expected to decrease following the evolution of alternative female strategies, such as female transfer and association with "hired gun" male protectors? Finally, how exactly is "female-bonded" defined in this context, and how does this reduce the costs of male coercion? Female-bondedness usually refers to female philopatry accompanied by well-defined dominance relationships and at least occasional alliances among females (van Schaik 1989; Sterck *et al.* 1997), yet the coercion defense hypothesis involves only female defense against male harassment that is "supported by kin selection and reciprocal altruism," which does not require dominance relationships among the females (or philopatry in the case of reciprocal altruism).

In examining the coercion–defense hypothesis, the issue boils down to how literally one takes the hypothesized evolutionary transitions (see Brereton 1995: 209). Can the proposed pathways of social evolution be reconstructed phylogenetically, as implied by the model, or are they just a heuristic device that can be used to generate more explicit (and hopefully testable) predictions regarding the constellations of traits expected at various points in the male–female "arms race"? Furthermore, the patterns and functions of male aggression are not likely to be constant across social systems, but may instead be modified by the socioecological context. For example, harassment in the form of herding is expected to be more common in one-male groups, as this herding would presumably help to prevent female transfer to other groups (e.g., Asian colobines: Bernstein 1968; Rudran 1973; Stanford 1991; Steenbeek 1999a). In contrast, male harassment in multimale groups might be aimed more explicitly at achieving immediate matings, including forced copulations, because in this case competition occurs among males within the group rather than simply between groups.

So what, then, can we say about harassment as an influence on primate social systems? Certainly harassment should be more common when males are much larger than females and when males compete intensely among themselves for mates, two factors that probably go hand-in-hand.

We therefore expect to find more harassment among large-bodied taxa because dimorphism increases with body size in primates (Clutton-Brock *et al.* 1977) and because large-bodied species experience a more skewed operational sex ratio owing to longer interbirth intervals (Mitani *et al.* 1996b). Furthermore, terrestrial species tend to be large-bodied, and coercion is probably further facilitated by terrestriality (Gowaty 1997b; van Schaik *et al.*, Chapter 15; see the Discussion). Nevertheless, all the major female social systems are found among the larger species, including resident-nepotistic (baboons), resident-egalitarian (some colobines), dispersal-egalitarian (gorillas, chimpanzees), and semi-solitary (orang-utans). Similarly, one-male (gorillas, hamadryas baboon) and multimale (chimpanzees) social systems are found in the large-bodied species. All of these social system categories are also found in small-bodied species that are less dimorphic, strongly suggesting that harassment risk alone does not explain variation in primate social systems.

One other pattern can be investigated: male "brotherhoods" have been hypothesized by Brereton (1995) and previously by others (Wrangham & Rubenstein 1986; Wrangham 1987a) to function in reducing male harassment. We considered four genera to have male brotherhoods: spider monkeys, muriquis, red colobus and chimpanzees (van Schaik 1996). Species with male brotherhoods are larger in body size, as might be expected if harassment is greater (mean female mass of 16.1 kg compared to 7.0 kg in species without brotherhoods, restricted to species with more than two females per group; $t_s = 2.71$, $N = 72$, $P = 0.004$, one-tailed). However, only two of the four contrasts in body mass were in the predicted direction in phylogenetic tests, including one large contrast between the gorilla and the two species of chimpanzees. Furthermore, harassment is common in many of the species with brotherhoods, and, contrary to the hypothesis, this harassment often comes from individuals within the group rather than outside it (e.g., Smuts & Smuts 1993; Treves 1998). It is also difficult to exclude ecology as the main factor in the evolution of these more flexible groupings with male brotherhoods. In particular, large-bodied frugivorous females may be selected to forage solitarily, which may result in male cooperative mate-guarding strategies that operate at a community rather than individual level. Hence, the more folivorous and large-bodied gorilla has gregarious females and no brotherhoods, but these gregarious females are thought to share the protection of a single male.

In conclusion, we do not yet know whether male harassment is a serious selective pressure on patterns of female sociality. However, harassment probably does not account for patterns of philopatry as proposed by Brereton (1995) because temporary alliances can exist in female-dispersal species through either reciprocal altruism or, more likely, mutualistic benefits (Dugatkin 1997). Furthermore, "hired guns" may function to reduce infanticide risk rather than harassment risk (see also Treves 1998). If counterstrategies to harassment exist it seems likely that they will operate within ecological constraints, as in female counterstrategies to infanticide. For example, females that already form alliances to defend food resources should readily form alliances against harassing males. Despite these criticisms it seems reasonable to expect females to prefer males that offer them the most protection from aggressive males. These preferences will again exist within ecological constraints, especially competition over resources and options for female transfer (e.g., folivores: Watts 1989; Steenbeek 1999a).

Discussion

These results suggest that ecological factors, such as predation risk and the strength of contest competition, explain more cross-species variation in primate social systems than do intersexual factors, such as risks associated with infanticide and harassment. Detailed intraspecific studies provide more clear-cut evidence for the effect of infanticide risk on primate social systems, and so we do not claim that social counterstrategies are not important. We can only conclude that these counterstrategies are not expressed in a consistent manner that can be identified in broad interspecific comparisons.

A major reason for the lack of effect at the interspecific level is probably that social counterstrategies to infanticide are highly flexible and thus can be tailored to the ecological factors that are in place. For example, infanticide may be countered with dispersal in species where female within-group competition is low (e.g., Thomas's langurs, Steenbeek 1999a), but where the latter is high and requires female philopatry, equally effective social tactics may be possible. Thus infanticide may lead to additional males (and promiscuous mating) in social groups when ecological factors allow larger groups of both males and females. In cases of resident-nepotistic species, female group size may overshoot its optimal value for ecological reasons, resulting in a multimale group (see Sterck et al. 1997). Females of these species may then opt for paternity

Table 16.2. *Interaction between ecological and intersexual risks*

Ecological risks (predation and competition)	Intersexual risks (infanticide and harassment)	
	Low	High
Low	(No clear predictions)	(2) Social system a function of infanticide risk, but probably also some effect of ecology
High	(1) Social system a function of ecological risks, not intersexual risks	(3) Social system a function of both ecological and intersexual risks; female sexual counterstrategies to infanticide most prominent because social strategies will be more constrained

confusion mechanisms coupled with promiscuous mating. In contrast, female group size can be more easily maintained in situations of low within-group contest; thus folivores will tend to experience added selection to limit female group size to prevent takeovers (Steenbeek 1999a; Crockett & Janson, Chapter 4). Under all types of competitive regimes and patterns of philopatry we expect female alliances against harassment and infanticide to exist (Brereton 1995; Treves 1998), as these alliances can be maintained through mutualistic benefits that do not depend on reciprocity or kinship (Dugatkin 1997). Male coercion is therefore not a likely explanation for patterns of female philopatry.

Table 16.2 summarizes how this framework can be used in future comparative or field studies on the social effects of ecological and intersexual conflict. (1) When intersexual risks are low but ecological risks are high, then social system variation should be explained mainly by ecological factors. (2) Conversely, when intersexual risks are high but ecological risks are low, then intersexual risks may influence the social system to a greater degree; however, female sexual behavior is still a likely target for selection because some ecological pressures will probably always exist and fewer options may be available for dealing with these pressures. (3) When both intersexual and ecological risks are high, then sexual behavior should be most elaborated (e.g., exaggerated swellings, pregnancy matings), as females will have less social flexibility to counter coercion. (The entry representing low ecological risks and low intersexual risks is ignored, as it makes no clear predictions to differentiate between the two categories of selective forces.)

Weaknesses of the interspecific approach to study infanticide and harassment

Using different methods, we have confirmed repeat patterns found in previous comparative studies of ecological factors. Hence our dataset is not in error and incorporating phylogenetic information does not overly weaken our statistical resolution. However, several other factors may have contributed to the non-significant results for intersexual pressures.

First, the tests may be insufficiently broad: if infanticide risk is roughly similar across anthropoid primates, then there will be insufficient variability to examine correlates of infanticide risk in these species. The analysis could be expanded to a broader taxonomic scale, although this may introduce additional confounding factors that are not easy to control (including variation in the solutions to ecological pressures; e.g., Wrangham & Rubenstein 1986). Even when the analysis is expanded to include other mammals, variation in the underlying reproductive mechanisms leads to ambiguous predictions for female counter-strategies (see van Noordwijk & van Schaik, Chapter 14). Yet such a broadened approach can still lead to some insights. For example, van Schaik (2000) noted that permanent groups of carnivore or primate females were more likely than ungulates to contain multiple males. Because infanticide risk is lower in ungulates, this observation is consistent with an infanticide avoidance function of multimale groups.

Second, we could not evaluate the effect of harassment because clear predictions are not yet available. When better hypotheses and predictions become available, several variables could be used to proxy the risk of harassment across species. One measure involves sexual dimorphism in body and canine size. The assumption is that when males are substantially larger than females it is easier for them to physically subdue and thus coerce females. In addition, substrate use may affect harassment risk. For example, harassment may be more difficult in an arboreal environment where males are constrained in their ability to harass females because they need more "points of contact" for arboreal locomotion (e.g., Gowaty 1997b; Nunn 1999c; van Schaik et al., Chapter 15). The three-dimensional structure of arboreal habitats may also provide more escape options for females and may explain why coercion is reduced in birds as compared with mammals (Gowaty 1997b), or why males have been observed to cooperatively (rather than singly) mob females in the three-dimensional habitats of dolphins (Connor et al. 1992) and New World monkeys (Fedigan & Baxter 1984; Boinski 1987b).

A final problem involves the measurement of infanticide risk. Estimates based on observed patterns of infanticide (e.g., Treves 1998) fail to separate rates from risks, yet when counterstrategies are being assessed, it is the risks that are most relevant (Hill & Dunbar 1998; Janson 1998). We used the ratio of lactation to gestation as our measure of risk, but lactation especially is very flexible and so this ratio is subject to extensive error when one is calculating species values. Hence our estimates of infanticide risk may be too variable to detect patterns and better measures of infanticide risk are needed. The real risk of infanticide is likely to be the rate of rapid turnover of dominant males (within the life history factors identified by van Schaik (2000)). However, this rate probably depends on numerous details of social organization, demography and population density that cannot be captured easily in cross-species analyses, although proxies can be developed for detailed intraspecific tests (e.g., Treves & Chapman 1996). While no immediate direction for a new proxy is apparent, effort placed in developing new measures of infanticide risk will no doubt lead to new insights.

Despite our emphasis on risk, observed rates may also provide insights to selective factors. This is because counterstrategies are rarely cost-free so that some residual level of risk remains, which will be estimated (see Janson 1998). One problem involves the separation of risk that can be reduced by further counterstrategies from the risk that cannot be eliminated. Another problem is that the level of residual risk is variable because of cross-species variation in the costs of implementing anti-predation strategies. Finally, in the present case, the high reproductive costs of predation and infanticide probably result in low levels of useful variation for statistical testing. In any case, we lack quantitative evidence for the most problematic variable in our study (infanticide), which is further complicated as a measure by problems of "negative evidence" in deciding which species do *not* exhibit infanticide (cf. van Schaik, Chapter 3; Blumstein, Chapter 8). A better cross-species measure of infanticide rate would involve quantitative measures of infant mortality due to infanticide, but this information is not presently available for enough species.

Despite these caveats, a quite reasonable interpretation of our results is that ecological factors modify the social strategies that females use to counter male sexual coercion. A likely reason is that more equally effective strategies exist to counter infanticide risk, as compared to strategies to deal with ecological pressures. This kind of complexity cannot be easily captured in a cross-species comparative analysis. Nevertheless, the value of cross-species comparative research is to identify the species that are most important for particular questions and to begin identifying the selective factors that can account for species differences.

Acknowledgments

We thank Roberto Delgado, Charles Janson, Adrian Treves and Maria van Noordwijk for comments on the manuscript. This research was supported by an NSF Dissertation Improvement Grant to C. L. N.

Appendix 16.1. Data used in the comparative tests[a]

Species	Substrate[b]	Habitat[c]	Activity period[d]	Predation risk[e]	Female foraging group size	Male foraging group size	Male population group size[f]	Regular female dispersal
Lemurs: Madagascar								
Avahi laniger	A	W	N					Yes
Eulemur coronatus	A	W	D	H				
Eulemur fulvus	A	W	D	M	3.5	3.2	3.2	Yes
Eulemur macaco	A	W	D	M	3.1	4.0	4.0	
Eulemur mongoz	A	W	D	H	1.3	1.3		Yes
Eulemur rubriventer	A	W	D	H				Yes
Hapalemur aureus	A	W	D	H				
Hapalemur griseus	A	W	D	H	1.0	1.0	1.0	
Hapalemur simus	A	W	D	H				
Indri indri	A	W	D	L	1.0	1.0	1.0	Yes
Lemur catta	T	W	D	H	6.4	5.9	5.9	No
Lepilemur mustelinus	A	W	N					Yes
Microcebus murinus	A	W	N					Yes
Mirza coquereli	A	W	N					Yes
Propithecus diadema	A	W	D	L	2.3	1.7		No
Propithecus tattersalli	A	W	D	M				
Propithecus verreauxi	A	W	D	M	2.8	3.0	3.0	No
Varecia variegata	A	W	D	M			2.5	Yes
Lorisids and tarsiers: Africa and Asia								
Galago alleni	A	W	N					No
Galago senegalensis	A	W	N					No
Galagoides demidoff	A	W	N					No
Galagoides zanzibaricus	A	W	N					No

Appendix 16.1. (cont.)

Species	Substrate[b]	Habitat[c]	Activity period[d]	Predation risk[e]	Female foraging group size	Male foraging group size	Male population group size[f]	Regular female dispersal
Otolemur crassicaudatus	A	W	N					No
Otolemur garnettii	A	W	N					No
Perodicticus potto	A	W	N					Yes
Phaner furcifer	A	W	N					Yes
Tarsius spectrum	A	W	N		1.4	1.1	1.1	
Platyrrhines: Neotropics								
Alouatta belzebul	A	W	D	M				Yes
Alouatta caraya	A	W	D	M	3.3	2.1		Yes
Alouatta fusca	A	W	D	M	1.9	1.1		Yes
Alouatta palliata	A	W	D	M	6.3	2.8	2.8	Yes
Alouatta pigra	A	W	D	M	1.6	1.4		
Alouatta seniculus	A	W	D	M	2.2	1.8	1.8	Yes
Aotus trivirgatus	A	W	N				1.1	Yes
Ateles belzebuth	A	W	D	M	4.2	2.8		Yes
Ateles fusciceps	A	W	D	L				Yes
Ateles geoffroyi	A	W	D	M	4.1	0.5	4.0	Yes
Ateles paniscus	A	W	D	L	2.3	0.9		Yes
Brachyteles arachnoides	A	W	D	L	7.1	7.0		Yes
Cacajao calvus	A	W	D	H				
Cacajao melanocephalus	A	W	D	H				
Callicebus brunneus	A	W	D	H				
Callicebus moloch	A	W	D	H	1.0	1.0		Yes
Callicebus personatus	A	W	D	H	1.0	1.0		Yes
Callicebus torquatus	A	W	D	H	1.0	1.0	1.0	Yes

Callimico goeldii	A	W	D	H	2.0	1.0	1.0	No
Callithrix argentata	A	W	D	H				Yes
Callithrix humeralifer	A	W	D	H	3.3	2.8		Yes
Callithrix jacchus	A	W	D	H	2.9	2.7	2.7	Yes
Cebuella pygmaea	A	W	D	H	1.0	1.5	1.5	Yes
Cebus albifrons	A	W	D	H	6.5	2.8	2.8	No
Cebus apella	A	W	D	H	4.9	3.2		No
Cebus capucinus	A	W	D	H	4.5	2.6		No
Cebus olivaceus	A	W	D	H	6.5	1.1		No
Chiropotes albinasus	A	W	D	H	9.0	8.0		
Chiropotes satanas	A	W	D	H	9.0	1.3		
Lagothrix flavicauda	A	W	D	L	4.0	3.0		
Lagothrix lagothricha	A	W	D	M	8.2	5.8	5.8	Yes
Leontopithecus chrysomelas	A	W	D	H				Yes
Leontopithecus chrysopygus	A	W	D	H				Yes
Leontopithecus rosalia	A	W	D	H				Yes
Pithecia hirsuta	A	W	D		1.0	1.0		
Pithecia monachus	A	W	D	H	1.0	1.3		
Pithecia pithecia	A	W	D	H	1.0	1.0	1.0	
Saguinus bicolor	A	W	D	H				Yes
Saguinus fuscicollis	A	W	D	H	1.5	1.5	1.5	Yes
Saguinus imperator	A	W	D	H				Yes
Saguinus inustus	A	W	D					Yes
Saguinus labiatus	A	W	D	H	1.6	2.2		Yes
Saguinus leucopus	A	W	D	H				Yes
Saguinus midas	A	W	D	H				Yes
Saguinus mystax	A	W	D	H	1.7	2.0		Yes
Saguinus nigricollis	A	W	D	H	1.2	1.2		Yes
Saguinus oedipus	A	W	D	H	1.9	2.6		Yes
Saguinus tripartitus	A	W	D					Yes

Appendix 16.1. (cont.)

Species	Substrate[b]	Habitat[c]	Activity period[d]	Predation risk[e]	Female foraging group size	Male foraging group size	Male population group size[f]	Regular female dispersal
Saimiri oerstedii	A	W	D	H	8.9	4.0		Yes
Saimiri sciureus	A	W	D	H	7.9	2.7	2.7	No
Catarrhines: Africa								
Allenopithecus nigroviridis	T	W	D	H				
Cercocebus albigena	A	W	D	M	7.8	4.8	4.8	No
Cercocebus aterrimus	A	W	D	M	5.0	3.0		
Cercocebus galeritus	T	W	D	M	9.5	2.5		
Cercocebus torquatus	T	W	D	M				
Cercopithecus aethiops	T	O	D	H	7.7	4.6	4.6	No
Cercopithecus ascanius	A	W	D	H	8.5	1.0	1.0	No
Cercopithecus campbelli	A	W	D	H	4.0	1.0		No
Cercopithecus cephus	A	W	D	H	2.0	1.0		
Cercopithecus diana	A	W	D	M	7.0	1.0		
Cercopithecus erythrogaster	A	W	D	H				
Cercopithecus erythrotis	A	W	D	H				
Cercopithecus hamlyni	A	W	D	M				
Cercopithecus lhoesti	T	W	D	H				
Cercopithecus mitis	A	W	D	M	8.8	1.1		No
Cercopithecus neglectus	T	W	D	H	1.9	1.0		No
Cercopithecus nictitans	A	W	D	M	2.0	1.0		
Cercopithecus petaurista	A	W	D	H				
Cercopithecus pogonias	A	W	D	H	3.0	1.0		
Cercopithecus preussi	T	W	D	H				
Cercopithecus solatus	T	W	D	H				

Cercopithecus wolfi	A	W	D	H	3.9	2.4		
Colobus angolensis	A	W	D	M	11.4	4.4		Yes
Colobus badius	A	W	D	M	2.9	1.3		No
Colobus guereza	A	W	D	L				
Colobus kirkii	A	W	D	M				
Colobus polykomos	A	W	D	M	4.8	2.8	2.8	No
Colobus satanas	A	W	D	M	5.5	2.2	2.2	No
Erythrocebus patas	T	O	D	H	10.4	1.2		Yes
Gorilla gorilla	T	W	D	L	4.4	1.8	1.8	No
Macaca sylvanus	T	W	D	M	7.0	5.2		No
Mandrillus sphinx	T	W	D	M		4.0		No
Miopithecus talapoin	A	W	D	H	21.3	13.0	13.0	Yes
Pan paniscus	T	W	D	L	3.4	1.9		Yes
Pan troglodytes	T	W	D	L	2.9	2.2	6.7	Yes
Papio anubis	T	O	D	H	17.5	9.1		No
Papio cynocephalus	T	O	D	H	19.4	6.6	6.6	No
Papio hamadryas	T	O	D	H	8.2	1.0	6.5	Yes
Papio papio	T	W	D	M				No
Papio ursinus	T	O	D	H	13.9	6.6		No
Procolobus verus	A	W	D	M	3.3	2.0		Yes
Theropithecus gelada	T	O	D	H	4.1	1.5	1.5	No
Catarrhines: Asia								
Hylobates agilis	A	W	D	M	1.0	1.0		Yes
Hylobates concolor	A	W	D	L	1.9	1.0		
Hylobates hoolock	A	W	D	L	1.0	1.0		
Hylobates klossi	A	W	D	M	1.0	0.9	0.9	Yes
Hylobates lar	A	W	D	M	1.0	1.0	1.0	Yes
Hylobates moloch	A	W	D	L				
Hylobates muelleri	A	W	D	M				

Appendix 16.1. (cont.)

Species	Substrate[b]	Habitat[c]	Activity period[d]	Predation risk[e]	Female foraging group size	Male foraging group size	Male population group size[f]	Regular female dispersal
Hylobates pileatus	A	W	D	M	1.0	1.0		
Hylobates syndactylus	A	W	D	L	1.0	1.0		Yes
Macaca arctoides	T	W	D	M				No
Macaca assamensis	A	W	D	L	6.5	3.0		
Macaca cyclopis	T	W	D	H	3.4	1.1		
Macaca fascicularis	A	W	D	M	10.6	4.6	4.6	No
Macaca fuscata	T	W	D	M	17.5	7.3	7.3	No
Macaca maurus	T	W	D	L	8.5	2.5		No
Macaca mulatta	T	W	D	M	23.8	9.1	9.1	No
Macaca nemestrina	T	W	D	M	14.9	2.4	2.4	No
Macaca nigra	T	W	D	L				No
Macaca ochreata	A	W	D	M				
Macaca radiata	T	W	D	H	9.9	6.9	6.9	No
Macaca silenus	A	W	D	L	7.1	3.1	3.1	No
Macaca sinica	A	W	D	M	6.2	2.6		No
Macaca thibetana	T	W	D	M	9.5	5.2		No
Macaca tonkeana	T	W	D	L				
Nasalis larvatus	A	W	D	L	3.7	1.0		Yes
Pongo pygmaeus	A	W	D	L	1.0	1.0	1.0	Yes
Presbytis aurata	A	W	D			1.0		
Presbytis comata	A	W	D	L				
Presbytis cristata	A	W	D	M	15.6	1.3		Yes
Presbytis entellus	T	W	D	M	11.5	2.4	2.4	Yes
Presbytis françoisi	A	W	D	L	6.5	1.0		Yes
Presbytis frontata	A	W	D	M		1.0		Yes

Presbytis geei	A	W	D	L	5.3		1.1	
Presbytis johnii	A	W	D	L	3.1		1.3	Yes
Presbytis melalophos	A	W	D	L	5.2		1.1	Yes
Presbytis obscura	A	W	D	L	7.0		2.0	
Presbytis phayrei	A	W	D	L				
Presbytis pileatus	A	W	D	L	3.6		1.1	Yes
Presbytis potenziani	A	W	D	L	0.9		1.0	
Presbytis rubicunda	A	W	D	L	2.3		1.0	Yes
Presbytis vetulus	A	W	D	M	3.8		1.0	Yes
Pygathrix avunculus	A	W	D	L			1.0	
Pygathrix nemaeus	A	W	D	L				Yes
Pygathrix roxellana	T	W	D	M				
Simias concolor	A	W	D	L	1.9		1.0	

Notes:

[a] Information provided only on datasets compiled by the authors from the literature (see also Nunn 1999a; Nunn & van Schaik 2001). For other variables: female body mass was taken from Smith & Jungers (1997), competition levels from Plavcan et al. (1995), social system categories from Sterck et al. (1997) subject to modifications described in Methods, lactation/gestation ratios from van Schaik (Chapter 3), and female sexual behaviors (duration of receptivity and pregnancy matings) from van Schaik et al. (1999). Blanks indicate that no information was available.

[b] A, arboreal; T, terrestrial, following Nunn & van Schaik (2001).

[c] W, wooded; O, open, following Nunn & van Schaik (2001).

[d] N, nocturnal; D, diurnal, following Nunn & van Schaik (2001).

[e] L, low; M, medium; H, high. See Methods and Table 16.1 for categorization of species. Information on predation risk and corresponding variables tested comparatively (e.g., foraging group size) provided only for diurnal species. Blanks for other species indicate that information on body mass was not available.

[f] Data provided only for those species with corresponding information on lactation/gestation ratios.

Part IV

Infanticide by females

17

Infanticide by female mammals: implications for the evolution of social systems

The complexity and richness inherent in the social networks female primates forge for themselves has, too often, obscured a vital fact of their lives: that competition among females is central to primate social organization.

(HRDY 1981: 96)

Introduction

Female competition, especially female reproductive competition, is likely to play an important role in shaping the social systems of all mammals (Altmann 1997; Gowaty 1997b; Hrdy 1981). While this competition can often be subtle, it also can lead to what is perhaps the most extreme form of reproductive competition: infanticide.

The chapters in this volume focus primarily, if not exclusively, on the phenomenon of infanticide by males and how it may be influencing the evolution of social systems. Generally, threats from conspecific males intensify under a specific context: the presence of an unrelated male, often following immigration into an established group (e.g., Hrdy 1974). But females and their young also face threats from conspecific females, including both unfamiliar intruders and fellow group mates. The threat of infanticide by females is likely to be taxonomically more widespread and, for group-living females, potentially a more constant threat than other forms of infanticide.

The purpose of this chapter is to examine the phenomenon of infanticide by females other than the mother (hereafter "infanticide by females") in a variety of mammalian taxa. How similar is this behavior to infanticide by unrelated males? What makes infants vulnerable to attack from female conspecifics, and what are the contexts in which female attacks on infants occur? By investigating the nature and extent of infanticide by females in a variety of social contexts, we can make preliminary speculations about the impact of this phenomenon on the evolution of mammalian social systems. Ultimately, I hope this research will lead to a re-examination of our perception of mammalian mothers and the extent to which they will go both to protect and to provide for their young.

Definitions and hypotheses

For the purposes of this chapter, we need a broader definition of infanticide than that used elsewhere in this volume. To encompass a variety of different threats, I am defining infanticide as an act by an individual (or group of individuals) that makes a direct or significant contribution to the immediate or imminent death of conspecific young (modified from Hrdy & Hausfater (1984) and Mock (1984) to exclude cases of gamete destruction and feticide). This definition includes cases where infants die as the result of physical aggression (direct infanticide) as well as cases where enforced neglect or harassment of the mother results ultimately in an infant's death (indirect infanticide). Many cases of enforced neglect, especially those involving "kidnapping," are often excluded from discussions of infanticide, in part because the behavior (a type of overzealous allomaternal care; see Hrdy 1976) appears contrary to the broader context of infant-killing. Nevertheless, these behaviors do occasionally result in infant death. It is the overall pattern of both direct and indirect infanticide that will determine the effect of conspecific threats on the evolution of social systems.

Hrdy (1979) originally proposed five hypotheses to explain the phenomenon of infanticide: sexual selection, resource competition, exploitation, parental manipulation and social pathology. The first four of these provide adaptive explanations for infanticide in particular contexts and have often been used to describe "classes" of infanticide. The fifth hypothesis, social pathology, allows for the possibility that not all cases are adaptive. These hypotheses have been discussed in depth elsewhere (Hrdy 1979; Sherman 1981; Hrdy & Hausfater 1984; Labov et al. 1985; Parmigiani & vom Saal 1994; Hoogland 1995; Ebensperger 1998a), I will therefore briefly discuss each (with the exception of parental manipulation, which is, by definition, not relevant to infanticide by non-parental individuals) only in terms of infanticide by females other than the mother.

Sexual selection

Of the five hypotheses suggested by Hrdy (1979), the sexual selection hypotheses is the one most closely associated with infanticide by males. The hypothesis proposes that the limited sex (i.e., the sex whose reproductive success is limited by access to the opposite sex) will benefit from killing the dependent young of the limiting sex, thus freeing the victim's caregiver to mate with the perpetrator (Hrdy 1974, 1979). In mammals, it is typically the males that are limited in their access to females and thus are predicted to be the more frequent perpetrators of sexually selected infanticide. Nevertheless, it is possible to have reversals where females are the limited sex, and thus the sexual selection hypothesis could be invoked to explain infanticide by females (Hrdy 1979). Intriguing examples of this have so far been described only for birds (e.g., jacanas (*Jacana jacana*), Emlen *et al.* 1989; and reed warblers (*Acrocephalus arundinaceus*), Hansson *et al.* 1997). In these cases, male care is a limiting factor for infant survival, and females compete for access to mates. In mammals, such "role reversal" scenarios are unlikely, given the female mammal's high energetic investment in each young compared with the investment of even the most attentive male (e.g., Trivers 1972). Infanticide by female mammals differs from most cases of infanticide by males simply because these females are unlikely to gain from killing the offspring of potential mates.

Traditionally applied, the sexual selection hypothesis refers to a type of *intra*sexual selection. Voland (see Voland & Stephan, Chapter 18) has introduced a new approach to the sexual selection hypothesis, suggesting that some females may kill their own offspring (maternal infanticide) in order to make themselves more "attractive" to a potential mate (i.e., *inter*sexual selection). This approach appears to be applicable only to maternal infanticide, and therefore is not discussed further here.

Resource competition

In contrast to the sexual selection hypothesis, the resource competition hypothesis addresses a factor more closely linked to female reproductive success: access to resources that will increase survivorship of young (e.g., Darwin 1871; Trivers 1972; Hrdy 1979; Wrangham 1980). The hypothesis predicts that by killing the young of other females, the infanticidal female and her young will have fewer competitors for limited resources, both now and in the future (Hrdy 1979; Sherman 1981; Hoogland 1995). Typically, the resource in question is food. Unfortunately, measuring the effect of infanticide on competition over food is difficult, especially when one is trying to determine changes in the reproductive success of the perpetrator (Hrdy & Hausfater 1984; Pierotti 1991). In addition, where females live in groups, *all* group members would presumably benefit from the reduced competition, thus diluting the effect of the infanticide for the perpetrator and her young (i.e., it presents a collective action problem: Nunn & Lewis 2001). Some of these problems can be circumvented by investigating what direct gains the perpetrator accrues following an infanticide.

Changes in access to such limited resources as nest sites or burrows (Sherman 1981), delineated foraging areas (Hoogland 1995), and access to helpers (Creel & Waser 1991; Emlen 1991; Digby 1995a; Hoogland 1995) as a consequence of infanticide can be directly measured. In most cases, the forfeiture of the resource by one female is the direct result of the death of her young. The reasons for forfeiture vary according to the resource that is being given up. For example, black-tailed prairie dog (*Cynomys ludovicianus*) females defend foraging territories only during the period between conception and weaning, presumably owing to her need for additional food during reproduction (Hoogland 1995). If the offspring die or are killed, the prairie dog female ceases to be territorial. In Belding's ground squirrels (*Spermophilus beldingi*), females abandon their burrows following the premature death of their young and subsequently search for "safer" sites (Sherman 1981). In the case of species that cooperatively rear their young, the time and energy of helpers is generally divided among the various infants in the group (e.g., Emlen 1991). With fewer offspring needing care following an infanticide, helpers are then available to spend more energy on the offspring of the perpetrator (Creel & Waser 1991; Digby 1995a). By killing the dependent young of other females living in either the adjacent territories or even the same group or coterie, infanticidal females and their young therefore gain direct (and generally unshared) access to these limited resources.

An alternative form of resource competition is actually the avoidance of a resource cost: the avoidance of misdirected parental care (e.g., Hrdy 1976; Sherman 1981; LeBoeuf & Campagna 1994; Hoogland 1995). In many species, females may end up caring for or "accidentally" adopting unrelated infants (Pierotti 1991). When maternal resources (e.g., milk) are limited, this misdirected care could cause a significant decrease in a female's ability to successfully raise her own young. Competition over limited resources (or the avoidance of resource theft) may have led to a feedback loop where females may commit infanticide in order to avoid becoming victims themselves. Following the death of their own offspring, the mothers of the victims are no longer in need of extra resources and thus are no longer infanticidal (Hoogland 1995). Such a feedback loop, however, is dependent on other factors (e.g., resource competition) to instigate the initial infanticidal behavior.

Exploitation

The exploitation hypothesis suggests that infanticidal females benefit from "use" of the victim itself (Hrdy 1979). Typically, this refers to cannibalism where the perpetrator benefits in terms of the nutritional value of the infant as a food item (also referred to as the predation hypothesis; Ebensperger 1998a; Sherman 1981). The original formulation of this hypothesis also included cases of infanticide where perpetrators "exploited" a live infant in order to avoid aggression or gain maternal experience (the death is thus an indirect consequence of this behavior). Although not often included in discussions of infanticide, such cases of "live use" may be important influences in the shaping of gregarious social systems. It is worth noting that the exploitation hypothesis is not mutually exclusive of the other adaptive hypotheses put forward by Hrdy (1979). Indeed, both male and female perpetrators may gain from "preying" on (or gaining experience from) their victims while simultaneously reducing competition for mates or resources.

In summary, of the four adaptive hypotheses originally proposed by Hrdy (1979), the resource competition and exploitation hypotheses are most relevant to infanticide by females (e.g., Sherman 1981; Hoogland 1995). Both are best divided into subhypotheses that address specific benefits (or avoidance of costs) to the perpetrator (Table 17.1). In addition to the general removal of future competitors, resource competition can be broken down into: (1) competition over a limited food or foraging territory; (2) competition over nest/burrow sites; (3) competition over helpers (helper coercion); and (4) avoidance of misdirected parental care. The exploitation hypothesis can be divided into (1) conspecific predation and (2) exploitation of live infants ("live use": deaths are a by-product of this exploitation). In most cases, these hypotheses fit under the broad concept of reproductive competition where the perpetrator is acting to increase the survivorship of her young, while decreasing the reproductive success of other females.

It should be noted that not all cases of infanticide by females necessarily have an adaptive explanation. Deaths of infants may be the result of "social pathologies" or accidents (e.g., accidental crushing of infants in

Table 17.1. *Hypotheses explaining infanticide by females*

Hypothesis	Subhypothesis	Predictions: perpetrator	Predictions: victim[a]	Examples
Resource Competition	Removal of future competitor (general)	Distantly related or unrelated	Victim is philopatric and likely to compete for same resource	Baboons (Wasser & Starling 1986)
	Competition over limited food or foraging territory	Distantly related or unrelated	Presence of infant is creating or increasing the limiting nature of the resource	Black-tailed prairie dogs (Hoogland 1995)
	Competition over nest sites or burrow	Perpetrator gains access nest/burrow, is generally from distant (site-limited) home range	Victim is pre-emergence and nests are "unshareable"	Belding's ground squirrels (Sherman 1981)
	Helper coercion	Perpetrator gains increased access to helpers	Victim is dependent on helpers and is of similar age to that of perpetrators young	Common marmosets (Digby 1995); wild dogs (van Lawick 1974)
	Avoidance of mis-directed parental care	Perpetrator avoids the loss of reproductive energy, most often milk and caregiving	Victim is attempting or may attempt to "steal" milk or has become separated from its mother and thus is attempting to be "adopted"	Northern elephant seals (LeBoeuf et al. 1972)
Exploitation	Predation	Perpetrator consumes the victim	Victim is young enough to be easily attacked	Chimpanzees (Goodall 1986)
	"Live use"	Perpetrator "uses" victim while still alive	Victim typically born to subordinate or less aggressive mother; mother unable to retrieve infant from perpetrator	Old World monkeys (Brain 1992; Collins et al. 1984)
Sexual selection	Intrasexual selection	Perpetrator has limited access to mates due to presence of second female and her young	Victim is limiting female from acquiring additional mates and/or care for additional offspring	No cases have yet been described for mammals

Notes:
[a] In each case, infants are predicted to be altricial and unable to defend themselves.

elephant seals, LeBoeuf *et al.* 1972; or infant lemurs injured during attacks on their mothers, Vick & Pereira 1989). Just how "pathological" or "accidental" such cases truly are is difficult to determine. Even in these cases it is important to investigate whether such deaths occur in a pattern that may be playing a role in the evolution of mammalian social systems.

Implications of infanticide by females for the evolution of mammalian social systems

Infanticide by females has been documented in over 50 species (300+ cases) in 5 mammalian orders (rodents, lagomorphs, pinnipeds, carnivores and primates; L. Digby *et al.,* unpublished data). The extent of infanticide varies between species. In the black-tailed prairie dog, up to 22% of litters are partially or completely killed by conspecific females (Hoogland 1995). In Columbian ground squirrels (*Spermophilus columbianus*) 6% of all newly emerged young are killed by lactating females (Stevens 1998), and in northern elephant seals (*Mirounga angustirostris*), necropsies and behavioral observations indicate that more than half of all infant deaths are at least partially the result of attacks by lactating adult females (LeBoeuf *et al.* 1972; LeBoeuf & Briggs 1977). While there are several species in which infanticide by females is a major mortality factor, there are many populations in which infanticide is relatively rare (e.g., yellow baboons (*Papio cynocephalus*): Shopland & Altmann 1987; Wasser & Starling 1986) and whole taxonomic groups for which the behavior has never been reported (e.g., ungulates, marsupials, edentates).

The variation in the extent of recorded infanticides may be, in part, due to methodological difficulties (e.g., attacks are difficult to document in den- or burrow-living species; for review, see Labov *et al.* 1985; Packer & Pusey 1984) or due to a lack of intensive behavioral observations in a given species . But real differences in the nature and extent of infanticide are also likely to be the result of differences in the costs and benefits of the behavior in diverse contexts. In many species, the killing of another female's offspring will result in little or no gain to the perpetrator. For example, most ungulate females will not benefit significantly from competing over access to relatively abundant food sources and, being herbivorous, they are unlikely to cannibalize infants (e.g., Elgar & Crespi 1992; Polis *et al.* 1984). Social circumstances will also influence potential costs. For example, infants born into social groups with female kin are less likely to be the targets of within-group attacks. This is due to the costs of eliminating close kin (and thus decreasing inclusive fitness) out-weighing the potential benefits of infanticide in most contexts (note: there are some exceptions to the pattern; e.g., cooperatively breeding groups, see below). In some cases, females may benefit from the mainte-nance of larger group sizes (whether due to foraging efficiency or preda-tor protection; Wrangham 1980; van Schaik 1983), again counterbalancing any potential benefit to infanticide. The vulnerability of offspring to infanticide will also vary between species (see Treves, Chapter 10). Precocial young, for example, are less vulnerable as mobility will make infants more costly to target (e.g., most ungulates). Young that are con-stantly carried (e.g., most marsupials, where offspring are carried in pro-tected pouches until they are mobile) also makes attacks on infants more costly or impractical (Wolff & Peterson 1998). Thus life history traits in many species can be used to predict the likelihood of infanticide by females.

Costs and benefits of infanticidal behavior by females will also vary according to behavioral counterstrategies against attacks on infants. Such counterstrategies may include broad-scale changes in social system (e.g., female philopatry resulting in groups of close female kin), habitual behaviors such as the carrying of infants (e.g., primates, Treves, Chapter 10), and direct defense of young. The presence of counterstrategies is of particular interest as they are indicative of strong selective pressure for infanticide avoidance. For species in which counterstrategies are present, whether or not high mortality rates due to infanticide have been reported, the *threat* of infanticide is likely to be shaping reproductive strategies, selecting for such behaviors as aggressive defense of young, female territoriality, and intolerance of unfamiliar female conspecifics (e.g, Christenson & LeBoeuf 1978; Maestripieri 1992; Hoogland 1995; Wolff & Peterson 1998).

For those species in which infanticide is a threat, the ways in which the behavior may impact on social relationships and overall organization will differ depending on whether females are dispersed (thus facing threats from unfamiliar intruders) or gregarious (facing additional threats from fellow group members). I present several case studies for each general social category (dispersed/territorial females, female aggregations, multi-female groupings, and cooperative breeding groups), discussing varia-tion in infant vulnerability, potential benefits to the perpetrator (in light of the above hypotheses), and maternal counterstrategies. I argue that

Table 17.2. *Dispersed/territorial females*

Vulnerability to infanticide by females
Altricial young
Young occasionally left alone/undefended

Benefits to perpetrator
Nutrition from consumption of infant/cannibalism (predation hypothesis)
Increases access to foraging area and/or nest sites (resource competition)
Elimination of future competitors (resource competition)

Counterstrategies and implications for social system
Selection for maternal aggression (from parturition until independence)
Selection for female territoriality (especially from parturition until independence)

Notes:
For examples, see Hoogland 1995; Hrdy 1979; Maestripieri 1992, Sherman 1981; Stevens 1998; Wolff 1985, 1997.

threats from conspecific females have probablly shaped the reproductive strategies and social systems of these species. It should be noted that this is a preliminary analysis of a large body of data. I review both previously discussed hypotheses alongside new, often speculative, hypotheses. Both, I hope, will instigate further investigations into the impact of infanticide by female mammals.

Dispersed/territorial females

Background and vulnerability

Among the documented cases of infanticide, only among the rodents do females exclude other females from their home range during birth and lactation (L. Digby *et al.*, unpublished data. Note: I am including here only those species where all conspecific adult females are excluded. Group-living species that are territorial are discussed in a later section). Infants born to solitary females or into monogynous social groups would, presumably, be at low risk of being killed by conspecific females. Infanticide in this context is typically opportunistic, with one female invading the territory of another and, in many cases, partially or totally consuming infants (e.g., Columbian ground squirrels, Balfour 1983, Stevens 1998). Infants in these species are altricial and thus helpless against intruders and vulnerable to attacks (Table 17.2). Although observations of infanticide are often hampered by the fact that most attacks occur inside the den

of the victim, field experiments, den excavations and captive manipulations have demonstrated that infanticide is indeed occurring, and can be
invoked under appropriate conditions (for reviews, see Labov *et al.* 1985;
Blumstein, Chapter 8; see also Hoogland 1995). While dens are typically
defended by mothers, even brief departures to feed or patrol territory
present opportunities for marauders to eat infants or drag them away
(e.g., Columbian ground squirrel, Stevens 1998; white-footed and deer
mice (*Peromyscus leucopus* and *P. maniculatus*), Wolff 1985; bank voles
(*Clethrionomys glareolus*), Ylönen *et al.* 1997).

Potential benefits to the perpetrator

Most cases of "opportunistic" infanticide involve some degree of
cannibalism (see above) and are therefore best explained in terms of the
predation (exploitation) hypothesis with infants serving as a food source
(for a review, see Elwood 1992). Additional cases involve competition over
a limited resource or the avoidance of misdirected parental care as illustrated in the case of the black-tailed prairie dog (for a review, see
Hoogland 1995).

In black-tailed prairie dogs, normally group-living females will isolate
themselves soon after conceiving and maintain an exclusive territory
until the time when infants emerge from their burrows (Hoogland 1995).
Although normally protected by the presence of their mother in the den,
infants left alone while the mother is foraging may be attacked by fellow
members of the coterie, often kin (Hoogland 1985, 1994, 1995). Seventy-
eight percent of the perpetrators are lactating when they commit infanticide, and, because females breed fairly synchronously within a given
coterie, the infants of the perpetrator (when she has them) are generally
the same age as her victims (Hoogland 1995). This pattern, combined with
the common occurrence of communal nursing/milk theft, suggests that
infanticidal females may be avoiding misdirected parental care. In addition, infanticide in black-tailed prairie dogs may result in nutritional
benefits to the perpetrator (predation hypothesis), increased access to foraging area (resource competition), increased access to the victim's mother's help (in terms of nursing and defense of the coterie territory; resource
competition; helper coercion), and decreasing the chances of infanticide
directed toward the perpetrator (Hoogland 1995).

Interestingly, Hoogland's prairie dog study is one of the few that has
been able to determine whether or not infanticidal females have a higher

reproductive success than non-infanticidal females. However, there is no significant difference between the two types of female (Hoogland 1985, 1995). While there is no significant increase in the number of offspring produced by infanticidal females, there is also no decrease, suggesting that this behavioral pattern may have attained equilibrium in this population, with counterstrategies preventing the perpetrator strategy from becoming more widespread than it already is. This "equilibrium" hypothesis remains to be tested.

Counterstrategies and implications

Infants in monogynous social systems are at the greatest risk when non-maternal females invade their territory and are able to access their den or burrow (Wolff 1985; Hoogland 1995; Stevens 1998). Two counterstrategies to such a threat would be maternal aggression (to prevent access to the den/burrow) and, on a broader scale, territoriality (Ostermeyer 1983; Wolff 1985; Maestripieri 1992; Wolff & Peterson 1998; Table 17.2). It has been argued that maternal aggression is primarily a response to con-specific threats of infanticide, rather than the often assumed threat by predators (Maestripieri 1992). Indeed, maternal aggression against con-specifics is most often described for the same taxa in which infanticide (by both males and females) has been observed (primates, carnivores, pin-nipeds, and rodents; Maestripieri 1992). Interestingly, in many rodent species conspecific threats come more often from females than from males (for a review, see Ebensperger 1998a). Aggression against females (and to some degree males) increases during the perinatal and lactational periods in many species, and ceases following the experimental removal of young (Ebensperger 1998a).

Infanticide may also be the driving force behind territoriality in small mammals (Wolff 1993; see also Blumstein, Chapter 8). Wolff & Peterson (1998) argued that territoriality is best explained in terms of a deterrent to infanticide. Evidence in support of this hypothesis includes a suite of characteristics more commonly associated with infanticide risk (altricial young, parking of the infants in nests or burrows) than with the defend-ability of food resources (limited, clumped foods) (Wolff 1993, 1997; Wolff & Peterson 1998). Lone females (whether solitary or in monogynous groups) may represent a baseline social system where maternal aggres-sion and territoriality can be used to avoid the threat of infanticide by conspecifics (in addition to avoiding the costs of food competition).

Species that form female aggregations

Background and vulnerability

In some species, the formation of large aggregations of females can increase the risk of infanticide due, in part, to the close proximity of aggressive female conspecifics. Examples of this phenomenon come primarily from the pinnipeds (nine species), with reports of over 100 pups dying at least in part due to wounds inflicted by adult females (LeBoeuf & Campagna 1994; for a review, see L. Digby *et al.*, unpublished data). The extent of this phenomenon varies between species and between years, explaining 6% to 73% of infant deaths (LeBoeuf *et al.* 1972; LeBoeuf & Briggs 1977).

The northern elephant seal, one of the best known of the pinniped species, is a good illustration of the reproductive costs facing females living in aggregations. During the breeding season, female elephant seals form large one-male harem groups (up to several hundred females on a single beach) that are maintained for 4–5 weeks (LeBoeuf & Briggs 1977). The benefits of such a gathering may include protection from predators, increased chance of breeding with a high-quality male, and protection from harassment by non-dominant males (LeBoeuf & Briggs 1977; LeBoeuf & Campagna 1994). But the costs are also very high, with infant mortality increasing with female density (LeBoeuf *et al.* 1972; Christenson & LeBoeuf 1978; Riedman & LeBoeuf 1982). Attacks from females are, at least in part, responsible for 58% to 73% of all infant deaths in this species (LeBoeuf & Briggs 1977).

In general, elephant seal pups are well protected from other females by their close proximity to their mothers (Christenson & LeBoeuf 1978; Table 17.3). But if pups become separated from their mothers, they will eventually attempt to suckle on other cows. In response, an adult female will attack an unfamiliar pup, biting it, shaking it and/or tossing the infant away from herself. Such attacks are often repeated as the infant again attempts to suckle on other females (e.g., LeBoeuf & Briggs 1977; Christenson & LeBoeuf 1978). Even if the infant does not become separated, it may be attacked by females that are more aggressive than its mother (Christenson & LeBoeuf 1978; McCann 1982). In these cases, less aggressive mothers simply fail to defend their young and infants are eventually driven away (thus subjecting them to further attacks). Similar examples have been seen in southern elephant seals (*M. leonina,* McCann 1982), gray seals (*Halichoerus grypus,* Anderson *et al.,* 1979; Coulson &

Table 17.3. *Female aggregations*

Vulnerability to infanticide by females
Altricial young
Separation from mother
Close proximity of potentially infanticidal females
Mother unwilling to defend young/less aggressive mother
Benefits to perpetrator
Avoidance of misdirected parental care (resource competition)
Counterstrategies and implications for social system
Selection for maternal aggression (from parturition until independence)
Selection for female territoriality (transient and on small scale)
Benefits of aggregations (e.g., predator avoidance, access to high-quality mates, avoidance of harassment from males) must outweigh the cost of female–female aggression and statistical probability of losing infant

Notes:
For examples, see Christenson & LeBoeuf 1978; LeBoeuf *et al.* 1972, LeBoeuf & Campagna 1994; LeBoeuf & Briggs 1977.

Hickling 1964), and Stellar sea lions (*Eumetopias jabatus*, Bruemmer 1994) and together have been labeled "trauma/separation syndrome", with most victims dying from a combination of physical injury and dehydration (LeBoeuf & Briggs, 1977: 193). In species where the mothers leave infants to forage, unprotected infants again are vulnerable to fatal attacks (e.g., South American fur seal (*Arctocephalus australis*), R. Harcourt 1992).

Potential benefit to the perpetrator

Most cases of infanticide by female pinnipeds appear to be motivated by the avoidance of misdirected parental care (Table 17.3). Milk is an expensive and highly limited resource in many pinnipeds (e.g., northern elephant seal infants gain three to five times their birth weight over the course of a 1-month suckling period; LeBoeuf *et al.* 1972; Reiter *et al.* 1978). As the females of many species fast (or go for long periods between foraging) during this period, milk theft could have significant negative effects on a female's ability to raise her own young successfully (LeBoeuf & Campagna 1994). The avoidance of misdirected parental care is therefore an important reproductive strategy in this species.

Counterstrategies and implications

A key counterstrategy to avoiding misdirected parental care is strong pup recognition. In addition to visual and olfactory cues, both mother and infant use calls to locate and identify one another (LeBoeuf *et al.* 1972; Reidman & LeBoeuf 1982). This recognition also allows aggressive females to attack unrelated infants without risking accidental injury to their own young. As with solitary females, threats from conspecific females appear to be the driving force behind maternal aggression and "territoriality" (elephant seals defend areas immediately surrounding themselves; Christenson & LeBoeuf 1978). This pattern also suggests that aggression from conspecifics, especially other females, make aggregations a costly strategy. While larger territories might appear to be an appropriate counterstrategy to threats from conspecific females, beaches that are protected from potential terrestrial predators are limited, resulting in high densities of females (which increase during high tides and storms; Christenson & LeBoeuf 1978). In addition, the morphology of these females limits their mobility on land, preventing them from economically defending areas much larger than a few body lengths in radius. Presumably, the benefits of living in such aggregations (e.g., predator protection, increased mate quality, and reduction in male harassment, see above) outweigh the costs of possible attacks by conspecific females.

Group-living species: infanticide by female immigrants and intruders

Infants born into permanent multifemale groups are at risk from two kinds of infanticidal female: intruders from outside the group (similar to that seen in dispersed/territorial females) and females within their own group (discussed in a later section).

Background and vulnerability

Opportunistic infanticide from intruders is relatively rare in group-living species (approximately 10% of all infanticide cases in group-living species, $N = 26$; L. Digby *et al.*, unpublished data). Examples include both a female chimpanzee (*Pan troglodytes*) seen eating an infant from a neighboring troop (Pusey *et al.* 1997), and nomadic female lions who attacked and cannibalized infants while crossing through territories (Schaller 1972; Bygott & Handby, pers. comm., as cited in Packer & Pusey 1984). Such cases are unusual at least in part because immigration into groups by females is relatively rare owing to intolerance of unfamiliar females (e.g., black-tailed prairie dogs, 2 infanticides by immigrant females out of 591 litters; Hoogland 1995). Thus group-living and the potential for

Table 17.4. *Multifemale groups*

Vulnerability to infanticide by females
Altricial young
Infants occasionally left alone/undefended
Close proximity of potentially infanticidal females
Allomaternal care/passing of infants between group members
Variance in power (social or physical) among group members

Benefits to perpetrator
Removal of future competitors (resource competition)
Increase/maintenance of social status (resource competition?)
Avoidance of misdirected parental care (resource competition)
Occasional nutrition from consumption of infant/cannibalism (predation hypothesis)

Counterstrategies and implications for social system
Selection for maternal aggression (from parturition until independence)
Selection for territoriality (exclusion of unrelated females)
Selection for limited female dispersal (high cost of encountering/living with unrelated
    females)
Selection for coalitions with either male or female conspecifics (for protection) benefits
    of living in a group (e.g., predator avoidance, access to food resources, defensive
    coalitions) must outweigh the cost of female–female aggression and statistical
    probability of losing infant (as well as food competition)

Notes:
For examples, see Hrdy 1979; Hoogland 1985; Maestripieri 1992; Wasser & Starling 1986;
Wolff & Peterson 1998; L. Digby *et al.*, unpublished data.

defensive coalitions reduces the risk of infanticide. Where female
immigration is more common, so are opportunistic killings. For
example, Belding's ground squirrel females who have lost their young
will emigrate and attempt to kill the young in another burrow and subsequently take it over (Sherman 1981, 1984). Again, infants in den-living
species become vulnerable when left unattended (Table 17.4). Infant primates should be less vulnerable than many mammalian infants because
they are habitually carried during the early stages of development
(Treves, Chapter 10). Infant primates become vulnerable only when their
mothers can be overpowered (e.g., by coalitions of females, or by more
socially dominant females) and their infants taken from them (see within-
group cases below).

Potential benefits to the perpetrator and
counterstrategies

These cases are best explained either by the predation (exploitation)
hypothesis or by competition over a limited resource (e.g., nest sites;
Table 17.4). Although infanticide by outside females is infrequent, it is

reasonable to assume that such threats may still be influencing female reproductive strategies. Maternal aggression and protectiveness is common in group-living species (Maestripieri 1992), and among many primates and social carnivores, females tend to focus territorial efforts against other females (e.g., gibbons, Mitani 1984; lion tamarins, French & Inglett 1991; marmosets and titi monkeys, Anzenberger 1993). Previously, such observations in the wild, backed up by numerous intruder introduction studies under captive conditions, have been interpreted to reflect female competition over food or mates (for reviews, see Cheney 1987; Silk 1993). The data on infanticide suggest that intrasexual aggression may also reflect the threat of attacks on infants (Maestripieri 1992; Wolff & Peterson 1998).

Similar to the patterns described as a result of male infanticide (van Schaik & Dunbar 1990; van Schaik & Kappeler 1997), we might speculate that reproductive competition among females may, in turn, be influencing the costs of female dispersal (female immigrants are a potential threat, and therefore are not tolerated) and, ultimately, may be influencing social structure (e.g., the lack of female immigration results in groups comprising female relatives). A preliminary glimpse at primate species in which females do disperse (e.g., *Alouatta* spp. *Colobus badius, Gorilla* spp.; for a review, see Pusey & Packer 1987a), suggests that female food competition in those species may be reduced because of primarily folivorous diets (but see Watts 1991b), and thus the gains resulting from infanticide by females would be limited. Nevertheless, there are exceptions to this pattern (e.g., *Papio hamadryas, Pan troglodytes*) that future analyses need to explain.

Within-group infanticide

In group-living species, threats of infanticide can also come from the female members of the victim's own group. Within-group infanticide can take very different forms, from unmistakable physical attacks (direct infanticide), to enforced neglect or repeated harassment of the mother, which eventually leads to cessation of lactation and starvation of the infant (indirect infanticide). Both types appear to be the result of a skew in power (both physical and social) among females in a given group, and can sometimes be interpreted as part of an overall pattern of harassment of subordinate females (e.g., Silk *et al.* 1981; Table 17.4). Direct and indirect cases of infanticide occur in a variety of mammalian taxa, including rodents, carnivores and primates (for a review, see L. Digby *et al.*, unpublished data; see also Ebensperger 1998a), with more extreme cases occurring in cooperatively breeding species (discussed in a later section).

Direct cases: background and vulnerability

Among the primates, direct forms of infanticide tend not to involve closely related females. Perhaps the best known case of direct infanticide in the primates involves a mother and daughter pair of common chimpanzees who were observed to attack new mothers, pull off their newborn infants and subsequently consume them (Goodall 1986). The pair was known to kill at least three infants, and are suspected of killing as many as seven more infants over the course of 4 years. These females were only successful while they worked together to overpower the mother of the victim. Following the death of the older female, the younger female did attempt at least one attack but failed to gain access to the infant. Although intraspecific predation may, in part, explain within-group infanticide in chimpanzees, cannibalism of infants by female primates (under free-ranging conditions) is rare (Hiraiwa-Hasegawa 1992), reported only for chimpanzees (Goodall 1986) and black lemurs (Andrews 1998).

Most cases of direct within-group infanticide reflect general patterns of dominance relations between female group members. For example, in a population of Tonkean macaques (*Macaca tonkeana*), a female coalition involving the matriarch of the dominant matriline attacked an unrelated infant of a more subordinate female (jaw injuries eventually resulted in the death of the infant; Muroyama & Thierry 1996). Similar cases have been described for yellow baboons (Wasser & Starling 1986) and common black lemurs (*Eulemur macaco macaco*; Andrews 1998). In the latter case, the attack occurred following the death of the group's dominant female and the perpetrator subsequently rose to the top ranking position (Andrews 1998).

Direct cases: potential benefits to the perpetrator

These cases of infanticide are somewhat difficult to explain in terms of Hrdy's (1979) four hypotheses. Although general resource competition may be playing a role (e.g., limited food availability for recently weaned infants: Wasser & Starling 1986), it is difficult to quantify food competition and thus the impact of infanticide on the perpetrators' reproductive success. Another possible benefit to infanticide by females in these species may be the maintenance of dominance status, which itself has potential benefits in terms of general resource and reproductive competition.

Although striking, cases of direct infanticide within groups are relatively unusual among primates, reported for only five species resulting in 12 deaths (direct infanticide is far more common in cooperatively breeding species, see below).

Indirect cases: background and vulnerability

In comparison with direct cases of infanticide, indirect cases of infanticide have been described for at least 10 species in 15 populations of primates, including prosimians, New World monkeys and Old World monkeys (27+ infants killed; additional cases are described for cooperatively breeding species below; L. Digby *et al.*, unpublished data). Cases of indirect infanticide are quite consistent between species, with a dominant female "kidnapping" a non-weaned infant from a more subordinate female and preventing the mother from retrieving her young (e.g., Collins *et al.* 1984, Brain 1992; for a review, see Hrdy 1976). These cases, by definition, do not involve any direct injury inflicted upon the infant by the perpetrator. Most infants that are "used" by adult females (most often simply carried by the females; Hrdy 1976) are eventually returned to their mothers unharmed, but there are multiple cases where kidnapped infants die of starvation or dehydration as a result of prolonged separation from their mothers (e.g., squirrel monkeys (*Saimiri sciureus*), Rosenblum 1971 cited in Hrdy 1976; Campbell's guenon (*Cercopithecus campbelli lowei*), Bourlière *et al.* 1970; yellow baboons, Shopland & Altmann 1987; chacma baboon (*Papio ursinus*), Collins *et al.* 1984; rhesus macaque (*Macaca mulatta*), Maestripieri & Carroll 1998). Additional cases of indirect infanticide result from dominant females repeatedly harassing or attacking subordinate females. Although the infants are not the focus of the harassment, infants may fall from their mothers, or be injured during attacks on the females resulting in their deaths (e.g., ringtailed lemurs (*Lemur catta*), Andrews 1998; Jolly *et al.* 2000; Vick & Pereira 1989).

Indirect cases: potential benefits to the perpetrator

Both types of indirect infanticide appear to be the result of highly skewed social and physical power within groups, with the infants of subordinate females being at greater risk of infanticide than those infants born to dominant females. As with the direct cases outlined above, benefits to the perpetrator may accrue simply from the maintenance of dominance within these groups (e.g., higher infant survival, shorter interbirth intervals; Silk 1993) in addition to possible benefits accrued from the infant handling itself (e.g., maternal experience, buffer from aggression; Hrdy 1976). Alternatively, indirect infanticides may simply be the by-product of infant "attractiveness" in these species (Paul 1999; Silk 1999). Again, further study of these indirect cases of infanticide is needed to determine the overall influence of these events on the reproductive success of the perpetrators.

Direct and indirect cases: counterstrategies and implications

Both direct and indirect infanticide by females within groups could be influencing the evolution of reproductive strategies and grouping patterns in these species. Given the differential threat to offspring born to dominant versus subordinate females, we predict maternal strategies will vary between females of different ranks. To limit infanticidal threats, subordinate females should be more protective of their young, not allowing them to wander and preventing other group members from getting too close (e.g., marmosets (see below), Digby 1995a; yellow baboons, Altmann 1980; bonnet macaques (*Macaca radiata*), Silk *et al.* 1981). Habitual infant-carrying would also be an effective way to increase protection for young. Whether infant-carrying is a specific counterstrategy or the result of other selective factors, it should result in primate infants being less vulnerable to attacks than many other species of mammals (Treves, Chapter 10). In addition, females may deter attacks from conspecifics by forming coalitions or "friendships" with either males or females from their own group (e.g., Smuts 1985).

Threats from within groups tend to go across matrilineal lines (avoiding inclusive fitness costs); thus there are benefits to maintaining groups with female kin via female philopatry (see above). Again, this is a speculative hypothesis that needs to be tested further, but, with one exception, infanticide by female primates has not been observed in species in which family groups (rather than multiple matrilines) are the norm. The exception, the marmosets, are cooperative breeders, a social system that may give rise to intense female reproductive competition.

Cooperative breeding species

Background and vulnerability

Some of the clearest examples of how infanticide may be influencing and shaping social structures come from cooperatively breeding species. A subset of these species restricts reproduction to a single, behaviorally dominant pair ("singular breeders"; Solomon & French 1997: 3; for a review, see Emlen 1991), and in several species subordinate females may be physiologically inhibited from ovulation (e.g., Abbott 1984, 1993; Creel & Waser 1991). Thus social status can have a profound effect on female reproductive opportunities. But, even in those species where singular breeding is the norm, subordinate females do occasionally attempt reproduction (e.g., carnivores, Creel & Waser 1991, Doolan & Macdonald 1997b, Macdonald & Moehlman 1982; marmosets, Digby 1995a). When this occurs, resources, especially the limited time and energy of helpers, have to be shared among multiple litters. The higher the degree of overlap between litters, the more limited group resources will become.

Examples of infanticide in cooperative breeders come from the New World primate subfamily Callitrichinae, as well as several species of social carnivores. Infants in these groups are vulnerable primarily because they are often passed among group members (e.g., carrying in marmosets) or are fed by a number of different group members (e.g., food sharing with both mother and infants in carnivores; Table 17.5). Among the marmosets, cases are remarkably similar, with socially dominant females killing offspring born to subordinate females (e.g., Alonso 1986; Digby 1995a; Kirkpatrick-Tanner et al. 1996; Yamamoto et al. 1996). In the one case where a subordinate female was observed killing the offspring of the dominant female, the females subsequently reversed social status, with the perpetrator attaining the top-ranking position (Roda & Roda 1988; Roda & Mendes Pontes 1998).

Infanticide in social carnivores follow a similar pattern (for reviews, see Packer & Pusey 1984; Rasa 1994). In wild dogs (*Lycaon pictus*), dingos (*Canis familiaris dingo*), brown hyaenas (*Hyaena brunnea*), and meerkats (*Suricata suricatta*) dominant females have been observed to harass and attack subordinate females that give birth (Frame et al. 1979; Owens & Owens 1984; Corbett 1988; Clutton-Brock et al. 1998). In addition to direct attacks, infants may die as a result of enforced neglect (indirect infanticide) due to dominant females preventing group members from sharing food with subordinates, ultimately resulting in cessation of lactation and starvation of pups (e.g., Frame et al. 1979; Owens & Owens 1984; Macdonald 1996). As subordinate females are forced to hunt for themselves, their pups are left alone in their dens, making them vulnerable to attack (e.g., van Lawick 1974). One exception to this general pattern has been described for meerkats, where both dominant and subordinate females have been observed attacking and killing the young of other

Table 17.5. *Cooperative breeding groups*

Vulnerability to infanticide by females
Altricial young
Close proximity of potentially infanticidal females
Allomaternal care/passing of infants between group members
Occasional caching of infants/infants left alone in den
Variance in power (primarily social) between females

Benefits to perpetrator
Helper coercion (resource competition)
Group energy/attention focused on perpetrators young (resource competition)
Removal of future competitors (resource competition)
Increase/maintenance of social status (resource competition?)
Avoidance of misdirected parental care (resource competition)
Occasional nutrition from consumption of infant/cannibalism (predation hypothesis)

Counterstrategies and implications for social system
Selection for reproductive suppression in subordinate females (to avoid wasted
    reproductive effort)
Maintenance of single breeding female under most circumstances
Subordinates breed only when the benefits/chance of success outweighs costs to future
    reproductive success (e.g., when females are older or when dispersal opportunities
    very limited or when birth spacing limits overlap between litters)
Selection for maternal aggression (from parturition until independence)
Selection for territoriality (exclusion of unrelated females)
Selection for limited female dispersal (high cost of encountering/living with unrelated
    females)
Benefits of living in a group (e.g., predator avoidance, access to food resources,
    defensive coalitions) must outweigh the cost of female–female aggression and
    statistical probability of losing infant (as well as food competition)

Notes:
For examples, see Creel & Waser 1991; Digby 1995a; L. Digby *et al.*, unpublished data;
Hrdy 1979; Macdonald & Moehlman 1982; Packer & Pusey 1984; Wasser & Barash 1983.

female group members (Clutton-Brock *et al.* 1998). But even in this
species, indirect evidence suggests that dominant females commit infanticide more frequently than do subordinates (Clutton-Brock *et al.* 1998).

Potential benefits to the perpetrator
In almost all cases of infanticide in cooperative breeders, perpetrators are
either lactating or in the late stages of pregnancy (e.g., van Lawick 1974;
Digby 1995a; Clutton-Brock *et al.* 1998). By committing infanticide, the
perpetrator achieves sole access to the group's resources for its young,
including increased attention from helpers and increased access to foods.
Particularly striking is that victimized females often go on to suckle the
young of the perpetrator (marmosets, Digby 1995a; carnivores, Pusey &

Packer 1984). Together, the pattern suggests general resource and reproductive competition. As many infanticidal females are closely related to their victims, competition over resources is predicted to be severe enough that the benefits accrued from killing closely related young are greater than the cost to the perpetrator's inclusive fitness. Future studies could be aimed at investigating these fitness trade-offs. In addition, females may be avoiding misdirected parental care and the threat of infanticide.

Counterstrategies and implications

The consequences of the threat of infanticide by female conspecifics in cooperatively breeding species are probably playing a key role in the social structure and organization of these groups. Subordinate females may forego current reproduction (in many cases physiologically inhibiting ovulation) in response to threats from more dominant females. By delaying reproduction, they avoid wasted reproductive effort (Wasser & Barash 1983; Creel & Wasser 1991; Digby 1995a). Group size, in turn, will be determined by reproductive opportunities. Females that are third or more in line for the breeding position in a group may opt to emigrate. Only those females that have little chance of successfully reproducing (often older females) will attempt to reproduce as a subordinate, perhaps making the best of a poor situation (e.g., Creel & Wasser 1991). As subordinates may sometimes be a threat (either because of the risk of misdirected parental care or because of infanticide risk), they may be expelled from their group by more dominant females (e.g., meerkats, Clutton-Brock et al. 1998). As such, females are controlling both group composition and the breeding opportunities within a given group.

Conclusions

One goal of this chapter is to review the variety of ways in which infanticide can manifest itself and its potential impact on the evolution of social systems. Infanticide by females other than the mother has been observed in a wide variety of species (L. Digby et al., unpublished data). Unlike the threat of infanticide by males, conspecific females may present an almost constant threat for breeding females, with potentially infanticidal females coming from both outside and within groups.

Within groups, the threat of infanticide may be shaping social relations and dispersal patterns. In several species infants born to subordinate females are more vulnerable to attacks, thus favoring more protective maternal styles, or self-imposed isolation following a birth. In species where competition reaches an extreme, such as cooperative breeding systems, the threat of infanticide may be so great that a subordinate female will forego reproduction while in the presence of a more dominant female, waiting for a future opportunity when she may become the sole breeding female (Creel & Wasser 1991; Digby 1995a). Even in those species in which dominance does not play a major role in infanticide (e.g., species with dispersed females), maternal aggression is likely to be a response to threats from conspecific females (Maestripieri 1992). In most social groups, subordinate females or matrilines may be responding to conspecific threats by emigrating and forming new groups where the threat of infanticide by females, at least initially, is reduced. When formation of new groups is not possible, the best option may be to group with closely related females (female philopatry), where the potential benefits of inclusive fitness effects are more likely to outweigh the costs of reproductive competition between females. Unfamiliar females appear to be a threat to both solitary and group-living females. Our increased understanding of infanticide by females is beginning to suggest that territoriality may be a counterstrategy more to threats from conspecific females than it is to food competition (Wolff 1993).

The threat of infanticide increases the costs of grouping for many species of mammals. In addition to estimating the costs and benefits of feeding competition and foraging efficiency, group size models should also factor in the cost of threats from conspecific females. Such threats may have presented an unsurmountable cost to many species, making social grouping untenable. For those taxa that are gregarious, the potential cost of infanticide needs to be balanced by clear-cut benefits, most likely in terms of predator protection, foraging efficiency, cooperation among kin, or infant protection (group defense) from possible marauders.

While all these elements of social behavior and social systems may be impacted on in some way by the threat of infanticide by females, it is clearly only *one* factor influencing the costs and benefits of sociality. In formulating models of the evolution of social systems we need to take into account the costs and benefits of as many factors as possible, including food availability and distribution, predator threats, food competition from other species sharing the habitat, climatic variability, social status, and the threats from both male and female conspecifics. Increased reporting of infanticide, and of the *threat* of infanticide by females other than the mother will increase our understanding of the phenomenon and the role it plays in the evolution of reproductive strategies and social systems.

The risk of infanticide by conspecific females is determined by the life history, ecology and social system of a given species as well as the current circumstances. Those infants that are precocial (thus able to evade potential attackers) or well protected during their early stages of development will be less vulnerable to infanticide. The extent of infanticide within a given species will also be determined by the potential benefits to the perpetrator. Of the five hypotheses originally put forth by Hrdy (1979), exploitation (including both "predation" and "live use") and resource competition (including general competition, direct access to food or foraging territory, helper coercion and avoidance of misdirected parental care) are most applicable to infanticide by females. Counterstrategies to the threat of infanticide by females include maternal aggression, infant recognition, female territoriality, female philopatry, intolerance for unrelated females, and, in extreme cases, suppression of reproduction (e.g., ovulation suppression in some cooperatively breeding species). Each of these counterstrategies has probably had a significant influence both on female reproductive strategies (e.g., maternal aggression, coalitions and timing of reproduction) as well as on the evolution of mammalian social systems.

Acknowledgments

I thank Carel van Schaik, Charlie Janson, Sarah Hrdy, Dario Maestripieri, Susan Alberts, Elena Davis, Michelle Merrill, Burney LeBoeuf, Charles Pell, and the members of "BEAST" (Biological Anthropology and Anatomy's behavior discussion group) for their help on various aspects of this manuscript.

18

"The hate that love generated" – sexually selected neglect of one's own offspring in humans

Introduction

Emelina was a great beauty. She was proud, passionate and extremely ambitious. She forced all of the men who asked for her to their knees and only gave herself to the one who had money and power. She had three children. One day a prince galloped into the village on a golden horse. His face had the beauty of a god. He had a generous heart and a noble character. It is said that the man drove the whole village crazy with his magnanimity and that he was able to seduce every woman with a single glance or a simple gesture. He loved women and women loved him. And so Emelina came to know him. He was not just any man, he was a powerful leader and had many men under his command. And he had money. He owned so many sacks of grain that they could have filled the storehouse of the whole village. They met. They made love. They floated in the realm of dreams on the fragile wings of passion. Nevertheless, they were unable to attain perfect happiness because the man was married and polygamous, while Emelina was married and had three children. Their mutual dream was to live together until death parted them, but this dream was unattainable. Emelina compared her husband with her lover. Separating from a husband was simple, but how was she to separate herself from her children? She had to find a way to free herself of them. One day the village was attacked, it was one of those minor, harmless attacks, but serious enough to allow her to implement her devilish plan. During the attack, she locked the three children into the straw hut and set it on fire. Then she began to scream and call the neighbors, but only after she was sure that the children were dead. As she lay with her lover, she breathed a sigh of relief and whispered:

"Now I am free for love". And the man replied: "I will give you new children in whom your beauty will be united with my courage". And they made love with the greatest pleasure in the world.

This story of a triple infanticide is told by Mozambique writer Paulina Chiziane in her novel *Ventos do Apocalipse* (1993). What makes this case so interesting and distinguishes it from the overwhelming majority of ethnologically described infanticides (Daly & Wilson 1984) is the rather unusual combination of perpetrator and motive: the biological mother kills her children because of her sexual desire. If this tale is a poetic interpretation of an evolved human behavioral tendency, we would have a case of sexually selected maternal infanticide – a behavior for which, to the best of our knowledge, no parallels have been described to date among non-human primates. Here (and in other species), sexually selected infanticide is known to be an adaptive strategy of males (or females in species with a sex role reversal, respectively, as in some birds and callitrichids; Digby 1995a, Chapter 17; Veiga, Chapter 9) to redirect female (or male) reproductive effort from offspring by others to their own offspring (this subject will be dealt with by the majority of the other contributions to this volume) and not an affair of the mothers.

Sexually selected infanticide is also not unknown to humans, of course. Studies of historical, traditional and modern societies all document a widespread folk wisdom, namely that in comparison with genetic children, stepchildren are exposed to a noticeably higher risk of being abandoned, unloved, neglected, exploited, or even killed. For example, in Canada the risk (controlled for poverty, family size and maternal age) for preschool-aged stepchildren being killed is 40 to 100 times greater than the comparable value for children who live with their biological parents (Daly & Wilson 1994). In historical Ditfurt (Germany) living with a stepmother in the 17th to 19th century reduced the life expectancy on average by approximately 7 years (Stephan 1992). Numerous other findings on parental discrimination against non-related offspring have meanwhile become available (for reviews, see Daly & Wilson 1984, 1994, 1996). The adaptive logic of this human inclination is easy to understand. Stepparents attempt to direct the parental effort of their partners toward their own offspring and away from the offspring of others. Although this applies equally to stepmothers and stepfathers, there are varying consequences for the evolution of counterstrategies.

Even if the risk associated with "the wicked stepmother" is widely known and has been referred to in many myths and fairy tales, men are relatively powerless when faced with this risk. What chances would men have to increase the nursing care of their partners for children from a first marriage? A certain possibility exists in marriage with the first wife's sister, for whom a neglect of her husband's children from a first marriage would be associated with costs, due to the effect of kin selection. Apart from this measure, there are hardly any possibilities for defense against a stepmother's aggressive tendencies. In contrast, women have more possibilities. Because the risk associated with infanticidal males depends to a decisive degree on their interpretation of paternity, women can have a manipulative impact on the male psyche. They can attempt to convince their partners that they were actually also the fathers of their offspring and thus throttle their aggressive motivation. Studies in Canada (Daly & Wilson 1982) and Mexico (Regalski & Gaulin 1993) show that mothers (and their kin) frequently try to dispel obvious male doubts about paternity by stressing the baby's similarities to the mother's partner. This is especially obvious with first children, i.e., when the parental partnership has not been in existence for very long. Approximately 80% of the mothers then claim that the similarity of the newborn is larger with the father than with themselves.

One can only speculate whether babies systematically do look more similar to their fathers than to their mothers, because historically it would be advantageous for them to signal their genetic parentage in order to elicit parental investment. In actual fact there are indications along these lines (Christenfeld & Hill 1995), which, however, could not be replicated (Brédart & French 1999). At the moment therefore, the extremely exciting question of whether or not the evolutionary history of male infanticide could possibly have led to genetic counterstrategies on the part of the babies (or their mothers?) is still open. The evolution of menopause might perhaps even be understood as a spectacular outflow of evolutionary "arms race" between infanticidal tendencies of males and protective counterstrategies of females. Under historical mating conditions, if the men are much older than their wives on average, and accordingly were likely to be the first to die, the surviving wives attracted other men. Therefore, the wives' children, especially the youngest, were exposed to a special risk. Infecundity would doubtless have been able to reduce this risk, which in turn might have favored the evolution of menopause (Turke 1997).

Maternal infanticide originates from other contexts, for example in connection with economic–ecological fluctuations, high economic or reproductive opportunity costs of raising offspring, specific life history constraints and from several additional scenarios, when a negative cost–benefit analysis argues against the raising of a child (Daly & Wilson 1994). The majority of these cases is to be understood against the background of a maternal psychology whose adaptive features have been shaped by natural selection: in view of the predominant economic, social and biographical conditions, maternal underinvestment can best be understood as a strategic measure to optimally allocate the always limited reproductive potential in the interests of as large a lifetime reproductive success as possible. This happens then at the expense of individual children, who – for whatever reasons – have the wrong sex, too young a mother, too many siblings, or do not appear to be vital enough or were simply born at the wrong time. As such children have a reduced reproductive value for extrinsic or intrinsic reasons, investment in these children is reduced or completely discontinued.

Even though human infanticide is a very heterogeneous phenomenon with respect to both personal motives and triggering socioecological framework conditions, at first glance it nevertheless appears as if the diversity of the phenomena could be pressed into a simple causal explanatory scheme through the simple equation of "male infanticide = sexually selected" and "female infanticide = naturally selected". This is a much too simple categorization, however, which, though perhaps applicable to other species, does not do justice to human conditions, because human offspring can also be neglected because they have the wrong father. Thus sexual selection could also form the functional background for differential maternal solicitude, something that has been less systematically considered to date. To be explicit, the sexual selection hypothesis of infanticide has conventionally focused on intrasexual selection, i.e., mainly male–male competition (Hrdy 1979), while we would like to add the idea that intersexual selection, i.e., mate choice, may also play a significant role in explaining human infanticide.

In European social history there was (and still is, albeit in somewhat changed form; Daly & Wilson 1994) a life context in which the "problem of the wrong father" was particularly noticeable, namely the situation of unmarried mothers. Apart from the clerical and social ostracism and mental anguish, to which the respective women were subjected, and which – as is known – has often ended tragically (Richter 1998), these women were also faced with a dilemma that can be expressed in the sober terminology of life history theory. In view of the relatively low chances of marriage that unmarried mothers have, should they continue to remain unmarried and concentrate their life effort on raising their illegitimate children, or should they instead attempt to improve their marriage chances by getting rid of their children? In other words, should they attempt to maximize their maternal effort or their mating effort? Both options are cost intensive. To remain unmarried meant having to lead a life without any reliable male support, therefore to be more directly exposed to ecological and economic fluctuations and crises and at best to leave behind children with below-average social chances. On the other hand, infanticide or child neglect could permanently ruin a woman's social reputation and counteract the actual purpose, namely improving one's mate value while creating the risk of total reproductive failure. In the following case study, we wish to pursue the social reality of this dilemma and to first inquire which mating and investment decisions unmarried mothers made in a German parish (Ditfurt). On the basis of a second set of data (Krummhörn), we then pursue the question of with which life history consequences these decisions coincided.

Unmarried mothers and their offspring in Ditfurt (Sachsen-Anhalt, Germany, 1655–1939)

The basis for this analysis is the family reconstitution study of Ditfurt, a relatively large Protestant rural community northeast of the Harz Mountains. In 1804, the community had 1790 inhabitants and by 1905 this number had increased to 2316. Particulars of the political, cultural, agricultural and population history of Ditfurt as well as on the sources evaluated, their scope, their quality and the nature of the data collection are found in great detail in Stephan (1983). It may be sufficient here to indicate that the results depicted below are based on the evaluation of the church registers and records (1655–1939).

Table 18.1 indicates the prevalence of illegitimate births. Families which were established by marriage in Ditfurt and for which the deaths of both spouses are documented in Ditfurt are likely to have been permanent residents, which is why their reproductive histories can be considered to be fully "under observation" (=completely known families). Then there are the immigrant, emigrant and transit families, from whom either the beginning of their marriage, the end of their marriage or both are unknown, as well as unmarried mothers with their children.

The remarkable rise in illegitimacy in the 19th century, as is also

Table 18.1. *Relative frequencies of unmarried mothers and their offspring in Ditfurt (1655–1939), according to P. Stephan (unpublished data)*

1	2	3	4	5	6	7
	All families		Completely known families		Unmarried mothers	
Cohort	No. of families	No. of children	No. of families % of 2	No. of children % of 3	No. of mothers % of 2	No. of children % of 3
1655–1684	235	1097	144 61.3	864 78.8	9 3.8	9 0.8
1685–1714	350	1447	232 66.3	1230 85.0	25 7.1	31 2.1
1715–1744	451	1573	319 70.7	1266 80.5	47 10.4	63 4.0
1745–1774	409	1491	334 81.7	1257 84.3	29 7.1	31 2.1
1775–1804	449	1828	367 81.7	1629 89.1	20 4.5	24 1.3
1805–1834	489	2063	383 78.3	1822 88.3	50 10.2	63 3.1
1835–1864	608	2341	392 64.5	1923 82.1	89 14.6	103 4.4
1865–1894	691	2554	401 58.0	1881 73.6	138 20.0	165 6.5
1895–1924	786	1559	439 55.8	972 62.3	156 19.8	183 11.7
1925–1939	449	615	215 47.9	435 70.7	86 19.2	98 15.9

known for other European regions (e.g., Sweden, Brändström 1996; Bavaria, Lee 1977; Scotland, Blaikie 1993) is traced by P. Stephan (unpublished data) to a change in attitudes toward premarital sex, which was probably triggered by economic changes. This aspect is not discussed any further in this chapter. What is more interesting for our subject is, first of all, a comparison of the infant and child mortality of legitimate and illegitimate offspring. Unfortunately, this comparison cannot be done straightforwardly. While mortality can be measured exactly in completely known families, this is not so for illegitimate children: frequently, unmarried mothers left their homes more or less quickly after giving birth in order to avoid social degradation so that their future fate and that of their children remain unknown (cf. column 10 in Table 18.2). The mortality rates in Table 18.2 are, therefore, only very conservative minimum estimates for illegitimate children. Despite this data-related restriction, infant mortality among illegitimate children in the three cohorts is higher than that of legitimate children (Z-test, $Z = 10\,312$; $P < 0.0001$). The actual difference is most likely to be considerably larger in view of the unknown number of infants who died elsewhere.

The increased mortality of illegitimate infants in Ditfurt is not surprising, because this result fits very well into what could be expected from comparable studies in other regions of Europe (e.g., Sweden, Brändström 1996; Lombada, Portugal, Abade & Bertranpetit 1995; Galicia, Spain, Fuster 1984; England, Wrigley *et al.* 1997). We now look at the background to this finding.

We compared the fate of those illegitimate children whose mothers later married, and restricted ourselves for reasons of data density to the time frame 1805–1939. The father of the illegitimate child was named in the baptismal register if he was known and if he acknowledged paternity. On the other hand, if the father was unknown or denied paternity, the corresponding entries are missing in the baptismal register. Table 18.3 documents how eminently important this question was for the survival of the children affected. If the mother married the man named in the baptismal register (on average 1.1 years after giving birth), significantly fewer children died (on average 3.6 years after giving birth) in comparison with if the mother married a man other than the acknowledged sire. Infant mortality then rose to approximately six times the value that is measured when father and mother did marry after the child's birth ($Z = 7276$; $P < 0.0001$).

The cohort 1895–1924 permits an even stricter test, because, owing to an administrative innovation, during this time, a subsequent entry was made to the baptismal register if the later husband legitimized the child during the course of a civil wedding ceremony. Here too, the risk of dying as an infant was almost six times higher for children with unknown fathers than for children whose mothers married the fathers who acknowledged paternity (Table 18.4; $Z = 4121$, $P < 0.0001$). Infant deaths of illegitimate children from mothers marrying a man other than the acknowledging father occurred before later marriage. Hence, stepfather violence plays no role here.

What is remarkable, however, is not only this still rather robust difference, despite the lower number of cases, but also the fact that the mortality of the legitimized children (only 69.0‰), is below the infant mortality of legitimate children (171.8‰). This permits, in our view, only one

Table 18.2. *Infant and child mortality of legitimate and illegitimate offspring in Ditfurt (1655–1939), according to P. Stephan (unpublished data)*

1	2	3	4	5	6	7	8	9	10
	Children from completely known families				Illegitimate children				
Cohort	N	‰ infant mortality	‰ deceased during childhood (age 1–18 years)	‰ survived 18th birthday	N	‰ infant mortality (minimum)	‰ deceased during childhood (age 1–18) (minimum)	‰ survived 18th birthday (minimum)	‰ unknown cases
1655–1804	6246	199	183	618	158	222	120	133	525
1805–1894	5626	232	191	576	331	317	145	314	224
1895–1939	1407	150	46	804	281	249	29	544	178

Table 18.3. *Mortality of illegitimate offspring whose unmarried mothers later married the putative father of the child named at baptism or another man in Ditfurt (1805–1939)*

	Unmarried mothers and their offspring No. of mothers = 312, no. of children = 356	
	Mother marries the named father of her child No. of children = 138	Mother marries, but not the named father of her child No. of children = 218
Infant mortality (N)	6	55
(‰)	43.8	252.3

reasonable interpretation. The later marriage with the child's father frequently only occurred if the child survived. If the child died, obviously the selection of a mate was reconsidered. Against this background, it cannot be ruled out that some women nursed their child with very special care, because the child's survival may have been a guarantee for the planned marriage – a speculation, that unfortunately cannot be confirmed empirically.

A second hypothesis comes to mind. The above-average mortality of the non-legitimized children could be related to the eventual marriage of their mothers. Although maternal manipulation may be considered as the probable cause of overmortality, there may be other explanations. There could be a confounded effect, because unmarried mothers, to the extent that they come from a prosperous natal home, were both able to experience better support while raising the child and also better able to bind the father of the child to them due to their "better" origins. However, a table of the social backgrounds of the natal families of unmarried mothers (Table 18.5) contradicts this possibility – the profile of the social background of both groups of women is too similar. Thus there remains only one plausible explanation for the overmortality of the illegitimate children of those mothers who married a man other than the father of the child: the mothers have not been willing to nurture their illegitimate offspring.

Depending on whether the father of an illegitimate child acknowledged his paternity or not, the survival of the child was advantageous or disadvantageous to the unmarried mother, and the statistics document that maternal solicitude of the child depended to a very crucial degree on

Table 18.4. Mortality of legitimate children from completely known families and of illegitimately born children whose mothers later married the father who acknowledged paternity or another man in Ditfurt (1895–1924), according to P. Stephan (unpublished data)

| | | Completely known families No. of families = 389, No. of children = 972 | Unmarried mothers and their offspring No. of mothers = 95, No. of children = 111 | |
			Mother marries acknowledging father of her child No. of children = 58	Mother marries, but not the acknowledging father of the child No. of children = 53
Infant mortality	N	167	4	21
	‰	171.8	69.0	396.2
Deceased during childhood (age 1–18 years)	N	49	4	1
	‰	50.4	69.0	18.9
Survived 18th birthday	N	756	50	31
	‰	777.8	862.1	584.9

Table 18.5. *Distribution (%) of the fathers of unmarried mothers by occupation (Ditfurt, 1895–1924), according to P. Stephan (unpublished data)*

	Men whose illegitimate grandchildren were legitimized by the son-in-law	Men whose illegitimate grandchildren were not legitimized by the son-in-law
Laborers	55	43
Craftsmen	18	13
Master craftsmen	8	8
Farmers	6	6
Others	5	6
Unknown	8	24

the answer to this question. No final clarification of by what differences in behavior differential maternal investment was expressed is possible, however, as the church records obviously do not contain any information on this subject. An analysis of the causes of death does not provide any help either, because for almost half of all of the children who died (legitimate and illegitimate) between 1768 and 1834 the cause of death was given as "*Jammer*" (literally misery or wretchedness). Between 1835 and 1864 in 42% of the cases, and between 1865 and 1894 in 68% of the cases "*Krämpfe*" (literally convulsion) was given as the cause of death. Although "*Jammer*" or "*Krämpfe*" describes the misery of the dying child, it does not designate a cause, which would be even halfway verifiable in view of current medical knowledge. The frequency with which this very vague indication appears in the registers ultimately allows the surmise that the population at that time had no lasting interest in really understanding the reasons for the high infant and child mortality. This widespread lack of knowledge, perhaps even paired with not wanting to know, which denies responsibility for what happened, has surely facilitated maternal manipulations.

Does it pay to abandon the bastard? – evidence from the Krummhörn

Can maternal neglect increase a woman's reproductive fitness despite the costs initially incurred? This might happen if the early death of an illegitimate child actually increased the chances of marriage for the child's mother. Unfortunately, this question can not be fully clarified on the basis

Table 18.6. *Infant and child deaths of illegitimate offspring and the fate of the mothers (Krummhörn, 1720–1874)*

	Mother marries	Mother remains unmarried	Total
Child dies before 15th birthday	19	5	24
Child survives 15th birthday	25	30	55
Total	44	35	79

of the Ditfurt material, because owing to the high mobility of unmarried mothers, not enough life histories have been able to be completely reconstituted. The data from the neighboring parishes of Ditfurt are not available in a form that permits an evaluation. Here, however, data from the family reconstitution study of the Krummhörn (Ostfriesland, Germany, 1720–1874) can assist us. This set of data now comprises church record entries from 19 neighboring parishes, so that the small distance mobility of the population, including that of unmarried mothers, is "under observation". Details on the background of this study, as well as a summary of the main results to date may be found in Voland (1995).

In order to pursue the question of the consequences of the early death of illegitimate children for the fitness of their mothers, we have extracted the life histories of a total of 79 unmarried mothers from the Krummhörn data base according to the following criteria:

> Only those women who did not marry the named father of the child or whose marriage did not occur within a very tight time frame, i.e., within 1 month after giving birth, were considered.

> The illegitimate children had to have been born in the parishes of the Krummhörn under observation and their mothers had to have died there. The combination of these two criteria makes it probable that the life histories of the women after the birth of the illegitimate child and the further fate of the child are completely known. Of course, it cannot be ruled out that children were given to the care of the maternal parents (or other relatives) who lived outside these parishes and died there without the Krummhörn church records containing any entries of such deaths. Therefore, the data in Table 18.6 perhaps slightly underestimate the actual infant and child mortality.

> Only those women were considered who gave birth to a child while unmarried and not the women who gave birth to an illegitimate child as a widow.

Table 18.7. *Social background of the fathers of deserted mothers according to whether the women remained single or later married (Krummhörn, 1720–1874)*

	Laborers	Tradesmen and smallholders	Farmers	Unknown	Total
Mothers married	27	8	2	7	44
Mothers remained single	22	9	1	3	35

> The women had to have lived for at least 5 years after the birth of their illegitimate child.
>
> Women with stillbirths were not considered.

Table 18.6 shows the relation between the early death of the illegitimate offspring and the marriage probability of their mothers. Interestingly enough, all of the deceased children died prior to the marriage of their mothers. Women whose children attained the age of adulthood remained unmarried to a significantly higher degree than women whose children died ($Z = 2773$; $P < 0.01$). The death of the child increased the marriage probability from $P = 0.454$ to $P = 0.792$ in the Krummhörn. This corresponds to an increase of approximately 75%. Now it would be conceivable, of course, that unmarried mothers who later married were forced to do so out of poverty (and being poor meant being more likely to lose one's offspring), as compared with those who could afford to remain single, as they were well supported by their kin. Table 18.7 shows, however, that this alternative hypothesis seems rather unlikely. Deserted mothers were largely found within the poorer group of society anyway. The 35 mothers who remained single do not appear to have enjoyed a richer family background on average than those 44 mothers who decided to marry. Thus raising an illegitimate child and marriage with a man other than the putative father of the child were antagonistic options, and a motivation for underinvesting in the child with the goal of being able to marry would have had a good prospect of success.

The Krummhörn data also disclosed that an increase in one's chances of marriage through an early child loss on average did increase the reproductive success of the women. The 19 women who married after the death of their children subsequently gave birth to a total of 37 legitimate children. Purely arithmetically, the subsequent 1.9 children per woman compensated for the loss of the one illegitimate child. Three of the 19 women remained childless in their subsequent marriage and thus were unable to compensate for the loss. The 30 unmarried mothers whose children survived and who remained unmarried throughout their lives gave birth to a total of another 18 children, from which an average of a total of 1.6 children per woman is calculated. Such a comparison can, of course, not be much more than a very rough initial estimate pertaining to the direction of the effect. For a final balance of fitness, additional essential cost–benefit relevant factors would have to be considered, the quantitative significance of which we have not been able to estimate very well. Above all, we have to consider second-generation effects, i.e., we would have to consider the different social and reproductive chances of legitimate and illegitimate offspring, such as different dispersal and marriage probabilities, differential adult mortality and different physical fitness (Low & Clarke 1991; Abade & Bertranpetit 1995; Baten & Murray 1997, 1998).

A second function of maternal neglect could have promoted its evolution. Even in the unlikely case that the survival of an illegitimate child would have been no obstacle for a subsequent marriage, the mere presence of a stepchild might have reduced the psychological well-being and ultimately the reproductive success of such a marriage, since stepfathers are known to be more ready to maltreat their wives (Daly & Wilson 1996). If children fathered by previous partners constitute a risk factor for violence against women, infanticide of one's own offspring prior to marriage might become a kind of self-protection. But of course, church records do not contain any entries on intrafamilial violence, which leaves the hypothesis of maternal neglect as an evolved means of self-protection in the realm of speculation.

To sum up, everything that we are able to draw from the entries in the church records from Ditfurt and the Krummhörn point to the fact that the neglect of illegitimate offspring via a maximization of mating success on average led to reproductive benefits for their unmarried mothers. We think that the empirical evidence suggests that some manifestations of maternal neglect of one's own offspring have been sexually selected.

Upgrading one's mate value: abandoning children to get a man

While divorce in order to become paired with a presumably "better" mate is not uncommon among socially monogamous species, especially among birds (Choudhury 1995), this does not coincide with abandoning current offspring. For seasonal breeders with a relatively short period of parental investment, as most birds are, it does not pay for females to abandon current offspring in favor of another mate. It is more promising to change mates – if need be – between reproductive bouts. Even in primates with their relatively long periods of parental investment, females rarely if ever abuse or kill their own offspring under natural conditions (Hrdy 1995). For organisms with individual phases of parental investment for each off-spring largely overlapping, as in humans, divorce between reproductive bouts is not available, which creates a specific human trade-off problem for females. Should they continue investing in current reproduction or go for increasing their mating success?

The fact that children in whom no father invests (because he has died or left the mother) have a higher risk of being killed or neglected is well known from several and very different cultural contexts (Daly & Wilson 1984; Voland 1988; Hurtado & Hill 1992; Bhuiya & Chowdhury 1997). However, there are two functional backgrounds to be considered for this phenomenon, which must be kept separate analytically. On one hand, the loss of parental investment automatically leads to a drastic reduction in the child's reproductive value in many societies, because mothers (or their kin) are unable to compensate for the father's investment due to their ecological circumstances. If this is why the child's reproductive value is reduced for extrinsic reasons, the benefit of continued reproductive effort is reduced for the mother according to current parental investment theory. Maternal investment promises little amortization and is therefore expensive.

On the other hand, continued investment can incur opportunity costs for the mothers, regardless of to what extent the lack of a father actually lowers the child's reproductive value. It would be conceivable, after all, that in ecologically halfway buffered societies, such as in Ditfurt or in the Krummhörn, the survival chances of illegitimate offspring due to kin support, from the grandparents for example, would not have to be reduced to any significant degree. Abandoning these children, even though their reproductive value does not have to be noticeably reduced due to extrinsic factors, can nevertheless be understood as an adaptive reaction, if the high opportunity costs, namely the factual exclusion of the mothers from the marriage market, can be avoided.

Maternal underinvestment in one's own offspring as a reaction to the lack of a paternal contribution to child care can, consequently, be adaptive for two reasons. First, maternal neglect may be shaped by natural selec-tion, which favors a reduction of parental investment in offspring with reduced reproductive value, provided the mother has children or other relatives of higher reproductive value to invest in, or is likely to acquire such children in the future. If, however, the current children are her only vehicles for genetic propagation, their reproductive values should be irrelevant to the mother's investment decision. Second, maternal neglect may be shaped by sexual selection, which favors a reduction of investment in offspring, even if such offspring have an average reproductive value, if mating success is thus increased.

Sexually selected maternal neglect of one's own offspring is only expected, of course, if biological reproduction occurs in cooperative relationships (pair bonds), in which males invest not only their germ cells but also resources into their reproduction. This explains the lack of evidence for sexually selected maternal infanticide in non-human primates, because it looks very much as if the interest of non-human primate males in the offspring of their mates has to be understood as mating effort, instead of paternal investment (van Schaik & Paul 1996–7). If, however, as in many human societies, fathers develop a continued interest in the well-being of their own offspring and if therefore paternity concerns become of crucial importance for the social transactions between the sexes, then a very important prerequisite is given for the possibility that maternal solicitude also becomes a target of sexual selection. Because men are hardly willing, on one hand, to invest in stepchildren (Daly & Wilson 1994) and, on the other, to expose their own children to competition for maternal investment with older stepsiblings, maternal infanticide can initially be understood as an adaptive reflex to male interests, while nevertheless also ultimately serving the mothers' own interests in a dialectic knot. The varying prevalence of sexually selected maternal neglect of one's own offspring is most likely to depend primarily on the importance of male parental investment for the reproductive success of the women.

Moreover, the probability of sexually selected infanticide being exercised is likely to depend on two additional factors, namely on the degree of social stratification of a society and also on the opportunity structures of the local mating/marriage markets. The more strongly stratified is a society, the more different are the mate values of the men. Under otherwise equal conditions, child neglect is worth it if potential mates are more resource rich or have a higher social rank. In societies where there is hardly any variance in the male resource holding potential, for example because the socioecological living circumstances do not permit any accumulation of property, mothers – other things being equal – will on average be less willing to abandon children in favor of another man.

The costs/benefit structure for maternal neglect is crucially influenced by the variance in mate quality, because high-quality mates with a high investment potential provide positive incentives to women to reconsider their mate selection. At the same time, however, a biological market, as defined by Noë & Hammerstein (1995), must be in operation, so that the abandonment of already born children can lead to an upgrading of market value and ultimately to an increase in mating success. The prerequisite for this is a combination of female competition and male choice, which allows more privileged men, because a supply/demand structure of the mating/marriage market that is more favorable to them, to assert their mate selection standards rather uncompromisingly. If men hardly have to accept any compromises with respect to their mate selection standards, women with dependent offspring will hardly have any chance with them. Interestingly enough, Emelina was not the only one aware of this correlation, since Waynforth & Dunbar (1995) were able to show by their analysis of "lonely-hearts" advertisements that British and North American women with dependent offspring make much more modest demands of the mate they seek than women without dependent offspring (the same applies to males, by the way). Obviously, they also know that their market value is reduced because of their children. And the remarriage chances of divorced women with children in the USA are actually reduced in comparison with those without offspring (Buckle *et al.* 1996). The same also applies to the 19th century, even if the end of the first marriage was usually constituted by the decease of one's mate and only rarely by divorce. The remarriage probabilities for widows declined with increasing number of children (Netherlands, Van Poppel 1995).

What makes the behavioral strategies of Emelina in Mozambique so comparable with those of the women in England and in the USA who place marriage advertisements and with that of unmarried mothers in historical Germany is an evolved, species-specific psychology. Darwinian algorithms weigh current reproduction against probable future reproduction by considering also benefits possibly arising from an enhancement of one's mating success. Since mating success plays an eminent role as a proximate parameter, when asserting different reproductive options for maximizing lifetime reproductive success, we must conclude that differential maternal investment has been shaped not only by natural selection, but also by sexual selection.

The end of the story

"Emelina experienced moments of triumph like a grand lady" it goes on to say in Paulina Chiziane's novel. "The man, who was even more crazy about her than before, satisfied her every mood. In her madness she demanded more and more impossible evidence of his love. 'I want you only for myself', said Emelina. 'Against my will, you still have two wives. Kill her, like I killed my children.' The man was close to keeping this promise when his conscience made him see reason. Suddenly he understood that his love to Emelina would turn him into a criminal. He decided to flee from this disaster. When she saw that her castle in the air had dissolved, Emelina roamed about in despair. The man had taken her soul with him, she was unable to find herself again. After one year she returned to the village in the hope that the people had forgotten the scandal. But the community does not forgive her. Emelina does not grasp why common people can maintain feelings of revenge for such a long time. She does not feel responsible for the death of her children at all, on the contrary, she is full of hate for those people who do not want to forgive her.

This is the story which Emelina tells Danila. Danila remains silent, then she asks with a cold, cutting voice: 'And this daughter comes from that man?' 'Yes. He took the others away and gave me this one. She looks a lot like him.' 'But how is this all possible, Emelina?' 'I don't know, I . . .' Danila mentally searches for a name for this crazy, unbelievable romance. Perhaps the title could be 'The hate that love generated'".

Conclusions

We show that in rural regions of historical Germany, infant mortality of illegitimate children varied according to whether or not the unmarried mother married the putative father of the child. If a man other than the father who publicly acknowledged paternity was married, infant mortality rose by a factor of approximately 6. We cannot see any other reason for this increase apart from maternal manipulation. The probability of marrying a man other than the father of her illegitimate child increased for unmarried mothers with the early death of her illegitimate children by about 75%. Thus maternal neglect increased mating success. Moreover, the loss of the bastard was on the average more than compensated for by subsequent legitimate children. We interpret the willingness of women to underinvest in their illegitimate children as an adaptive outcome of a sexually selected maternal algorithm that weighs current reproduction against future mating success. While, in non-human primates, sexually selected maltreatment of one's own offspring seems to be non-existent (Hrdy 1995), humans appear to be the only primates that have evolved a psychology that allows them to abandon their own dependent offspring from a former relationship in order to upgrade their mate value.

Acknowledgments

We thank Annette Scheunpflug for drawing our attention to the work of Paulina Chiziane. We are grateful to Carel van Schaik and three anonymous reviewers for their thorough inspection of the chapter and their valuable comments, which helped to strengthen our line of argument.

Part V

Conclusion

19

The behavioral ecology of infanticide by males

Introduction

The chapters in this volume point to four major conclusions. First, infanticide by males in many kinds of animal is most reasonably interpreted as a reproductive strategy, although not every observed case fits neatly all the criteria specified by this hypothesis (van Schaik, Chapter 2; Palombit *et al.*, Chapter 6; Blumstein, Chapter 8; Veiga, Chapter 9). As further evidence accumulates, the patterns appear to reinforce the sexual selection hypothesis (e.g., Borries *et al.* 1999b) rather than non-adaptive interpretations (see Sommer, Chapter 1). Second, there are predictable correlates of infanticide risk among primates and other mammal species, including life history factors such as long infant dependency relative to gestation, large litter size, altriciality and social factors such as the loss of protectors, in particular the rate of breeding male replacement (van Schaik, Chapters 2 and 3; Borries & Koenig, Chapter 5; Blumstein, Chapter 8; van Noordwijk & van Schaik, Chapter 14; Nunn & van Schaik, Chapter 16). Third, females in species with infanticide by males are not just passive recipients of male aggression, but have developed a broad array of behavioral and physiological strategies to reduce infanticide risk. These traits are a major focus of this volume and will be developed more fully later.

A fourth conclusion to emerge is that females may also kill infants, in some species more frequently than do males (Blumstein, Chapter 8; Veiga, Chapter 9; Digby, Chapter 17; Voland & Stephan, Chapter 18). Although the fitness benefits for females primarily involve resource competition and infant exploitation (Blumstein, Chapter 8; Digby, Chapter 17), there are a few instances of sexually selected infanticide in females (Veiga, Chapter 9; Voland & Stephan, Chapter 18). Whatever the evolutionary causes of infanticide by females, it may have selected for particular counterstrategies in females, just as infanticide by males has. Because of the overall emphasis on infanticide by males in this volume, we will not dwell in detail on female infanticide, which deserves a more complete separate treatment (for an overview of work to date, see Blumstein, Chapter 8; Digby, Chapter 17).

This chapter addresses each of the three remaining major themes by posing a question and trying to develop at least a preliminary answer. Before we do so, we should re-emphasize the importance of distinguishing infanticide *risk* from infanticide *rate*. Risk as used here is analogous to the idea of "intrinsic" risk, that is the probability of an event holding all other things equal except some variable of interest (see Janson 1998). In this case, the event is infanticide and the variable of interest may be morphological (body size, sexual dimorphism) or extrinsic (diet, density of solitary males). Although ideally we would prefer to present a wide-ranging compilation across all mammals, in practice quantitative estimates of infanticide rates and their possible social and ecological correlates in wild populations are available primarily for primates, so our data analysis focuses on this group, with qualitative comparisons drawn from studies in other taxa.

The first question we address is: why does infanticide risk occur at all? Why do not all female mammals have postpartum estrus and small litter sizes? We shall explore the ecological, social, and life-history correlates of major risk factors for infanticide in primates. One result, expanding previous work of van Schaik (2000, Chapter 3), is that infanticide risk should be present in many species of mammals where little or no actual infanticide is observed.

This observation brings up the second question: why are rates of infanticide so variable among primate species? We examine this variation and the social and ecological factors that appear to explain it (see also Borries & Koenig, Chapter 5). Interestingly, some traits that appear most important in explaining broad-scale patterns of infanticide risk among primate species have little power to explain differences in rates of infanticide. Thus other factors must operate to modulate observed rates of infanticide. In addition to variation in male mating strategies, there may be differences in female counterstrategies.

This variation leads to our third question: why are not female counterstrategies more effective? Presumably, the answer is that they must be limited by ecological circumstances (see Nunn & van Schaik, Chapter 16).

We ask how ecology constrains female behaviors with respect to infanticide in primates. In particular, it is interesting to ask when anti-infanticidal behaviors complement, and when they conflict with, behaviors of females to increase feeding success or reduce predation risk, the main themes of primate socioecology prior to the last decade of the twentieth century.

In the Discussion, we ask why infanticide has remained hidden as a potential selective force on social systems (particularly in primates) for so long, and how future studies may advance beyond those of this volume to elucidate the causes and consequences of infanticide by males. Finally, we consider the conservation impact of especially high rates of infanticide on population survival in the context of human disturbance.

Why cannot female mammals avoid infanticide risk altogether?

The major proximate correlate of infanticide risk in primates and other mammals is a long infant dependency period relative to gestation (van Schaik, Chapter 3). A long infant dependency may be brought about by intrinsically slow infant growth or development, large litter size, or strong breeding seasonality. Any of these factors is subject to change by natural selection, so the question arises why females have not evolved values of these traits that minimize or eliminate infanticide risk altogether. There are two major, non-exclusive possibilities: (1) intrinsic physiological constraints on growth or development make it difficult or impossible to avoid long dependency periods; and (2) the costs of modifying values of these risk-related traits is greater than the benefit of reduced infant mortality.

What are the patterns of infant dependency relative to gestation length, the major predictive correlate of infanticide risk among mammalian orders and primate families (van Schaik, Chapter 3)? In a preliminary analysis, van Schaik (2000) suggests that body size is an important determinant of lactation/gestation ratio, and thus of infanticide risk, in both carnivores and primates. This effect is to be expected, since allometric studies consistently show shallower slopes for the body-size dependence of gestation length than lactation length (Lee 1999). Previous studies have also documented that frugivores have slow development relative to folivores (Ross 1988; Ross & Jones 1999). Thus we may expect lactation/gestation ratios to be affected by diet, once body mass is controlled.

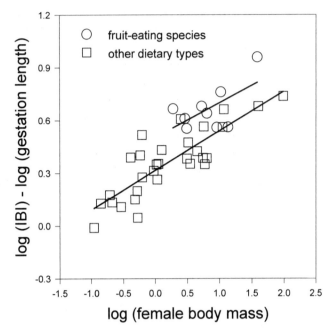

Figure 19.1. The length of the interbirth interval (IBI) relative to gestation period in primates increases with female body mass ($t = 15.63$, $P < 0.001$) and is greater in frugivores than other dietary types of a given body mass in primates ($F_{[1,36]} = 14.83$, $P < 0.001$). The other dietary types (folivores, frugivore-insectivores, insectivores) did not differ among each other after accounting for body mass ($F_{[2,23]} = 0.28$, $P = 0.76$). Data based on Ross & Jones (1999), except for population-specific values included in Table 19.2.

For our analysis, we used interbirth interval (IBI) as a widely available measure of infant dependency. Similar to the results of van Schaik (2000), we found the IBI/gestation risk ratio increases markedly with body mass among primates, exceeding 2 for all species greater than 1.1 kg (Figure 19.1). The relationship is fairly tight, with a range of only about 2.5-fold in the risk ratio for any given value of body mass. Thus it seems that it may be physiologically difficult (or pleiotropically too costly) for a female primate above about 1.5 kg to reduce her risk ratio below 1.0.

In addition, we examined the effect of diet after controlling for body mass (since diet and body mass are well known to covary in primates; e.g., Kay 1984). Each species was assigned to a diet category based on the food on which it spent the greatest fraction of time feeding (leaves, fruits, gums, insects) and on its major protein source or emergency food (leaves, seeds, insects), yielding four categories: folivores, frugivore-omnivores (including the use of seeds or leaves as an emergency or protein source), frugivore-insectivores (with little or no folivory), and insectivores (including use of gums). Analysis of covariance of the risk ratio using body mass as the covariate and diet as the factor revealed significant differences among dietary categories. In particular, frugivore-omnivores had higher risk ratios for their body mass than any other diet category, while the other categories did not differ significantly from each other (Figure 19.1). Thus the longer dependency periods in largely frugivorous primates should increase the per capita risk of dying from infanticide, all other things being equal.

These strong effects of body size and diet show that changes in life history to reduce the intrinsic risk of infanticide by males will be very costly. We therefore expect the strongest counterstrategies against infanticide in primate species with the greatest intrinsic risk: large-bodied frugivores. In addition, we should expect more changes in life history in the highly vulnerable species than in those at lower risk. Treves (Chapter 10) has provided an initial exploration of these important issues for primates, but much more remains to be done.

Mammals other than primates show other methods to reduce infanticide risk by changes in reproductive life histories (van Noordwijk & van Schaik, Chapter 14). Delayed implantation effectively increases the gestation length, so that a female can have postpartum estrus without reducing lactation to her current offspring. In many carnivores and rodents, females nest in relatively secure dens where they cache their offspring, and, at least in primates, this behavior is correlated with rapid infant development, short lactation and thus low risk of infanticide. At least in the latter case, the scarcity of this mechanism in most primates is probably related to a combination of an arboreal lifestyle and long daily travel distances.

Ecological and social correlates of observed rates of infanticide

What toll does infanticide take on infants? Even a cursory glance at the literature reveals tremendous variation in rates of infant death due to infanticide by males. However, infanticide is likely to be seriously underestimated if only the directly observed cases are included in an estimate. For example, mortality risk of red howler infants was almost 200 times higher within 1 month after male status changes than during non-change periods of the same groups, yet fewer than 5% of suspected infant killings by newly dominant males were actually observed (Crockett & Sekulic 1984). Likewise, Sommer (1994) noted that 93% of known and suspected cases of infanticidal assaults happened soon after takeovers, even though only a small number was directly observed.

There are various ways to express loss of infants to infanticide. Assume that, in a cohort of N infants, d infants die before becoming juveniles, i of which are killed by infanticide. Thus, i/N is the proportion of infants killed by infanticide in the cohort. This should be a relevant measure of natural selection to females, because it directly affects a female's reproductive rate (fecundity × a newborn's probability of survival to adulthood). However, it could be argued that a given level of infanticide is less important in a species in which many other sources of mortality reduce infant survival to low values, because any infant that escapes infanticide is likely to die of other causes. Because of interspecific differences in infant mortality, the proportion of infant mortality that can be attributed to infanticide (i/d) may be compared more easily across species. In what follows, we mostly analyze variation in i/N, which we shall call "infanticide rate" or "infanticide rate per infant born". We note where results for i/N differ markedly from those for i/d, which we shall label the "*relative infanticide rate*".

Table 19.1 provides realistic (not conservative) estimates of infanticide rates in natural primate populations. We base our estimates on the assessments of the original authors (who used a variety of evidence; but see also footnotes to Table 19.1), and, though it contains an element of arbitrariness, it should be noted that infant disappearances labeled as suspected infanticides tend to be far more likely during the periods when they are theoretically expected. To reduce the chance that a major fraction of infanticides could have been missed between periodic censuses, we used data only from studies in which the population was monitored nearly continuously or censused at no more than 3-month intervals.

One conclusion can be drawn immediately. While estimates of infanticide as a source of infant mortality vary, from close to zero in some studies to some 60% in others, infanticide is at least occasionally a major source of mortality, and therefore exerts strong selective pressure on the evolution of counter-measures. Hence, even though these are estimated mortality rates rather than directly observed ones, it is difficult to dismiss them as "genetically inconsequential", as suggested by Bartlett *et al.* (1993).

However, some of the rates presented in Table 19.1 are probably subject to serious biases. First, populations with high i/d ratios may be subject to low rates of natural predation relative to the species norm, probably because they are often in modified habitats. Second, the observed rate for gorillas may be exceptionally high because it involved a period where several adult males were killed by poachers and thus produced male-less groups that were subsequently taken over by extra-group males (Fossey 1984; cf. Steenbeek 1996). Third, a serious problem is that populations in which infanticide is rare or absent are unlikely to be included because detailed reports are not available. Conversely, the rare instances of infanticide that are published in these populations may lead to overestimates of the infanticide rate if comparable demographic data from non-infanticidal periods are not published. The latter problem may affect estimated infanticide rates in baboons (*Papio cynocephalus*) and Japanese macaques (*Macaca fuscata*) in Table 19.1.

These limitations should be kept in mind when one is interpreting the results of further analysis. To identify the possible underlying causes for the observed variability, we divide variables into two major groups: those that affect variation in (1) opportunity for infanticide or (2) incentive or benefits gained by the infanticidal male. Opportunity-related variables include male replacement rate, sexual dimorphism in body size and female dispersal. Incentive-related variables include the degree of monopolization of matings by the dominant male and relative advancement of the female's return to receptivity. Other variables, such as diet and social structure, are expected to exert their influence through these more proximate variables.

Opportunity and infanticide

Male replacement rate

Analyses of within-population variation in infanticide rates should provide the best available control for phylogenetic factors, including life history and physiological factors that may be important in explaining between-species variation. They also control for extrinsic socioecological variables, such as food availability, density of extra-group males and presence or absence of female dispersal, which have been argued to influence differences in infanticide rates between populations (Butynski, 1982; Borries 1997; Treves & Chapman 1996). The most obvious immediate factor affecting rates of infanticide *within populations* is the rate of

Table 19.1. Rates of infanticide (i/N and i/d) in various species of primates. See text for definitions for rates of infanticide and male replacement rate (RR). Sexual dimorphism based on data in Plavcan & van Schaik (1997). All other values taken from diverse literature sources on the same populations as the calculated infanticide rates (reference list available from senior author on request)

Species	i/N	i/d	Male RR	IBI	Relative male RR	Sexual dimorphism	Female dispersal	One-male mating?	Mean no. males/group	Male takeover?	Diet	Ref.
Alouatta caraya	0.16	0.47	0.15	1.32	0.198	0.1847	Y	Y	1.4	Y	L	1
Alouatta palliata	0.14	0.32	0.40	1.88	0.752	0.1072	Y	Y	3	Y	L	2
Alouatta seniculus	0.09	0.42	0.25	1.42	0.355	0.1004	Y	Y	1.932	Y	L	3
Cebus apella	0.08	0.2	0.19	1.61	0.307	0.1399	N	N	2.9	Y	F	4
Cercopithecus mitis, Kanyawara	0	0	0.29	5.00	1.450	0.2148	N	Y	1	Y	F	5
Cercopithecus mitis, Ngogo	0.3	0.47	2.07	2.38	4.927	0.2148	N	Y	1	Y	F	5
Colobus badius	0.05	0.3	0.67	2.13	1.427	0.1761	Y	N	4	N	L	6
Gorilla gorilla	0.13	0.37	0.07	4.00	0.280	0.2122	Y	Y	1	Y	L	7
Macaca fascicularis	0.04	0.2	0.55	1.33	0.732	0.1643	N	N	4.5	N	F	8
Macaca fuscata	0.04	0.13	0.20	1.96	0.392	0.1072	N	N	13.5	N	F	9
Pan troglodytes, Gombe	0	0	0.18	5.20	0.936	0.1139	Y	N	7.8	N	F	10
Pan troglodytes, Mahale	0.1	0.21	0.24	5.40	1.296	0.1139	Y	N	8.4	N	F	11
Papio anubis	0.07	0.21	0.27	1.61	0.435	0.2504	N	N	9.17	N	F	12
Papio cynocephalus	0.04	0.13	0.86	1.75	1.505	0.2480	N	N	11	N	F	13
Papio ursinus	0.14	0.34	1.82	1.61	2.930	0.2765	N	N	10.5	N	F	14
Presbytis thomasi	0.12	0.33	0.22	2.10	0.467	0.0379	Y	Y	1	Y	L	15
Propithecus diadema	0.18	0.36	0.20	2.08	0.417	0.0000	N	N	1.565	N	L	16
Semnopithecus entellus, Jodhpur	0.14	0.41	0.45	1.24	0.558	0.2014	Y	Y	1	Y	L	17

S. entellus, Kanha	0.2	0.5	0.33	1.66	0.548	0.2014	Y	Y	Y	1	Y	L	18
S. entellus, Ramnagar	0.15	0.31	1.03	2.15	2.215	0.2014	N	N	N	2.5	N	L	19

Notes:

IBI, interbirth interval; Y, Yes; N, no; L, leaves; F, fruit.

References for i/N and i/d values: (1) Zunino *et al.* 1986; Rumiz 1990 (mean of range between 5 and 9 infant deaths due to infanticides). (2) Clarke 1983; Clarke & Glander 1984; at least 6 infanticides; 19 dead on 44 infants born. (3) Crockett & Rudran 1987b; from their Table 2; this is an underestimate; if possible cases are added, midvalues become 0.12 and 0.56, respectively. (4) C. H. Janson & M. S. DiBitetti, unpublished data. (5) Butynski 1982, 1990 · (6) Struhsaker & Leland 1985; i/d is mid-range of two estimates. (7) Watts 1989; perhaps high due to recent poaching. (8) M. A. van Noordwijk & C. P. van Schaik, unpublished data; 6 estimated infanticides; 30 infants died in total on 156 born. (9) For 1994–6, T. Tanaka (unpublished data); for 1997–8: Soltis *et al.* 2000. (10) Goodall 1977; including only infants lost to females within community due to infanticide by males. (11) Hamai *et al.* 1992 for i/N; 2 deaths due to attacks by extra-community males; rest intra-community. Fossey 1984 for i/d, mid-range of two estimates. (12) Collins *et al.* 1984; estimating 12 cases of infanticide by males from tables. (13) Shopland 1982; Pereira 1983. (14) Palombit *et al.* 1997; Palombit *et al.* Chapter 6. (15) Steenbeek 1999. (16) P. C. Wright, pers. comm. (17) Sommer 1994, reanalysis of original data by A. Koenig & C. Borries. (18) Newton 1987. (19) Borries 1997.

breeding male replacement. In one-male groups, the breeding male is assumed to be the resident male (but see Cords 1987) whereas in multi-male groups, it is assumed to be the top dominant male because mating success is typically correlated with dominance rank (Cowlishaw & Dunbar 1991; de Ruiter & van Hooff 1993; Paul 1997).

Several studies have estimated infanticide rates for female groups of different sizes (Packer *et al.* 1988; Borries 1997; Crockett & Janson, Chapter 4; Steenbeek, Chapter 7). In all these studies, the rate of breeding male replacement and the infanticide rate per infant varied in parallel across groups. In howling monkeys and both langur species, infanticide rates increased with group size, but in hanuman langurs, rates decreased again in the largest groups, producing a curvilinear relationship. In lions, infanticide rates were lowest in groups of intermediate size.

Does the rate of breeding male replacement also help to explain differences in infanticide rate between populations and species? For the populations in Table 19.1, we calculated rates of breeding male replacement, either as the number of male replacements divided by the number of male-years (the sum of observation years per male summed for all males monitored) of observation, or as the inverse of the mean breeding male tenure length. In multimale species, we used the mean tenure of males at the top dominant position, as even in large multimale groups, dominance and reproductive success are related (Cowlishaw & Dunbar 1991). However, we cannot use male replacement rate as a simple measure of infanticide risk for comparisons between species because socioecological and life history variables will differ among species and even populations. The most obvious confounding factor is that inter-birth intervals (IBIs) vary greatly among species. A given rate of male replacement might present little threat of infant death when IBIs are much shorter than the mean male tenure, but could produce a high rate when IBIs are substantially longer than male tenure. We therefore multiplied the rate of breeding male change (number per year) by the mean IBI (years per birth) to yield the expected number of breeding male replacements per birth or "relative male replacement rate" (RMRR). If all males were infanticidal and perfectly successful in killing vulnerable offspring, observed infanticide rates should nearly equal RMRR, saturating as RMRR $>$ 1.0.

Despite much variation in ecological and social variables among species in our sample, RMRR is positively correlated with infanticide rate, albeit weakly (Table 19.1, $r=0.44$, $P=0.026$, one-tailed). However, this significant result depends entirely on including values for blue monkeys (*Cercopithecus mitis*), for which two populations provided the most extreme values of i/N in the entire sample. Excluding these populations, there is little effect of RMRR alone ($r = 0.015$) and it is in the wrong direction (i/N decreases with increasing RMRR)! With or without blue monkeys, RMRR did not correlate significantly with relative infanticide rate. We did not perform phylogenetically adjusted tests, as the greatest variation in infant death rates from infanticide were found between populations of the same species, suggesting that phylogenetic constraints are not likely to be important. As we shall see below, RMRR is important, but its effects are confounded with those of other variables, particularly the number of competing males.

If male replacement is a major proximate cause of infanticide rate, then we are left with the problem of explaining variation in male replacement rate. Butynski (1982, 1990) showed that infant mortality due to infanticide varied markedly between two populations of blue monkeys only 10 km apart and depended directly on takeover rates. Takeover rates, in turn, were well correlated with the density of non-group solitary males, who are continuously monitoring groups for possible takeovers. Ironically, the population with the better food supply, probably because it could easily harbor the solitary males, had a much lower population density, suggesting that the high rate of infanticide was almost certainly not due to overcrowding. It is even possible that the low density of the population in the area of high food density was caused by the impact of the anomalously high rate of male takeovers. Differences in predation risk are not likely to be important in this case, as the two study areas are only 10 km apart. Differences in takeover rate between populations and species have also been related to one-male versus multimale social structure (see below).

Sexual dimorphism

The bigger a male is relative to a female, the easier it may be for him to challenge her and gain possession of her infant. So we would expect higher rates of infanticide to occur where males are increasingly larger than females. To test this idea, we used the dimorphism values (log-transformed) from Plavcan & van Schaik (1997), using population-specific values when available, but otherwise using the most reliable estimates (see Plavcan & van Schaik 1997). However, after controlling for male replacement rate, this index of sexual dimorphism was not correlated with either measure of infanticide rate in our sample (Table 19.2). This result suggests that female defense does not impose a large cost on infanticidal males.

Table 19.2. *Results of statistical tests discussed in text. Values in parentheses are the test and probability values when both populations of blue monkeys (Cercopithecus mitis) in Table 19.1 are excluded; if exclusion does not affect test, this is indicated as "ditto". All probability values are two-tailed*

Dependent variable	Main independent variable	Covariates/conditions	Overall test	Test for main variable
i/N	Sexual dimorphism	RMRR	$F_{[2,17]} = 2.47, P = 0.11$ ($F_{[2,15]} = 0.01, P = 0.99$)	$t = -0.82, P = 0.43$ ($t = -0.10, P = 0.92$)
i/d	Sexual dimorphism	RMRR	$F_{[2,17]} = 0.12, P = 0.89$ ($F_{[2,15]} = 0.30, P = 0.74$)	$t = -0.13, P = 0.90$ ($t = 0.54, P = 0.60$)
i/N	Female dispersal (Y, N)	RMRR	$F_{[2,17]} = 2.91, P = 0.08$ ($F_{[2,15]} = 0.28, P = 0.76$)	$t = -1.18, P = 0.26$ ($t = -0.74, P = 0.47$)
i/d	Female dispersal (Y, N)	RMRR	$F_{[2,17]} = 1.98, P = 0.17$ ($F_{[2,15]} = 1.32, P = 0.30$)	$t = -1.92, P = 0.07$ ($t = -1.51, P = 0.15$)
i/N	Separate slopes for RMRR according to female dispersal	RMRR, female dispersal	$F_{[3,16]} = 4.56, P = 0.02$ ($F_{[3,14]} = 1.71, P = 0.21$)	$t = 2.47, P = 0.025$ ($t = 2.11, P = 0.053$)
i/d	Separate slopes for RMRR according to female dispersal	RMRR, female dispersal	$F_{[3,16]} = 4.12, P = 0.02$ ($F_{[3,14]} = 3.38, P < 0.05$)	$t = 2.65, P = 0.02$ ($t = 2.56, P = 0.02$)
i/N	RMRR	Female dispersal, Y	$t = -1.92, P = 0.09$ (ditto)	
i/N	RMRR	Female dispersal, N	$t = 2.93, P = 0.02$ ($t = 1.01, P = 0.35$)	
i/d	RMRR	Female dispersal, Y	$t = -2.11, P = 0.07$ (ditto)	
i/d	RMRR	Female dispersal, N	$t = 2.13, P = 0.07$ ($t = 1.22, P = 0.27$)	
i/N	Male takeover (Y, N)	RMRR	$F_{[2,17]} = 5.28, P = 0.02$ ($F_{[2,15]} = 3.29, P = 0.06$)	$t = -2.28, P = 0.04$ ($t = -2.57, P = 0.02$)
i/d	Male takeover (Y, N)	RMRR	$F_{[2,17]} = 2.63, P = 0.10$ ($F_{[2,15]} = 5.97, P = 0.01$)	$t = -2.23, P = 0.04$ ($t = -3.38, P < 0.01$)
i/N	Number of males/group	RMRR	$F_{[2,17]} = 6.50, P < 0.01$ ($F_{[2,15]} = 5.18, P = 0.02$)	$t = -2.68, P = 0.02$ ($t = -3.22, P < 0.01$)
i/d	Number of males/group	RMRR	$F_{[2,17]} = 3.86, P = 0.04$ ($F_{[2,15]} = 8.25, P < 0.01$)	$t = -2.72, P = 0.01$ ($t = -3.98, P < 0.01$)

i/N	RMRR	No. of males/group	$F_{[2,17]} = 6.50, P < 0.01$ ($F_{[2,15]} = 5.18, P = 0.02$)	$t = 2.56, P = 0.02$ ($1.20, P = 0.25$)
i/d	RMRR	No. of males/group	$F_{[2,17]} = 3.86, P = 0.04$ ($F_{[2,15]} = 8.25, P < 0.01$)	$t = 0.70, P = 0.49$ ($t = 0.84, P = 0.41$)
i/N	Log (female body mass)	RMRR	$F_{[2,17]} = 2.053, P = 0.16$ ($F_{[2,15]} = 0.0, P = 0.99$)	$t = -0.06, P = 0.95$ ($t = 0.04, P = 0.97$)
i/d	Log (female body mass)	RMRR	$F_{[2,17]} = 0.156, P = 0.86$ ($F_{[2,15]} = 0.28, P = 0.76$)	$t = -0.31, P = 0.763$ ($t = -0.50, P = 0.626$)
i/N	Diet (fruit, leaf)	–	Wilcoxon, $P = 0.02$ ($P < 0.01$)	
i/d	Diet (fruit, leaf)	–	Wilcoxon, $P < 0.01$ ($P < 0.01$)	
i/N	Diet (fruit, leaf)	RMRR	$F_{[2,17]} = 9.18, P < 0.01$ ($F_{[2,15]} = 6.31, P = 0.01$)	$t = -3.39, P < 0.01$ ($t = -3.55, P < 0.01$)
i/d	Diet (fruit, leaf)	RMRR	$F_{[2,17]} = 11.70, P < 0.01$ ($F_{[2,15]} = 12.62, P < 0.01$)	$t = -4.78, P < 0.01$ ($t = -4.94, P < 0.01$)
i/N	Diet (fruit, leaf)	RMRR, no. of males	$F_{[3,16]} = 6.25, P < 0.01$ ($F_{[3,14]} = 4.30, P = 0.02$)	$t = -1.92, P = 0.07$ ($t = -1.38, P = 0.19$)
i/d	Diet (fruit, leaf)	RMRR, no. of males	$F_{[3,16]} = 7.44, P < 0.01$ ($F_{[3,14]} = 8.42, P < 0.01$)	$t = -3.22, P < 0.01$ ($t = -2.17, P < 0.05$)
No. of males	Diet (fruit, leaf)	–	Wilcoxon, $P = 0.01$ ($P < 0.01$)	

Notes:
RMRR, relative male replacement rate; Y, yes; N, no.

Female dispersal

Female dispersal could increase infanticide risk by reducing the likelihood of female alliances against male attackers. Conversely, female dispersal could decrease infanticide risk by allowing females to move among groups to choose the best male protectors (Steenbeek, Chapter 7) or reduce group size to make the group a less attractive target for male takeovers (Crockett & Janson, Chapter 4). Which of these effects dominates? A comparison of langur populations (Sterck 1998) suggested that routine female dispersal (defined here as >0.1 migration per female per year) is associated with reduced infanticide rates. Overall, however, Sterck & Korstjens (Chapter 13) conclude that infanticide avoidance is not a major cause of female dispersal decisions, though it is probably important in some species. Instead, females in most primate species disperse apparently to avoid inbreeding, though such dispersal might well affect rates of infanticide.

In our sample, there is no significant effect of female dispersal alone on infanticide rates once RMRR is controlled statistically in an analysis of covariance, whether or not blue monkey populations are included (Table 19.2). However, female dispersal does affect the relationship between RMRR and i/N or i/d. The relationships are positive, as expected, where females are resident, but negative where females disperse. The slopes are significantly different from each other in each analysis (Table 19.2), but most of the slopes for each dispersal type separately do not differ from zero (Figure 19.2, Table 19.2). This trend suggests that female dispersal might limit the impact of frequent male replacement on infant death rates, as hypothesized by van Schaik (1996).

Differences in incentive: competition from other males

One- versus multimale groups

In populations that typically contain a single male per social group, the density of males excluded from breeding should be high, thus leading to high rates of dominant male replacement and consequent infanticide. Robbins (1995) showed that among mountain gorillas all known cases of infanticide occurred in single-male groups, and that infants in these groups were significantly more likely to die from infanticide than those in multimale groups. Newton (1986) noted that the great majority of reported cases of observed or strongly suspected infanticide among hanuman langurs was from populations in which one-male groups predominated (but see Borries 1997). Among red colobus, infanticide by

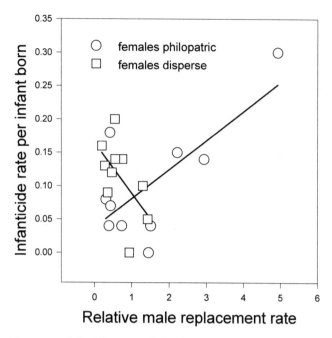

Figure 19.2. Infanticide rate per infant born increases with relative male replacement
rate in primate species in which females remain in their natal group, but
decreases in species in which females disperse from their natal group. The
slope shown for females that disperse is significantly different from 0 ($P =$
0.019), as is the difference in slopes between the two dispersal strategies (Table
19.2).

males is common in the one-male groups found at the Tana River (Marsh
1979b) but rare in the multimale groups seen in most of its range (Leland
et al. 1984; Struhsaker & Leland 1985). Crockett & Janson (Chapter 4) found
no effect of the number of males in red howler groups on infanticide rates,
but in red howlers, as in howlers in general, only a single male in the
group mates. Thus, on the whole, infants in multimale groups may suffer
a lower rate of infanticide, provided there is polyandrous mating, which
lowers the paternity likelihood for the dominant male. However, some
multimale groups may show little polyandrous mating, as in howlers,
whereas groups with a single resident male may show high levels of polyandrous mating during seasonal male influxes (e.g., Cords 1987). More
studies of genetic mating success by male primates are needed to improve
our understanding of male incentives to commit infanticide.

The differences in infanticide rates between one-male and multimale
groups were argued to be driven by parallel differences in replacement

rate, but it is also possible that the number of competing males affects infanticide rates via disincentives to a potentially infanticidal male. First, if the previous dominant male remains in the group, he is likely to defend any infants he has sired and thus increase the cost of infanticidal behavior to the new male. There is extensive observational evidence for this (see van Schaik 1996; Palombit, Chapter 11; Paul *et al.*, Chapter 12). Also, the new dominant male is sometimes an insider who would be less inclined to commit infanticide because he has mated with group females before and may have sired some offspring by them. This effect should be revealed by dividing populations into those with takeovers, where the previous dominant male is removed upon entry of a new dominant, versus those in which coresidence is typical. For instance, Sterck (1998) found that within a single subfamily, the colobine monkeys, estimated infanticide rate is much higher where takeovers occur than where they do not occur (but for an exception, see Borries & Koenig, Chapter 5). Second, if many males are present in the group when a female returns to estrus after losing her infant, the dominant male may have less assurance of inseminating the female, especially if she mates promiscuously. Male dominance and siring rate are less strongly correlated in large primate social groups (Cowlishaw & Dunbar 1991), especially where breeding is seasonal (Paul 1997). If this second effect is important, we expect infanticide rates to decrease with the number of adult males residing in the group *after* controlling for the rate of male replacement.

We tested the importance of these effects in our dataset. When RMRR is controlled for in an analysis of covariance, populations with habitual male takeovers have higher rates of i/N and i/d than those with habitual male co-residence, as predicted (Table 19.2). A much stronger effect was found for the number of males in a group. In a multiple regression, the effects of RMRR and number of males per group were both significant and in the expected directions (Figure 19.3, Table 19.2). The model explained 41% to 43% of the variance in rates of i/N, depending on whether blue monkeys were excluded or included, respectively. No other single variable analyzed above (sexual dimorphism, female dispersal, male takeovers, one-male mating) was statistically significant, either singly or in combination, when added to the effects of RMRR and number of males per group, as revealed by a stepwise regression. We take this as strong support for reduced incentive to commit infanticide when future matings with newly receptive females are less assured. The lack of additional effect of male takeover status on i/N after controlling for number of

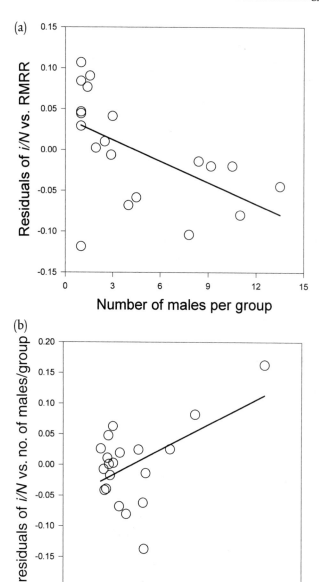

Figure 19.3. (a) When relative male replacement rate (RMRR) is controled statistically, infanticide rate per infant born is negatively related to the number of males per group (Table 19.2). (b) When the number of males per group is held constant statistically, infanticide rate is positively related to relative male replacement rate (Table 19.2).

males per group and RMRR suggests that the protection of previous, but now defeated, dominant males does little to slow the rate of infanticide.

The significant effect of numbers of males in reducing infanticide rates prompts the further question of what affects variation in male numbers (see Nunn & van Schaik, Chapter 16). Numbers of males are known to increase with female group size (Andelman 1986), which in turn varies with the level of food competition and predation risk (Janson & Goldsmith 1995). In addition, there are more males per group where predation risk is higher, independent of the number of females in the group (Nunn & van Schaik, Chapter 16) or their breeding synchrony (Nunn 1999b). If the predation effect is driven by a reluctance of males to roam alone where predators are common or predation risk is high, then we would expect that infanticide is most common in habitats that lack predators naturally or where disturbance has removed predators recently, or in larger-bodied species (see Isbell 1994). However, infanticide rates in our sample are not correlated with body mass (Table 19.2) or disturbance (despite the trend in langurs). It remains possible, therefore, that infanticide risk also affects the number of males in a group, but so far it has been impossible to demonstrate such an effect.

Diet

Hrdy (1979) was the first to note the apparent preponderance of infanticide in folivorous primate species relative to frugivorous ones, even though the longer infant dependence predicts a higher intrinsic risk for frugivores (see above). Following up on this suggestion, Crockett & Janson (Chapter 4) suggest that infanticide may limit group size in folivores more frequently than in frugivores (cf. Steenbeek, Chapter 7). Indeed, overall rates of infanticide per infant born are significantly higher in folivores than frugivores (Table 19.2), and the same is true for relative infanticide rate (Table 19.2). Thus, even if infanticide increased with group size in both frugivores and folivores, the quantitative effect in favoring smaller groups might be important only in the latter. Unfortunately, we have no good data on group-size-related rates of infanticide in any frugivorous species.

The higher rate of infanticide in folivores might be the result of diet-related changes in numbers of males per group or RMRR. We therefore tested for an effect of diet after controlling for these other variables. In this case, diet remained a significant predictor of relative infanticide rate

(Table 19.2), but not of i/N (Table 19.2). In either case, the effect of number of males became non-significant when diet was included. Because diet and number of males are highly related (Table 19.2), it is premature to speculate which of the two is the driving variable. However, if the diet effect is upheld in future analyses, its origin must lie in some variable other than the effects controlled for here. Dominant males in folivorous species may monopolize mating more than they do in similar-sized groups of frugivorous species, thus providing a greater benefit to infanticide in multimale groups. This will become a testable proposition as increasing numbers of studies of genetic paternity in primates become available. Males in folivorous species may monopolize matings more easily because their time budgets are less constrained than in frugivorous species. Most folivores spend the large majority of each day digesting rather than ingesting or searching for food, as do most frugivores, especially frugivore-insectivores (e.g., Terborgh 1983). Another possibility is that females in groups of folivores are more cohesive than those in groups of frugivores, thus making it easier for a male to monitor their locations.

Female responses to male infanticidal threat and their ecological implications

The particular female features highlighted in this volume as possible mechanisms to reduce the infanticidal threat from males include: (1) association with a male capable of defending infants from aggression by infanticidal males (Palombit *et al.*, Chapter 6; Steenbeek, Chapter 7); (2) confusion of paternity by polyandrous mating combined with unpredictable ovulation, increasing the likelihood that a male might kill its own infant and thereby imposing a cost on infanticidal behavior (Borries & Koenig, Chapter 5; van Noordwijk & van Schaik, Chapter 14; van Schaik *et al.*, Chapter 15; Nunn & van Schaik, Chapter 16); (3) avoidance of unfamiliar males during times of infant vulnerability (Treves, Chapter 10); (4) social defense by groups of related females (lions, Packer & Pusey, 1983; langurs, Treves & Chapman, 1996); (5) dispersal (Sterck 1998; Sterck & Korstjens, Chapter 13); and (6) caching vulnerable offspring (Treves, Chapter 10). When female counterstrategies are not effective, females associated with infanticidal males should cease investing in current offspring, leading to induced abortion or embryo resorption (Bruce effects: Blumstein, Chapter 8; van Noordwijk & van Schaik, Chapter 14).

How effective are female counterstrategies? Although female defenses are not perfect, the mortality of pre-weaned infants during male take-overs in primates is usually far from 100%, even though most vulnerable infants are attacked (van Schaik, Chapter 2; Borries & Koenig, Chapter 5; Palombit *et al.* Chapter 6; Steenbeek, Chapter 7). In some cases, the survival of an infant can be attributed rather directly to intervention by male or female protectors (e.g., Palombit *et al.*, Chapter 6), but it is more difficult to assess the effectiveness of paternity confusion. The significant decline in infanticide rates in primate groups with increasing numbers of males (Table 19.2, Figure 19.3a) suggests that paternity confusion via pro-miscuous mating is an effective female tactic to reduce infanticide risk. Nevertheless, there is little direct evidence based on differences among individual females in their use of promiscuous mating. Janson (1984) reported a possible example in one group of capuchin monkeys. The dominant female cycled many times and mated promiscuously during a portion of each cycle, whereas the subordinate females tended to cycle only during the peak period of conception and rarely mated with any male other than the dominant one. It may be coincidence, but the only two cases of known infant death associated with male takeovers in this species were both infants of the most subordinate female (M. S. DiBitetti & C. H. Janson, unpublished data).

It is not obvious at first why the various female counterstrategies are not developed fully in all species subject to infanticide risk. The answer must be that (1) some aspects are not fully under a female's control, (2) some aspects carry considerable hidden costs, or (3) some aspects may conflict directly with other aspects. Let us explore these possibilities in turn. Females may not be able to avoid infanticide risk altogether because selection on females to avoid infanticide may be balanced by equally strong selection on males to carry it out. Thus simple avoidance of males during periods of infant vulnerability may not be possible because males may seek out the infants.

Female counterstrategies with substantial ecological costs include communal female defense against males, female dispersal and infant caching. Communal defense may mean increasing female group size, which carries costs of either increased food competition (Janson & Goldsmith 1995) or increased infanticide risk through increased risk of male takeover (Crockett & Janson, Chapter 4; Steenbeek, Chapter 7). Moreover, coordinating female defense against infanticidal males may be difficult as males typically have less constrained time budgets (C. P. van

Schaik, unpublished data) and so can choose opportune moments to kill dependent offspring. The second alternative, female dispersal, has important costs (for a review, see Sterck & Korstjens, Chapter 13), which may limit its use by females in many species. Finally, infant caching is likely to be effective only if the female can leave the infant for long periods (so a male cannot track her to the cache site), but this may increase the infant's risk of predation or starvation.

The remaining two female counterstrategies, use of protector males and paternity confusion, may not always be compatible. Although the likely sire of an offspring should be interested in seeing it survive, males should be selected to adjust the risks they take to protect offspring in parallel with their likelihood of paternity in the infant, whether retrospectively or prospectively (cf. van Schaik & Paul 1996–7). Thus, if a female is successful in confusing paternity of her offspring, she provides little incentive for any one male to defend her infants, while reducing the benefit for any of those males to commit infanticide themselves. The use of a protector male versus paternity confusion should therefore depend on whether the major infanticidal threat comes from outside or inside the group (van Schaik et al. 1999). If most infanticide occurs because of male incursions from outside the group, then females should choose a protector male and forego paternity confusion, as defense in this case may require very high risks on the part of the defending male. Conversely, when most of the infanticide risk occurs because of status changes among resident males in a group, then females should prefer a strategy of paternity confusion to reduce the risk that any newly dominant male benefits from killing existing offspring in the group. In howler monkeys, the extent of infanticidal threat from outside versus inside the group changes with group size (Crockett & Janson, Chapter 4), and we might expect female strategies to change in parallel. Apparently, however, female red howlers preferentially mate with the group's dominant male in all groups (Pope 1990).

This overview shows that female counterstrategies are not perfect, and that both lack of control and incompatibility with other strategies are responsible for this. However, females could also manipulate infanticide rates by affecting male dispersal options, a possibility that remains almost entirely unexplored. For instance, in brown capuchin monkeys, females tend to support the existing dominant against all immigrant challengers, perhaps leading to the observed system wherein new males enter at the bottom of the hierarchy (C. H. Janson & M. S. DiBitetti, unpublished data). Conversely, in rhesus macaques, new immigrants appear to be favored mating partners even though they are of low status (Manson 1995; Berard 1999). Both these situations would lead to reduced infanticide by making newly dominant males likely to have sired some infants in the group. We urge workers to consider the potential for females to modulate the risk of takeovers by influencing male dispersal strategies.

Conclusions: on the causes of variation in rates of infanticide

Within and across primate populations and species, the relative rate of male replacement is a major first predictor of infanticide rates and must be included in any analysis of other contributing factors. Of the several theoretically justifiable social and ecological factors that should influence infanticide rates, only three produced suggestive results after male replacement rate was held constant statistically. First, infanticide rates decline as the number of males per group increases and this effect is stronger than that of sexual dimorphism or any social variable we examined (female dispersal, presence of male takeovers, one- versus multimale mating). This result supports the idea that promiscuous mating in multimale groups dramatically reduces a newly dominant male's incentive to commit infanticide. Second, female dispersal may affect how male replacement rate affects observed infanticide rates. Where females are philopatric, male replacement rates and infanticide rates vary in parallel, but where females disperse, infanticide rates paradoxically appear to decline as male replacement rates increase! Although this result becomes statistically non-significant when the number of males per group is controlled for, it is intriguing enough to warrant further examination. Third, the effect of diet on infanticide rates mimics the effect of number of males, with infanticide rates lower in frugivorous (but also multimale) primate species. Although the two results may not be separable at present, it is worth noting that the lower rates of infanticide by males in frugivorous species precisely contradicts the trends in underlying risk of infanticide based on relative infant dependency periods (Figure 19.1). Indeed, we predicted above that large-bodied frugivores should manifest the strongest females tactics against infanticide. Perhaps encouraging multimale grouping is the main tactic used by females to reduce infanticide risk, and the results of the statistical analysis suggest it is so effective that females in frugivorous species end up with lower overall infanticide rates than females in folivorous species.

Discussion

Why was infanticide ignored for so long as a significant selection pressure on social systems?

Although infanticide in primates and other mammals has been known for over 25 years, it is only in the past decade that it has been widely recognized as a potential source of selection on social systems (but see Hrdy 1979). There are several reasons that the scientific community has been slow in accepting infanticide as a selective force on a par with ecology or population structure. First, observed cases of infanticide are still relatively uncommon, and constitute a minority of the infant deaths that can reasonably be attributed to infanticide because of their association with male status changes (cf. Crockett & Sekulic 1984). In most frugivorous species, where absolute rates tend to be low (Table 19.1), it would be easy for infanticide to be overlooked entirely except in very intensive and long-term studies.

Second, the idea that infanticide might be adaptive was controversial (and still is: Hrdy *et al.* 1995; Sussman *et al.* 1995) and came to dominate interest in this phenomenon, at least in primates (cf. Pusey & Packer 1994). Thus scientists were slow to recognize that infanticide might be important to females regardless of its adaptive value to males. Because mortality rates of primate infants tend to be low compared with those in many other mammals, even relatively uncommon instances of infanticide can represent an avoidable and appreciable fraction of infant deaths (Table 19.1).

Finally, many of the results and counterstrategies associated with infanticide mimic those of other selection pressures that are either easier to document (food competition, e.g., Janson 1985) or more dramatic (predation). The finding by van Schaik (1983) that infant/female ratios declined with group size in many primate populations was taken as consistent with an effect of food competition in reducing birth rates in larger female groups. However, for red howler monkeys, this inference may be wrong, as increased infanticide rates in larger groups cause reduced infant/female ratios even though births rates are not markedly affected by group size (Crockett & Janson, Chapter 4). Similarly, early studies of vigilance or scanning behavior assumed that most outwardly directed scanning was directed against predators (e.g., de Ruiter 1986), but several recent studies have found evidence that such vigilance is directed against both predators and conspecifics (Cowlishaw 1998; Steenbeek, Chapter 7). Monogamy in primates was once thought to be a simple consequence of female territoriality, combined with an inability of the male to defend a territory large enough to encompass more than one female (Leighton 1987). More recently, the limiting factor on males has been suggested to be guarding of infants against infanticidal males rather than the energetics of territory defense (van Schaik & Dunbar 1990).

Although the new studies suggest that infanticide can no longer be ignored as a selective force in primate social evolution, this does not mean that the "old" socioecological model for explaining variation in primate social systems should be discarded (Nunn & van Schaik, Chapter 16). On the contrary, it seems that ecological factors still explain much of the interspecific variation. Nunn & van Schaik (Chapter 16) noted two possible reasons for this. First, the anti-infanticide responses are very flexible, and more or less equally effective counterstrategies may exist in a variety of ecologically imposed social organizations. Second, it may be more difficult to assess the impact of infanticide risk in broad interspecific analyses because of the difficulty of devising accurate proxy measures for intrinsic infanticide risk.

Assuming that anti-infanticide strategies are indeed more flexible, we are most likely to fathom their importance by comparisons of closely related species or populations of the same species. There may be many different ways of responding socially to the single selection pressure of infanticide, and which one is used by a given species may depend on ecology and phylogenetic history. Indeed, some of the ideas that appear to work well in explaining infanticidal risk and counterstrategies in primates are less successful when applied across mammals as a whole (van Noordwijk & van Schaik, Chapter 14; van Schaik 2000). To create a general theory of socioecology that includes infanticide will require distilling the common features shared by very distinct taxa in their responses to infanticide.

A methodological innovation that should be useful in future comparative analyses is to define *a priori* levels of infanticide risk using theoretical criteria (van Schaik, Chapter 2; Nunn & van Schaik, Chapter 16). Such *a priori* evaluations of risk reduce the problem of multiple correlated variables and have proven very valuable in understanding the impact of another rare behavior, predation, on primate social and ecological behaviors (Cowlishaw 1997, 1998). Another alternative to disentangle the web of intercorrelated variables is to reduce the correlations between them by examining a broader range of taxa, each of which will have distinct ranges of life history and ecological traits. Ideally, future theories of infanticide and its impacts can be tested across mammals and birds as well as primates.

Comparisons across species are not the only way to support or document the importance of infanticide to social evolution. Field researchers need to consider explicitly infanticide risk as a possible selective force on social structure as they take data in the field. Detailed and innovative studies on single species like that of Steenbeek (Chapter 7) can reveal the subtle but often pervasive effects of infanticide avoidance behaviors on daily social relationships and activity allocation. In addition, there are many other aspects of social organization peculiar to smaller taxonomic groups (such as triadic male–infant–male relationships) that may provide additional important avenues of investigation (Paul *et al.*, Chapter 12).

A final avenue for future research concerns experimental approaches both in captivity and the field. Using existing and novel hypotheses, it is possible to develop tests that distinguish clearly between ecological and infanticidal causes of particular social traits, all other things being equal. To make all other things equal, experimental manipulation of one variable is required. Although some possible manipulations (removal of an existing dominant male, for instance) might not be ethical or desirable, other experiments (use of playbacks to simulate the presence of intruding males) are both feasible and non-injurious.

Infanticide rates and population disturbance

We would like to end with a note on conservation. Populations with extremely high rates of infanticide suggest that female counterstrategies are not functioning well, and this in turn may be due to disequilibrium caused by disturbance (Sterck 1998; van Schaik, Chapter 2). Thus, even though female counterstrategies normally reduce infanticide to low rates, under certain circumstances, this male behavior can have serious demographic consequences.

An example may be Butynski's (1982, 1990) blue monkeys in Kibale forest, in which the Ngogo population with more food had far more solitary males, perhaps because these males were attracted to (or preferred to remain in) this area, compared to resource-poor regions nearby. These solitary males caused high rates of infanticide, so high that they may have depressed the population density of the Ngogo population below its carrying capacity. In this case, local heterogeneity in food supply for non-group males may have led to major differences in infanticide rates against which females had little defense as they may not have been selected to respond to such high rates in the past.

In any case, it is worth considering the possibility that higher rates of infanticide may be one symptom of disturbance in mammal populations subjected to habitat removal or hunting. This result would be expected if the disturbance is such that it favors high rates of male turnover (e.g., selective hunting of socially dominant or territorial males, which often are the individuals that confront hunters or are the prized trophy animals; cf. Swenson *et al.* 1997), high densities of extra-group males (habitat disturbance that concentrates vagrant males into novel home ranges), or restricted female counterstrategies (reduced dispersal for lack of habitat, reduced polyandry because of altered habitat structure).

Extremely high rates of infanticide (>15% of all infants dying of infanticide or >35% of infant mortality attributable to infanticide, see Table 19.1) may endanger population viability. This can emerge directly because high infant mortality may reduce population growth rate to levels where the population has difficulty recovering from any external disturbance. There may also be indirect effects, wherein the response of females to reduce infanticide risk results in population characteristics that make population growth difficult. Thus, if gibbons are socially monogamous to reduce infanticide risk (van Schaik & Dunbar 1990, but see Palombit 1999, Chapter 11), the resulting dispersion and low population density of females makes them vulnerable to population crashes and local extinction even if actual infanticide rates are so low as to be negligible. Habitat disturbance and hunting might therefore lead to an extinction "vortex" (Soulé 1986), in which infanticide rate is increased, population growth suffers, and the population is not able to recover from occasional years with poor survival (e.g., Foster 1982) or from increased mortality pressure exerted by human activities.

Acknowledgments

We thank the many contributors of field data cited in Table 19.1 for access to unpublished data and for clarification of published results. The manuscript benefited by a careful reading by Maria van Noordwijk. This is contribution number 1073 from the Graduate Program in Ecology and Evolution at the State University of New York at Stony Brook.

References

Abade, A. & Bertranpetit, J. (1995). Birth, marriage and death in illegitimacy: a study in Northern Portugal. *Journal of Biosocial Science*, **27**, 443–55.

Abbott, D. H. (1984). Behavioral and physiological suppression of fertility in subordinate marmoset monkeys. *American Journal of Primatology*, **6**, 169–86.

Abbott, D. H. (1993). Social conflict and reproductive suppression in marmoset and tamarin monkeys. In *Primate Social Conflict*, ed. W. A. Mason & S. P. Mendoza, pp. 331–72. New York: State University of New York Press.

Abegglen, J.-J. (1976). On socialization in Hamadryas baboons. Ph.D. thesis, Universität Zürich.

Agoramoorthy, G. & Rudran, R. (1995). Infanticide by adult and subadult males in free-ranging red howler monkeys of Venezuela. *Ethology*, **99**, 75–88.

Agoramoorthy, G., Mohnot, S. M., Sommer, V. & Srivastava, A. (1988). Abortions in free ranging Hanuman langurs (*Presbytis entellus*) – a male induced strategy? *Human Evolution*, **3**, 297–308.

Agrell, J. (1995). A shift in female social organization independent of relatedness: an experimental study on the field vole (*Microtus agrestis*). *Behavioral Ecology*, **6**, 182–91.

Agrell, J., Wolff, J. O. & Ylönen, H. (1998). Counterstrategies to infanticide in mammals: costs and consequences. *Oikos*, **83**, 507–17.

Alberts, S. C. & Altmann, J. (1995). Balancing costs and opportunities: dispersal in male baboons. *American Naturalist*, **145**, 279–306.

Alberts, S., Altmann, J. & Wilson, M. L. (1996). Mate guarding constrains foraging activity of male baboons. *Animal Behaviour*, **51**, 1269–77.

Alcorn, J. R. (1940). Life history notes on the piute ground squirrel, *Journal of Mammalogy*, **21**, 160–70.

Alderton, D. (1993). *Wild Cats of the World*. New York: Facts on File Publications.

Alderton, D. (1994). *Foxes, Wolves and Dogs of the World*. New York: Facts on File Publications.

Alderton, D. (1996). *Rodents of the World*. New York: Facts on File Publications.

Alexander, R. D. (1974). The evolution of social behavior. *Annual Review of Ecology and Systematics*, **5**, 325–83.

Allen, R. W. & Nice, M. M. (1952). A study of the breeding biology of the purple martin (*Progne subis*). *American Midland Naturalist*, **47**, 606–45.

Alonso, C. (1986). Fracasso na inibição da reprodução de uma fêmea subordinada e troca de hierarquia em um grupo familiar de *Callithrix jacchus jacchus*. In *A Primatologia No Brasil – 2*, ed. M. Thiago de Mello, p. 203. Campinas, SP: Sociedade Brasileira de Primatologia.

Altmann, J. (1980). *Baboon Mothers and Infants*. Cambridge, MA: Harvard University Press.

Altmann, J. (1990). Primate males go where the females are. *Animal Behaviour*, **39**, 193–5.

Altmann, J. (1997). Mate choice and intrasexual reproductive competition: contributions to reproduction that go beyond acquiring more mates. In *Feminism and Evolutionary Biology: Boundaries, Insections and Frontiers*, ed. P. A. Gowaty, pp. 320–33. New York: Chapman & Hall.

Altmann, J., Alberts, S. C., Haines, S. A., Dubach, J., Muruthi, P., Coote, T., Geffen, E., Cheesman, D. J., Mututua, R. S., Saiyalel, S. N., Wayne, R. K., Lacy, R. C. & Bruford, M. W. (1996). Behavior predicts genetic structure in a wild primate group. *Proceedings of the National Academy of Sciences, USA*, **93**, 5797–801.

Altmann, J., Altmann, S. A. & Hausfater, G. (1978). Primate infant's effects on mother's future reproduction, *Science*. **201**, 1028–9.

Altmann, J., Hausfater, G. & Altmann, S. A. (1988). Determinants of reproductive success in savannah baboons, *Papio cynocephalus*. In *Reproductive Success: Studies of Individual Variation in Contrasting Breeding System,* ed. T. H. Clutton-Brock, pp. 403–18. Chicago: University of Chicago Press.

Amos, W., Twiss, S., Pomeroy, P. P. & Anderson, S. (1993). Male mating success and paternity in the grey seal, *Halichoerus grypus*: a study using DNA fingerprinting. *Proceedings of the Royal Society, London B*, **252**, 199–207.

Andelman, S. J. (1986). Ecological and social determinants of cercopithecine mating systems. In *Ecological Aspects of Social Evolution: Birds and Mammals*, ed. D. I. Rubenstein & R. W. Wrangham, pp. 201–16. Princeton, NJ: Princeton University Press.

Andelman, S. J. (1987). Evolution of concealed ovulation in vervet monkeys (*Cercopithecus aethiops*). *American Naturalist*, **129**, 785–99.

Anderson, C. M. (1983). Levels of social organization and male–female bonding in the genus *Papio*. *American Journal of Physical Anthropology*, **60**, 15–22.

Anderson, C. M. (1986). Predation and primate evolution. *Primates*, **27**, 15–39.

Anderson, C. M. (1987). Female transfer in baboons. *American Journal of Primatology*, **73**, 241–50.

Anderson, C. M. (1992). Male investment under changing conditions among chacma baboons at Suikerbosrand. *American Journal of Physical Anthropology*, **87**, 479–96.

Anderson, S. (1985). Taxonomy and systematics. In *Biology of New World Microtus,* ed. R. H. Tamarin, pp. 52–83. Washington, DC: Special Publication of the American Society of Mammalogists.

Anderson, S. S., Baker, J. R., Prime, J. H. & Baird, A. (1979). Mortality in grey seal pups: incidence and causes. *Journal of Zoology, London*, **189**, 407–17.

Andersson, M. (1994). *Sexual Selection*. Princeton, NJ: Princeton University Press.

Andersson, M., Wiklund, G. & Rundgren, H. (1980). Parental defense of offspring: a model and an example. *Animal Behaviour*, **28**, 536–42.

Andrews, J. (1998). Infanticide by a female black lemur, *Eulemur macaco*, in disturbed habitat on Nosy Be, North-Western Madagascar. *Folia Primatologica*, **69**, 14–17.

Angst, W. & Thommen, D. (1977). New data and a discussion of infant killing in old world monkeys and apes. *Folia Primatologica*, **27**, 198–229.

Anonymous (1995). Infanticide observed in black howlers. *Community Conservation Consultants*, **6**, 1.

Anthony, L. L. & Blumstein, D. T. (2000). Integrating behaviour into wildlife conservation: the central role of N_e. *Biological Conservation,* in press.

Anzenberger, G. (1993). Social conflict in two monogamous New World primates: pairs and rivals. In *Primate Social Conflict,* ed. W. A. Mason & S. P. Mendoza, pp. 291–329. New York: State University of New York Press.

Armitage, K. B. (1981). Sociality as a life-history tactic of ground squirrels. *Oecologia*, **48**, 36–49.

Armitage, K. B., Johns, D. & Andersen, D. C. (1979). Cannibalism among yellow-bellied marmots. *Journal of Mammalogy*, **60**, 205–7.

Arnold, W. (1993). Social evolution in marmots and the adaptive value of joint hibernation. *Verhandlungen der Deutschen Zoologischen Gesellschaft*, **86**, 79–93.

Aronson, D. (1995). Infant killing among primates more myth than reality. Press Release (*July 1995*). St Louis, MO: Washington University.

Auerbach, G. & Taub, D. M. (1979). Paternal behavior in a captive 'harem' group of cynomolgus macaques (*Macaca fascicularis*). *Laboratory Primate Newsletter,* **18**, 7–11.

Aujard, F., Heistermann, M., Thierry, B. & Hodges, J. K. (1998). Functional significance of behavioral, morphological, and endocrine correlates across the ovarian cycle in semifree ranging female tonkean macaques. *American Journal of Primatology*, **46**, 285–309.

Aureli, F. & de Waal, F. B. M. (eds.) (2000). *Natural Conflict Resolution*. Berkeley, CA: University of California Press.

Avise, J. C. (1996). Three fundamental contributions of molecular genetics to avian ecology and evolution. *Ibis*, **138**, 16–25.

Ayer, M. L. & Whitsett, J. M. (1980). Aggressive behaviour of female prairie deer mice in laboratory populations. *Animal Behaviour*, **28**, 763–71.

Baker-Dittus, A. (1985). Infant and juvenile-directed care behaviors in adult toque macaques, *Macaca sinica*. Ph. D. dissertation, University of Maryland, Washington, DC.

Baldellou, M. & Henzi, S. P. (1992). Vigilance, predator detection and the presence of supernumerary males in vervet monkey troops. *Animal Behaviour*, **43**, 451–61.

Balfour, D. (1983). Infanticide in the Columbian ground squirrel, *Spermophilus columbianus*. *Animal Behaviour,* **31**, 949–50.

Banbura, J. & Zielinski, P. (1995). A clear case of sexually selected infanticide in the swallow *Hirundo rustica. Journal für Ornithologie*, **136**, 299–301.

Bart, J. & Tornes, A. (1989). Importance of monogamous male birds in determining reproductive success: evidence for house wrens and a review of male-removal experiments. *Behavioral Ecology and Sociobiolology*, **24**, 109–16.

Bartlett, J. (1988). Male mating success and paternal care in *Necrophorus vespilloides* (Coleoptera: Silphidae). *Behavioral Ecology and Sociobiolology*, **23**, 297–303.

Bartlett, T. Q., Sussman, R. W. & Cheverud, J. M. (1993). Infant killing in primates: a review of observed cases with specific reference to the sexual selection hypothesis. *American Anthropologist,* **95**, 958–90.

Barton, R. A., Byrne, R. W. & Whiten, A. (1996). Ecology, feeding competition and social structure in baboons. *Behavioral Ecology and Sociobiology*, **38**, 321–9.

Bartos, L. & Madlafousek, J. (1994). Infanticide in a seasonal breeder: the case of the red deer. *Animal Behaviour*, **47**, 217–19.

Baten, J. & Murray, J. E. (1997). Bastardy in south Germany revisited: an anthropometric synthesis. *Journal of Interdisciplinary History*, **28**, 47–56.

Baten, J. & Murray, J. E. (1998). Women's stature and marriage markets in preindustrial Bavaria. *Journal of Family History*, **23**, 124–35.

Bauers, K. A. & Hearn, J. P. (1994). Patterns of paternity in relation to male social rank in the stumptailed macaque, *Macaca arctoides*. *Behaviour*, **129**, 149–76.

Bearder, S. K. (1987). Lorises, bushbabies, and tarsiers: diverse societies in solitary foragers. In *Primate Societies*, ed. B. B. Smuts, D. L. Cheney, R. M. Seyfarth, R. W. Wrangham and T. T. Struhsaker, pp. 11–24. Chicago: University of Chicago Press.

Belles-Isles, J. C. & Picman, J. (1986). House wren nest-destroying behavior. *Condor*, **88**, 190–3.

Bennett, E. L. (1983). The Banded Langur: Ecology of a Colobine in West Malaysian Rain-forest, Ph. D. dissertation, University of Cambridge.

Bennett, E. L. & Sebastian, A. C. (1988). Social organization and ecology of proboscis monkeys (*Nasalis larvatus*) in mixed coastal forest in Sarawak. *International Journal of Primatology*, **9**, 233–55.

Bensch, S. & Hasselquist, D. (1994). Higher rate of nest loss among primary than secondary females: infanticide in great reed warbler? *Behavioral Ecology Sociobiology*, **35**, 309–17.

Berard, J. (1999). A four-year study of the association between male dominance rank, residency status, and reproductive activity in rhesus macaques (*Macaca mulatta*). *Primates*, **40**, 159–75.

Bercovitch, F. B. (1987). Reproductive success in male savanna baboons. *Behavioral Ecology and Sociobiology*, **21**, 163–72.

Bercovitch, F. B. (1988). Coalitions, cooperation and reproductive tactics among adult male baboons. *Animal Behaviour*, **36**, 1198–209.

Bercovitch, F. B. (1991). Mate selection, consortship formation, and reproductive tactics in adult female savanna baboons. *Primates*, **32**, 437–52.

Berger, J. (1986). *Wild Horses of the Great Basin: Social Competition and Population Size*. Chicago: Chicago University Press.

Berger, J. & Cunningham, C. (1994). *Bison: Mating and Conservation in Small Populations*. New York: Columbia University Press.

Berman, C. M. (1982). Demography and mother–infant relationships: implications for group structure. In *Ecology and Behavior of Food-Enhanced Primate Groups*, ed. J. Fa & C. Southwick, pp. 269–96. New York: Alan R. Liss.

Bernstein, I. S. (1968). The lutong of Kuala Selangor. *Behaviour*, **32**, 1–32.

Bernstein, I. S. (1976). Activity patterns in a sooty mangabey group. *Folia Primatologica*, **26**, 185–206.

Bernstein, I. S. (1980). Activity patterns in a stumptail macaque group (*Macaca arctoides*). *Folia Primatologica*, **33**, 20–45.

Bernstein, I. S. & Cooper, M. A. (1998). Ambiguities in the behavior of Assamese macaques. *American Journal of Primatology*, **45**, 170–1.

Bernstein, I., Williams, L. & Ramsay, M. (1983). The expression of aggression in Old World monkeys. *International Journal of Primatology*, **4**, 113–25.

Bertram, B. C. R. (1975). Social factors influencing reproduction in wild lions. *Journal of Zoology, London*, **177**, 463–82.

Bertram, B. C. R. (1978). Living in groups: predators and prey. In *Behavioural Ecology*, ed. J. R. Krebs & N. B. Davies, pp. 64–96. Oxford: Blackwell.

Betts, B. J. (1976). Behaviour in a population of Columbian ground squirrels, *Spermophilus columbianus columbianus*. *Animal Behaviour*, **24**, 652–80.

Betzig, L. (1992). Of human bonding: cooperation or exploitation? *Social Science Information*, **31**, 611–42.

Bhuiya, A. & Chowdhury, M. (1997). The effect of divorce on child survival in a rural area of Bangladesh. *Population Studies*, **51**, 57–61.

Biben, M., Symmes, D. & Bernhards, D. (1989). Vigilance during play in squirrel monkeys. *American Journal of Primatology*, **17**, 41–9.

Bibikov, D. I. & Berendaev, S. A. (1978). The Altai marmot. In *Marmots: their Distribution and Ecology*, ed. R. P. Zimina, pp. 39–78. Moscow: Nauka.

Bielert, C. & Girolami, L. (1986). Experimental assessments of behavioral and anatomical components of female chacma baboon (*Papio ursinus*) sexual attractiveness. *Psychoneuroendocrinology*, **11**, 75–90.

Bingham, L. R. & Hahn, T. C. (1974). Observations on the birth of a lowland gorilla, *Gorilla g. gorilla*. *International Zoo Yearbook*, **14**, 113–15.

Birkhead, T. R., Atkin, L. & Møller, A. P. (1986). Copulation behaviour of birds. *Behaviour*, **101**, 101–138.

Birkhead, T. R. & Møller, A. P. (1992). *Sperm Competition in Birds: Evolutionary Causes and Consequences*. London: Academic Press.

Black, C. C. (1972). Holarctic evolution and dispersal of squirrels (Rodentia: Sciuridae). *Evolutionary Biology*, **6**, 305–22.

Blaikie, A. (1993). *Illegitimacy, Sex, and Society – Northeast Scotland 1750–1900*. Oxford: Clarendon.

Blumstein, D. T. (1997). Infanticide among golden marmots (*Marmota caudata aurea*). *Ethology, Ecology, and Evolution*, **9**, 169–73.

Blumstein, D. T. & Armitage, K. B. (1997). Does sociality drive the evolution of communicative complexity? A comparative test with ground-dwelling sciurid alarm calls. *American Naturalist*, **150**, 179–200.

Blumstein, D. T. & Armitage, K. B. (1998). Life history consequences of social complexity: a comparative study of ground-dwelling sciurids. *Behavioral Ecology*, **9**, 8–19.

Blumstein, D. T. & Armitage, K. B. (1999). Cooperative breeding in marmots. *Oikos*, **84**, 369–82.

Blumstein, D. T. & Arnold, W. (1998). Ecology and social behavior of golden marmots (*Marmota caudata aurea*). *Journal of Mammology*, **79**, 873–86.

Böer, M. & Sommer, V. (1992). Evidence for sexually selected infanticide in captive *Cercopithecus mitis, Cercocebus torquatus* and *Mandrillus leucophaeus*. *Primates*, **33**, 557–63.

Boesch, C. & Boesch, H. (2000). *The Chimpanzees of Taï*. Oxford: Oxford University Press.

Boggess, J. (1979). Troop male membership changes and infant killing in langurs (*Presbytis entellus*). *Folia Primatologica*, **32**, 65–107.

Boggess, J. E. (1984). Infant killing and male reproductive strategies in langurs (*Presbytis entellus*). In *Infanticide. Comparative and Evolutionary Perspectives*, ed. G. Hausfater & S. B. Hrdy, pp. 283–310. New York: Aldine de Gruyter.

Boinski, S. (1987a). Birth synchrony in squirrel monkeys (*Saimiri oerstedi*): a strategy to reduce neonatal predation. *Behavioral Ecology and Sociobiology*, **21**, 393–400.

Boinski, S. (1987b). Mating patterns in squirrel monkeys (*Saimiri oerstedi*). *Behavioral Ecology and Sociobiology*, **21**, 13–21.

Boinski, S. & Mitchell, C. L. (1994). Male residence and association patterns in Costa Rican squirrel monkeys (*Saimiri oerstedi*). *American Journal of Primatology*, **34**, 157–69.

Boness, D. J. & James, H. (1979). Reproductive behaviour of the grey seal (*Halichoerus grypus*) on Sabel Island, Nova Scotia. *Journal of Zoology, London*, **188**, 477–500.

Bonhomme, F., Iskander, D., Thaler, L. & Petter, F. (1985). Electromorphs and phylogeny in muroid rodents. In *Evolutionary Relationships among Rodents: a Multidisciplinary Analysis*, ed. W. P. Luckett & J. L. Hartenberger, pp. 671–83. New York: Plenum Press.

Boonstra, R. (1978). Effect of adult Townsend voles (*Microtus townsendii*) on survival of young. *Ecology*, **59**, 242–8.

Boonstra, R. (1980). Infanticide in microtines: importance in natural populations. *Oecologia*, **46**, 262–5.

Borries, C. (1997). Infanticide in seasonally breeding multimale groups of Hanuman langurs (*Presbytis entellus*) in Ramnagar (South Nepal). *Behavioral Ecology and Sociobiology*, **41**, 139–50.

Borries, C. (2000). Male dispersal and mating season influxes in Hanuman langurs living in multimale groups. In *Primate Males*, ed. P. M. Kappeler, pp. 146–58. Cambridge: Cambridge University Press.

Borries, C., Launhardt, K., Epplen, C., Epplen, J. T. & Winkler, P. (1999a). Males as infant protectors in Hanuman langurs (*Presbytis entellus*) living in multimale groups: defense pattern, paternity, and sexual behaviour. *Behavioral Ecology and Sociobiology*, **46**, 350–6.

Borries, C., Launhardt, K., Epplen, C., Epplen, J. T. & Winkler, P. (1999b). DNA analyses support the hypothesis that infanticide is adaptive in langur monkeys. *Proceedings of the Royal Society, London B*, **266**, 901–4.

Borries, C., Sommer, V. & Srivastava, A. (1994). Weaving a tight social net: allogrooming in free-ranging female langurs (*Presbytis entellus*). *International Journal of Primatology*, **15**, 421–44.

Bortolotti, G. (1986). Influence of sibling competition on nestling sex ratios of sexually dimorphic birds. *American Naturalist*, **127**, 495–50.

Bourlière, F., Hunkeler, C. & Bertrand, M. (1970). Ecology and behavior of Lowe's guenon (*Cercopithecus campbelli lowei*) in the Ivory Coast. In *Old World Monkeys: Evolution, Systematics and Behavior*, ed. J. R. Napier & P. H. Napier, pp. 297–350. New York: Academic Press.

Brain, C. (1992). Deaths in a desert baboon troop. *International Journal of Primatology*, **13**, 593–9.

Brändström, A. (1996). Life history of single parents and illegitimate infants in nineteenth-century Sweden. *History of the Family*, **1**, 205–26.

Brédart, S. & French, R. M. (1999). Do babies resemble their fathers more than their mothers? A failure to replicate Christenfeld and Hill (1995). *Evolution and Human Behavior*, **20**, 129–35.

Brereton, A. (1995). Coercion-defence hypothesis: the evolution of primate sociality. *Folia Primatologica*, **64**, 207–14.

Brockelman, W. Y., Reichard, U., Treesucon, U. & Raemakers, J. J. (1998). Dispersal, pair formation, and social structure in gibbons (*Hylobates lar*). *Behavioral Ecology and Sociobiology*, **42**, 329–39.

Brockman, D. K. & Whitten, P. L. (1996). Reproduction in free-ranging *Propithecus verrauxi*: estrus and the relationship between multiple partner matings and fertilization. *American Journal of Physical Anthropology*, **100**, 57–69.

Brody, A. K. & Melcher, J. (1985). Infanticide in yellow-bellied marmots. *Animal Behaviour*, **33**, 673–4.

Brooks, R. J. (1984). Causes and consequences of infanticide in populations of rodents. In *Infanticide: Comparative and Evolutionary Perspectives*, ed. G. Hausfater & S. B. Hrdy, pp. 331–48. New York: Aldine de Gruyter.

Brotherton, P. N. M. & Manser, M. B. (1997). Female dispersion and the evolution of monogamy in the dik-dik. *Animal Behaviour*, **54**, 1413–24.

Bruce, H. M. (1960). A block to pregnancy in the house mouse caused by the proximity of strange males. *Journal of Reproduction and Fertility*, **1**, 96–103.

Bruce, K. E., Estep, D. Q. & Baker, S. C. (1988). Social interactions following parturition in stumptail macaques. *American Journal of Primatology*, **15**, 247–61.

Bruemmer, F. (1994). Rough rookeries. *Natural History*, **103**, 26–32.

Bryant, A. A. & Janz, D. W. (1996). Distribution and abundance of Vancouver Island marmots (*Marmota vancouverensis*). *Canadian Journal of Zoology*, **74**, 667–77.

Buckle, L., Gallup Jr, G. G. & Rodd, Z. A. (1996). Marriage as a reproductive contract: Patterns of marriage, divorce, and remarriage. *Ethology and Sociobiology*, **17**, 363–77.

Bulger, J. B. (1993). Dominance rank and access to estrous females in male savanna baboons. *Behaviour*, **127**, 67–103.

Bulger, J. B. & Hamilton III, W. J. (1987). Rank and density correlates of inclusive fitness measures in a natural chacma baboon (*Papio ursinus*) population. *International Journal of Primatology*, **8**, 635–50.

Bulger, J. B. & Hamilton III, W. J. (1988). Inbreeding and reproductive success in a natural chama baboon, *Papio cynocephalus ursinus*. *Animal Behaviour*, **36**, 574–8.

Burger, J. & Gochfeld, M. (1994). Vigilance in African mammals: differences among mothers, other females, and males. *Behaviour*, **131**, 153–69.

Burns, R. J. (1968). The role of agonistic behavior in regulation of density in Uinta ground squirrels (*Citellus armatus*). M. S. thesis. Utah State University, Logan.

Burt, A. (1992). 'Concealed ovulation' and sexual signals in primates. *Folia Primatologica*, **58**, 1–6.

Busch, R. H. (1996). *The Cougar Almanac. A Complete Natural History of the Mountain Lion*. New York: Lyons and Burford.

Buskirk, W. H., Buskirk, R. E. & Hamilton III, W. J. (1974). Troop-mobilizing behavior of adult male chacma baboons. *Folia Primatologica*, **22**, 9–18.

Busse, C. D. (1984a). Tail raising by baboon mothers toward immigrant males. *American Journal of Physical Anthropology*, **64**, 255–62.

Busse, C. (1984b). Triadic interactions among male and infant chacma baboons. In *Primate Paternalism*, ed. D. M. Taub, pp. 186–212. New York: Van Nostrand Reinhold.

Busse, C. & Gordon, T. P. (1983). Attacks on neonates by a male mangabey (*Cercocebus atys*). *American Journal of Primatology*, **5**, 345–56.

Busse, C. D. & Gordon, T. P. (1984). Infant carrying by adult male mangabeys (*Cercocebus atys*). *American Journal of Primatology*, **6**, 133–41.

Busse, C. & Hamilton, W. J. III. (1981). Infant carrying by male chacma baboons. *Science*, **212**, 1281–3.

Butynski, T. M. (1982). Harem-male replacement and infanticide in the blue monkey

(*Cercopithecus mitis stuhlmanni*) in the Kibale Forest, Uganda. *American Journal of Primatolology*, **3**, 1–22.

Butynski, T. (1990). Comparative ecology of blue monkeys (*Cercopithecus mitis*) in high and low density subpopulations. *Ecological Monographs*, **60**, 1–26.

Byers, J. A. (1997). *American Pronghorn. Social Adaptations and the Ghosts of Predators Past.* Chicago: University of Chicago Press.

Byrne, R. & Whiten, A. (ed.) (1988). *Machiavellian Intelligence.* Oxford: Clarendon.

Caley, J. & Boutin, S. (1985). Infanticide in wild populations of *Ondatra zibethicus* and *Microtus pennsylvanicus. Animal Behaviour*, **33**, 1036–7.

Calhoun, J. B. (1962). Population density and social pathology. *Scientific American*, **206**, 139–48.

Campagna, C. & LeBoeuf, B. J. (1988). Reproductive behaviour of Southern sea lions. *Behaviour*, **104**, 233–61.

Campbell, C. S. & Turek, F. W. (1981). Cyclic function of the mammalian ovary. In *Handbook of Behavioral Neurobiology,* ed. J. A. Schoff, pp. 523–45. Plenum Press, New York.

Camperio Ciani, A. (1984). A case of infanticide in a free-ranging group of rhesus monkeys (*Macaca mulatta*) in the Jackoo Forest, Simla, India. *Primates,* **25**, 372–7.

Cantoni, D. & Brown, R. E. (1997). Paternal investment and reproductive success in the California mouse, *Peromyscus californicus. Animal Behaviour*, **54**, 377–386.

Carleton, M. D. (1980). *Phylogenetic Relationships in Neotomine-Peromyscine Rodents (Muroidea) and a Reappraisal of the Dichotomy within New World Cricetinae.* Ann Arbor, MI: University of Michigan Museum of Zoology,.

Carleton, M. D. (1984). Introduction to rodents. In *Orders and Families of Recent Mammals of the World,* ed. S. Anderson & J. K. Jones Jr, pp. 255–65. New York: John Wiley and Sons.

Carleton, M. D. & Musser, G. G. (1984). Muroid rodents. In *Orders and Families of Recent Mammals of the World,* ed. S. Anderson & J. K. Jones Jr, pp. 289–379. New York: John Wiley and Sons.

Carpenter, C. C. (1942). Societies of monkeys and apes. *Biological Symposia*, **8**, 177–204.

Cartmill, M. (1998). Oppressed by evolution. *Discover,* March, 78–83.

Catzeflis, F. M., Dickerman, A. W., Michaux, J. & Kirsch, J. A. W. (1992). DNA hybridization and rodent phylogeny. In *Mammal phylogeny,* ed. F. S. Szalay, M. J. Novacek & M. C. McKenna, pp. 159–72. New York: Springer-Verlag.

Cavallini, P. (1998). Differential investment in mating by red foxes. *Journal of Mammology*, **79**, 215–21.

Chalmers, N. R. (1968). The social behaviour of free living mangabeys in Uganda. *Folia Primatologica*, **8**, 263–81.

Chanin, P. (1985). *The Natural History of Otters.* New York: Facts on File Publications.

Chapais, B. (1983). Reproductive activity in relation to male dominance and the likelihood of ovulation in rhesus monkeys. *Behavioral Ecology and Sociobiology*, **12**, 215–28.

Chapais, B. (1995). Alliances as a means of competition in primates: evolutionary, developmental, and cognitive aspects. *Yearbook of Physical Anthropology*, **38**, 115–36.

Chapman, M. & Hausfater, G. (1979). The reproductive consequences of infanticide in langurs: a mathematical model. *Behavioral Ecology and Sociobiology*, **5**, 227–40.

Charles-Dominique, P. (1995). Food distribution and reproductive constraints in the evolution of social structure: nocturnal primates and other mammals. In *Creatures of the Dark: The Nocturnal Prosimians*, ed. L. Alterman, G. A. Doyle & M. K. Izard, pp. 425–38. New York: Plenum Press.

Charles-Dominique, P. & Bearder, S. K. (1979). Field studies of lorisid behavior: methodological aspects. In *The Study of Prosimian Behavior*, ed. G. A. Doyle & R. D. Martin, pp. 567–629. New York: Academic Press.

Check, A. A. & Robertson, R. J. (1991). Infanticide in female tree swallows: a role for sexual selection, *Condor*, **93**, 454–7.

Cheney, D. L. (1987). Interactions and relationships between groups. In *Primate Societies*, ed. B. B. Smuts, D. L. Cheney, R. M. Seyfarth, R. W. Wrangham & T. T. Struhsaker, pp. 267–81. Chicago: University of Chicago Press.

Cheney, D. L. (1992). Intragroup cohesion and intergroup hostility: the relation between grooming distributions and intergroup competition among female primates. *Behavioral Ecology*, **3**, 334–45.

Cheney, D. L. & Seyfarth, R. M. (1990). *How Monkeys See the World*. Chicago: University of Chicago Press.

Cheney, D. L., Seyfarth, R. M., Andelman, S. J. & Lee, P. C. (1988). Reproductive success in vervet monkeys. In *Reproductive Success*, ed. T. H. Clutton-Brock, pp. 384–402. Chicago: University of Chicago Press.

Cheney, D. L. & Wrangham, R. W. (1987). Predation. In *Primate Societies*, ed. B. B. Smuts, D. L. Cheney, R. M. Seyfarth, R. W. Wrangham & T. T. Struhsaker, pp. 227–39. Chicago: University of Chicago Press.

Chiarello, A. G. (1995). Grooming in brown howler monkeys, *Alouatta fusca*. *American Journal of Primatology*, **35**, 73–81.

Chivers, D. J. & Raemaekers, J. J. (1980). Long-term changes in behaviour. In *Malayan Forest Primates: Ten Years' Study in Tropical Rain Forest*, ed. D. J. Chivers, pp. 209–58. New York: Plenum Press.

Chiziane, P. (1997). *Wind der Apokalypse*. Frankfurt/ M., Brandes and Apsel and Wien: Südwind. Original: *Ventos do Apocalipse*, 1993. (Our translation is from the German edition.)

Choudhury, S. (1995). Divorce in birds: a review of the hypotheses. *Animal Behaviour*, **50**, 413–29.

Christenfeld, N. & Hill, E. (1995). Whose baby are you? *Nature*, **378**, 669.

Christenson, T. E. & LeBoeuf, B. J. (1978). Aggression in the female northern elephant seal, *Mirounga angustirostris*. *Behaviour*, **64**, 158–72.

Cichy, K. A. (1996). Male–infant interactions in *Macaca fascicularis*. *IPS/ASP Congress Abstracts*, no. 634.

Cicirello, D. M. & Wolff, J. O. (1990). The effects of mating on infanticide and pup discrimination in white-footed mice. *Behavioral Ecology and Sociobiology*, **26**, 275–9.

Clapham, P. (1997). *Whales of the World*. Stillwater, MN: Voyageur Press.

Clarke, M. R. (1983). Infant-killing and infant disappearance following male takeovers in a group of free-ranging howling monkeys (*Alouatta palliata*). *American Journal of Primatology*, **5**, 241–7.

Clarke, M. R. & Glander, K. E. (1984). Female reproductive success in a group of free-ranging howling monkeys (*Alouatta palliata*) in Costa Rica. In *Female Primates: Studies by Women Primatologists*, ed. M. F. Small, pp. 111–26. New York: Alan R. Liss.

Clarke, M. R. & Zucker, E. L. (1994). Survey of the howling monkey population at La Pacifica: a seven-year follow-up. *International Journal of Primatology,* **15,** 61–73.

Clarke, M. R., Zucker, E. & Glander, K. (1994). Group takeover by a natal male howling monkey (*Alouatta palliata*) and associated disappearance and injuries of immatures. *Primates,* **35,** 435–42.

Clulow, F. V. & Clarke, J. R. (1968). Pregnancy-block in *Microtus agrestis*: an induced ovulator. *Nature,* **219,** 511.

Clulow, F. V., Franchetto, E. A. & Langford, P. E. (1982). Pregnancy failure in the red-backed vole, *Clethrionomys gapperi*. *Journal of Mammalogy,* **63,** 499–500.

Clulow, F. V. & Langford, P. E. (1971). Pregnancy block in the meadow vole, *Microtus pennsylvanicus*. *Journal of Reproduction and Fertility,* **24,** 275–7.

Clutton-Brock, T. H. (1989a). Female transfer and inbreeding avoidance in social mammals. *Nature,* **337,** 70–1.

Clutton-Brock, T. H. (1989b). Mammalian mating systems. *Proceedings of the Royal Society of London B,* **236,** 339–72.

Clutton-Brock, T. H. (1991). *The Evolution of Parental Care*. Princeton, NJ: Princeton University Press.

Clutton-Brock, T. H., Brotherton, N. M., Smith, R., McIlrath, G. M., Kansky, R., Gaynor, D., O'Riain, M. J. & Skinner, J. D. (1998). Infanticide and expulsion of females in a cooperative mammal. *Proceedings of the Royal Society: Biological Sciences,* **265,** 2291–5.

Clutton-Brock, T. H., Guinness, F. E. & Albon, S. D. (1982). *Red Deer: Behavior and Ecology of Two Sexes*. Chicago: University of Chicago Press.

Clutton-Brock, T. H. & Harvey, P. H. (1977). Primate ecology and social organization. *Journal of Zoology,* **183,** 1–39.

Clutton-Brock, T. H., Harvey, P. H. & Rudder, B. (1977). Sexual dimorphism, socionomic sex ratio and body weight in primates. *Nature,* **269,** 797–800.

Clutton-Brock, T. H. & Parker, G. A. (1992). Potential reproductive rates and the operation of sexual selection. *Quarterly Review of Biology,* **67,** 437–56.

Clutton-Brock, T. H. & Parker, G. A. (1995). Sexual coercion in animal societies. *Animal Behaviour,* **49,** 1345–65.

Collins, D. A. (1986). Interactions between adult male and infant yellow baboons (*Papio c. cynocephalus*) in Tanzania. *Animal Behaviour,* **34,** 430–43.

Collins, D. A., Busse, C. D. & Goodall, J. (1984). Infanticide in two populations of savanna baboons. In *Infanticide: Comparative and Evolutionary Perspectives*, ed. S. B. Hrdy & G. Hausfater, pp. 193–215. New York: Aldine de Gruyter Publishing Company.

Colmenares, F., East, M. L. & Hofer, H. (2000). Greeting ceremonies in baboons and hyenas. In *Natural Conflict Resolution*, ed. F. Aureli & F. B. M. de Waal. Berkeley, CA: University of California Press, in press.

Conaway, C. H. (1971). Ecological adaptation and mammalian reproduction. *Biology of Reproduction,* **4,** 239–47.

Connor, R. C., Richards, A. F., Smolker, R. A. & Mann, J. (1996). Patterns of female attractiveness in Indian Ocean bottlenose dolphins. *Behaviour,* **133,** 37–69.

Connor, R. C., Smolker, R. A. & Richards, A. F. (1992). Two levels of alliance formation among male bottlenose dolphins (*Tursiops sp.*). *Proceedings of the National Academy of Sciences, USA,* **89,** 987–90.

Converse, L. J., Carson, A. A., Ziegler, T. E. & Snowdon, C. T. (1995). Communication of ovulatory state to mates by female pygmy marmosets. *Animal Behaviour*, **49**, 615–21.

Corbet, G. B. (1978). *The Mammals of the Palaearctic Region: A Taxonomic Review*. Ithaca, NY: Cornell University Press.

Corbett, L. K. (1988). Social dynamics of a captive dingo pack: population regulation by dominant female infanticide. *Ethology*, **78**, 177–98.

Cords, M. (1987). Forest guenons and patas monkeys: male–male competition in one-male groups. In *Primate Societies*, ed. B. B. Smuts, D. L. Cheney, R. M. Seyfarth, R. W. Wrangham & T. T. Struhsaker, pp. 98–111. Chicago: University of Chicago Press.

Coulon, J., Graziani, L., Allainé, D., Bel, M. C. & Pouderoux, S. (1995). Infanticide in the alpine marmot (*Marmota marmota*). *Ethology, Ecology, and Evolution*, **7**, 191–4.

Coulson, J. C. & Hickling, G. (1964). The breeding biology of the grey seal, *Halichoerus grypus* (FAB.), on the Farne Islands, Northumberland. *Journal of Animal Ecology*, **33**, 485–512.

Cowlishaw, G. (1995). Behavioural patterns in baboon group encounters: the role of resource competition and male reproductive strategies. *Behaviour*, **132**, 75–86.

Cowlishaw, G. (1997). Trade-offs between foraging and predation risk determine habitat use in a desert baboon population. *Animal Behaviour*, **53**, 667–86.

Cowlishaw, G. (1998). The role of vigilance in the survival and reproductive strategies of desert baboons. *Behaviour*, **135**, 431–52.

Cowlishaw, G. & Dunbar, R. I. M. (1991). Dominance rank and mating success in male primates. *Animal Behaviour*, **41**, 1045–56.

Cox, C. R. & LeBoeuf, B. J. (1977). Female incitation of male competition: a mechanism in sexual selection. *American Naturalist*, **111**, 317–35.

Creel, S. R. & Waser, P. M. (1991). Failures of reproductive suppression in dwarf mongooses (*Helogale parvula*): accident of adaptation? *Behavioral Ecology*, **2**, 7–15.

Crockett, C. M. (1984). Emigration by female red howler monkeys and the case for female competition. In *Female Primates: Studies by Women Primatologists*, ed. M. F. Small, pp. 159–73. New York: Alan R. Liss.

Crockett, C. M. (1985). Population studies of red howler monkeys (*Alouatta seniculus*). *National Geographic Research*, **1**, 264–73.

Crockett, C. M. (1996). The relation between red howler monkey troop size and population growth in two habitats. In *Adaptive Radiations of Neotropical Primates*, ed. M. A. Norconk, A. L. Rosenberger & P. A. Garber, pp. 489–510. New York: Plenum Press.

Crockett, C. M. & Eisenberg, J. F. (1987). Howlers: variations in group size and demography. In *Primate Societies*, ed. B. B. Smuts, D. L. Cheney, R. M. Seyfarth, R. W. Wrangham & T. T. Struhsaker, pp. 54–68. Chicago: University of Chicago Press.

Crockett, C. M. & Pope, T. (1988). Inferring patterns of aggression from red howler monkey injuries. *American Journal of Primatology*, **15**, 289–308.

Crockett, C. M. & Pope, T. R. (1993). Consequences of sex differences in dispersal for juvenile red howler monkeys. In *Juvenile Primates: Life History, Development and Behavior*, ed. M. E. Pereira & L. A. Fairbanks, pp. 104–18. New York: Oxford University Press.

Crockett, C. M. & Rudran, R. (1987a). Red howler monkey birth data. I: Seasonal variation. *American Journal of Primatology*, **13**, 347–68.

Crockett, C. M. & Rudran, R. (1987b). Red howler monkey birth data. II: Interannual, habitat, and sex comparisons. *American Journal of Primatology*, **13**, 369–84.

Crockett, C. M. & Sekulic, R. (1982). Gestation length in red howler monkeys. *American Journal of Primatology*, **3**, 291–4.

Crockett, C. M. & Sekulic, R. (1984). Infanticide in red howler monkeys (*Alouatta seniculus*). In *Infanticide: Comparative and Evolutionary Perspectives*, ed. G. Hausfater & S. B. Hrdy, pp. 173–91. New York: Aldine de Gruyter.

Crook, J. H. & Gartlan, J. C. (1966). Evolution of primate societies. *Nature*, **210**, 1200–3.

Crook, J. R. & Shields, W. M. (1985). Sexually selected infanticide by adult male barn swallows. *Animal Behaviour*, **33**, 754–61.

Curtin, R. A. & Dolhinow, P. J. (1979). Infanticide among langurs – a solution to overcrowding? *Science Today*, **13**, 35–41.

D'Amato, F. R. (1993). Effect of familiarity with the mother and kinship on infanticidal and alloparental behaviour in virgin house mice. *Behaviour*, **124**, 313–26.

Da Silva, J., Macdonald, D. W. & Evans, P. G. H. (1993). Net costs of group living in a solitary forager, the European badger. *Behavioral Ecology*, **5**, 151–8.

Dagg, A. I. (1999a). Infanticide by male lions hypothesis: a fallacy influencing research into human behavior. *American Anthropologist*, **100**, 940–50.

Dagg, A. I. (1999b). Sexual selection is debatable. *Anthropology News*, (December), 20.

Dagg, A. I. & Foster, J. B. (1976). *The Giraffe. Its Biology, Behavior, and Ecology*. New York: Van Nostrand Reinhold Company.

Dahlgren, J. (1990). Female choose vigilant males: an experiment with the monogamous grey partridge, *Perdix perdix. Animal Behaviour*, **39**, 646–51.

Daly, M. & Wilson, M. (1982). Whom are newborn babies said to resemble? *Ethology and Sociobiology*, **3**, 69–78.

Daly, M. & Wilson, M. (1984). A sociobiological analysis of human infanticide. In *Infanticide: Comparative and Evolutionary Perspectives,* ed. G. Hausfater & S. B. Hrdy, pp. 487–502. New York: Aldine de Gruyter.

Daly, M. & Wilson, M. (1988). *Homicide*. New York: Aldinede Gruyter.

Daly, M. & Wilson, M. (1994). Stepparenthood and the evolved psychology of discriminative parental solicitude. In *Infanticide and Parental Care*, ed. S. Parmigiani & F. S. vom Saal, pp. 121–34. Chur (Switzerland): Harwood.

Daly, M. & Wilson, M. (1996). Evolutionary psychology and marital conflict: the relevance of stepchildren. In *Sex, Power, Conflict – Evolutionary and Feminist Perspectives*, ed. D. M. Buss & N. Malamuth, pp. 9–28. Oxford: Oxford University Press.

Daniel, J. C. (1970). Dormant embryos of mammals. *Bioscience*, **20**, 411–15.

Darwin, C. (1871). *The Descent of Man, and Selection in Relation to Sex*. Princeton, NJ: Princeton University Press (1981 reprint).

David, J. H. M. (1975). Observations on mating behaviour, parturition, suckling and the mother–young bond in the Bontebok (*Damaliscus dorcas dorcas*). *Journal of Zoology London*, **177**, 203–23.

Davidov, G. S., Neranov, I. M., Usachev, G. P. & Yakovlev, E. P. (1978). The long-tailed marmot. In *Marmots: Their Distribution and Ecology,* ed. R. P. Zimina, pp. 117–25. Moscow: Nauka.

Davies, A. G. (1987). Adult male replacement and group formation in *Presbytis rubicunda. Folia Primatologica*, **49**, 111–4.

Davies, N. (1992). *Dunnock Behaviour and Social Evolution*. Oxford: Oxford University Press.

Dawkins, R. (1976). *The Selfish Gene*. New York: Oxford University Press.

de la Maza, H. M., Wolff, J. O. & Lindsey, A. (1999). Exposure to strange adults does not cause pregnancy disruption or infanticide in the gray-tailed vole. *Behavioral Ecology and Sociobiology*, **45**, 107–13.

de Ruiter, J. R. (1986). The influence of group size on predator scanning and foraging behavior of wedge-capped capuchin monkeys *Cebus olivaceus*. *Behaviour*, **98**, 240–58.

de Ruiter, J. R., & van Hooff, J. A. R. A. M. (1993). Male dominance rank and reproductive success in primate groups. *Primates,* **34**, 513–23.

de Ruiter, J. R., van Hooff, J. A. R. A. M. & Scheffrahn, W. (1994). Social and genetic aspects of paternity in wild long-tailed macaques (*Macaca fascicularis*). *Behaviour*, **129**, 203–24.

de Villiers, D. J. (1986). Infanticide in the tree squirrel, *Paraxerus cepapi*. *South African Journal of Zoology*, **21**, 183–4.

de Waal, F. B. M. (1982). *Chimpanzee Politics*. London: Jonathan Cape.

de Waal, F. B. M. (1989). Dominance 'style' and primate social organizaton. In *Comparative Socioecology: The Behavioural Ecology of Humans and Other Mammals*. ed. V. Standen & R. A. Foley, pp. 243–63. Oxford: Blackwell.

de Waal, F. B. M. (1992). Coalitions as part of reciprocal relations in the Arnhem chimpanzee colony. In *Coalitions and Alliances in Humans and Other Animals*, ed. A. H. Harcourt & F. B. M. de Waal, pp. 233–57. New York: Oxford University Press.

de Waal, F. B. M. (1996). *Good Natured. The Origins of Right and Wrong in Humans and Other Animals*. Cambridge, MA, and London: Harvard University Press.

de Waal, F. B. M. (1998). Survival of the kindest. *The Chronicle of Higher Education*, 7 August, B4–B5.

de Waal, F. B. M., van Hooff, J. A. R. A. M. & Netto, W. J. (1976). An ethological analysis of types of agonistic interaction in a captive group of Java-monkeys (*Macaca fascicularis*). *Primates*, **17**, 257–90.

Deag, J. M. (1974). A study of the social behaviour and ecology of the wild Barbary macaque, *Macaca sylvanus* L. Ph.D. dissertation, University of Bristol.

Deag, J. M. (1980). Interactions between males and unweaned Barbary macaques: testing the agonistic buffering hypothesis. *Behaviour*, **75**, 54–81.

Deag, J. M. & Crook, J. H. (1971). Social behaviour and 'agonistic buffering' in the wild Barbary macaque, *Macaca sylvanus* L. *Folia Primatologica*, **15**, 183–200.

DeBry, R. W. (1992). Biogeography of New World taiga-dwelling *Microtus* (Mammalia: Arvicolidae): a hypothesis that accounts for phylogenetic uncertainty. *Evolution*, **46**, 1347–57.

Deng, Z. & Zhao, Q. (1987). Social structure in a wild group of *Macaca thibetana* at Mount Emei, China. *Folia Primatologica*, **49**, 1–10.

Derix, R. R. W. M. & van Hooff, J. A. R. A. M. (1995). Male and female partner preferences in a captive wolf pack (*Canis lupus*) – specificity versus spread of sexual attention. *Behaviour*, **132**, 127–49.

Derrida, J. (1970). Structure, sign and play in the discourse of the human sciences. In *The Languages of Criticism and the Sciences of Man: The Structuralist Controversy*, ed. R. Macksey & E. Donato, pp. 247–72. Baltimore, MD: Johns Hopkins University Press.

Digby, L. J. (1995a). Infant care, infanticide, and female reproductive strategies in polygynous groups of common marmosets (*Callithrix jacchus*). *Behavioral Ecology and Sociobiology,* **37,** 51–61.

Digby, L. J. (1995b). Social organization in a wild population of *Callithrix jacchus*. II. Intragroup social behavior. *Primates* **36,** 361–75.

Digby, L. J. & Barreto, C. E. (1993). Social organization in a wild population of *Callithrix jacchus*. I. Group composition and dynamics. *Folia Primatologica,* **61,** 123–34.

Dittus, W. P. J. (1975). The ecology and behavior of the toque monkey, *Macaca sinica*. Ph.D. dissertation, University of Maryland.

Dittus, W. P. J. (1980). The social regulation of primate populations: a synthesis. In *The Macaques: Studies in Ecology, Behavior and Evolution*, ed. D. G. Lindburg, pp. 263–86. New York: van Nostrand Reinhold Co.

Dittus, W. P. J. (1988). Group fission among wild toque macaques as a consequence of female resource competition and environmental stress. *Animal Behaviour,* **36,** 1626–45.

Dixson, A. F. (1983). The hormonal control of sexual behaviour in primates. *Oxford Reviews of Reproductive Biology,* **5,** 131–219.

Dixson, A. (1998). *Primate Sexuality*. Oxford: Oxford University Press.

Djojosudharmo, S. & van Schaik, C. P. (1992). Why are orang utans so rare in the highlands? Altitudinal changes in a Sumatran forest. *Tropical Biodiversity,* **1,** 11–22.

Dobson, F. S. (1990). Environmental influences on infanticide in Columbian ground squirrels. *Ethology,* **84,** 3–14.

Dobson, F. S. & Jones, W. T. (1985). Multiple causes of dispersal. *American Naturalist,* **126,** 855–8.

Dolhinow, P. J. (1977). Normal monkeys? *American Scientist,* **65,** 266.

Dolhinow, P. (1994). Social systems and the individual. *Evolutionary Anthropology,* **3,** 73–4.

Dolhinow, P. (1999). Understanding behavior: a langur monkey case study. In *The Nonhuman Primates*, ed. P. Dolhinow & A. Fuentes, pp. 189–95. Mountain View, CA, London, Toronto: Mayfield.

Dolhinow, P. & Fuentes, A. (eds.) (1999). *The Nonhuman Primates*. Mountain View, CA, London, Toronto: Mayfield.

Doolan, S. P. & Macdonald, D. W. (1997a). Breeding and juvenile survival among slender-tailed meerkats (*Suricata suricatta*) in the south-western Kalahari: ecological and social influences. *Journal of Zoology, London,* **242,** 309–27.

Doolan, S. P. & Macdonald, D. W. (1997b). Band structure and failures of reproductive suppression in a cooperatively breeding carnivore, the slender-tailed meerkat (*Suricata suricatta*). *Behaviour,* **134,** 827–48.

Dugatkin, L. A. (1997). *The Evolution of Cooperation*. New York: Oxford University Press.

Dukelow, W. R. (1977). Ovulatory cycle characteristics in *Macaca fascicularis*. *Journal of Medical Primatology,* **6,** 33–42.

Dunbar, R. I. M. (1980). Determinants and evolutionary consequences of dominance among female gelada baboons. *Behavioral Ecology Sociobiology,* **7,** 253–65.

Dunbar, R. I. M. (1984a). Infant-use by male gelada in agonistic contexts: agonistic buffering, progeny protection or soliciting support? *Primates,* **25,** 28–35.

Dunbar, R. I. M. (1984b). *Reproductive Decisions. An Economic Analysis of Gelada Baboon Social Strategies*. Princeton, NJ: Princeton University Press.

Dunbar, R. I. M. (1987). Habitat quality, population dynamics, and group composition in colobus monkeys. *International Journal of Primatology*, **8**, 299–329.

Dunbar, R. I. M. (1988). *Primate Social Systems*. London: Croom Helm.

Dunbar, R. I. M. (1995). The mating system of callitrichid primates. I. Conditions for the coevolution of pair bonding and twinning. *Animal Behaviour*, **50**, 1057–70.

Dunbar, R. I. M. & Dunbar, E. P. (1974). Ecology and population dynamics *of Colobus guereza* in Ethiopia. *Folia Primatologica*, **21**, 188–208.

Duncan, P. (1982). Foal killing by stallions. *Applied Animal Ethology*, **8**, 567–70.

Eaton, R. L. (1978). Why some felids copulate so much: a model for the evolution of copulation frequency. *Carnivore*, **1**, 42–51.

Ebensperger, L. A. (1998a). Strategies and counterstrategies to infanticide in mammals. *Biological Reviews*, **73**, 321–46.

Ebensperger, L. A. (1998b). Do female rodents use promiscuity to prevent male infanticide? *Ethology Ecology and Evolution*, **10**, 129–41.

Eberhard, W. G. (1996). *Female Control: Sexual Selection by Cryptic Female Choice*. Princeton, NJ: Princeton University Press.

Edwards, H. E., Reburn, C. J. & Wynne-Edwards, K. E. (1995). Daily patterns of pituitary prolactin secretion and their role in regulating maternal serum progesterone concentrations across pregnancy in the Djungarian hamster (*Phodopus campbelli*). *Biology of Reproduction*, **52**, 814–23.

Eggert, A.-K. & Müller, J. K. (1997). Biparental care and social evolution in burying beetles: lessons from the larder. In *The Evolution of Social Behavior in Insects and Arachnids,* ed. J. C. Choe & B. J. Crespi, pp. 216–36. Cambridge: Cambridge University Press.

Eggert, A.-K., Reinking, M. & Müller, J. K. (1998). Parental care improves offspring survival and growth in burying beetles. *Animal Behaviour*, **55**, 97–101.

Eggert, A. K. & Sakaluk, S. K. (1995). Female-coerced monogamy in burying beetles. *Behavioral Ecology and Sociobiology*, **37**, 147–53.

Ehrlich, A. & MacBride, L. (1989). Mother–infant interactions in captive slow lorises (*Nycticebus coucang*). *American Journal of Primatology*, **19**, 217–28.

Eibl-Eibesfeldt, I. (1965). Warum sich Tiere nicht töten. *Das Tier*, **6**, 14–16.

Eibl-Eibesfeldt, I. (1984). *Die Biologie des menschlichen Verhaltens. Grundriß der Humanethologie*. Munich, Zurich: Piper.

Eibl-Eibesfeldt, I. (1997). *Die Biologie des menschlichen Verhaltens. Grundriß der Humanethologie*, 3rd edn. Weyarn: Seehammer.

Eisenberg, J. F. (1981). *The Mammalian Radiations*. Chicago: University of Chicago Press.

Elgar, M. A. & Crespi, B. J. (1992). Ecology and evolution of cannibalism. In *Cannibalism: Ecology and Evolution among Diverse Taxa*, ed. M. A. Elgar & B. J. Crespi, pp. 1–12. Oxford: Oxford University Press.

Elwood, R. W. (1980). The development, inhibition and disinhibition of pup-cannibalism in the Mongolian gerbil. *Animal Behaviour*, **28**, 1188–94.

Elwood, R. W. (1986). The inhibition of infanticide and the onset of paternal care in male mice, *Mus musculus. Journal of Comparative Psychology*, **99**, 457–67.

Elwood, R. W. (1989). Female aggression and the inhibition of infanticidal tendencies in male mice. *Aggressive Behavior*, **15**, 54–5.

Elwood, R. W. (1991). Ethical implications of studies on infanticide and maternal aggression in rodents. *Animal Behaviour*, **42**, 841–9.

Elwood, R. (1992). Pup-cannibalism in rodents: causes and consequences. In

Cannibalism: Ecology and Evolution among Diverse Taxa, ed. M. A. Elgar & B. J. Crespi, pp. 299–322. Oxford: Oxford University Press.

Elwood, R. W. (1994). Temporal-based kinship recognition: a switch in time saves mine. *Behavioral Processes*, **33**, 15–24.

Elwood, R. W. & Kennedy, H. F. (1994). Selective allocation of parental and infanticidal responses in rodents: a review of mechanisms. In *Infanticide and Parental Care*, ed. S. Parmigiani & F. S. vom Saal, pp. 397–425. Chur (Switzerland): Harwood.

Elwood, R. W. & Ostermeyer, M. C. (1984). Infanticide by male and female mongolian gerbils: ontogeny, causation, and function. In *Infanticide: Comparative and Evolutionary Perspectives*, ed. G. Hausfater & S. B. Hrdy, pp. 367–86. New York: Aldine de Gruyter.

Emlen, S. T. (1991). Evolution of cooperative breeding in birds and mammals. In *Behavioural Ecology: An Evolutionary Approach*, ed. J. R. Krebs & N. B. Davies, pp. 301–37. London: Blackwell Scientific Publications.

Emlen, S. T., Demong, N. J. & Emlen, D. J. (1989). Experimental induction of infanticide in female wattled jacanas. *The Auk*, **106**, 1–7.

Emlen, S. T. & Oring, L. W. (1977). Ecology, sexual selection, and the evolution of mating systems. *Science*, **197**, 215–23.

Endler, J. A. (1986). *Natural Selection in the Wild*. Princeton, NJ: Princeton University Press.

Engel, S. R., Hogan, K. M., Taylor, J. F. & Davis, S. K. (1998). Molecular systematics and paleobiogeography of the South American Sigmodontine rodents. *Molecular Biology and Evolution*, **15**, 35–49.

Epple, G. (1986). *Communication by Chemical Signals*. New York: Alan R. Liss.

Erhart, E. M. & Overdorff, D. J. (1998). Infanticide in *Propithecus diademia edwardsi*: an evaluation of the sexual selection hypothesis. *International Journal of Primatology*, **19**: 73–81.

Errington, P. L. (1963). *Muskrat Populations*. Ames: Iowa State University Press.

Estes, J. A. (1989). Adaptations for aquatic living by carnivores. In *Carnivore Behavior, Ecology, and Evolution*, ed. J. L. Gittleman, pp. 242–82. Ithaca, NY: Cornell University Press.

Estes, R. D. (1991). *The Behavior Guide to African Mammals. Including Hoofed Mammals, Carnivores, Primates*, Berkeley, CA: University of California Press.

Estrada, A. (1984). Male–infant interactions among free-ranging stumptail macaques. In *Primate Paternalism*, ed. D. M. Taub, pp. 56–87. New York: Van Nostrand Reinhold.

Estrada, A. & Sandoval, J. M. (1977). Social relations in a free-ranging troop of stumptail macaques (*Macaca arctoides*): male-care behaviour I. *Primates*, **18**, 793–813.

Fairbanks, L. A. (1990). Reciprocal benefits of allomothering for female vervet monkeys. *Animal Behaviour*, **40**, 553–62.

Fairbanks, L. A. & McGuire, M. T. (1987). Mother–infant relationships in vervet monkeys: response to new adult males. *International Journal of Primatology*, **8**, 351–66.

Fairgrieve, C. (1995). Infanticide and infant eating in the blue monkey (*Cercopithecus mitis stuhlmanni*) in the Budongo Forest Reserve, Uganda. *Folia Primatologica*, **64**, 69–72.

Fedigan, L. (1993). Sex differences and intersexual relations in adult white-faced capuchins (*Cebus capucinus*). *International Journal of Primatology*, **14**, 853–77.

Fedigan, L. M. & Baxter, M. J. (1984). Sex differences and social organization in free-ranging spider monkeys (*Ateles geoffroyi*). *Primates*, **25**, 279–94.

Feh, C. (1999). Alliances and reproductive success in Camargue stallions. *Animal Behaviour*, **57**, 705–13.

Felsenstein, J. (1985). Phylogenies and the comparative method. *American Naturalist*, **125**, 1–25.

Fetterolf, P. M. (1983). Infanticide and non-fatal attacks on chicks by ring-billed gulls. *Animal Behaviour*, **31**, 1018–28.

Fisher, H. (1992). *Anatomy of Love. The Natural History of Monogamy, Adultery, and Divorce.* New York: W. W. Norton.

Fitzgerald, J. P. & Lechleitner, R. R. (1974). Observations on the biology of Gunnison's prairie dog in central Colorado. *American Midland Naturalist*, **92**, 146–63.

FitzGibbon, C. D. (1997). The adaptive significance of monogamy in the golden-rumped elephant-shrew. *Journal of Zoology, London*, **242**, 167–77.

Fleagle, J. G. (1999). *Primate Adaptation and Evolution*, 2nd edn. San Diego, CA: Academic Press.

Fleming, A. S. (1979). Maternal nest defense in the desert woodrat *Neotoma lepida lepida*. *Behavioral and Neural Biology*, **26**, 41–63.

Flowerdew, J. R. (1987). *Mammals: Their Reproductive Biology and Population Ecology.* London: Edward Arnold.

Foerg, R. (1982a). Reproduction in *Cheirogaleus medius*. *Folia Primatologica*, **39**, 49–62.

Foerg, R. (1982b). Reproductive behavior in *Varecia variegata*. *Folia Primatologica*, **38**, 108–21.

Fossey, D. (1983). *Gorillas in the Mist*. Boston, MA: Houghton Mifflin Company.

Fossey, D. (1984). Infanticide in mountain gorillas (*Gorilla gorilla beringei*) with comparative notes on chimpanzees. In *Infanticide. Comparative and Evolutionary Perspectives*, ed. G. Hausfater & S. B. Hrdy, pp. 217–35. New York: Aldine de Gruyter.

Foster, R. (1982). Famine on Barro Colorado Island. In *The Ecology of a Tropical Forest: Seasonal Rhythms and Long-term Changes*, ed. E. G. Leigh, A. S. Rand & D. M. Windsor, pp. 201–12. Washington, DC: Smithsonian Institution Press.

Fox, E. A. (1998). The function of female mate choice in the Sumatran orangutan (*Pongo pygmaeus abelii*). Ph. D. thesis, Duke University.

Frame, L. H., Malcohm, J. R., Frame, G. W. & van Lawick, H. (1979). Social organization of African wild dogs (*Lycaon pictus*) on the Serengeti Plains, Tanzania 1967–78. *Zeitschrift für Tierpsychologie*, **50**, 225–49.

Freed, L. A. (1986). Territory takeover and sexually selected infanticide in tropical house wrens. *Behavioral Ecology and Sociobiology*, **19**, 197–206.

Freed, L. A. (1987). Prospective infanticide and protection of genetic paternity in tropical house wrens. *American Naturalist*, **130**, 948–54.

Freeman-Gallant, C. R. (1996). DNA-fingerprinting reveals female preference for male parental care in savanna sparrows. *Proceedings of the Royal Society of London B*, **263**, 157–60.

Freeman-Gallant, C. R. (1997). Parentage and paternal care: consequences of intersexual selection in savanna sparrows. *Behavioral Ecology and Sociobiology*, **40**, 395–400.

French J. A. & Inglett, B. J. (1991) Responses to novel social stimuli in callitrichid monkeys: a comparative perspective. In *Primate Responses to Environmental Change*, ed. H. Box, pp. 275–94. London: Chapman & Hall.

Fuentes, A. (1999). Variable social organization: what can looking at primate groups tell us about the evolution of plasticity in primate societies? In *The Nonhuman Primates*, ed. P. Dolhinow & A. Fuentes, pp. 183–8. Mountain View, CA, London, Toronto: Mayfield.

Fuentes, A. & Tenaza, R. R. (1995). Infant parking in the pig-tailed langur (*Simias concolor*). *Folia Primatologica*, **65**, 172–3.

Furuichi, T. (1989). Social interactions and the life history of female *Pan paniscus* in Wamba. *International Journal of Primatology*, **10**, 173–97.

Furuichi, T. & Ihobe, H. (1994). Variation in male relationships in bonobos and chimpanzees. *Behaviour*, 130, 211–28.

Fuster, V. (1984). Extramarital reproduction and infant mortality in rural Galicia (Spain). *Journal of Human Evolution*, **13**, 457–63.

Gagneux, P., Woodruff, D. S. & Boesch, C. (1997). Furtive mating in female chimpanzees. *Nature*, **387**, 358–9.

Gajda, A. M. T. & Brooks, R. J. (1993). Paternal care in collared lemmings (*Dicrostonyx richardsoni*) – artifact or adaptation. *Arctic*, **46**, 312–15.

Galat-Luong, A. & Galat, G. (1979). Conséquences comportementales des perturbations sociales repetées sure une troupe de Mones de Lowe, *Cercopithecus campbelli lowei*, de Côte d'Ivoire. *Terre et Vie*, **33**, 4–57.

Galetti, M., Pedroni, F. & Paschoal, M. (1994). Infanticide in the brown howler monkey, *Alouatta fusca*. *Neotropical Primates*, **2**(4), 6–7.

Garber, P. A. (1997). One for all and breeding for one: cooperation and competition as a tamarin reproductive strategy. *Evolutionary Anthropology*, **5**, 187–99.

Garland, T., Dickerman A. W., Janis, C. M. & Jones, J. A. (1993). Phylogenetic analysis of covariance by computer simulation. *Systematic Biology*, **42**, 265–92.

Garland, T. J., Harvey, P. H. & Ives, A. R. (1992). Procedures for the analysis of comparative data using phylogenetically independent contrasts. *Systematic Biology*, **41**, 18–32.

Gehrt, S. D. & Fritzell, E. K. (1996). Second estrus and late litters in raccoons. *Journal of Mammalogy*, **77**, 388–93.

Gehrt, S. D. & Fritzell, E. K. (1999). Behavioural aspects of the raccoon mating system: determinants of consortship success. *Animal Behaviour*, **57**, 593–601.

Getz, U. & Carter, C. S. (1998). Inbreeding avoidance in the prairie vole, *Microtus ochrogaster*. *Ethology, Ecology, Evolution*, **10**, 115–27.

Gibber, J. R., Piontkewitz, Y. & Terkel, J. (1984). Response of male and female Siberian hamsters towards pups. *Behavior and Neural Biology*, **42**, 177–82.

Gilbert, D., Packer, C., Pusey, A. E., Stephens, J. C. & O'Brien, S. J. (1991). Analytical DNA fingerprinting in lions: parentage, genetic diversity and kinship. *Journal of Heredity*, **82**, 378–86.

Girman, D. J., Mills, M. G. L., Geffen, E. & Wayne, R. K. (1997). A molecular genetic analysis of social structure, dispersal and interpack relationships of the African wild dog (*Lycaon pictus*). *Behavioral Ecology and Sociobiology*, **40**, 187–98.

Gittleman, J. L. (1986). Carnivore life history patterns: allometric, phylogenetic, and ecological associations. *American Naturalist*, **127**, 744–71.

Gittleman, J. L. & Luh, H.-K. (1992). On comparing comparative methods. *Annual Review of Ecology and Systematics*, **23**, 383–404.

Gittleman, J. L. & Thompson, S. D. (1988). Energy allocation in mammalian reproduction. *American Zoologist*, **28**, 863–75.

Gjershaug, J. O., Järvi, T. & Røskaft, E. (1989). Marriage entrapment by 'solitary' mothers: a study on male deception by female pied flycatchers. *American Naturalist*, **133**, 273–6.

Glander, K. E. (1980). Reproduction and population growth in free-ranging mantled howling monkeys. *American Journal of Physical Anthropology*, **53**, 25–36.

Glander, K. E. (1992). Dispersal patterns in Costa Rican mantled howling monkeys. *International Journal of Primatology*, **13**, 415–36.

Goldizen, A. W. (1987). Tamarins and marmosets: communal care of offspring. In *Primate Societies*, ed. B. Smuts, D. L. Cheney, R. M. Seyfarth, R. W. Wrangham & T. T. Struhsaker, pp. 34–43. Chicago: University of Chicago Press.

Goldizen, A. W., Mendelson, J., van Vlaardingen, M. & Terborgh, J. (1996). Saddle-back tamarin (*Saguinus fuscicollis*) reproductive strategies: evidence from a thirteen year study of a marked population. *American Journal of Primatology*, **38**, 57–83.

Goldizen, A. W. & Terborgh, J. (1989). Demography and dispersal patterns of a tamarin population: possible causes of delayed breeding. *American Naturalist*, **134**, 208–24.

Goldstein, H., Eisikovitz, D. & Yom-Tov, Y. (1986). Infanticide in the Palestine sunbird. *Condor*, **88**, 528–9.

Gomendio, M. & Colmenares, F. (1989). Infant killing and infant adoption following the introduction of new males to an all-female colony of baboons. *Ethology*, **80**, 223–44.

Gomendio, M. & Roldan, E. R. S. (1993). Mechanisms of sperm competition: linking physiology and behavioural ecology. *Trends in Ecology and Evolution*, **8**, 95–100.

Gompper, M. E., Gittleman, J. L. & Wayne, R. K. (1997). Genetic relatedness, coalitions and social behaviour of white-nosed coatis, *Nasua narica*. *Animal Behaviour*, **53**, 781–97.

Goodall, J. (1977). Infant killing and cannibalism in free-living chimpanzees. *Folia Primatologica*, **28**, 259–82.

Goodall, J. (1986). *The Chimpanzees of Gombe. Patterns of Behavior*. Cambridge, MA: Harvard University Press.

Goodall, J., Bandora, A., Bergmann, E., Busse, C., Matama, H., Mpongo, E., Pierce, A. & Riss, D. (1979). Intercommunity interactions in the chimpanzee population of the Gombe National Park. In *The Great Apes*, ed. D. A. Hamburg & E. R. McCown, pp. 13–54. Menlo Park, CA: Benjamin Cummmings.

Goossens, B., Graziani, L., Waits, L. P., Farand, E., Magnolon, S., Coulon, J., Bel, M-C., Taberlet, P. & Allain, D. (1998). Extra-pair paternity in the monogamous Alpine marmot revealed by nuclear DNA microsatellite analysis. *Behavioral Ecology and Sociobiology*, **43**, 281–8.

Gori, D. F., Rohwer, S. & Caselle, J. (1996). Accepting unrelated broods helps replacement male yellow-headed blackbirds attract mates. *Behavioral Ecology*, **7**, 49–54.

Gould, S. J. (1999). Dorothy, It's really Oz. *Time*, 23 August, 59.

Gouzoules, H. (1975). Maternal rank and early social interactions of infant stumptail macaques, *Macaca arctoides*. *Primates*, **16**, 405–18.

Gouzoules, H. (1984). Social relations of males and infants in a troop of Japanese monkeys: a consideration of causal mechanisms. In *Primate Paternalism*, ed. D. M. Taub, pp. 127–45, New York: Van Nostrand Reinhold.

Gowaty, P. A. (1996a). Multiple mating by females selects for males that stay: another hypothesis for social monogamy in passerine birds. *Animal Behaviour*, **51**, 482–4.

Gowaty, P. A. (1996b). Battles of the sexes and origins of monogamy. In *The Study of Monogamy. Partnerships in Birds,* ed. J. M. Black, pp. 21–52. Oxford: Oxford University Press.

Gowaty, P. (ed.) (1997a). *Feminism and Evolutionary Biology: Boundaries, Insections and Frontiers*. New York: Chapman & Hall.

Gowaty, P. (1997b). Sexual dialectics, sexual selection, and variation in reproductive behavior. In *Feminism and Evolutionary Biology: Boundaries, Insections and Frontiers*, ed. P. A. Gowaty, pp. 351–84. New York: Chapman & Hall.

Gowaty, P. A. & Buschhaus, N. (1998). Ultimate causation of aggressive and forced copulation in birds: female resistance, the CODE hypothesis, and social monogamy. *American Zoologist*, **38**, 207–25.

Grafen, A. (1991). Modelling in behavioural ecology. In *Behavioural Ecology, an Evolutionary Approach,* 3rd edn, ed. J. R. Krebs & N. B. Davies, pp. 5–31. Oxford: Blackwell Scientific Publications.

Greenwood, P. J. (1980). Mating systems, philopatry and dispersal in birds and mammals. *Animal Behaviour*, **28**, 1140–62.

Gubernick, D. J. (1994). Biparental care and male–female relations in mammals. In *Infanticide and Parental Care,* ed. S. Parmigiani & F. S. vom Saal, pp. 427–63. Chur (Switzerland): Harwood.

Gubernick, D. J., Schneider, K. A. & Jeannotte, L. A. (1994). Individual differences in the mechanisms underlying the onset and maintenance of paternal behavior and the inhibition of infanticide in the monogamous biparental California mouse, *Peromyscus californicus. Behavioral Ecology and Sociobiology*, **34**, 225–31.

Gubernick, D. J., Winslow, J. T., Jensen, P., Jeanotte, L. & Bowen, J. (1995). Oxytocin changes in males over the reproductive cycle in the monagomous, biparental California mouse, *Peromyscus californicus. Hormones and Behavior*, **29**, 59–73.

Gurmaya, K. J. (1986). Ecology and behavior of *Presbytis thomasi* in northern Sumatra. *Primates*, **27**, 151–72.

Gust, D. A. (1994a). A brief report on the social behavior of the crested mangabey (*Cercocebus galeritus galeritus*) with a comparison to the sooty mangabey (*C. torquatus atys*). *Primates*, **35**, 375–83.

Gust, D. A. (1994b). Alpha-male sooty mangabeys differentiate between females' fertile and their postconception maximal swellings. *International Journal of Primatology*, **15**, 289–302.

Gust, D. A., Gordon, T. P. & Gergits, W. (1995). Proximity at birth relates to a sire's tolerance of his offspring among sooty mangabeys. *Animal Behaviour*, **49**, 1403–5.

Gust, D. A., McCaster, T., Gordon, T. P., Gergits, W. F., Casna, N. J. & McClure, H. M. (1998). Paternity in sooty mangabeys. *International Journal of Primatology*, **19**, 83–94.

Hackländer, K. & Arnold, W. (1999). Male-caused failure in female reproduction and its adaptive value in alpine marmots (*Marmota marmota*). *Behavioral Ecology and Sociobiology,* **10**, 592–7.

Hadidian, J. M. (1979). Allo- and auto-grooming in a captive black ape colony (*Macaca nigra*). Ph.D. dissertation, Pennsylvania State University.

Hall, K. R. L. & DeVore, I. (1965). Baboon social behavior. In *Primate Behavior. Field Studies of Monkeys and Apes*, ed. I. DeVore, pp. 53–110. New York: Holt, Rinehart and Winston.

Haltenorth, Th., Diller, H. & Smeenk, C. (1979). *Elseviers Gids van de Afrikaanse Zoogdieren*. Amsterdam: Elsevier.

Hamai, M., Nishida, T., Takasaki, H. & Turner, L. A. (1992). New records of within-group infanticide and cannibalism in wild chimpanzees. *Primates, 33*, 151–62.

Hamilton, W. D. (1964a). The genetical evolution of social behaviour. I. *Journal of Theoretical Biology, 7*, 1–16.

Hamilton, W. D. (1964b). The genetical evolution of social behaviour. II. *Journal of Theoretical Biology, 7*, 17–52.

Hamilton III, W. J. (1984). Significance of paternal investment by primates to the evolution of male–female associations. In *Primate Paternalism*, ed. D. M. Taub, pp. 309–35. New York: Van Nostrand Reinhold.

Hamilton III, W. J. & Bulger, J. B. (1990). Natal male baboon rank rises and successful challenges to resident alpha males. *Behavioral Ecology and Sociobiology, 26*, 357–63.

Hamilton III, W. J. & Bulger, J. B. (1992). Facultative expression of behavioral differences between one-male and multimale savanna baboon groups. *American Journal of Primatology, 28*, 61–71.

Hamilton III, W. J., Buskirk, R. E. & Buskirk, W. H. (1976). Defense of space and resources by chacma (*Papio ursinus*) baboon troops in an African desert and swamp. *Ecology, 57*, 1264–72.

Hamilton III, W. J., Busse, C. D. & Smith, K. S. (1982). Adoption of infant orphan chacma baboons. *Animal Behaviour, 30*, 29–34.

Hannon, S. J. (1984). Factors limiting polygyny in the willow ptarmigan. *Animal Behaviour, 32*, 153–61.

Hansson, B. Bensch, S. & Hasselquist, D (1997). Infanticide in great reed warblers: secondary females destroy eggs of primary females. *Animal Behavior, 54*, 297–304.

Haraway, D. J. (1989). *Primate Visions: Gender, Race, and Nature in the World of Modern Science*. New York: Routledge.

Harcourt, A. H. (1978). Strategies of emigration and transfer by primates, with particular reference to gorillas. *Zeitschrift für Tierpsychologie, 48*, 401–20.

Harcourt, A. H. (1979a). Contrasts between male relationships in wild gorilla groups. *Behavioral Ecology and Sociobiology, 5*, 39–49.

Harcourt, A. H. (1979b). Social relationships between adult male and female mountain gorillas in the wild. *Animal Behaviour, 27*, 325–42.

Harcourt, A. H. (1992). Coalitions and alliances: are primates more complex than nonprimates? In *Coalitions and Alliances in Humans and Other Animals*, ed. A. H. Harcourt & F. B. M. de Waal, pp. 445–71. Oxford: Oxford University Press.

Harcourt, A. H. & de Waal, F. B. M. (1992). Coalitions and alliances: a history of ethological research. In *Coalitions and Alliances in Humans and Other Animals*, ed. A. H. Harcourt & F. B. M. de Waal, pp. 1–19. New York: Oxford University Press.

Harcourt, A. H., Fossey, D., Stewart, K. & Watts, D. P. (1980). Reproduction in wild gorillas and some comparisons with chimpanzees. *Journal of Reproductive Fertility Supplement, 28*, 59–70.

Harcourt, A. H., Purvis, A. & Liles, L. (1995). Sperm competition: mating system, not breeding season, affects testes size of primates. *Functional Ecology, 9*, 468–76.

Harcourt, A. H., Stewart, K. S. & Fossey, D. (1976). Male emigration and female transfer in wild mountain gorilla. *Nature, 263*, 226–7.

Harcourt, C. (1986). Seasonal variation in the diet of South African galagos. *International Journal of Primatology*, **7**, 491–506.

Harcourt, R. (1992). Factors affecting early mortality in the South American fur seal *Arctocephalus gazella* at South Georgia. *Journal of Zoology London,* **202**, 449–60.

Haring, D. M. & Wright, P. C. (1989). Hand-raising a Phillipine tarsier (*Tarsius syrichta*). *Zoo Biology*, **8**, 265–74.

Hartwig, W. C. (1996). Perinatal life history traits in New World monkeys. *American Journal of Primatology,* **40**, 99–130.

Harvey, P. & Pagel, M. (1991). *The Comparative Method in Evolutionary Biology*. Oxford: Oxford University Press.

Hasegawa, T. (1989). Sexual behavior of immigrant and resident female chimpanzees at Mahale. In *Understanding Chimpanzees*, ed. P. G. Heltne & L. A. Marquardt, pp. 90–103. Cambridge, MA: Harvard University Press.

Hasegawa, T. & Hiraiwa, M. (1980). Social interactions of orphans observed in a free-ranging troop of Japanese monkeys. *Folia Primatologica*, **33**, 129–58.

Hauser, M. D. (1986). Male responsiveness to infant distress calls in free-ranging vervet monkeys. *Behavioral Ecology and Sociobiology*, **19**, 65–71.

Hauser, M. D. (1988). Variation in maternal responsiveness in free-ranging vervet monkeys A response to infant mortality risk? *American Naturalist*, **131**, 573–87.

Hausfater, G. (1975). *Dominance and Reproduction in Baboons: A Quantitative Analysis*. Basel: Karger.

Hausfater, G. (1984). Infanticide in langurs: strategies, counterstrategies, and parameter values. In *Infanticide: Comparative and Evolutionary Perspectives*, ed. G. Hausfater & S. B. Hrdy, pp. 257–81. New York: Aldine de Gruyter.

Hausfater, G. & Hrdy, S. B. (eds.) (1984). *Infanticide: Comparative and Evolutionary Perspectives*. Foundations of Human Behavior Series. New York: Aldine de Gruyter.

Hausfater, G., Saunders, C. D. & Chapman, M. (1981). Some applications of computer models to the study of primate mating and social systems. In *Natural Selection and Social Behavior*, ed. R. D. Alexander & D. W. Tinkle, pp. 345–60. New York: Chiron.

Hawkes, K. (1993). Why hunter–gatherers work; an ancient version of the problem of public goods. *Current Anthropology,* **34**, 341–61.

Hawkes, K., Rogers, A. R. & Charnov, E. L. (1995). The male's dilemma: increased offspring production is more paternity to steal. *Evolutionary Ecology*, **9**, 662–77.

Hayssen, V., van Tienhoven, A. & van Tienhoven, A. (1993). *Asdell's Patterns of Mammalian Reproduction; a Compendium of Species-Specific Data*. Ithaca, NY: Comstock Publishing Associates.

Heise, S. & Lippke, J. (1997). Role of female aggression in prevention of infanticidal behavior in male common voles, *Microtus arvalis* (Pallas, 1779). *Aggressive Behavior* **23**, 293–8.

Heistermann, M., Finke, M. & Hodges, J. K. (1995). Assessment of female reproductive status in captive-housed hanuman langurs (*Presbytis entellus*) by measurement of urinary and fecal steroid excretion patterns. *American Journal of Primatology,* **37**, 275–84.

Heistermann, M., Möhle, U., Vervaecke, H., van Elsacker, L. & Hodges, J. K. (1996).

Application of urinary and fecal steroid measurements for monitoring ovarian function and pregnancy in the bonobo (*Pan paniscus*) and evaluation of perineal swelling patterns in relation to endocrine events. *Biology of Reproduction*, **55**, 844–53.

Heltne, P. G., Wojcik, J. F. & Pook, A. G. (1981). Goeldi's monkey, genus *Callimico*. In *Ecology and Behavior of Neotropical Primates*, ed. A. F. Coimbra-Filho & R. A. Mittermeier, pp. 184–210. Brasil: Academia Brasileira de Ciencias.

Hendy-Neely, H. & Rhine, R. J. (1977). Social development of stumptail macaques (*Macaca arctoides*): momentary touching and other interactions with adult males during the infants' first 60 days of life. *Primates*, **18**, 589–600.

Henzi, S. P. & Lucas, J. W. (1980). Observations on the inter-troop movement of adult vervet monkeys (*Cercopithecus aethiops*). *Folia Primatologica*, **33**, 220–35.

Heske, E. J. (1987). Pregnancy interruption by strange males in the California vole. *Journal of Mammalogy*, **68**, 406–10.

Heske, E. J. & Nelson, R. J. (1984). Pregnancy interruption in *Microtus ochrogaster*: laboratory artifact or field phenomenon? *Biology of Reproduction*, **31**, 97–103.

Hewlett, B. S. (1991). *Intimate Fathers. The Nature and Context of Aka Pygmy Paternal Infant Care*. Ann Arbor, MI: University of Michigan Press.

Hill, D. A. (1986). Social relationships between adult male and immature rhesus macaques. *Primates*, **27**, 425–40.

Hill, D. A. & van Hooff, J. A. R. A. M. (1994). Affiliative relationships between males in groups of nonhuman primates: a summary. *Behaviour*, **130**, 143–9.

Hill, R. A. & Dunbar, R. I. M. (1998). An evaluation of the roles of predation rate and predation risk as selective pressures on primate grouping behaviour. *Behaviour*, **135**, 411–30.

Hill, R. A. & Lee, P. C. (1998). Predation risk as an influence on group size in cercopithecoid primates: implications for social structure. *Journal of Zoology London*, **245**, 447–56.

Hinde, R. A. (1975). The concept of function. In *Function and Evolution in Behavior: Essays in Honour of Professor Niko Tinbergen*, ed. G. Baerends, C. Beer & A. Manning, pp. 3–15. Oxford: Clarendon.

Hinde, R. A. (ed.) (1983). *Primate Social Relationships: An Integrated Approach*. Sunderland, MA: Sinauer.

Hiraiwa, M. (1981). Maternal and alloparental care in a troop of free-ranging Japanese monkeys. *Primates*, **22**, 309–29.

Hiraiwa-Hasegawa, M. (1988). Adaptive significance of infanticide in primates. *Trends in Ecology and Evolution*, **3**, 102–5.

Hiraiwa-Hasegawa, M. (1990). Maternal investment before weaning. In *The Chimpanzees of the Mahale Mountains*, ed. T. Nishida, pp. 257–66. Tokyo: University of Tokyo Press.

Hiraiwa-Hasegawa, M. (1992). Cannibalism among non-human primates. In *Cannibalism: Ecology and Evolution among Diverse Taxa*, ed. M. A. Elgar & B. J. Crespi, pp. 322–38. Oxford: Oxford University Press.

Hiraiwa-Hasegawa, M. & Hasegawa, T. (1994). Infanticide in nonhuman primates: Sexual selection and local resource competition. In *Infanticide and Parental Care*, ed. S. Parmigiani & F. S. vom Saal, pp. 137–54, Chur (Switzerland): Harwood.

Hodges, J. K. (1992). Detection of oestrous cycles and timing ovulation. *Symposium of the Royal Society London*, **64**, 73–88.

Hodges, J. K., Czekala, N. M. & Lasley, B. L. (1979). Estrogen and luteinizing hormone secretion in diverse primate species from simplified urinary analysis. *Journal of Medical Primatology*, **8**, 349–64.

Hofshi, H., Gersani, M. & Katzir, G. (1987). A case of infanticide among Tristram's grackles, *Onychognathus tristramii*. *Ibis*, **129**, 389–90.

Hogg, J. T. & Forbes, S. H. (1997). Mating in bighorn sheep: frequent male reproduction via a high-risk 'unconventional' tactic. *Behavioral Ecology and Sociobiology*, **41**, 33–48.

Honacki, J. H., Kinman, K. E. & Koeppl, J. W. (1982). *Mammal Species of the World*. Lawrence, Kansas: Allen Press, Inc.

Hood, L. C. (1994). Infanticide among ringtailed lemurs (*Lemur catta*) at Berenty reserve, Madagascar. *American Journal of Primatology*, **33**, 65–9.

Hoogland, J. L. (1985). Infanticide in prairie-dogs: lactating females kill offspring of close kin. *Science*, **230**, 1037–40.

Hoogland, J. L. (1994). Nepotism and infanticide among prairie dogs. In *Infanticide and Parental Care*, ed. S. Parmigiani & F. S. vom Saal, pp. 321–37. Chur (Switzerland): Harwood.

Hoogland, J. L. (1995). *The Black-Tailed Prairie Dog: Social Life of a Burrowing Mammal*. Chicago: University of Chicago Press.

Hoogland, J. L. (1998). Why do female Gunnison's prairie dogs copulate with more than one male? *Animal Behaviour*, **55**, 351–9.

Horne, T. J. & Ylönen, H. (1996). Female bank voles (*Clethrionomys glareolus*) prefer dominant males; but what if there is no choice? *Behavioral Ecology and Sociobiology*, **38**, 401–5.

Hotta, M. (1994). Infanticide in little swifts taking over costly nests. *Animal Behaviour*, **47**, 491–3.

Hrdy, S. B. (1974). Male–male competition and infanticide among the langurs (*Presbytis entellus*) of Abu, Rajasthan. *Folia Primatologica*, **22**, 19–58.

Hrdy, S. B. (1976). Care and exploitation of nonhuman primate infants by conspecifics other than the mother. In *Advances in the Study of Behavior*, vol. 6, ed. L. Rosenblatt, R. A. Hinde, R. Shaw & C. Beer, pp. 101–58. New York: Academic Press.

Hrdy, S. B. (1977). *The Langurs of Abu – Female and Male Strategies of Reproduction*. Cambridge, MA: Harvard University Press.

Hrdy, S. B. (1979). Infanticide among animals: a review, classification, and examination of the implications for the reproductive strategies of females. *Ethology and Sociobiology*, **1**, 13–40.

Hrdy, S. B. (1981). *The Woman that Never Evolved*. Cambridge, MA: Harvard University Press.

Hrdy, S. B. (1982). Positivist thinking encounters field primatology, resulting in agonistic behaviour. *Social Science Information*, **21**, 245–50.

Hrdy, S. B. (1995). Natural-born mothers. *Natural History*, **104**(12), 30–43.

Hrdy, S. B. & Hausfater, G. (1984). Comparative and evolutionary perspectives on infanticide: introduction and overview. In *Infanticide: Comparative and Evolutionary Perspectives*, ed. G. Hausfater & S. B. Hrdy, pp. xiii–xxxv. New York: Aldine de Gruyter.

Hrdy, S. B., Janson, C. H. & van Schaik, C. P. (1995). Infanticide: let's not throw out the baby with the bath water. *Evolutionary Anthropology*, **3**, 151–4.

Hrdy, S. B. & Whitten, P. L. (1987). Patterning of sexual activity. In *Primate Societies*, ed. B. B. Smuts, D. L. Cheney, R. M. Seyfarth, R. W. Wrangham & T. T. Struhsaker, pp. 370–84. Chicago: University of Chicago Press.

Huck, U. W. (1982). Pregnancy block in laboratory mice as a function of male social status. *Journal of Reproductive Fertility*, **66**, 181–4.

Huck, U. W. (1984). Infanticide and the evolution of pregnancy block in rodents. In *Infanticide. Comparative and Evolutionary Perspectives*, ed. G. Hausfater & S. B. Hrdy, pp. 349–65. New York: Aldine de Gruyter.

Huck, U. W., Lisk, R. D., Miller, K. S. & Bethel, A. (1988). Progesterone levels and socially-induced implantation failure and fetal resorption in golden hamsters (*Mesocricetus auratus*). *Physiology and Behavior*, **44**, 321–6.

Huck, U. W., Quinn, R. P. & Lisk, R. D. (1985). Determinants of mating success in the golden hamster (*Mesocricetus auratus*): IV. Sperm competition. *Behavioral Ecology and Sociobiology*, **17**, 239–52.

Huck, U. W., Soltis, R. L. & Coopersmith, C. B. (1982). Infanticide in male laboratory mice: effects of social status, prior sexual experience, and basis for discrimination between related and unrelated young. *Animal Behaviour*, **30**, 1158–65.

Hurtado, A. M. & Hill, K. R. (1992). Paternal effect on offspring survivorship among Ache and Hiwi hunter-gatherers: implications for modeling pair-bond stability. In *Father–Child Relations – Cultural and Biosocial Contexts*, ed. B. Hewlett, pp. 31–55, New York: Aldine de Gruyter.

Idani, G. (1991). Social relationships between immigrant and resident bonobo (*Pan paniscus*) females at Wambe. *Folia Primatologica*, **57**, 83–95.

Ihobe, H. (1992). Male–male relationships among wild bonobos (*Pan paniscus*) at Wamba, Republic of Zaïre. *Primates*, **33**, 163–79.

Isbell, L. A. (1991). Contest and scramble competition: patterns of female aggression and ranging behavior among primates. *Behavioral Ecology*, **2**, 143–55.

Isbell, L. A. (1994). Predation on primates: ecological patterns and evolutionary consequences. *Evolutionary Anthropology*, **3**, 61–71.

Isbell, L. A., Cheney, D. L. & Seyfarth, R. M. (1990). Costs and benefits of home range shifts among vervet monkeys (*Cercopithecus aethiops*) in Amboseli National Park, Kenya. *Behavioral Ecology and Sociobiology*, **27**, 351–8.

Isbell, L. A. & Pruetz, J. D. (1998). Differences between vervets (*Cercopithecus aethiops*) and patas monkeys (*Erythrocebus patas*) in agonistic interactions between adult females. *International Journal of Primatology*, **19**, 837–55.

Isbell, L. A. & van Vuren, D. (1996). Differential costs of locational and social dispersal and their consequences for female group-living primates. *Behaviour*, **133**, 1–36.

Itani, J. (1959). Paternal care in the wild Japanese monkey, *Macaca fuscata fuscata*. *Primates*, **2**, 61–93.

Izard, M. K. & Simons, E. L. (1986). Isolation of females prior to parturition reduces neonatal mortality in Galago. *American Journal of Primatology*, **10**, 249–55.

Izawa, K. (1978). A field study of the ecology and behavior of the black-maned tamarin (*Saguinus nigricollis*). *Primates*, **19**, 241–74.

Izawa, K. & Lozano, M. H. (1991). Social changes within a group of red howler monkeys

(*Alouatta seniculus*). III. *Field Studies of New World Monkeys, La Macarena, Colombia,* **5**, 1–16.

Izawa, K. & Lozano M. H. (1994). Social changes within a group of red howler monkeys (*Alouatta seniculus*). V. *Field Studies of New World Monkeys, La Macarena, Colombia,* **9**, 33–9.

Jakubowski, M. & Terkel, J. (1982). Infanticide and caretaking in non-lactating *Mus musculus*: influence of genotype, family group and sex. *Animal Behaviour,* **30**, 1029–35.

Jakubowski, M. & Terkel, J. (1985). Transition from pup killing to parental behavior in male and virgin female albino rats. *Physiology and Behavior,* **34**, 683–6.

Janson, C. H. (1984). Female choice and mating system of the brown capuchin monkey *Cebus apella* (Primates: Cebidae). *Zeitschrift für Tierpsychologie,* **65**, 177–200.

Janson, C. H. (1985). Aggressive competition and individual food intake in wild brown capuchin monkeys. *Behavioral Ecology and Sociobiology,* **18**, 125–38.

Janson, C. H. (1988). Intra-specific food competition and primate social structure: a synthesis. *Behaviour,* **105**, 1–17.

Janson, C. H. (1997). Ecological and social determinants of male coalitions in New World primates. *Primate Report,* **48–2**, 22.

Janson, C. H. (1998). Testing the predation hypothesis for vertebrate sociality: prospects and pitfalls. *Behaviour,* **135**, 389–410.

Janson, C. H. & Goldsmith, M. (1995). Predicting group size in primates: foraging costs and predation risks. *Behavioral Ecology,* **6**, 326–36.

Jay, P. C. (1963). The social behavior of the langur monkey. Ph.D. thesis, University of Chicago.

Jensen, P. M. & Gustafsson, T. O. (1984). Evidence for pregnancy failure in young female *Lemmus lemmus* and *Microtus oeconomus. Canadian Journal of Zoology,* **62**, 2568–70.

Jeppson, B. (1986). Mating by pregnant water voles (*Arvicola terrestris*): a strategy to counter infanticide by males? *Behavioral Ecology and Sociobiology,* **19**, 293–6.

Johnson, L. S. & Albrecht, D. J. (1993). Does the cost of polygyny in house wrens include reduced male assistance in defending offspring? *Behavioral Ecology and Sociobiology,* **33**, 131–6.

Johnson, L. S. & Kermott, L. H. (1993). Why is reduced male parental assistance detrimental to the reproductive success of secondary female house wrens? *Animal Behaviour,* **46**, 1111–20.

Johnson, L. S., Merkle, M. S. & Kermott, L. H. (1992). Experimental evidence for importance of male parental care in monogamous house wrens. *Auk,* **109**, 662–4.

Jolly, A. (1998). Pair-bonding, female aggression and the evolution of lemur societies. *Folia Primatologica,* **69**, 1–13.

Jolly, A., Caless, S., Cavigelli, S., Gould, L., Pitts, A., Pereira, M. E., Pride, R. E., Rabenandrasana, H. D., Walker, J. D. & Zafison, T. (2000). Infant killing, wounding, and predation in *Eulemur* and *Lemur. International Journal of Primatology,* **21**, 21–40.

Jones, C. B. (1981). The evolution and socioecology of dominance in primate groups: a theoretical formulation, classification and assessment. *Primates,* **22**, 70–83.

Jones, C. B. (1997). Subspecific differences in vulva size between *Alouatta palliata palliata* and *A. p. mexicana*: implications for assessment of female receptivity. *Neotropical Primates,* **5**, 46–47.

Kappeler, P. M. (1991). Patterns of sexual dimorphism in body weight among prosimian primates. *Folia Primatologica*, **57**, 132–46.

Kappeler, P. M. (1993). Variation in social structure: the effects of sex and kinship on social interactions in three lemur species. *Ethology*, **93**, 125–45.

Kappeler, P. M. (1995). Life history variation among nocturnal prosimians. In *Creatures of the Dark: The Nocturnal Prosimians*, ed. L. Alterman, G. A. Doyle & M. K. Izard, pp. 75–92. New York: Plenum Press.

Kappeler, P. M. (1998). Nests, tree holes, and the evolution of primate life histories. *American Journal of Primatology*, **46**, 7–33.

Kappeler, P. M. (1999). Lemur social structure and convergence in primate socioecology. In *Comparative Primate Socioecology*, ed. P. C. Lee, pp. 273–99. Cambridge: Cambridge University Press.

Kawai, M. (1963). *Nihonzaru No Seitai* [in Japanese]. Tokyo: Kawade Shobo.

Kawamura, S., Norikoshi, K. & Azuma, N. (1991). Observations of Formosan monkeys (*Macaca cyclopis*) in Taipingshan, Taiwan. In *Primatology Today*, ed. A. Ehara, T. Kimura, O. Takenaka & M. Iwamoto, pp. 97–100. Amsterdam: Elsevier.

Kawanaka, K. (1981). Infanticide and cannibalism in chimpanzees, with special reference to the newly observed case in the Mahale mountains. *Africa Studies Monographs*, **1**, 69–91.

Kay, C. E. (1994). Aboriginal overkill. The role of native Americans in structuring western ecosystems. *Human Nature*, **5**, 359–98.

Kay, R. F. (1984). On the use of anatomical features to infer foraging behavior in extinct primates. In *Adaptations for Foraging in Nonhuman Primates,* ed. P. Rodman & J. Cant, pp. 21–53. New York: Columbia University Press.

Keil, A. & Sachser, N. (1998). Reproductive benefits from female promiscuous mating in a small mammal. *Ethology*, **104**, 897–903.

Keller, L. & Reeve, H. K. (1995). Why do females mate with multiple males? The sexually selected sperm hypothesis. *Advances in the Study of Behavior*, **24**, 291–315.

Kelley, J. (1997). Paleobiological and phylogenetic significance of life history in Miocene hominoids. In *Function, Phylogeny, and Fossils: Miocene Hominoid Evolution and Adaptations*, ed. D. R. Begun, C. V. Ward & M. D. Rose, pp. 173–208. New York: Plenum Press.

Kendrick, K. M. & Dixson, A. F. (1983). The effect of the ovarian cycle on the sexual behaviour of the common marmoset (*Callithrix jacchus*). *Physiology and Behavior*, **30**, 735–42.

Kenney, A. M., Evans, R. L. & Dewsbury, D. A. (1977). Postimplantation pregnancy disruption in *Microtus ochrogaster*, *M. pennsylvanicus*, and *Peromyscus maniculatus*. *Journal of Reproduction and Fertility*, **49**, 365–7.

Kermott, L. H. & Johnson, L. S. (1990). Brood adoption and apparent infanticide in a north-temperate house wren population. *Wilson Bulletin*, **102**, 333–6.

Kermott, L. H., Johnson, L. S. & Merkle, M. S. (1991). Experimental evidence for the function of mate replacement and infanticide by males in a north-temperate population of house wrens. *Condor,* **93**, 630–6.

Keverne, E. B. (1976). Sexual receptivity and attractiveness in monkeys. *Advances in the Study of Behavior,* **7**, 155–200.

King, C. (1989) *Weasels and Stoats*. Ithaca, NY: Cornell University Press.

King, W. J. & Allainé, D. (1998). Copulatory behaviour of alpine marmots (*Marmota marmota*). *Mammalia*, **62**, 439–41.

Kinnaird, M. F. (1990). Behavioral and demographic responses to habitat change by the Tana River crested mangabey (*Cercocebus galeritus galeritus*). Ph. D. thesis, University of Florida, Gainesville.

Kinzey, W. G. & Wright, P. C. (1982). Grooming behavior in the titi monkey (*Callicebus torquatus*). *American Journal of Primatology*, **3**, 267–75.

Kirkpatrick-Tanner, M., Aeschlimann, C. & Anzenberger, G. (1996). Occurrence of an infanticide within a captive polygynous group of common marmosets, *Callithrix jacchus*. *Folia Primatologica*, **67**, 52–8.

Kitchener, A. (1991). *The Natural History of the Wild Cats*. New York: Cornell University Press.

Kizilov, V. A. & Berendaev, S. A. (1978). The long-tailed marmot. In *Marmots: Their Distribution and Ecology*, ed. R. P. Zimina, pp. 79–116. Moscow: Nauka.

Kleiman, D. G. (1977). Monogamy in mammals. *Quarterly Review of Biology*, **52**, 39–69.

Kleiman, D. G. & Malcolm, J. R. (1981). The evolution of male parental investment in mammals. In *Parental Care in Mammals*, ed. D. J. Gubernick & P. H. Klopfer, pp. 347–87. New York: Plenum Press.

Klopfer, P. H. & Boskoff, K. J. (1979). Maternal behavior in prosimians. In *The Study of Prosimian Behavior*, ed. G. A. Doyle & R. D. Martin, pp. 123–56. New York: Academic Press.

Koenig, A. (1995). Group size, composition, and reproductive success in wild common marmosets (*Callithrix jacchus*). *American Journal of Primatology*, **35**, 311–17.

Koenig, A., Beise, J., Chalise, M. K. & Ganzhorn, J. U. (1998). When females should contest for food – testing hypotheses about resource density, distribution, size, and quality with Hanuman langurs (*Presbytis entellus*). *Behavioral Ecology and Sociobiology*, **42**, 225–37.

Koenig, A., Borries, C., Chalise, M. K. & Winkler, P. (1997). Ecology, nutrition, and timing of reproductive events in an Asian primate, the Hanuman langur (*Presbytis entellus*). *Journal of Zoology, London*, **243**, 215–35.

Koenig, W. D. (1990). Opportunity of parentage and nest destruction in polygynadrous acorn woodpeckers, *Melanerpes formicivorus*. *Behavioral Ecology*, **1**, 55–61.

Koenig, W. D., Mumme, R. L. & Pitelka, F. A. (1983). Females roles in cooperatively breeding acorn woodpeckers. In *Social Behavior of Female Vertebrates*, ed. S. K. Wasser, pp. 235–6. New York: Academic Press.

Komers, P. E. & Brotherton, P. N. M. (1997). Female space is the best predictor of monogamy in mammals. *Proceedings of the Royal Society, London B*, **264**, 1261–70.

König, B. (1989). Kin recognition and maternal care under restricted feeding in house mice (*Mus domesticus*). *Ethology*, **82**, 328–43.

König, B., Riester, J. & Markl, H. (1988). Maternal care in house mice (*Mus musculus*): II. the energy cost of lactation as a function of litter size. *Journal of Zoology, London*, **216**, 195–210.

Koskela, E., Horne, T. J., Mappes, T. & Ylönen, H. (1996). Does risk of small mustelid predation affect the oestrous cycle in the bank vole, *Clethrionomys glareolus*. *Animal Behaviour*, **51**, 1159–63.

Krebs, J. R. & Davies, N. B. (1987). *An Introduction to Behavioural Ecology*, 2nd edn. Sunderland, MA: Sinauer.

Kruuk, H. (1972). *The Spotted Hyena. A Study of Predation and Social Behavior*. Chicago: University of Chicago Press.

Kuester, J. & Paul, A. (1992). Influence of male competition and female mate choice on male mating success in Barbary macaques (*Macaca sylvanus*). *Behaviour*, **120**, 192–217.

Kuester, J. & Paul, A. (2000). Triadic male–infant interactions and male–male aggression in Barbary macaques (*Macaca sylvanus*). In *Natural Conflict Resolution*, ed. F. Aureli & F. B. M. de Waal. Berkeley: University of California Press, in press.

Kumar, A. (1987). The ecology and population dynamics of the lion-tailed macaque (*Macaca silenus*) in South India. Ph.D. thesis, Cambridge University.

Kumar, A. & Kurup, G. U. (1981). Infant development in the lion-tailed macaque, *Macaca silenus* (Linnaeus): The first eight weeks. *Primates*, **22**, 512–22.

Kummer, H. (1967). Tripartite relations in hamadryas baboons. In *Social Communication among Primates*, ed. S. A. Altmann, pp. 63–71. Chicago: The University of Chicago Press.

Kummer, H. (1968). *Social Organization of Hamadryas Baboons*. Chicago: University of Chicago Press.

Kummer, H., Banaja, A. A., Abo-Khatwa, A. N. & Ghandour, A. M. (1985). Differences in social behavior between Ethiopian and Arabian hamadryas baboons. *Folia Primatologica*, **45**, 1–8.

Kummer, H., Götz, W. & Angst, W. (1974). Triadic differentiation: an inhibitory process protecting pair bonds in baboons. *Behaviour*, **59**, 62–87.

Künkele, J. & Hoeck, H. N. (1989). Age-dependent discrimination of unfamiliar pups in *Galea musteloides* (Mammalia, Caviidae). *Ethology*, **83**, 316–19.

Kurland, J. A. (1977). Kin selection in the Japanese monkey. In *Contributions to Primatology*, vol. 12, ed. F. S. Szalay, pp. 1–145. Basel: Karger.

Kuroda, S. (1992). Paternal care in the pygmy chimpanzees. In *XIVth Congress of the International Primatological Society*, p. 222, Strasbourg.

Kyes, R. C., Rumawas, R. E., Sulistiawati, E. & Budiarsa, N. (1995). Infanticide in a captive group of pig-tailed macaques (*Macaca nemestrina*). *American Journal of Primatology*, **36**, 135–6.

Labov, J. B. (1981). Pregnancy blocking in rodents: adaptive advantages for females. *American Naturalist*, **118**, 361–71.

Labov, J. B. (1984). Infanticidal behavior in male and female rodents: sectional introduction and directions for future research. In *Infanticide. Comparative and Evolutionary Perspectives*, ed. G. Hausfater & S. B. Hrdy, pp. 323–9. New York: Aldine de Gruyter.

Labov, J. B., Huck, U. W., Elwood, R. W. & Brooks, R. J. (1985). Current problems in the study of infanticidal behavior in rodents. *Quarterly Review of Biology*, **60**, 1–20.

Lack, D. (1954). *The Natural Regulation of Animal Numbers*. Oxford: Clarendon Press.

Lack, D. (1968). *Ecological Adaptations for Breeding in Birds*. London: Methuen.

Lamb, M. E., Pleck, J. H., Charnov, E. L. & Levine, J. A. (1987). A biosocial perspective on paternal behavior and involvement. In *Parenting Across the Life Span. Biosocial Dimensions*, ed. J. B. Lancaster, J. Altmann, A. E. Rossi & L. R. Sherrod, pp. 111–42. New York: Aldine de Gruyter.

Lancaster, J. (1971). Play-mothering: the relation between juvenile females and young infants among free-ranging vervet monkeys (*Cercopithecus aethiops*). *Folia Primatologica*, **15**, 161–82.

Landeau, L. & Terborgh, J. (1986). Oddity and the 'confusion effect' in predation. *Animal Behaviour*, **34**, 1372–80.

Lasley, B. L. & Kirkpatrick, J. F. (1991). Monitoring ovarian function in captive and free-ranging wildlife by means of urinary and fecal steroids. *Journal of Zoo and Wildlife Medicine,* **22**, 23–31.

Launhardt, K. (1998). *Paarungs- und Reproduktionserfolg männlicher Hanuman-Languren* (Presbytis entellus) *in Ramnagar/Südnepal.* Göttingen: Cuvillier.

Launhardt, K., Borries, C., Epplen, C., Epplen, J. T. & Winkler, P. (2000). Male reproductive success in Ramnagar langurs (*Semnopithecus*) – the influence of social organization and dominance rank. *Animal Behaviour,* in press.

Launhardt, K., Epplen, C., Epplen, J. T. & Winkler P. (1998). Amplification of microsatellites adapted from human systems in faecal DNA of wild Hanuman langurs (*Presbytis entellus*). *Electrophoresis,* **19**, 1356–61.

LeBoeuf, B. J. (1972). Sexual behavior in the northern elephant seal *Mirounga angustirostris. Behaviour,* **41**, 1–26.

LeBoeuf, B. J. & Briggs, K. T. (1977). The cost of living in a seal harem. *Mammalia,* **41**, 167–95.

LeBoeuf, B. J. & Campagna, C. (1994). Protection and abuse of young in pinnipeds. In *Infanticide and Parental Care,* ed. S. Parmigiani & F. S. vom Saal, pp. 257–76. Chur (Switzerland): Harwood.

LeBoeuf, B. J., Whiting, R. J. & Gantt, R. F. (1972). Perinatal behavior of northern elephant seal females and their young. *Behaviour,* **43**, 121–56.

Lee, P. C. (1999). Comparative ecology of postnatal growth and weaning among haplorhine primates. In *Comparative Primate Socioecology,* ed. P. C. Lee, pp. 111–36. Cambridge: Cambridge University Press.

Lee, W. R. (1977). Bastardy and the socioeconomic structure of south Germany. *Journal of Interdisciplinary History,* 7, 403–45.

Leighton, D. R. (1987). Gibbons: Territoriality and monogamy. In *Primate Societies,* ed. B. B. Smuts, D. L. Cheney, R. M. Seyfarth, R. W. Wrangham & T. T. Struhsaker, pp. 135–45. Chicago: University of Chicago Press.

Leighton-Shapiro, M. E. (1986) . Vigilance and the costs of motherhood to rhesus monkeys. *American Journal of Primatology,* **10**, 414–15.

Leland, L., Struhsaker, T. T. & Butynski, T. M. (1984). Infanticide by adult male in three primate species of the Kibale Forest, Uganda: a test of hypotheses. In *Infanticide: Comparative and Evolutionary Perspectives,* ed. G. Hausfater and S. B. Hrdy, pp. 151–72. New York: Aldine de Gruyter.

Lewison, R. (1998). Infanticide in the hippopotamus: evidence for polygynous ungulates. *Ethology Ecology and Evolution,* **10**, 177–286.

Liberg, O. & Schantz, T. von (1985). Sex-biased philopatry and dispersal in birds and mammals: the Oedipus hypothesis. *American Naturalist,* **126**, 129–35.

Lidicker, W. Z. Jr (1979). Analysis of two freely-growing enclosed populations of the California vole. *Journal of Mammalogy,* **60**, 447–66.

Liedtke, M. (1999). Die Entwicklung von Wertvorstellungen. Genetische Voraussetzungen und der "naturalistische Fehlschluß". In *Die Natur der Moral. Evolutionäre Ethik und Erziehung,* ed. D. Neumann, A. Schöppe & A. K. Treml, pp. 159–76. Stuttgart, Leipzig: Hirzel.

Lindburg, D. G. (1971). The rhesus monkey in North India: an ecological and behavioral study. In *Primate Behavior: Developments in Field and Laboratory Research,* vol. 2, ed. L. A. Rosenblum, pp. 2–106. New York: Academic Press.

Lindburg, D. G., Lyles, A. M. & Czekala, N. M. (1989). Status and reproductive potential of lion-tailed macaques in captivity. *Zoo Biology Supplement,* **1**, 5–16.

Lockhart, A. B., Thrall, P. H. & Antonovics, J. (1996). Sexually transmitted diseases in animals: ecological and evolutionary implications. *Biological Reviews*, **71**, 415–71.

Lorenz, K. (1943). Die angeborenen Formen möglicher Erfahrung. *Zeitschrift für Tierpsychologie*, **5**, 235–409.

Lorenz, K. (1955). Über das Töten von Artgenossen. *Jahrbuch der Max-Planck-Gesellschaft Göttingen*, pp. 105–40.

Lorenz, K. (1963). *Das sogenannte Böse. Zur Naturgeschichte der Aggression*. Wien: Borotha-Schoeler.

Lorenz, K. (1973). *Die Rückseite des Spiegels. Versuch einer Naturgeschichte menschlichen Erkennens*. Munich, Zurich: Piper.

Low, B. S. & Clarke, A. L. (1991). Family patterns in nineteenth-century Sweden: impact of occupational status and landownership. *Journal of Family History*, 16, 117–38.

Loy, J. (1981). The reproductive and heterosexual behaviours of adult patas monkeys in captivity. *Animal Behaviour*, **29**, 714–26.

Macdonald, D. W. (1996). Social behaviour of captive bush dogs. *Journal of the Zoological Society, London*, **239**, 525–43.

Macdonald, D. W. & Moehlman, P. D. (1982). Cooperation, altruism, and restraint in the reproduction of carnivores. In *Perspectives in Ethology*, vol. 5 *Ontogeny*, ed. P. P. G. Bateson & P. H. Klopfer, pp. 433–47. New York: Plenum Press.

Maddison, W. P. (1990). A method for testing the correlated evolution of two binary characters: are gains or losses concentrated on certain branches of a phylogenetic tree. *Evolution*, **44**, 539–57.

Maddison, W. P. & Maddison, D. R. (1992). *MacClade. Analysis of Phylogeny and Character Evolution*. Sunderland, MA: Sinauer Associates.

Maestripieri, D. (1992). Functional aspects of maternal aggression in mammals. *Canadian Journal of Zoology*, **70**, 1069–77.

Maestripieri, D. (1993). Vigilance costs of allogrooming in macaque mothers. *American Naturalist*, **141**, 744–53.

Maestripieri, D. (1994). Mother–infant relationships in three species of macaques (*Macaca mulatta, M. nemestrina, M. arctoides*). II. The social environment. *Behaviour*, **131**, 97–113.

Maestripieri, D. (1998). The evolution of male–infant interactions in the tribe Papionini (Primates: Cercopithecidae). *Folia Primatologica*, **69**, 247–51.

Maestripieri, D. & Alleva, E. (1991). Do male mice use parental care as a buffering strategy against maternal aggression? *Animal Behaviour*, **41**, 904–6.

Maestipieri, D. & Carroll, K. A. (1998). Risk factors for infant abuse and neglect in group-living rhesus monkeys. *Psychological Science*, **9**, 143–5.

Makin, J. W. & Porter, R. H. (1984). Paternal behavior in the spiny mouse (*Acomys cahirinus*). *Behavior and Neural Biology*, **41**, 135–51.

Mallory, F. F. & Brooks, R. J. (1978). Infanticide and other reproductive strategies in the collared lemming, *Dicrostonyx groenlandicus*. *Nature*, **273**, 144–6.

Mallory, F. F. & Brooks, R. J. (1980). Infanticide and pregnancy failure: reproductive strategies in the female collared lemming (*Dicrostonyx groenlandicus*). *Biology of Reproduction*, **22**, 192–6.

Mallory, F. F. & Clulow, F. V. (1977). Evidence of pregnancy failure in the wild meadow vole, *Microtus pennsylvanicus*. *Canadian Journal of Zoology*, **55**, 1–17.

Manson, J. H. (1995). Do female rhesus macaques choose novel males? *American Journal of Primatology*, **37**, 285–96.

Manson, J. H. (1999). Infant handling in wild *Cebus capucinus*: testing bonds between females? *Animal Behaviour*, **57**, 911–21.

Markman, S., Yom-Tov, Y. & Wright, J. (1996). The effect of male removal on female parental care in the orange-tufted sunbird. *Animal Behav*iour, **52**, 437–44.

Marler, P., Evans, C. S. & Hauser, M. D. (1992). Animal signals: Motivational, referential, or both? In *Nonverbal Vocal Communication: Comparative and Developmental Approaches*, ed. H. Papoušek, U. Jürgens & M. Papoušek, pp. 66–86. New York: Cambridge University Press.

Marlowe, F. (1999). Male care and mating effort among Hadza foragers. *Behavioral Ecology and Sociobiology*, **46**, 57–64.

Marsh, C. W. (1979a). Comparative aspects of social organization in the Tana River red colobus, *Colobus badius rufomitratus*. *Zeitschrift für Tierpsychologie*, **51**, 337–62.

Marsh, C. W. (1979b). Female transference and mate choice among Tana River red colobus. *Nature*, **281**, 568–9.

Martin, R. D. (1992). Female cycles in relation to paternity in primate societies. In *Paternity in Primates: Genetic Tests and Theories*, ed. R. D. Martin, A. F. Dixson & E. J. Wickings, pp. 238–274. Basel: Karger.

Martin, R. D. (1996). Scaling of the mammalian brain: the maternal energy hypothesis. *News in Physiological Sciences*, **11**, 149–53.

Martins, E. P. & Garland, T. (1991). Phylogenetic analyses of the correlated evolution of continuous characters: a simulation study. *Evolution*, **45**, 534–57.

Massey, A. (1977). Agonistic aids and kinship in a group of pigtail macaques. *Behavioral Ecology and Sociobiology*, **2**, 31–40.

Matsumoto-Oda, A. (1998). Injuries to the sexual skin of female chimpanzees at Mahale and their effect on behaviour. *Folia Primatologica*, **69**, 400–4.

Matsumura, S. (1991). A preliminary report on the ecology and social behavior of moor macaques (*Macaca maurus*) in Sulawesi, Indonesia. *Kyoto University Overseas Research Report of Studies on Asian Non-Human Primates*, **8**, 27–41.

Matsumura, S. (1997). Mothers in a wild group of moor macaques (*Macaca maurus*) are more attractive to other group members when holding their infants. *Folia Primatologica*, **68**, 77–85.

Matsumura, S. (1999). The evolution of 'egalitarian' and 'despotic' social systems among macaques. *Primates*, **40**, 23–31.

Matteo, S. & Rissman, E. F. (1984). Increased sexual activity during the midcycle portion of the human menstrual cycle. *Hormones and Behavior*, **18**, 249–55.

Maynard-Smith, J. (1982). *Evolution and the Theory of Games*. Cambridge: Cambridge University Press.

McCann, T. S. (1982). Aggressive and maternal activities of female southern elephant seals (*Mirounga leonina*). *Animal Behaviour*, **30**, 268–76.

McCarthy, M. M., Bare, J. E. & vom Saal, F. (1986). Infanticide and parental behavior in wild female house mice: effects of ovariectomy, adrenalectomy and administration of oxytocin and prostaglandin F2. *Physiology and Behavior*, **36**, 17–23.

McLean, I. G. (1983). Paternal behaviour and killing of young in arctic ground squirrels. *Animal Behaviour*, **31**, 32–44.

Mead, R. A. (1989). The pysiology and evolution of delayed implantation. In *Carnivore Behavior, Ecology, and Evolution*, ed. J. L. Gittleman, pp. 437–64. Ithaca, NY: Cornell University Press.

Mead, R. A. (1994). Reproduction in *Martes*. In *Martens, Sables, and Fishers. Biology and Conservation*, ed. S. W. Buskirk, A. S. Harestad, M. G. Raphael & R. A. Powell, pp. 404–23. Ithaca, NY: Cornell University Press.

Megantara, E. N. (1989). Ecology, behavior and sociality of *Presbytis femoralis* in East central Sumatra. *Comparative Primatology Monographs, Padjadjaran University of Padjadjaran*, **2**, 171–301.

Meikle, D. B. & Vessey, S. H. (1981). Nepotism among rhesus monkey brothers. *Nature*, **294**, 160–1.

Melnick, D. J. & Pearl, M. C. (1987). Cercopithecines in multimale groups: genetic diversity and population structure. In *Primate Societies*, ed. B. Smuts, D. L. Cheney, R. M. Seyfarth, R. W. Wrangham & T. T. Struhsaker, pp. 121–34. Chicago: University of Chicago Press.

Ménard, N., Scheffrahn, W., Vallet, D. & Reber, C. (1992). Application of blood protein electrophoresis and DNA fingerprinting to the analysis of paternity and social characteristics of wild Barbary macaques. In *Paternity in Primates. Genetic Tests and Theories*, ed. R. D. Martin, A. F. Dixson & E. J. Wickings, pp. 155–74. Basel: Karger.

Mendes Pontes, A. R. & Monteiro da Cruz, M. A. O. (1995). Home range, intergroup transfers, and reproductive status in common marmosets *Callithrix jacchus* in a forest fragment in North-eastern Brazil. *Primates*, **36**, 335–47.

Merkel, F. W. (1982). Sozialverhalten von individuell markierten Staren *Sturnus vulgaris*, in enier kleinen Nistkastenkolonie. 4. Fortpflantzungstrategien *Luscinia*, **44**, 239–54.

Mesnick, S. L. (1997). Sexual alliances: evidence and evolutionary implications. In *Feminism and Evolutionary Biology: Boundaries, Intersections, and Frontiers*, ed. P. A. Gowaty, pp. 207–60. New York: Chapman & Hall.

Michener, G. R. (1973). Maternal behaviour in Richardson's ground squirrel (*Spermophilus richardsonii richardsonii*): retrieval of young by non-lactating females. *Animal Behaviour*, **21**, 157–9.

Michener, G. R. (1982). Infanticide in ground squirrels. *Animal Behaviour*, **30**, 936–8.

Michener, G. R. (1983). Kin identification, matriarchies, and the evolution of sociality in ground-dwelling sciurids. In *Advances in the Study of Mammalian Behavior*, ed. J. F. Eisenberg & D. G. Kleiman, pp. 528–72. Shippensburg, PA: The American Society of Mammalogists.

Millesi, E., Huber, S., Dittami, J., Hoffman, I. & Daan, S. (1998). Parameters of mating effort and success in male European ground squirrels, *Spermophilus citellus*. *Ethology*, **104**, 298–313.

Milligan, S. R. (1976). Pregnancy blocking in the vole, *Microtus agrestis*. I. Effect of the social environment. *Journal of Reproduction and Fertility*, **46**, 91–5.

Milton, K. (1985). Mating patterns of woolly spider monkeys, *Brachyteles arachnoides*: implications for female choice. *Behavioral Ecology and Sociobiology*, **17**, 53–9.

Mitani, J. C. (1984). The behavioral regulation of monogamy in gibbons (*Hylobates muelleri*). *Behavioral Ecology and Sociobiology*, **15**, 225–9.

Mitani, J. C. (1985). Gibbon song duets and intergroup spacing. *Behaviour*, **92**, 59–96.

Mitani, J. C. (1987). Territoriality and monogamy among agile gibbons (*Hylobates agilis*). *Behavioral Ecology and Sociobiology*, **20**, 265–9.

Mitani, J. C. (1990). Demography of agile gibbons (*Hylobates agilis*). *International Journal of Primatology*, **11**, 411–24.

Mitani, J. C., Gros-Louis, J. & Manson, J. (1996a). Number of males in primate groups: comparative tests of competing hypotheses. *American Journal of Primatology*, **38**, 315–32.

Mitani, J., Gros-Louis, J. & Richards, A. F. (1996b). Sexual dimorphism, the operational sex ratio, and the intensity of male competition in polygynous primates. *American Naturalist*, **147**, 966–80.

Mitchell, C. L., Boinski, S. & van Schaik, C. P. (1991). Competitive regimes and female bonding in two species of squirrel monkeys (*Saimiri oerstedi* and *S. sciureus*). *Behavioral Ecology and Sociobiology*, **28**, 55–60.

Mittermeier, R. A., Tattersall, I., Konstant, W. R., Meyers, D. M., Mast, R. B. & Nash, S. D. (1994). *Lemurs of Madagascar*. Washington, DC: Conservation International.

Mock, D. W. (1984). Infanticide, siblicide, and avian nestling mortality. In *Infanticide: Comparative and Evolutionary Perspectives*, ed. G. Hausfater & S. B. Hrdy, pp. 2–30. New York: Aldine de Gruyter.

Mock, D. W. & Fujioka, M. (1990). Monogamy and long-term pair bonding in vertebrates. *Trends in Ecology and Evolution*, **5**, 39–43.

Moehlman, P. D. (1989). Intraspecific variation in canid social systems. In *Carnivore Behavior, Ecology, and Evolution*, ed. J. L Gittleman, pp. 143–63. Ithaca, NY: Cornell University Press.

Mohnot, S. M. (1971). Some aspects of social changes and infant killing in the Hanuman langur (*Presbytis entellus*) (Primates: Cercopithecidae) in Western India. *Mammalia*, **35**, 175–98.

Møller, A. P. (1986). Mating systems among European passerines: a review. *Ibis*, **128**, 34–250.

Møller, A. P. (1988). Infanticidal and anti-infanticidal strategies in the swallow *Hirundo rustica*. *Behavioral Ecology and Sociobiology*, **22**, 365–71.

Montgomery, W. I., Wilson, W. L. & Elwood, R. W. (1997). Spatial regulation and population growth in the wood mouse *Apodemus sylvaticus*: experimental manipulations of males and females in natural populations. *Journal of Animal Ecology*, **66**, 755–68.

Moore, J. (1984). Female transfer in primates. *International Journal of Primatology*, **5**, 537–89.

Moore, J. (1993). Inbreeding and outbreeding in primates: what's wrong with 'the dispersing sex'? In *The Natural History of Inbreeding and Outbreeding. Theoretical and Emperical Perspectives*, ed. N. W. Thornhill, pp. 392–426. Chicago: University of Chicago Press.

Moore, J. (1999). Population density, social pathology, and behavioral ecology. *Primates*, **40**, 5–26.

Moos, R., Rock, J. & Salzert, W. (1985). Infanticide in gelada baboons (*Theropithecus gelada*). *Primates*, **26**, 497–500.

Morales, J. C. & Melnick, D. J. (1998). Phylogenetic relationships of the macaques (Cercopithecidae, *Macaca*), as revealed by high resolution restriction site mapping of mitochondrial ribosomal genes. *Journal of Human Evolution*, **34**, 1–23.

Mori, U. & Dunbar, R. I. M. (1985). Changes in the reproductive condition of female gelada baboons following the takeover of one-male units. *Zeitschrift für Tierpsychologie*, **67**, 215–24.

Mori, A., Iwamoto, T. & Bekele, A. (1997). A case of infanticide in a recently found gelada population in Arsi, Ethiopia. *Primates*, **38**, 79–88.

Morin, P. A., Moore, J. J., Chakraborti, R., Jin, L., Goodall, J. & Woodroffe, D. S. (1994). Kin selection, social structure, gene flow, and the evolution of chimpanzees. *Science*, **265**, 1193–201.

Morland, H. S. (1993). Reproductive activity of ruffed lemurs (*Varecia variegata variegata*) in a Madagascar rain forest. *American Journal of Physical Anthropology*, **91**, 71–82.

Morris, D. (1967). *The Naked Ape*. London: Cape.

Moss, C. J. (1983). Oestrous behaviour and female choice in the African elephant. *Behaviour*, **86**, 176–96.

Mühlmann, W. E. (1984). *Geschichte der Anthropologie*. Wiesbaden: Aula.

Müller, J. K., Eggert, A. K. & Sakaluk, S. K. (1998). Carcass maintenance and biparental brood care in burying beetles: are males redundant? *Ecological Entomology*, **23**, 195–200.

Müller-Hill, B. (1984). *Tödliche Wissenschaft*. Reinbek: Rowohlt.

Mumme, R. L., Koenig, W. D. & Pitelka, F. A. (1983). Reproductive competition in the communal acorn woodpecker: sisters destroy each others' eggs. *Nature*, **306**, 583–4.

Muroyama, Y. & Thierry, B. (1996). Fatal attack on an infant by an adult female Tonkean macaque. *International Journal of Primatology*, **17**, 219–27.

Musser, G. G. & Carleton, M. D. (1993). Family Muridae. In *Mammal Species of the World: A Taxonomic and Geographic Reference*, 2nd edn, ed. D. E. Wilson & D. M. Reeder, pp. 501–755. Washington, DC: Smithsonian Institution Press.

Myllymaki, A. (1977). Demographic mechanisms in the fluctuating populations of the field vole *Microtus agrestis*. *Oikos*, **29**, 468–93.

Nadler, R. D. (1992). Sexual behavior and the concept of estrus in the great apes. In *Topics in Primatology*, ed. N. Itoigawa, Y. Sugiyama, G. P. Sackett & R. K. R. Thompson, pp. 191–206. Tokyo University Press, Tokyo.

Nagy, K. A. & Milton, K. (1979). Energy metabolism and food consumption by wild howler monkeys (*Alouatta palliata*). *Ecology*, **60**, 475–80.

Neal, E. (1986). *The Natural History of Badgers*. New York: Facts on File Inc.

Newton, P. N. (1986). Infanticide in an undisturbed forest population of hanuman langurs, *Presbytis entellus*. *Animal Behaviour*, **34**, 785–9.

Newton, P. N. (1987). The social organization of forest hanuman langurs (*Presbytis entellus*). *International Journal of Primatology*, **8**, 199–232.

Newton, P. N. (1988). The variable social organization of hanuman langurs (*Presbytis entellus*), infanticide, and the monopolization of females. *International Journal of Primatology*, **9**, 59–77.

Newton, P. N. & Dunbar, R. I. M. (1994). Colobine monkey society. In *Colobine Monkeys: Their Ecology, Behaviour and Evolution*, ed. A. G. Davies & J. F. Oates, pp. 311–46. Cambridge: Cambridge University Press.

Nicolson, N. A. (1987). Infants, mothers and other females. In *Primate Societies*, ed. B. B. Smuts, D. L. Cheney, R. M. Seyfarth, R. W. Wrangham & T. T. Struhsaker, pp. 330–42. Chicago: University of Chicago Press.

Nicolson, N. & Demment, M. W. (1982). The transition from suckling to independent feeding in wild baboon infants. *International Journal of Primatology*, **3**, 318.

Nieuwenhuijsen, K., de Neef, K. J. & Slob, A. K. (1986). Sexual behaviour during ovarian cycles, pregnancy and lactation in group-living stumptail macaques (*Macaca arctoides*). *Human Reproduction*, **1**, 159–69.

Nishida, T. (1983a). Alloparental behaviour of wild chimpanzees (*Pan troglodytes schweinfurthii*). *Folia Primatologica*, **41**, 1–33.

Nishida, T. (1983b). Alpha status and agonistic alliance in wild chimpanzees (*Pan troglodytes schweinfurthii*). *Primates*, **24**, 318–36.

Nishida, T. & Hiraiwa-Hasegawa, M. (1987). Chimpanzees and bonobos: cooperative relationships among males. In *Primate Societies*, ed. B. B. Smuts, D. L. Cheney, R. M. Seyfarth, R. W. Wrangham & T. T. Struhsaker, pp. 165–77. Chicago: University of Chicago Press.

Nishida, T., Hiraiwa-Hasegawa, M., Hasegawa, T. & Takahata, Y. (1985). Group extinction and female transfer in wild chimpanzees in the Mahale national park, Tanzania. *Zeitschrift für Tierpsychologie*, **67**, 284–301.

Nishida, T., Takasaki, H. & Takahata, Y. (1990). Demography and reproductive profiles In *The Chimpanzees of the Mahale Mountains. Sexual and Life History Strategies,* ed. T. Nishida, pp. 64–97. Tokyo: University of Tokyo Press.

Noë, R. & Hammerstein, P. (1995). Biological markets. *Trends in Ecology and Evolution,* **10**, 336–9.

Noë, R. & Sluijter, A. A. (1990). Reproductive tactics of male savanna baboons. *Behaviour*, **113**, 117–70.

Noë, R. & Sluijter, A. A. (1995). Which adult male savanna baboons form coalitions? *International Journal of Primatology*, **16**, 77–105.

Nowak, R. M. (1999). *Walker's Mammals of the World*. Baltimore, MD: Johns Hopkins University Press.

Nunn, C. L. (1999a). A comparative approach to primate socioecology and intersexual conflict. Ph.D., dissertation, Duke University.

Nunn, C. L. (1999b). The number of males in primate social groups: a comparative test of the socioecological model. *Behavioral Ecology and Sociobiology*, **46**, 1–13.

Nunn, C. L. (1999c). The evolution of exaggerated sexual swellings in primates and the graded signal hypothesis. *Animal Behaviour*, **58**, 229–46.

Nunn, C. L. & Lewis, R. J. (2001). Cooperation and collective action in animal behavior. In *Economics in Nature,* ed. R. Noë, J. A. R. A. M. van Hooff & P. Hammerstein. Cambridge: Cambridge University Press, in press.

Nunn, C. L. & van Schaik, C. P. (2001). Reconstructing the behavioral ecology of extinct primates. In *Reconstructing Behavior in the Fossil Record*, ed. J. M. Plavcan, R. F. Kay, W. L. Jungers & C. P. van Schaik. New York: Plenum Press, in press.

O'Brien, T. G. (1991). Female–male social interactions in wedge-capped capuchin monkeys: benefits and costs of group living. *Animal Behaviour*, **41**, 555–67.

O'Brien, T. G. & Kinnaird, M. F. (1997). Behavior, diet, and movements of the Sulawesi crested black macaque (*Macaca nigra*). *International Journal of Primatology*, **18**, 321–51.

O'Connell, S. & Cowlishaw, G. (1994). Infanticide avoidance, sperm competition and mate choice: the function of copulation calls in female baboons. *Animal Behaviour*, **48**, 687–94.

O'Connor, R. J. (1978). Brood reduction in birds: selection for fratricide, infanticide, and suicide. *Animal Behaviour*, **26**, 79–96.

Oates, J. F. (1977). The social life of a black-and-white colobus monkey, *Colobus guereza*. *Zeitschrift für Tierpsychologie*, **45**, 1–60.

Oates, J. F., Davies, A. G. & Delson, E. (1994). The diversity of living colobines. In *Colobine Monkeys: Their Ecology, Behaviour and Evolution*, ed. A. G. Davies & J. F. Oates, pp. 45–73. Cambridge: Cambridge University Press.

Ogawa, H. (1995a). Bridging behavior and other affiliative interactions among male Tibetan macaques (*Macaca thibetana*). *International Journal of Primatology*, **16**, 707–29.

Ogawa, H. (1995b). Recognition of social relationships in bridging behavior among Tibetan macaques (*Macaca thibetana*). *American Journal of Primatology*, **35**, 305–10.

Oi, T. (1990). Patterns of dominance and affiliation in wild pig-tailed macaques (*Macaca nemestrina nemestrina*) in West Sumatra. *International Journal of Primatology*, **11**, 339–56.

Oi, T. (1996). Sexual behaviour and mating system of the wild pig-tailed macaque in West Sumatra. In *Evolution and Ecology of Macaque Societies*, ed. J. E. Fa & D. G. Lindburg, pp. 342–68. Cambridge: Cambridge University Press.

Omland, K. (1997). Examining two standard assumptions of ancestral reconstructions: repeated loss of dichromatism in dabbling ducks (Anatini). *Evolution*, **51**, 1636–46.

Ostermeyer, M. C. (1983). Maternal aggression. In *Parental Behaviour of Rodents*, ed. R. W. Elwood, pp 151–79. New York: John Wiley and Sons.

Overdorff, D. J. (1991). Ecological correlates to social structure in two prosimian primates in Madagascar. Ph.D. Thesis, Duke University.

Overdorff, D. J. (1998). Are *Eulemur* species pair-bonded? Social organization and mating strategies in *Eulemur fulvus rufus* from 1988–1995 in southeast Madagascar. *American Journal of Physical Anthropology*, **105**, 153–66.

Owens, D. D. & Owens, M. J. (1984). Helping behaviour in brown hyenas. *Nature*, **308**, 843–5.

Packer, C. (1977). Reciprocal altruism in olive baboons. *Nature*, **265**, 441–3.

Packer, C. (1979). Inter-troop transfer and inbreeding avoidance in *Papio anubis*. *Animal Behaviour*, **27**, 1–36.

Packer, C. (1980). Male care and exploitation of infants in *Papio anubis*. *Animal Behaviour*, **28**, 512–20.

Packer, C. (1986). The ecology of sociality in felids. In *Ecological Aspects of Social Evolution: Birds and Mammals*, ed. D. I. Rubenstein & R. W. Wrangham, pp. 429–51. Princeton, NJ: Princeton University Press.

Packer, C., Herbst, L., Pusey, A. E., Bygott, J. D., Hanby, J. P., Cairns, S. J. & Mulder, M. B. (1988). Reproductive success of lions. In *Reproductive Success: Studies of Individual Variation in Contrasting Breeding Systems*, ed. T. H. Clutton-Brock, pp. 363–83. Chicago: University of Chicago Press.

Packer, C. & Pusey, A. E. (1983). Adaptations of female lions to infanticide by incoming males. *American Naturalist*, **121**, 716–28.

Packer, C. & Pusey, A. E. (1984). Infanticide in carnivores. In *Infanticide: Comparative and Evolutionary Perspectives*, ed. S. B. Hrdy & G. Hausfater, pp. 31–42. New York: Aldine de Gruyter.

Packer, C. & Pusey, A. (1985). Asymmetric contests in social mammals: respect, manipulation and age-specific effects. In *Evolution. Essays in Honour of John Maynard Smith*, ed. P. J. Greenwood, P. H. Harvey & M. Slatkin, pp. 173–86, Cambridge: Cambridge University Press.

Pagel, M. (1993). Seeking the evolutionary regression coefficient: an analysis of what comparative methods measure. *Journal of Theoretical Biology*, **164**, 191–205.

Pagel, M. (1997). Desperately concealing father: a theory of parent–infant resemblance. *Animal Behaviour*, **53**, 973–81.

Palanza, P., Re, L., Mainardi, D., Brain, P. F. & Parmigiani, S. (1996). Male and female competitive strategies of wild house mice pairs (*Mus musculus domesticus*) contronted with intruders of different sex and age in artificial territories. *Behaviour*, **133**, 863–82.

Pallaud, B. (1984). Conséquences d'un changement de hiérarchie et de territoire dans un groupe de Macaques crabiers (*Macaca fascicularis*). *Biology of Behaviour*, **9**, 89–99.

Palombit, R. A. (1994a). Dynamic pair bonds in hylobatids: implications regarding monogamous social systems. *Behaviour*, **128**, 65–101.

Palombit, R. A. (1994b). Extra-pair copulations in a monogamous ape. *Animal Behaviour*, **47**, 721–3.

Palombit, R. A. (1995). Longitudinal patterns of reproduction in wild female siamang (*Hylobates syndactylus*) and white-handed gibbons (*Hylobates lar*). *International Journal of Primatology*, **16**, 739–60.

Palombit, R. A. (1996). Pair bonds in monogamous apes: a comparison of the siamang (*Hylobates syndactylus*) and the white-handed gibbon (*Hylobates lar*). *Behaviour*, **133**, 321–56.

Palombit, R. A. (1999). Infanticide and the evolution of pair bonds in nonhuman primates, *Evolutionary Anthropology*, **7**, 117–29.

Palombit, R. A., Seyfarth, R. M. & Cheney, D. L. (1997). The adaptive value of 'friendship' to female baboons: experimental and observational evidence. *Animal Behaviour*, **54**, 599–614.

Palombit, R. A., Cheney, D. L. & Seyfarth, R. M., (2000), Female–female competition for male "friends" in wild chacma baboons (*Papio cynocephalus ursinus*). *Animal Behaviour*, in press.

Parish, A. R. (1996). Female relationships in bonobos (*Pan paniscus*): evidence for bonding, cooperation, and female dominance in a male philopatric system. *Human Nature*, **7**, 61–96.

Parker, G. A. (1979). Sexual selection and sexual conflict. In *Sexual Selection and Reproductive Competition in Insects*, ed. S. M. Blum & N. A. Blum, pp. 123–66. New York: Academic Press.

Parmigiani, S., Palanza, P., Mainardi, D. & Brain, P. F. (1994). Infanticide and protection of young in house mice (*Mus domesticus*): female and male strategies. In *Infanticide and Parental Care,* ed. S. Parmigiani & F. S. vom Saal, pp. 341–63. Chur (Switzerland): Harwood.

Parmigiani, S. & vom Saal, F. S. (eds.) (1994). *Infanticide and Parental Care.* New York: Harwood.

Partridge, L. & Harvey, P. H. (1988). The ecological context of life history evolution. *Science*, **241**, 1449–55.

Patterson, I. A. P., Reid, R. J., Wilson, B., Ross, H. M. & Thompson, P. M. (1998). Evidence for infanticide in bottlenose dolphins: an explanation for violent interactions with harbour porpoises? *Proceedings of the Royal Society, London B,* **265**, 1167–70.

Paul, A. (1984). Zur Sozialstruktur und Sozialisation semi-freilebender Berberaffen (*Macaca sylvanus* L. 1758). Ph.D. dissertation. University of Kiel.

Paul, A. (1997). Breeding seasonality affects the association between dominance and reproductive success in non-human male primates. *Folia Primatologica*, **68**, 344–9.

Paul, A. (1999). The socioecology of infant handling in primates: is the current model convincing? *Primates*, **40**, 33–46.

Paul, A., Kuester, J. & Arnemann, J. (1992). DNA fingerprinting reveals that infant care by male Barbary macaques (*Macaca sylvanus*) is not paternal investment. *Folia Primatologica,* **58,** 93–8.

Paul, A., Kuester, J. & Arnemann, J. (1996). The sociobiology of male–infant interactions in Barbary macaques, *Macaca sylvanus. Animal Behaviour,* **51,** 155–70.

Penzhorn, B. L. (1984). A long-term study of social organization and behaviour of cape mountain zebras, *Equus zebra zebra. Zeitschrift für Tierpsychologie,* **64,** 97–146.

Pereira, M. E. (1983). Abortion following the immigration of an adult male baboon (*Papio cynocephalus*). *American Journal of Primatology,* **4,** 93–8.

Pereira, M. E. & Weiss, M. L. (1991). Female mate choice, male migration, and the threat of infanticide in ringtailed lemurs. *Behavioral Ecology and Sociobiology,* **28,** 141–52.

Perrigo, G., Belvin, L., Quindry, P., Kadir, T., Becker, J., van Look, C., Niewoehner, J. & vom Saal, F. S. (1993). Genetic mediation of infanticide and parental behavior in male and female domestic and wild stock house mice. *Behavioral Genetics,* **23,** 525–31.

Perrigo, G., Bryant, W. C. & vom Saal, F. S. (1990). A unique neural timing system prevents male mice from harming their own offspring. *Animal Behaviour,* **39,** 535–9.

Perrigo, G. & vom Saal, F. S. (1994). Behavioral cycles and the neural timing of infanticide and parental behavior in male house mice. In *Infanticide and Parental Care,* ed. S. Parmigiani & F. S. vom Saal, pp. 365–96. Chur (Switzerland): Harwood.

Perrin, C., Allainé, D. & Le Berre, M. (1994). Intrusion de males et possibilité d'infanticide chez la marmotte alpine. *Mammalia,* **58,** 150–3.

Petrie, M. (1994). Improved growth and survival of offspring of peacocks with more elaborate trains. *Nature,* **371,** 598–599.

Picman, J. (1977). Intraspecific nest destruction in the long-billed marsh wren, *Telmatodytes palustris palustris. Canadian Journal of Zoology,* **55,** 1997–2003.

Pierotti, R. (1991). Infanticide versus adoption: an intergenerational conflict. *American Naturalist,* **138,** 1140–58.

Pierotti, R., Brunton, D. & Murphy, E. C. (1988). Parent–offspring and sibling–sibling recognition in gulls. *Animal Behaviour,* **36,** 620–1.

Pinxten, R., Eens, M. & Verheyen, R. F. (1991). Responses of male starlings to experimental intraspecific brood parasitism. *Animal Behaviour,* **42,** 1028–30.

Plavcan, M. & van Schaik, C. P. (1992). Intrasexual competition and canine dimorphism in primates. *American Journal of Physical Anthropology,* **87,** 461–77.

Plavcan, J. M. & van Schaik, C. P. (1997). Intrasexual competition and body weight dimorphism in anthropoid primates. *American Journal of Physical Anthropology,* **103,** 37–68.

Plavcan, J. M., van Schaik, C. P. & Kappeler, P. M. (1995). Competition, coalitions and canine size in primates. *Journal of Human Evolution,* **28,** 245–76.

Podzuweit, D. (1994). Sozio-Ökologie weiblicher Hanuman Languren (*Presbytis entellus*) in Ramnagar, Südnepal. Doctoral thesis, Georg-August University.

Poglayen-Neuwall, I., Durrant, B. S., Swansen, M. L., Williams, R. C. & Barnes, R. A. (1989). Estrous cycle of the tayra, *Eira barbara. Zoo Biology,* **8,** 171–7.

Poirier, F. E. (1970). The nilgiri langur (*Presbytis johnii*) of South India In *Primate Behavior. Developments in Field and Laboratory Research,* ed. L. A. Rosenblum, pp. 251–383. New York: Academic Press.

Polis, G. A., Myers, C. A. & Hess, W. R. (1984). A survey of intraspecific predation within the class mammalia. *Mammalian Review*, **14**,187–98.

Pope, T. R. (1989). The influence of mating system and dispersal patterns on the genetic structure of red howler monkey populations. Ph. D. dissertation, University of Florida.

Pope, T. R. (1990). The reproductive consequences of male cooperation in the red howler monkey: paternity exclusion in multi-male and single-male troops using genetic markers. *Behavioral Ecology and Sociobiology*, **27**, 439–46.

Popp, J. L. (1978). Male baboons and evolutionary principles. Ph. D. thesis, Harvard University.

Popp, J. L. & DeVore, I. (1979). Aggressive competition and social dominance theory: a synopsis. In *The Great Apes*, ed. D. A. Hamburg & E. R. MacCown, pp. 317–38. Menlo Park, CA: Benjamin Cummings.

Preuschoft, S. & Paul, A. (2000). Dominance, egalitarianism, and stalemate: an experimental approach to male–male competition in Barbary macaques. In *Primate Males*, ed. P. M. Kappeler, pp. 205–16. Cambridge: Cambridge University Press.

Preuschoft, S., Paul, A. & Kuester, J. (1998). Dominance styles of female and male Barbary macaques (*Macaca sylvanus*). *Behaviour*, **135**, 731–55.

Preuschoft, S. & van Schaik, C. P. (2000). Is the hierarchy a conflict-management device? In *Natural Conflict Resolution*, ed. F. Aureli & F. B. M. de Waal. Berkeley, CA: University of California Press, in press.

Price, E. (1992). The costs of infant carrying in captive cotton-top tamarins. *American Journal of Primatology*, **26**, 23–33.

Price, T. (1997). Correlated evolution and independent contrasts. *Philosophical Transactions of the Royal Society, London B*, **352**, 519–29.

Purvis, A. (1995). A composite estimate of primate phylogeny. *Philosophical Transactions of the Royal Society, London B*, **348**, 405–21.

Purvis, A. & Rambaut, A. (1995). Comparative analysis by independent contrasts (CAIC): an Apple Macintosh application for analysing comparative data. *Computer Applications in the Biosciences*, **11**, 247–51.

Purvis, A. & Webster, A. J. (1999). Phylogenetically independent contrasts and primate phylogeny. In *Comparative Primate Socioecology*, ed. P. Lee, pp. 44–68. Cambridge: Cambridge University Press.

Pusey, A. (1979). Intercommunity transfer of chimpanzees in Gombe National Park. In *The Great Apes*, ed. D. A. Hamburg & E. R. McCown, pp. 465–79. Menlo Park, CA: Benjamin/Cummings Publ. Co.

Pusey, A. E. & Packer C. (1987a). Dispersal and philopatry. In *Primate Societies*, ed. B. B. Smuts, D. L. Cheney, R. M. Seyfarth, R. W. Wrangham & T. T. Struhsaker, pp. 250–66. Chicago: University of Chicago Press.

Pusey, A. E. & Packer, C. (1987b). The evolution of sex-biased dispersal in lions. *Behaviour*, **101**, 275–310.

Pusey, A. E. & Packer, C. (1994). Infanticide in lions: consequences and counterstrategies. In *Infanticide and Parental Care*, ed. S. Parmigiani & F. S. vom Saal, pp. 277–99. London: Harwood.

Pusey, A. Williams, J. & Goodall, J. (1997). The influence of dominance rank on the reproductive success of female chimpanzees. *Science*, **277**, 828–31.

Quanstrom, W. R. (1968). Some aspects of the ethoecology of Richardson's ground squirrel in eastern North Dakota. Ph. D. thesis, University of Oklahoma.

Quinn, J. F. & Dunham, A. E. (1983). On hypothesis testing in ecology and evolution. *American Naturalist*, **122**, 602–17.

Quinn, J. S., Whittingham, L. A. & Morris, R. D. (1994). Infanticide in skimmers and terns: side effects of territorial attacks or inter-generational conflict? *Animal Behaviour*, **47**, 363–7.

Quinn, M. S. & Holroyd, G. L. (1989). Nestling and egg destruction by house wrens. *Condor*, **91**, 206–7.

Radcliffe-Brown, A. R. (1952). *Structure and Function in Primitive Society*. Glencoe, IL: Free Press.

Rajpurohit, L. S. & Mohnot, S. M. (1988). Fate of ousted male residents of one-male bisexual troops of Hanuman langurs (*Presbytis entellus*) at Jodhpur, Rajasthan, India. *Human Evolution*, **3**, 309–18.

Rajpurohit, L. S. & Mohnot, S. M. (1991). The process of weaning in Hanuman langurs *Presbytis entellus entellus*. *Primates*, **32**, 213–18.

Ransom, T. W. & Ransom, B. S. (1971). Adult male–infant relations among baboons (*Papio anubis*). *Folia Primatologica,* **16**, 179–95.

Rasa, O. A. E. (1994). Altruistic infant care or infanticide: the dwarf mongooses' dilemma. In *Infanticide and Parental Care*, ed. S. Parmigiani & F. S. vom Saal, pp. 301–20. Chur (Switzerland): Harwood.

Rasmussen, D. R. (1981). Communities of baboon troops (*Papio cynocephalus*) in Mikumi National Park, Tanzania. *Folia Primatologica*, **36**, 232–42.

Rasmussen, K. L. R. (1985). Changes in the activity budgets of yellow baboons (*Papio cynocephalus*) during sexual consortships. *Behavioral Ecology and Sociobiology,* **17**, 161–70.

Reena, M. & Ram, M. (1991). Departure of juvenile male *Presbytis entellus* from the natal group. *International Journal of Primatology*, **12**, 39–43.

Reeves, R. R., Stewart B. S. & Leatherwood, S. (1992). *The Sierra Club Handbook of Seals and Sirenians*. San Francisco: Sierra Club Books.

Regalski, J. M. & Gaulin, S. J. C. (1993). Whom are Mexican infants said to resemble? Monitoring and fostering paternal confidence in the Yucatan. *Ethology and Sociobiology*, **14**, 97–113.

Reichard, U. (1995). Extra-pair copulations in a monogamous gibbon (*Hylobates lar*). *Ethology*, **100**, 99–112.

Reichard, U. & Sommer, V. (1997). Group encounters in white-handed gibbons (*Hylobates lar*): agonism, affiliation, and the concept of infanticide. *Behaviour*, **134**, 1135–74.

Reinking, M. & Müller, J. K. (1990). The benefit of parental care in the burying beetle, *Necrophorus vespilloides*. *Verhandlungen der Deutschen Zoologischen Gesellschaft*, **83**, 655–6.

Reiter, J., Stinson, N. L. & LeBoeuf, B. J. (1978). Northern elephant seal development: the transition from weaning to nutritional independence. *Behavioral Ecology and Sociobiology,* **3**, 337–67.

Rhine, R. & Hendy-Neely, H. (1978). Social development of stumptail macaques (*Macaca arctoides*): momentary touching, play and other interactions with aunts and immatures during the infants' first 60 days of life. *Primates*, **19**, 115–23.

Rhine, R. J., Wasser, S. K. & Norton, G. W. (1988). Eight-year study of social and ecological correlates of mortality among immature baboons of Mikumi National Park, Tanzania. *American Journal of Primatology*, **16**, 199–212.

Ribble, D. O. (1991). The monogamous mating system *Peromyscus californicus* as revealed by DNA fingerprinting. *Behavioral Ecology and Sociobiology,* **29**, 161–6.

Richard, A. F. (1992). Aggressive competition between males, female-controlled polygyny and sexual monomorphism in a Malagasy primate, *Propithecus verreauxi*. *Journal of Human Evolution,* **22**, 395–406.

Richard, A. F., Rakotomanga, P. & Schwartz, M. (1993). Dispersal by *Propithecus verrauxi* at Beza Mahafaly, Madagascar: (1984–1991). *American Journal of Primatology*, **30**, 1–20.

Richards, M. P. M. (1966). Maternal behaviour in the golden hamster: responsiveness to young in virgin, pregnant, and lactating females. *Animal Behaviour*, **14**, 310–13.

Richardson, P. R. K. (1991). Territorial significance of scent marking during the non-mating season in the aardwolf *Proteles cristatus* (Carnivora: Protelidae). *Ethology*, **87**, 9–27.

Richter, J. S. (1998). Infanticide, child abandonment, and abortion in imperial Germany, *Journal of Interdisciplinary History,* **28**, 511–51.

Ridley, M. (1978). Paternal care. *Animal Behaviour*, **26**, 904–32.

Riedman, M. L. & LeBoeuf, B. J. (1982). Mother–pup separation and adoption in northern elephant seals. *Behavioral Ecology and Sociobiology,* **11**, 203–15.

Rijksen, H. D. (1978). *A Field Study on Sumatran Orang Utans* (Pongo pygmaeus abelii, *Lesson 1827): Ecology, Behaviour and Conservation*. Wageningen: Veenman.

Rijksen, H. D. (1981). Infant killing: a possible consequence of a disputed leader role. *Behaviour,* **78**, 138–68.

Ripley, S. (1980). Infanticide in langurs and man: adaptive advantage or social pathology? In *Biosocial Mechanisms of Population Regulation*, ed. M. N. Cohan, R. S. Malpass & H. G. Klein, pp. 349–90. New Haven, CT: Yale University Press.

Rissman, E. F. (1995). An alternative animal model for the study of female sexual behavior. *Current Directions in Psychological Science*, **8**, 6–10.

Robbins, M. M. (1995). A demographic analysis of male life history and social structure of mountain gorillas. *Behaviour*, **132**, 21–47.

Roberts, M. (1994). Growth, development and parental care in the western tarsier (*Tarsius bancanus*) in captivity: evidence for a 'slow' life history and nonmonogamous mating system. *International Journal of Primatology*, **15**, 1–28.

Roberts, M. S. & Kessler, D. S. (1979). Reproduction in red pandas, *Ailurus fulgens* (Carnivora: Ailuropodidae). *Journal of Zoology, London*, **188**, 235–49.

Robertson, I. C. (1993). Nest intrusions, infanticide, and parental care in the burying beetle, *Necrophorus orbicollis* (Coleoptera, Silphidae). *Journal of Zoology, London*, **231**, 583–93.

Robertson, R. J. (1990). Tactics and countertactics of sexually selected infanticide in tree swallows. In *Population Biology of Passerine Birds: An Integrated Approach*, ed. J. Blondel, A. Gosler, J.-D. Lebreton & R. McCleery, pp. 381–90. Berlin: Springer-Verlag.

Robertson, R. J. (1991). Infanticide or adoption by replacement males : the influence of female behaviour. *Acta XX Congressus Internationalis Ornithologici,* pp. 974–83.

Robertson, R. J. & Stutchbury, B. J. (1988). Experimental evidence for sexually selected infanticide in tree swallow. *Animal Behaviour*, **36**,749–53.

Robinson, J. A. & Goy, R. W. (1986). Steroid hormones and the ovarian cycle. In *Comparative Primate Biology,* vol. 3 *Reproduction and Development*, ed. W. R. Dukelow & J. Erwin, pp. 63–91. New York: Alan R. Liss.

Roda, S. A. & Mendes Pontes, A. R. (1998). Polygyny and infanticide in common marmosets in a fragment of the Atlantic forest of Brazil. *Folia Primatologica,* **69**, 372–6.

Roda, S. A. & Roda, S. (1988). Infanticide in a natural group of *Callithrix jacchus* (Callitrichidae – Primates). In *Abstracts of the XIIth Congress of the International Primatological Society*, p. 497. Brasília, Brazil.

Rodseth, L., Wrangham, R. W., Harrigan, A. M. & Smuts, B. B. (1991). The human community as a primate society. *Current Anthropology*, **32**, 221–54.

Rogowitz, G. L. (1996). Trade-offs in energy allocation during lactation. *American Zoologist*, **36**, 197–204.

Rohwer, S. (1986). Selection for adoption versus infanticide by replacement "mates" in birds. In *Current Ornithology*, vol. 3, ed. L. S. Johnson, pp. 353–95. New York: Plenum Press.

Rood, J. P. (1972). Ecological and behavioural comparisons of three genera of Argentine cavies. *Animal Behaviour Monographs*, **5**, 1–83.

Rose, L. M. (1994). Benefits and costs of resident males to females in white-faced capuchins, *Cebus capucinus*. *American Journal of Primatology*, **32**, 235–48.

Rose, L. M. & Fedigan, L. M. (1995). Vigilance in white-faced capuchin, *Cebus capuchinus*, in Costa Rica. *Animal Behaviour*, **49**, 63–70.

Rosenblum, L. A. & Albert, S. (1977). Response to mother and stranger: a first step in socialization. In *Primate Bio-social Development: Biological, Social and Ecological Determinants*, ed. S. Chevalier-Skolnikoff & F. E. Poirier, pp. 463–78. New York: Garland Publishing.

Ross, C. (1988). The intrinsic rate of natural increase and reproductive effort in primates. *Journal of Zoology, London*, **214**, 199–219.

Ross, C. (1993). Take-over and infanticide in South Indian hanuman langurs (*Presbytis entellus*). *American Journal of Primatology*, **30**, 75–82.

Ross, C. & Jones, K. E. (1999). Socioecology and the evolution of primate reproductive rates. In *Comparative Primate Socioecology*, ed. P. C. Lee, pp. 73–110. Cambridge: Cambridge University Press.

Ross, K. (1987). *Okavango: Jewel of the Kalahari*. New York: Macmillan Publishing Co.

Rowell, T. E. (1974). Contrasting adult male roles in different species of nonhuman primates. *Archives of Sexual Behavior*, **3**, 143–9.

Rudran, R. (1973). Adult male replacement in one-male troops of purple-faced langurs (*Presbytis senex senex*) and its effects on population structure. *Folia Primatologica*, **19**, 166–92.

Rudran, R. (1979). The demography and social mobility of a red howler (*Alouatta seniculus*) population in Venezuela. In *Vertebrate Ecology in the Northern Neotropics*, ed. J. F. Eisenberg, pp. 107–26. Washington, DC: Smithsonian Institution.

Rumiz, D. I. (1990). *Alouatta caraya*: Population density and demography in northern Argentina. *American Journal of Primatology*, **21**, 279–94.

Ryden, H. (1981). *Bobcat Year*. New York: Viking Press .

Ryder, O. A. & Massena, A. R. (1988). A case of male infanticide in *Equus przewalskii*. *Applied Animal Behaviour Science*, **21**, 187–90.

Saayman, G. S. (1971). Behaviour of the adult males in a troop of free-ranging chacma baboons (*Papio ursinus*). *Folia Primatologica*, **15**, 36–57.

Sakaluk, S. K., Eggert, A. K. & Müller, J. K. (1998). The 'widow effect' and its consequences for reproduction in burying beetles, *Necrophorus vespilloides* (Coleoptera, Silphidae). *Ethology*, **104**, 553–64.

Sánchez, S., Peláez, F., Gil-Bürmann, C. & Kaumanns, W. (1999). Costs of infant-carrying in the cotton-top tamarin (*Saguinus oedipus*). *American Journal of Primatology*, **48**, 99–111.

Sandell, M. (1989). The mating tactics and spacing patterns of solitary carnivores. In *Carnivore Behavior, Ecology, and Evolution*, ed. J. L. Gittleman, pp.164–82. Ithaca, NY: Cornell University Press.

Sargent, R. C. & Gross, M. R. (1993). Williams' principle: an explanation of parental care in teleost fish. In *Behaviour of Teleost Fishes*, ed. T. J. Pitcher, pp. 333–61. London: Chapman & Hall.

Scanlon, C. E., Chalmers, N. R. & Monteiro da Cruz, M. A. O. (1988). Changes in size, composition, and reproductive condition of wild marmoset groups (*Callithrix jacchus jacchus*) in North East Brazil. *Primates*, **29**, 295–305.

Schadler, M. H. (1981). Postimplantation abortion in pine voles (*Microtus pinetorum*) induced by strange males and pheromones of strange males. *Biology of Reproduction*, **25**, 295–7.

Schaller, G. B. (1972). *The Serengeti Lion: A Study of Predator–Prey Relations*. Chicago: University of Chicago Press.

Schaller, G. B., Jinchu, H., Wenshi, P. & Jing, Z. (1985). *The Giant Pandas of Wolong*. Chicago: University of Chicago Press.

Schenk, A. & Kovacs, K. M. (1995). Multiple mating between black bears revealed by DNA fingerprinting. *Animal Behaviour*, **50**, 1483–90.

Schubert, G. (1982). Infanticide by usurper Hanuman langur monkeys: a sociobiological myth. *Social Science Information*, **21**, 199–244.

Schües, C. & Ostbomk-Fischer, E. (1993). Das Menschenbild im Schatten der Soziobiologie. *GwG (Gesellschaft für wissenschaftliche Gesprächstherapie) Zeitschrift*, **89**, 14–18.

Schultz, L. A. & Lore, R. K. (1993). Communal reproductive success in rats (*Rattus norvegicus*): effects of group composition and prior social experience. *Journal of Comparative Psychology*, **107**, 216–22.

Schürmann, C. L. (1982). Mating behaviour of wild orang utans. In *The Orang Utan: Its Biology and Conservation*, ed. L. E. M. de Boer, pp. 269–84. The Hague: Dr W. Junk, Publishers, .

Schwagmeyer, P. L. (1979). The Bruce effect: an evaluation of male/female advantages. *American Naturalist*, **114**, 932–8.

Schwagmeyer, P. L. (1984). Multiple mating and intersexual selection in thirteen-lined ground squirrels. In *The Biology of Ground-Dwelling Squirrels. Annual Cycles, Behavioral Ecology, and Sociality*, ed. J. O. Murie & G. R. Michener, pp. 275–93. Lincoln: University of Nebraska Press.

Scott, M. P. (1989). Male parental care and reproductive success in the burying beetle, *Necrophorus orbicollis*. *Journal of Insect Behavior*, **2**, 133–7.

Scott, M. P. (1990). Brood guarding and the evolution of male parental care in burying beetles. *Behavioral Ecology and Sociobiology*, **26**, 31–9.

Scott, M. P. (1994). Competition with flies promotes communal breeding in the burying beetle, *Necrophorus tomentosus*. *Behavioral Ecology and Sociobiology*, **34**, 367–73.

Scott, M. P. & Traniello, F. A. (1990). Behavioural and ecological correlates of male and female parental care and reproductive success in burying beetles (*Necrophorus* spp.). *Animal Behaviour*, **39**, 274–83.

Segerstråle, U. (1992). Reductionism, 'bad science', and politics: a critique of anti-reductionist reasoning. *Politics and the Life Sciences*, **11**, 199–214.

Segerstråle, U. (2000). *Defenders of the Truth. The Battle for Science in the Sociobiology Debate and Beyond*. Oxford: Oxford University Press.

Sekulic, R. (1982a). Behavior and ranging patterns of a solitary female red howler (*Alouatta seniculus*). *Folia Primatologica*, **38**, 217–32.

Sekulic, R. (1982b). Daily and seasonal patterns of roaring and spacing in four red howler *Alouatta seniculus* troops. *Folia Primatologica*, **39**, 22–48.

Sekulic, R. (1983a). Male relationships and infant deaths in red howler monkeys (*Alouatta seniculus*). *Zeitschrift für Tierpsycholgie*, **61**, 185–202.

Sekulic, R. (1983b). Spatial relationships between recent mothers and other troop members in red howler monkeys (*Alouatta seniculus*). *Primates*, **24**, 475–85.

Semb-Johansson, A., Wiger, R. & Eugh, C. E. (1979). Dynamics of freely growing, confined populations of the Norwegian lemming *Lemmus lemmus*. *Oikos,* **33**, 246–60.

Setiawan, E., Knott, C. D. & Budhi, S. (1996). Preliminary assessment of vigilance and predator avoidance behavior of orang utans in Gunung Palung National Park, West Kalimantan, Indonesia. *Tropical Biodiversity*, **3**, 269–79.

Seyfarth, R. M. (1978). Social relationships among adult male and female baboons. II. Behaviour throughout the female reproductive cycle. *Behaviour*, **64**, 227–47.

Shea, C. (1999). Motive for murder. *Lingua Franca,* **9**(6), 23–5.

Sherman, P. W. (1981). Reproductive competition and infanticide in Belding's ground squirrels and other animals. In *Natural Selection and Social Behavior: Recent Research and Theory,* ed. R. D. Alexander & D. Tinkle, pp. 311–31. New York: Chiron Press.

Sherman, P. W. (1982). Infanticide in ground squirrels. *Animal Behaviour*, **30**, 938–9.

Sherman, P. W. (1984). The role of kinship. In *The Encyclopedia of Mammals*, ed. D. Macdonald, pp. 624–5. New York: Facts on File Publications.

Shopland, J. M. (1982). An intergroup encounter with fatal consequences in yellow baboons. *American Journal of Primatology*, **3**, 263–6.

Shopland, J. M. & Altmann, J. (1987). Fatal intragroup kidnapping in yellow baboons. *American Journal of Primatology*, **13**, 61–5.

Short, R. V. (1984). Oestrous and menstrual cycles. In *Hormonal Control of Reproduction,* ed. C. R. Austin & R. V. Short, pp. 115–52. Cambridge: Cambridge University Press.

Sicotte, P. (1994). Effect of male competition on male–female relationships in bi-male groups of mountain gorillas. *Ethology,* **97**, 47–64.

Siegel, S. & Castellan, N. J. (1988). *Nonparametric Statistics for the Behavioral Sciences.* New York: McGraw-Hill.

Sigg, H., Stolba, A., Abegglen, J.-J. & Dasser, V. (1982). Life history of hamadryas baboons: physical development, infant mortality, reproductive parameters and family relationships. *Primates,* **23**, 473–87.

Silk, J. B. (1980). Kidnapping and female competition among captive bonnet macaques. *Primates*, **21**, 100–10.

Silk, J. B. (1983). Local resource competition and facultative adjustment of sex ratios in relation to competitive abilities. *American Naturalist*, **121**, 56–66.

Silk, J. B. (1992a). Patterns of intervention in agonistic contests among male bonnet macaques. In *Coalitions and Alliances in Humans and Other Animals,* ed. A. H. Harcourt & F. B. M. de Waal, pp. 215–32. New York: Oxford University Press.

Silk, J. B. (1992b). The patterning of intervention among male bonnet macaques: reciprocity, revenge, and loyalty. *Current Anthropology*, **33**, 318–25.

Silk, J. B. (1993). The evolution of social conflict among female primates. In *Primate Social Conflict,* ed. W. A. Mason & S. P. Mendoza, pp. 49–83. New York: State University of New York Press.

Silk, J. B. (1994). Social relationships of male bonnet macaques: male bonding in a matrilineal society. *Behaviour*, **130**, 271–91.

Silk, J. B. (1999). Why are infants so attractive to others? The form and function of infant handling in bonnet macaques. *Animal Behaviour*, **57**, 1021–32.

Silk, J., Clark-Wheatley, C. B., Rodman, P. S. & Samuels, A. (1981). Differential reproductive success and facultative adjustment of sex ratios among captive female bonnet macaques (*Macaca radiata*). *Animal Behaviour*, **29**, 1106–20.

Silk, J. B. & Samuels, A. (1984). Triadic interactions among *Macaca radiata*: passports and buffers. *American Journal of Primatology*, **6**, 373–6.

Silk, J. & Stanford, C. (1999). Infanticide article disputed. *Anthropology News*, (September), 27 – 8.

Sillén-Tullberg, B. & Møller, A. P. (1993). The relationship between concealed ovulation and mating systems in anthropoid primates: a phylogenetic analysis. *American Naturalist*, **141**, 1–25.

Sillero-Zubiri, C., Gottelli, D. & Macdonald, D. W. (1996). Male philopatry, extra-pack copulations and inbreeding avoidance in Ethiopian wolves (*Canis simensis*). *Behavioral Ecology and Sociobiology*, **38**, 331–40.

Simonds, P. E. (1974). Sex differences in bonnet macaque networks and social structure. *Archives of Sexual Behavior*, **3**, 151–66.

Slagsvold, T. & Lifjeld, J. T. (1997). Incomplete female knowledge of male quality may explain variation in extra-pair paternity in birds. *Behaviour*, **134**, 353–71.

Sliwa, A. & Richardson, P. R. K. (1998). Responses of aardwolves, *Proteles cristatus*, Sparrman (1983), to translocated scent marks. *Animal Behaviour*, **56**, 137–46.

Small, M. F. (1990). Promiscuity in Barbary macaques (*Macaca sylvanus*). *American Journal of Primatology*, **20**, 267–82.

Small, M. F. (1993). *Female Choices: Sexual Behavior of Female Primates*. Ithaca, NY: Cornell University Press.

Smith, R. J. & Leigh, S. R. (1997). Sexual dimorphism in primate neonatal body mass. *Journal of Human Evolution*, **34**, 173–201.

Smith, E. O. & Pfeffer-Smith, P. G. (1984). Adult male–immature interactions in captive stumptail macaques (*Macaca arctoides*). In *Primate Paternalism*, ed. D. M. Taub, pp. 88–112. New York: Van Nostrand Reinhold.

Smith, E. O. & Whitten, P. L. (1988). Triadic interactions in savanna-dwelling baboons. *International Journal of Primatology*, **9**, 409–24.

Smith, H. G., Ottoson, U. & Sandell, M. (1994). Intrasexual competition among polygynously mated female starlings (*Sturnus vulgaris*). *Behavioral Ecology*, **5**, 57–63.

Smith, H. G., Wennerberg, L. & von Schantz, T. (1996). Adoption or infanticide: options of replacement males in the European starling. *Behavioral Ecology and Sociobiology*, **38**, 191–7.

Smith, R. J. & Jungers, W. L. (1997). Body mass in comparative primatology. *Journal of Human Evolution*, **32**, 523–59.

Smuts, B. B. (1983). Special relationships between adult male and female olive baboons: advantages. In *Primate Social Relationships: An Integrated Approach*, ed. R. A. Hinde, pp. 262–6. London: Blackwell.

Smuts, B. B. (1985) *Sex and Friendship in Baboons*. New York: Aldine de Gruyter.

Smuts, B. B. (1987a). Sexual competition and mate choice. In *Primate Societies*, ed. B. B. Smuts, D. L. Cheney, R. M. Seyfarth, R. W. Wrangham & T. T. Struhsaker, pp. 385–99. Chicago: University of Chicago Press.

Smuts, B. B. (1987b). Gender, aggression, and influence. In *Primate Societies*, ed. B. B. Smuts, D. L. Cheney, R. M. Seyfarth, R. W. Wrangham & T. T. Struhsaker, pp. 400–12. Chicago: University of Chicago Press.

Smuts, B. B. (1992). Male aggression against women: an evolutionary perspective. *Human Nature*, **3**, 1–44.

Smuts, B. B., Cheney, D. L., Seyfarth, R. M., Wrangham, R. W. & Struhsaker, T. T. (eds.) (1987). *Primate Societies*. Chicago: University of Chicago Press.

Smuts, B. B. & Gubernick, D. J. (1992). Male–infant relationships in nonhuman primates: paternal investment or mating effort? In *Father–Child Relations. Cultural and Biosocial Contexts*, ed. B. S. Hewlett, pp. 1–30. New York: Aldine de Gruyter.

Smuts, B. B. & Smuts, R. W. (1993). Male aggression and sexual coercion of females in nonhuman primates and other mammals: evidence and theoretical implications. *Advances in the Study of Behavior*, **22**, 1–63.

Smuts, B. B. & Watanabe, J. M. (1990). Social relationships and ritualized greetings in adult male baboons (*Papio cynocephalus anubis*). *International Journal of Primatology*, **11**, 147–72.

Snowdon, C. T. (1996). Infant care in cooperatively breeding species. In *Parental Care Evolution, Mechanisms and Adaptive Significance*, ed. J. S. Rosenblatt & C. T. Snowdon, pp. 643–89. San Diego: Academic Press.

Snowdon, C. T. & Suomi, S. J. (1982). Paternal behavior in primates. In *Nurturance*, vol. 3 *Studies of Development in Nonhuman Primates*, ed. H. E. Fitzgerald, J. Mullins & P. Child, pp. 63–108. New York: Plenum Press.

Sober, E. & Wilson, D. S. (1998). *Unto Others. The Evolution and Psychology of Unselfish Behavior*. Cambridge, MA, London: Harvard University Press.

Sokal, A. & Bricmont, J. (1998). *Intellectual Impostures. Postmodern Philosophers' Abuse of Science*. London: Profile Books.

Solomon, N. G. & French, J. A. (1997). *Cooperative Breeding in Mammals*. Cambridge: Cambridge University Press.

Solomon, N. G. & Getz, L. L. (1997). Examination of alternative hypotheses for cooperative breeding in rodents. In *Cooperative Breeding in Mammals*, ed. N. G. Solomon & J. A. French, pp. 199–230. Cambridge: Cambridge University Press.

Soltis, J., Thomsen, R., Matsubayashi, K. & Takenaka. O. (2000). Infanticide by resident males and female counterstrategies in wild Japanese macaques (*Macaca fuscata*). *Behavioral Ecology and Sociobiology*, in press.

Sommer, S. (1997). Monogamy in *Hypogeomys antimena*, an endemic rodent of the deciduous dry forest in western Madagascar. *Journal of Zoology, London*, **241**, 301–14.

Sommer, V. (1987). Infanticide among free-ranging langurs (*Presbytis entellus*) at Jodhpur (Rajasthan/India): recent observations and a reconsideration of hypotheses. *Primates*, **28**, 163–97.

Sommer, V. (1989a). Sexual harassment in langur monkeys (*Presbytis entellus*): competition for ova, sperm, and nurture? *Ethology*, **80**, 205–17.

Sommer, V. (1989b). *Die Affen. Unsere wilde Verwandtschaft*. Hamburg: GEO.

Sommer, V. (1994). Infanticide among the langurs of Jodhpur: testing the sexual selection hypothesis with a long-term record. In *Infanticide and Parental Care*, ed. S. Parmigiani & F. S. vom Saal, pp. 155–98. Chur (Switzerland): Harwood Academic Publishers.

Sommer, V. (1996). *Heilige Egoisten. Die Soziobiologie indischer Tempelaffen*. Munich: C. H. Beck.

Sommer, V., Asrivastava, A. & Borries, C. (1992). Cycles, sexuality, and conception in free-ranging langurs (*Presbytis entellus*). *American Journal of Primatology*, **28**, 1–27.

Sommer, V. & Mohnot, S. M. (1985). New observations on infanticides among Hanuman langurs (*Presbytis entellus*) near Jodhpur, Rajasthan, India. *Behavioral Ecology and Sociobiology,* **16**, 245–8.

Sommer, V. & Rajpurohit, L. S. (1989). Male reproductive success in harem troops of Hanuman langurs (*Presbytis entellus*). *International Journal of Primatology*, **10**, 293–317.

Soroker, V. & Terkel, J. (1988). Changes in incidence of infanticidal and parental responses during the reproductive cycle in male and female wild mice, *Mus musculus*. *Animal Behaviour*, **36**,1275–81.

Soukup, S. S. & Thompson, C. F. (1997). Social mating system affects the frequency of extra-pair paternity in house wrens. *Animal Behaviour*, **54**, 1089–105.

Soulé, M. E. (ed.) (1986). *Conservation Biology: The Science of Scarcity and Diversity*. Sunderland, MA: Sinauer.

Sowls, L. K. (1948). The Franklin ground squirrel, *Citellus franklinii* (Sabine), and its relationship to nesting ducks. *Journal of Mammalogy*, **29**, 113–37.

Spijkerman, R. P., van Hooff, J. A. R. A. M. & Jens, W. (1990). A case of lethal infant abuse in an established group of chimpanzees. *Folia Primatologica,* **55**, 41–44.

SPSS (1990) *SPSS for the Macintosh 4.0*. Chicago: SPSS Inc.

Stacey, P. B. & Edwards, T. C. (1983). Possible cases of infanticide by immigrant females in a group-breeding bird. *Auk*, **100**, 731–3.

Stanford, C. B. (1991). Social dynamics of intergroup encounters in the capped langur (*Presbytis pileata*). *American Journal of Primatology*, **25**, 35–47.

Stanford, C. B. (1992). Costs and benefits of allomothering in wild capped langurs (*Presbytis pileata*). *Behavioral Ecology and Sociobiology*, **30**, 29–34.

Stanford, C. B. (1998). Predation and male bonds in primate societies. *Behaviour*, **135**, 513–33.

Stanford, C. B., Wallis, J., Matama, H. & Goodall, J. (1994). Patterns of predation by chimpanzees on red colobus monkeys in Gombe National Park, (1982 -1991). *American Journal of Physical Anthro*pology, **94**, 213–28.

Stanger, K. F., Coffman, B. S. & Izard, M. K. (1995). Reproduction in Coquerel's dwarf lemur (*Mirza coquereli*). *American Journal of Primatology, 36*, 223–37.

Starin, E. D. (1991). Socioecology of the red colobus monkey in the Gambia with particular reference to female–male differences and transfer patterns. Ph.D. Thesis, City University of New York.

Starin, E. D. (1994). Philopatry and affiliation among red colobus. *Behaviour*, **130**, 253–70.

Stearns, S. C. (1992). *The Evolution of Life Histories*. Oxford: Oxford University Press.

Steenbeek, R. (1996). What a maleless group can tell us about the constraints on female transfer in Thomas's langurs (*Presbytis thomasi*). *Folia Primatologica*, **67**, 169–81.

Steenbeek, R. (1999a). Female choice and male coercion in wild Thomas's langurs. Ph.D. dissertation, Utrecht University.

Steenbeek, R. (1999b). Tenure related changes in wild Thomas's langurs: I: Between-group interactions. *Behaviour*, **136**, 595–625.

Steenbeek, R., Assink, P. & Wich, S. A. (1999a). Tenure related changes in wild Thomas's langurs: II: Loud calls. *Behaviour*, **136**, 627–650.

Steenbeek, R., Piek, R. C., van Buul, M. & vanHooff, J. A. R. A. M. (1999b). Vigilance in wild Thomas's langurs (*Presbytis thomasi*): the importance of infanticide risk. *Behavioral Ecology and Sociobiology*, **45**, 137–50.

Steenbeek, R., Sterck, E. H. M., de Vries, H. & van Hooff, J. A. R. A. M. (2000). Costs and benefits of the one-male, age-graded, and all-male phase in wild Thomas's langur groups. In *Primate Males*, ed. P. M. Kappeler, pp. 130–45. Cambridge: Cambridge University Press.

Steenbeek, R. & van Schaik, C. P. (2000). Group size as a limiting factor in Thomas's langurs: the folivore paradox revisited. *Behavioral Ecology and Sociobiology*, in press.

Stehn, R. A. & Jannett, F. J. Jr (1981). Male-induced abortion in various microtine rodents. *Journal of Mammalogy, 62*, 369–72.

Stein, D. M. (1984a). *The Sociobiology of Infant and Adult Male Baboons*. Norwood, NJ: Ablex.

Stein, D. M. (1984b). Ontogeny of infant–adult male relationships during the first year of life for yellow baboons (*Papio cynocephalus*). In *Primate Paternalism*, ed. D. M. Taub, pp. 213–43. New York: Van Nostrand Reinhold.

Stein, D. M. & Stacey, P. B. (1981). A comparison of infant–adult male relations in a one-male group with those in a multi-male group for yellow baboons (*Papio cynocephalus*). *Folia Primatologica, 36*, 264–76.

Steiner, A. L. (1972). Mortality resulting from intraspecific fighting in some ground squirrel populations. *Journal of Mammalogy, 53*, 601–3.

Steklis, H. D. & Fox, R. (1988). Menstrual-cycle phase and sexual behavior in semi-free ranging stumptail macaques (*Macaca arctoides*). *International Journal of Primatology 9*, 443–56.

Steklis, H. D. & Whiteman, C. H. (1989). Loss of estrus in human evolution: too many answers, too few questions. *Ethology and Sociobiology, 10*, 417–34.

Stephan, P. (1983). Bevölkerungsbiologische Untersuchungen in Ditfurt, einem Dorf im Nordharzvorland, im 17. und 18. Jahrhundert. Ph. D. thesis. Akademie der Wissenschaften.

Stephan, P. (1992). Wie egoistisch sind menschliche Gene? – Versuch einer Deutung von Differenzen in der Lebenserwartung von Kindern mit und ohne Stiefmutter. *Wissenschaftliche Zeitschrift der Humboldt-Universität zu Berlin, Reihe Medizin, 41*, 75–6.

Stephens, M. L. (1982). Mate takeover and possible infanticide by a female northern jacana, *Jacana spinosa*. *Animal Behaviour, 30*, 1253–4.

Steppan, S. J., Akhverdyan, M. R., Lyapunova, E. A., Fraser, D. G., Vorontsov, N. N., Hoffmann, R. S. & Braun, J. (2000) Molecular phylogeny of the marmots (Rodentia: Sciuridae): tests of evolutionary and biogeographic hypotheses. *Systematic Biology*, in press.

Sterck, E. H. M. (1997). Determinants of female dispersal in Thomas's langurs. *American Journal of Primatology, 42*, 179–98.

Sterck, E. H. M. (1998). Female dispersal, social organization, and infanticide in langurs: are they linked to human disturbance? *American Journal of Primatology, 44*, 235–54.

Sterck, E. H. M. (1999). Variation in langur social organization in relation to the socioecological model, human habitat alteration, and phylogenetic constraints. *Primates, 40*, 201–15.

Sterck, E. H. M. & Steenbeek, R. (1997). Female dominance relationships and food competition in the sympatric Thomas's langur and long-tailed macaque. *Behaviour, 139*, 9–10.

Sterck, E. H. M., Watts, D. P. & van Schaik, C. P. (1997). The evolution of female social relationships in nonhuman primates. *Behavioral Ecology and Sociobiology*, **41**, 291–309.

Stevens, S. D. (1998). High incidence of infanticide by lactating females in a population of Columbian ground squirrels (*Spermophilus columbianus*). *Canadian Journal of Zoology*, **76**, 1183–7.

Stewart, K. J. & Harcourt, A. H. (1987). Gorillas: variation in female relationships. In *Primate Societies*, ed. B. B. Smuts, D. L. Cheney, R. M. Seyfarth, R. W. Wrangham & T. T. Struhsaker, pp. 155–64. Chicago: University of Chicago Press.

Stopka, P. & Macdonald, D. W. (1998). Signal interchange during mating in wood mouse (*Apodemus sylvaticus*): the concept of active and passive signalling. *Behaviour*, **135**, 231–49.

Storey, A. E., Bradbury, C. G. & Joyce, T. L. (1994). Nest attendance in male meadow voles: the role of the female in regulating male interactions with pups. *Animal Behaviour*, **47**, 1037–46.

Stribley, J. A., French, J. A. & Inglett, B. J. (1987). Mating patterns in the golden lion tamarin (*Leontopithecus rosalia*): continuous receptivity and concealed estrus. *Folia Primatologica*, **49**, 137–50.

Strier, K. B. (1991). Demography and conservation of an endangered primate, *Brachyteles arachnoides*. *Conservation Biology*, **5**, 214–18.

Strier, K. B. (1994). Brotherhoods among Atelines: kinship, affiliation, and competition. *Behaviour*, **130**, 151–67.

Strier, K. B. (2000). From binding brotherhoods to short-term sovereignty: the dilemma of male Cebidae. In *Primate Males*, ed. P. M. Kappeler, pp. 72–83. Cambridge: Cambridge University Press.

Strier, K. B., Mendes, F. D. C., Rímoli, J. & Rímoli, A. O. (1993). Demography and social structure of one group of muriquis (*Brachyteles arachnoides*). *International Journal of Primatology*, **14**, 513–26.

Strier, K. B. & Ziegler, T. E. (1997). Behavioral and endocrine characteristics of the reproductive cycle in wild Muriqui monkeys, *Brachyteles arachnoides*. *American Journal of Primatology*, **42**, 299–310.

Struhsaker, T. T. (1975). *The Red Colobus Monkey*. Chicago: University of Chicago Press.

Struhsaker, T. T. (1977). Infanticide and social organization in the redtail monkey (*Cercopithecus ascanius schmidti*) in the Kibale Forest, Uganda. *Zeitschrift für Tierpsycholologie*, **45**, 75–84.

Struhsaker, T. T. (1981). Polyspecific associations among tropical rain-forest primates. *Zeitschrift für Tierpsychologie*, **57**, 268–304.

Struhsaker, T. T. (2000). The effects of predation and habitat quality on the socioecology of African monkeys: lessons from the islands of Bioko and Zanzibar. In *Old World Monkeys*, ed. P. Whitehead & C. Jolly. pp. 393–430. Cambridge: Cambridge University Press.

Struhsaker, T. T. & Leland, L. (1979). Socioecology of five sympatric monkey species in the Kibale Forest, Uganda. *Advances in the Study of Behavior*, **9**, 159–228.

Struhsaker, T. T. & Leland, L. (1985). Infanticide in a patrilineal society of red colobus monkeys. *Zeitschrift für Tierpsychology*, **69**, 89–132.

Struhsaker, T. T. & Leland, L. (1987). Colobines: Infanticide by adult males. In *Primate Societies*, ed. B. B. Smuts, D. L. Cheney, R. M. Seyfarth, R. W. Wrangham & T. T. Struhsaker, pp. 83–97. Chicago: University of Chicago Press.

Struhsaker, T. T. & Pope, T. R. (1991). Mating system and reproductive success: a comparison of two African forest monkeys (*Colobus badius* and *Cercopithecus ascanius*). *Behaviour*, **117**, 182–205.

Strum, S. C. (1974). Life with the Pumphouse Gang: new insights into baboon behavior. *National Geographic*, **147**, 672–91.

Strum, S. C. (1982). Agonistic dominance in male baboons: an alternative view. *International Journal of Primatology*, **3**, 175–202.

Strum, S. C. (1983). Use of females by male olive baboons (*Papio anubis*). *American Journal of Primatolology*, **5**, 93–110.

Strum, S. C. (1984). Why males use infants. In *Primate Paternalism,* ed. D. M. Taub, pp. 146–85. New York: Van Nostrand Reinhold.

Strum, S. (1987). *Almost Human. A Journey into the World of Baboons*. London: Elm Tree.

Stubbe, A. & Janke, S. (1994). Some aspects of social behaviour in the vole *Microtus brandti* (Radde, 1861). *Polish Ecological Studies*, **20**, 449–57.

Sugiyama, Y. (1965). On the social change of hanuman langurs (*Presbytis entellus*) in their natural condition. *Primates*, **6**, 213–47.

Sugiyama, Y. (1966). An artificial social change in a hanuman langur troop (*Presbytis entellus*). *Primates, 7*, 41–72.

Sugiyama, Y. (1967). Social organization of hanuman langurs. In *Social Communication among Primates*, ed. S. A. Altmann, pp. 221–36. Chicago: University of Chicago Press.

Sugiyama, Y. (1971). Characteristics of the social life of bonnet macaques (*Macaca radiata*). *Primates*, **12**, 247–66.

Sussman, R. W. (1991). Demography and social organization of free-ranging *Lemur catta* in the Beza Mahafaly Reserve, Madagascar. *American Journal of Physical Anthropology*, **84**, 43–58.

Sussman, R. W. (1992). Male life history and intergroup mobility among ringtailed lemurs (*Lemur catta*). *International Journal of Primatology*, **13**, 395–413.

Sussman, R. W., Cheverud, J. M. & Bartlett, T. Q. (1995). Infant killing as an evolutionary strategy: reality or myth? *Evolutionary Anthropology*, **3**, 149–51.

Sussman, R. W. & Garber, P. A. (1987). A new interpretation of the social organization and mating system of the Callitrichidae. *International Journal of Primatology*, **8**, 73–92.

Svare, B., Broida, J., Kinsley, C. & Mann, M. (1984). Psychobiological determinants underlying infanticide in mice. In *Infanticide: Comparative and Evolutionary Perspectives*, ed. G. Hausfater and S. B. Hrdy, pp. 387–400. New York: Aldine de Gruyter.

Swenson, J. E., Sandegren, F., Söderberg, A., Bjärvall, A., Franzén, R. & Wabakken, P. (1997). Infanticide caused by hunting of male bears. *Nature, 386*, 450–51.

Symington, M. McF. (1988). Demography, ranging patterns, and activity budgets of black spider monkeys (*Ateles paniscus chamek*) in the Manu National Park, Peru. *American Journal of Primatology*,**15**, 45–67.

Tarara, E. B. (1987). Infanticide in a chacma baboon troop. *Primates* **28**, 267–70.

Tardif, S. D. & Bales, K. (1997). Is infant-carrying a courtship strategy in callitrichid primates? *Animal Behaviour*, **53**, 1001–7.

Taub, D. M. (1978). Aspects of the biology of the wild Barbary macaque (Primates, Cercopithecinae, *Macaca sylvanus* L. 1758): Biogeography, the mating system and male–infant associations. Ph. D. dissertation, University of California, Davis.

Taub, D. M. (1980). Testing the 'agonistic buffering' hypothesis. I. The dynamics of participation in the triadic interaction. *Behavioral Ecology and Sociobiology*, **6**, 187–97.

Taub, D. M. (1984). Male caretaking behavior among wild Barbary macaques (*Macaca sylvanus*). In *Primate Paternalism,* ed. D. M. Taub, pp. 20–55, New York: Van Nostrand Reinhold.

Taub, D. M. (1985). Male–infant interactions in baboons and macaques: a critique and reevaluation. *American Zoologist*, **25**, 861–71.

Taylor, H., Teas, J., Richie, T., Southwick, C. & Shresta, R. (1978). Social interactions between adult male and infant rhesus monkeys in Nepal. *Primates*, **19**, 343–51.

Terborgh, J. (1983). *Five New World Primates: A Study in Comparative Ecology.* Princeton: Princeton University Press.

Thierry, B. & Anderson, J. R. (1986). Adoption in anthropoid primates. *International Journal of Primatology*, **7**, 191–216.

Thierry, B., Heistermann, M., Aujard F. & Hodges, J. K. (1996). Long-term data on basic reproductive parameters and evaluation of endocrine, morphological, and behavioral measures for monitoring reproductive status in a group of semifree-ranging Tonkean macaques (*Macaca tonkeana*). *American Journal of Primatology,* **39**, 47–62.

Tilson, R. L. (1981). Family formation strategies of Kloss's gibbons. *Folia Primatologica*, **35**, 259–87.

Tinbergen, N. (1963). On aims and methods of ethology. *Zeitschrift für Tierpsychologie*, **20**, 410–33.

Tinley, K. L., (1966). *An Ecological Reconnaissance of the Moremi Wildlife Reserve.* Cape Town: Gothic Printing Co., Ltd.

Tokida, E. (1976). A case of infanticide by Kabo at Jigokudani monkey park, Nagano, in 1970 [in Japanese]. *Nihonzaru,* **2**, 124–8.

Tolonen, P. & Korpimäki, E. (1995). Parental effort of kestrels (*Falco tinnunculus*) in nest defense: effects of laying time, brood size, and varying survival prospects of offspring. *Behavioral Ecology*, **6**, 435–41.

Topping, M. G. & Millar, J. S. (1998). Mating patterns and reproductive success in the bushy-tailed woodrat (*Neotoma cinerea*) as revealed by DNA fingerprinting. *Behavioral Ecology and Sociobiology,* **43**, 115–24.

Trail, P. W., Strahl, S. D. & Brown, J. L. (1981). Infanticide in relation to individual and flock histories in a communally breeding bird, the Mexican jay (*Aphelocoma ultramarina*). *American Naturalist*, **118**, 72–82.

Treves, A. (1996). A preliminary analysis of the timing of infant exploration in relation to social structure in 17 species of primates. *Folia Primatologica*, **67**, 153–6.

Treves, A. (1997). Primate natal coats: a preliminary analysis of distribution and function. *American Journal of Physical Anthropology*, **104**, 47–70.

Treves, A. (1998). Primate social systems: conspecific threat and coercion-defense hypotheses. *Folia Primatologica*, **69**, 81–8.

Treves, A. (1999). Within-group vigilance in red colobus and redtail monkeys. *American Journal of Primatology*, **48**, 113–26.

Treves, A. & Chapman, C. A. (1996). Conspecific threat, predation avoidance, and resource defense: implications for grouping in langurs. *Behavioral Ecology and Sociobiology*, **39**, 43–53.

Trillmich, F. (1986). Attendance behavior of Galapagos fur seals. In *Fur seals. Maternal Strategies on Land and at Sea*, ed. R. L. Gentry & G. L. Kooyman, pp. 168–85. Princeton, NJ: Princeton University Press.

Trivers, R. L. (1972). Parental investment and sexual selection. In *Sexual Selection and the Descent of Man 1871–1971*, ed. B. Campbell, pp. 136–79. Chicago: Aldine de Gruyter.

Trivers, R. L. (1985). *Social Evolution*. Menlo Park, CA: Benjamin Cummings.

Trulio, L. A. (1987). Infanticide in a population of California ground squirrels: functional significance and comparative importance. Ph. D. thesis, University of California, Davis.

Trulio, L. A. (1996). The functional significance of infanticide in a population of California ground squirrels (*Spermophilus beecheyi*). *Behavioral Ecology and Sociobiology*, **38**, 97–103.

Trumbo, S. T. (1990a). Interference competition among burying beetles (Silphidae, *Necrophorus*). *Ecological Entomology*, **15**, 347–55.

Trumbo, S. T. (1990b). Reproductive benefits of infanticide in a biparental burying beetle *Necrophorus orbicollis*. *Behavioral Ecology and Sociobiology*, **27**, 269–73.

Trumbo, S. T. (1991). Reproductive benefits and the duration of paternal care in a biparental burying beetle, *Necrophorus orbicollis*. *Behaviour*, **117**, 82–115.

Trumbo, S. T. & Eggert, A. (1994). Beyond monogamy: territory quality influences sexual advertisement in male burying beetles. *Animal Behaviour*, **48**, 1043–7.

Turke, P. W. (1997). Hypothesis: menopause discourages infanticide and encourages continued investment by agnates. *Evolution and Human Behavior*, **18**, 3–13.

Tutin, C. E. G. (1979). Responses of chimpanzees to copulation, with special reference to interference by immature individuals. *Animal Behaviour*, **27**, 845–54.

Uehara, S., Nishida, T., Takasaki, H., Kitopeni, R., Kasagula, M. B., Noriksoshi, K., Tsukahara, T., Nyundo, R. & Hamai, M. (1994). A lone male chimpanzee in the wild: the survivor of a disintegrated unit-group. *Primates*, **35**, 275–81.

Valderrama, X., Srikosomatara, S. & Robinson, J. G. (1990). Infanticide in wedge-capped capuchin monkeys, *Cebus olivaceus*. *Folia Primatologica*, **54**, 171–6.

van den Berghe, P. L. (1981). Sociobiology. Several views. *Bio-Science*, **31**, 406.

van Hooff, J. A. R. A. M. & van Schaik, C. P. (1992). Cooperation in competition: the ecology of primate bonds. In *Coalitions and Alliances in Humans and Other Animals*, ed. A. H. Harcourt & F. B. M. de Waal, pp. 357–89. New York: Oxford University Press.

van Hooff, J. A. R. A. M. & van Schaik, C. P. (1994). Male bonds: affiliative relationships among nonhuman primate males. *Behaviour*, **130**, 309–37.

van Horn, R. N. & Eaton, G. (1979). Reproductive physiology and behavior in prosimians. In *The Study of Prosimian Behavior*, ed. G. A. Doyle & R. D. Martin, pp. 79–122. New York: Academic Press.

van Lawick, H. (1974). *Solo: The Story of an African Wild Dog*. Boston, MA: Houghton Mifflin Company.

van Noordwijk, M. A. (1985). Sexual behaviour of Sumatran long-tailed macaques (*Macaca fascicularis*). *Zeitschrift für Tiepsychologie*, **70**, 277–96.

van Noordwijk, M. A. & van Schaik, C. P. (1985). Male migration and rank acquisition in wild long-tailed macaques (*Macaca fascicularis*). *Animal Behaviour*, **33**, 849–61.

van Noordwijk, M. A. & van Schaik, C. P. (1988). Male careers in Sumatran long-tailed macaques (*Macaca fascicularis*). *Behaviour*, **107**, 24–43.

van Noordwijk, M. A. & van Schaik, C. P. (1999). The effects of dominance rank and group size on female lifetime reproductive succes in wild long-tailed macaques, *Macaca fascicularis*. *Primates*, **40**, 105–30.

Van Poppel, F. (1995). Widows, widowers and remarriage in nineteenth-century Netherlands. *Population Studies*, **49**, 421–41.

van Roosmalen, M. G. M. (1985). Habitat preferences, diet, feeding strategy and social organization of the black spider monkey (*Ateles paniscus paniscus* Linnaeus 1758) in Surinam. *Acta Amazonica*, **15**.

van Schaik, C. P. (1983). Why are diurnal primates living in groups? *Behaviour*, **87**, 120–44.

van Schaik, C. P. (1989). The ecology of social relationships amongst female primates. In *Comparative Socioecology: The Behavioral Ecology of Humans and Other Mammals*, ed. V. Standen & R. A. Foley, pp. 195–218. Oxford: Blackwell.

van Schaik, C. P. (1996). Social evolution in primates: the role of ecological factors and male behaviour. *Proceedings of the British Academy*, **88**, 9–31.

van Schaik, C. P. (2000). Social counterstrategies against infanticide by males in primates and other mammals. *Primate Males*, ed. P. M. Kappeler, pp. 34–52. Cambridge: Cambridge University Press.

van Schaik, C. P., Assink, P. R. & Salafsky, N. (1992). Territorial behavior in southeast Asian langurs: resource defense or mate defense? *American Journal of Primatology*, **26**, 233–42.

van Schaik, C. P. & Dunbar, R. I. M. (1990). The evolution of monogamy in large primates: a new hypothesis and some crucial tests. *Behaviour*, **115**, 30–62.

van Schaik, C. P. & Hörstermann, M. (1994). Predation risk and the number of adult males in a primate group: a comparative test. *Behavioral Ecology and Sociobiology*, **35**, 261–72.

van Schaik, C. P. & Kappeler, P. M. (1993). Life history, activity period and lemur social systems. In *Lemur Social Systems and Their Ecological Basis*, ed. P. M. Kappeler & J. U. Ganzhorn, pp. 241–60. New York: Plenum Press.

van Schaik, C. P. & Kappeler, P. M. (1996). The social systems of gregarious lemurs: lack of convergence with anthropoids due to evolutionary disequilibrium. *Ethology*, **102**, 915–41.

van Schaik, C. P. & Kappeler, P. M. (1997). Infanticide risk and the evolution of male–female association in primates. *Proceedings of the Royal Society of London B*, **264**, 1687–94.

van Schaik, C. P. & Mirmanto, E. (1985). Spatial variation in the structure and litterfall of a Sumatran rain forest. *Biotropica*, **17**, 196–205.

van Schaik, C. P. & Paul, A. (1996–7). Male care in primates: does it ever reflect paternity? *Evolutionary Anthropology*, **5**, 152–6.

van Schaik, C. P. & van Noordwijk, M. A. (1985). Evolutionary effect of the absence of felids on the social organization of the macaques on the island of Simeulue (*Macaca fascicularis fusca*, Miller 1903). *Folia Primatologica*, **44**, 138–47.

van Schaik, C. P. & van Noordwijk, M. A. (1988). Scramble and contest in feeding competition among female long-tailed macaques (*Macaca fascicularis*). *Behaviour*, **105**, 77–98.

van Schaik, C. P. & van Noordwijk, M. A. (1989). The special role of male *Cebus* monkeys in predation avoidance and its effects on group composition. *Behavioral Ecology and Sociobiology*, **24**, 265–76.

van Schaik, C. P., van Noordwijk, M. A. & Nunn, C. L. (1999). Sex and social evolution in primates. In *Comparative Primate Socioecology*, ed. P. C. Lee, pp. 204–40. Cambridge: Cambridge University Press.

van Tienhoven, A. (1983). *Reproductive Physiology of Vertebrates*. Ithaca, NY: Cornell University Press.

Vehrencamp, S. L. (1977). Relative fecundity and parental effort in communally nesting Anis, *Crotophaga sulcirostris. Science*, **197**, 403–5.

Veiga, J. P. (1990a). Sexual conflict in the house sparrow: interference between polygynously mated females versus asymmetric male investment. *Behavioral Ecology and Sociobiology*, **27**, 345–50.

Veiga, J. P. (1990b). Infanticide by male and female house sparrows. *Animal Behaviour*, **39**, 496–502.

Veiga, J. P. (1992). Why are house sparrows predominantly monogamous: a test of hypotheses, *Animal Behaviour*, **43**, 361–70.

Veiga, J. P. (1993). Prospective infanticide and ovulation retardation in free-living house sparrows. *Animal Behaviour*, **45**, 43–6.

Velleman, P. F. (1997). *Data Desk: The New Power of Statistical Vision*. Ithaca, NY: Data Description Inc.

Venkataraman, A. B. (1998). Male-biased adult sex ratios and their significance for cooperative breeding in Dhole, *Cuon alpinus*, packs. *Ethology*, **104**, 671–84.

Vessey, S. & Meikle, D. (1984). Free living rhesus monkeys: adult male interactions with infants and juveniles. In *Primate Paternalism,* ed. D. M. Taub, pp. 113–26. New York: Van Nostrand Reinhold.

Vestal, B. M. (1991). Infanticide and cannibalism by male thirteen-lined ground squirrels. *Animal Behaviour*, **41**, 1103–4.

Vick, L. G. & Pereira, M. E. (1989). Episodic targeting aggression and the histories of *Lemur* social groups. *Behavioral Ecology and Sociobiology,* **25**, 3–12.

Vogel, C. (1976). Ökologie, Lebensweise und Sozialverhalten der grauen Languren in verschiedenen Biotopen Indiens. *Fortschritte der Verhaltensforschung, Beiheft* **17**, *Zeitschrift für Tierpsychologie*. Berlin: Parey.

Vogel, C. (1979). Der Hanuman-Langur (*Presbytis entellus*), ein Parade-Exempel für die theoretischen Konzepte der Soziobiologie? *Verhandlungen der Deutschen Zoologischen Gesellschaft*, **72**, 73–89.

Vogel, C. (1984). Patterns of infant-transfer within two troops of common langurs (*Presbytis entellus*) near Jodhpur: testing hypotheses concerning the benefits and risks. In *Current Primate Researches*, ed. M. L. Roonwal, S. M. Mohnot & N. S. Rathore, pp. 361–79. Jodhpur: Jodhpur University Press.

Vogel, C. (1989). *Vom Töten zum Mord. Das wirkliche Böse in der Evolutions-geschichte*. Munich: Hanser.

Voland, E. (1988). Differential infant and child mortality in evolutionary perspective: data from late 17th to 19th century Ostfriesland (Germany). In *Human Reproductive Behaviour – A Darwinian Perspective*, ed. L. Betzig, M. Borgerhoff Mulder & P. Turke, pp. 253–61. Cambridge: Cambridge University Press.

Voland, E. (1995). Reproductive decisions viewed from an evolutionarily informed historical demography. In *Human Reproductive Decisions – Biological and Social Perspectives, ed.* R. I. M. Dunbar, pp. 137–59. Houndsmills & London: Macmillan;

New York: St. Martin's.

vom Saal, F. S. (1984). Proximate and ultimate causes of infanticide and parental behavior in male house mice. In *Infanticide: Comparative and Evolutionary Perspectives*, ed. G. Hausfater & S. B. Hrdy, pp. 401–24. New York: Aldine de Gruyter.

vom Saal, F. S., Franks, P., Boechler, M., Palanza, P. & Parmigiani, S. (1995). Nest defense and survival of offspring in highly aggressive wild Canadian female house mice. *Physiology and Behavior*, **58**, 669–78.

vom Saal, F. S. & Howard, L. S. (1982). The regulation of infanticide and parental behavior: implications for reproductive success in male mice. *Science*, **215**, 1270–2.

von Glasersfeld, E. (1992). Konstruktion der Wirklichkeit und des Begriffs der Objektivität. In *Einführung in den Konstruktivismus*, Carl Friedrich von Siemens Stiftung, pp. 9–41. Munich, Zurich: Piper.

Wagner, R. H. (1991). The use of extrapair copulations for mate appraisal by razorbills, *Alca torda*. *Behavioral Ecology*, **2**, 198–203.

Wallen, K. (1995). The evolution of female sexual desire. *Sexual Nature, Sexual Culture*, ed. P. R. Abramson & S. D. Pinkerton, pp. 57–79. Chicago: Chicago University Press.

Wallis, J. & Lemon, W. B. (1986). Social behavior and genital swelling in pregnant chimmpanzees (*Pan troglodytes*). *American Journal of Primatolology*, **10**, 171–83.

Wamboldt, M. Z., Gelhard, R. E. & Insel, T. R. (1988). Gender differences in caring for infant *Cebuella pygmaea*: the role of infant age and relatedness. *Developmental Psychobiology*, **20**, 7–18.

Waser, P. M., Austad, S. N. & Keane, B. (1986). When should animals tolerate inbreeding? *American Naturalist*, **128**, 529–37.

Waser, P. M. & Jones, W. T. (1983). Natal philopatry among solitary mammals. *Quarterly Review of Biology*, **58**, 355–90.

Wasser, S. K. & Barash, D. P. (1983). Reproductive suppression among female mammals: implications for biomedical and sexual selection theory. *Quarterly Review of Biology*, **58**, 513–38.

Wasser, S. K. & Starling, A. K. (1986). Reproductive competition among female yellow baboons. In *Primate Ontogeny, Cognition and Social Behavior*, ed. J. G. Else & P. C. Lee, pp. 343–54. Cambridge: Cambridge University Press.

Wasser, S. K. & Starling, A. K. (1988). Proximate and ultimate causes of reproductive suppression among female yellow baboons at Mikumi National Park, Tanzania. *American Journal of Primatology*, **16**, 97–121.

Watanabe, K. (1979). Alliance formation in a free-ranging troop of Japanese macaques. *Primates*, **20**, 459–74.

Watanabe, K. & Matsumura, S. (1996). Social organization of moor macaques, *Macaca maurus*, in the Karaent Nature Reserve, South Sulawesi, Indonesia. In *Variations in the Asian Macaques*, ed. T. Shotake, pp. 147–62, Tokyo: Tokai University Press.

Waterman, J. M. (1998) Mating tactics of male Cape ground squirrels, *Xerus inauris*: consequences of year-round breeding. *Animal Behaviour*, **56**, 459–66.

Watts, D. P. (1985). Relations between group size and composition and feeding competition in mountain gorilla groups. *Animal Behaviour*, **33**, 72–85.

Watts, D. P. (1989). Infanticide in mountain gorillas: new cases and a review of evidence. *Ethology*, **81**, 1–18.

Watts, D. P. (1990). Ecology of gorillas and its relation to female transfer in mountain gorillas. *International Journal of Primatology*, **11**, 21–45.

Watts, D. P. (1991a). Mountain gorilla reproduction and sexual behavior. *American Journal of Primatology*, **24**, 211–26.

Watts, D. P. (1991b). Harassment of immigrant female mountain gorillas by resident females. *Ethology*, **89**, 135–53.

Watts, D. P. (1992). Social relationships of immigrant and resident female mountain gorillas. I. Male–female relationships. *American Journal of Primatology*, **28**, 159–81.

Watts, D. P. (1994). Social relationships of immigrant and resident female mountain gorillas. II. Relatedness, residence, and relationships between females. *American Journal of Primatology*, **32**, 13–30.

Watts, D. P. (1996). Comparative socio-ecology of gorillas. In *Great Ape Societies*, ed. W. C. McGrew, L. F. Marchant & T. Nishida, pp. 16–28. Cambridge: Cambridge University Press.

Waynforth, D. & Dunbar, R. I. M. (1995). Conditional mate choice strategies in humans: evidence from 'Lonely Hearts' advertisements. *Behaviour*, **132**, 755–79.

Webster, A. B. & Brooks, R. J. (1981). Social behavior of *Microtus pennsylvanicus* in relation to seasonal changes in demography. *Journal of Mammalogy*, **62**, 738–51.

Webster, A. B., Gartshorne, R. G. & Brooks, R. J. (1981). Infanticide in the meadow vole, *Microtus pennsylvanicus*: significance in relation to social system and population cycling. *Behavioral and Neural Biology*, **31**, 342–47.

Weingrill, T. (2000). Infanticide and the value of male–female relationships in mountain chacma baboons (*Papio cynocephalus ursinus*), *Behaviour*, in press.

Wesche, K. (1997). A classification of a tropical *Shorea robusta* forest stand in southern Nepal. *Phytocoenologia*, **27**, 103–18.

Whitmore, M. J. (1986). Infanticide of nestling noisy miners, communally breeding honeyeaters. *Animal Behaviour*, **34**, 933–5.

Whitten, P. L. (1987). Infants and adult males. In *Primate Societies*, ed. B. B. Smuts, D. L. Cheney, R. M. Seyfarth, R. W. Wrangham & T. T. Struhsaker, pp. 343–57. Chicago: University of Chicago Press.

Wickler, W. (1976). The ethological analysis of attachment. *Zeitschrift für Tierpsychologie*, **42**, 12–28.

Wickler, W. & Seibt, U. (1983). Monogamy: an ambiguous concept. In *Mate Choice*, ed. P. Bateson, pp. 33–50. Cambridge: Cambridge University Press.

Wiley, E. O., Siegel-Causey, D., Brooks, D. R. & Funk, V. A. (1991). *The Complete Cladist*. Lawrence: University of Kansas.

Wiley, R. H. & Poston, J. (1996). Indirect mate choice, competition for mates, and the coevolution of the sexes. *Evolution*, **50**, 1371–81.

Williams, L., Gibson, S., McDaniel, M., Bazzel, J., Barnes, S. & Abee, C. (1994). Allomaternal interactions in the Bolivian squirrel monkey (*Saimiri boliviensis boliviensis*). *American Journal of Primatology*, **34**, 145–56.

Wilson, D. E. & Reeder, D. M. (1993). Introduction. In *Mammal Species of the World: A Taxonomic and Geographic Reference*, 2nd edn, ed. D. E. Wilson & D. M. Reeder, pp. 1–12. Washington: Smithsonian Institution Press.

Wilson, E. O. (1975). *Sociobiology: The New Synthesis*. Cambridge, MA: Harvard University Press.

Wilson, E. O. (1998). *Consilience. The Unity of Knowledge*. New York: Alfred A. Knopf.

Wilson, J. D. (1992). A probable case of sexually selected infanticide by a male dipper *Cinclus cinclus*. *Ibis*, **134**, 188–90.

Wilson, W. L., Elwood, R. W. & Montgomery, W. I. (1993). Infanticide and maternal defence in the wood mouse *Apodemus sylvaticus*. *Ethology, Ecology and Evolution*, **5**, 365–70.

Winn, R. M. (1994). Preliminary study of the sexual behaviour of three aye-ayes (*Daubentonia madagascariensis*) in captivity. *Folia Primatologica*, **62**, 63–73.

Wishner, L. (1982). *Eastern Chipmunks. Secrets of their Solitary Life*. Washington, DC: Smithsonian Institution Press.

Witt, R., Schmidt, C. & Schmitt, J. (1981). Social rank and Darwinian fitness in a multimale group of Barbary macaques (*Macaca sylvana* Linnaeus, 1758). *Folia Primatologica*, **36**, 201–11.

Wittenberger, J. F. & Tilson, R. L. (1980). The evolution of monogamy: hypotheses and evidence. *Annual Review of Ecology and Systematics*, **11**, 197–232.

Wolf, K. E. (1984). Reproductive competition among co-resident male silvered leaf monkeys (*Presbytis cristata*). Ph.D. Thesis, Yale University.

Wolf, K. E. & Fleagle, J. G. (1977). Adult male replacement in a group of silvered leaf-monkeys (*Presbytis cristata*) at Kuala Selangor, Malaysia. *Primates*, **18**, 949–55.

Wolf, N. G. (1985). Odd fish abandon mixed-species groups when threatened. *Behavioral Ecology and Sociobiology*, **17**, 47–52.

Wolff, J. O. (1985). Maternal aggression as a deterrent to infanticide in *Peromyscus leucopus* and *P. maniculatus*. *Animal Behaviour*, **33**, 117–23.

Wolff, J. O. (1986). Infanticide in white-footed mice, *Peromyscus leucopus*. *Animal Behaviour*, **34**, 1568.

Wolff, J. O. (1993). Why are female small mammals territorial? *Oikos*, **68**, 364–70.

Wolff, J. O. (1997). Population regulation in mammals: an evolutionary perspective. *Journal of Animal Ecology*, **66**, 1–13.

Wolff, J. O. & Cicirello, D. M. (1989). Field evidence for sexual selection and resource competition infanticide in white-footed mice. *Animal Behaviour*, **38**, 637–42.

Wolff, J. O. & Cicirello, D. M. (1991). Comparative paternal and infanticidal behavior of white-footed mice (*Peromyscus leucopus noveoboracensis*) and deermice (*P. maniculatus nubiterrae*). *Behavioral Ecology*, **2**, 238–45.

Wolff, J. O. & Peterson, J. A. (1998). An offspring-defense hypothesis for territoriality in female mammals. *Ethology, Ecology and Evolution*, **10**, 227–39.

Wolff, J. O. & Schauber, E. M. (1996). Space use and juvenile recruitment in gray-tailed voles in response to intruder pressure and food abundance. *Acta Theriologica*, **41**, 35–43.

Woodroffe, R. & Vincent, A. (1994). Mother's little helpers: patterns of male care in mammals. *Trends in Ecology and Evolution*, **9**, 294–7.

Wrangham, R. W. (1979). On the evolution of ape social systems. *Social Science Information*, **18**, 335–68.

Wrangham, R. W. (1980). An ecological model of female-bonded primate groups. *Behaviour*, **75**, 262–99.

Wrangham, R. W. (1986). Ecology and social relationships in two species of chimpanzee. In *Ecology and Social Evolution: Birds and Mammals*, ed. D. I. Rubenstein & R. W. Wrangham, pp. 352–78. Princeton, NJ: Princeton University Press.

Wrangham, R. W. (1987a). Evolution of social structure. In *Primate Societies*, ed. B. B. Smuts, D. L. Cheney, R. M. Seyfarth, R. W. Wrangham & T. T. Struhsaker, pp. 282–97. Chicago: University of Chicago Press.

Wrangham, R. W. (1987b). The significance of African apes for reconstructing human social evolution. In *The Evolution of Human Behavior: Primate Models*, ed. W. G. Kinzey, pp. 51–71. Albany: SUNY Press.

Wrangham, R. W., Gittleman, J. L. & Chapman, C. A. (1993). Constraints on group size in primates and carnivores: population density and day-range as assays of exploitation competition. *Behavioral Ecology and Sociobiology*, **32**, 199–210.

Wrangham, R. & Peterson, D. (1996). *Demonic Males. Apes and the Origins of Human Violence.* London: Bloomsbury.

Wrangham, R. W. & Rubenstein, D. I. (1986). Social evolution in birds and mammals. In *Ecological Aspects of Social Evolution*, ed. D. I. Rubenstein & R. W. Wrangham, pp. 452–70. Princeton, NJ: Princeton University Press.

Wright, P. (1984). Biparental care in *Aotus trivirgatus* and *Callicebus moloch*. In *Female Primates: Studies by Women Primatologists*, ed. M. Small, pp. 59–75. New York: Alan R. Liss.

Wright, P. C. (1990). Patterns of paternal care in primates. *International Journal of Primatology*, **11**, 89–102.

Wright, P. C. (1995). Demography and life history of free-ranging *Propithecus diadema edwardsi* in Ranomafana National Park, Madagascar. *International Journal of Primatology*, **16**, 835–54.

Wright, P. L. (1963). Variations in reproductive cycles in North American mustelids. In *Delayed Implantation,* ed. A. C. Enders, pp. 77–97. Chicago: University of Chicago Press.

Wrigley, E. A., Davies, R. S., Oeppen, J. E. & Schofield, R. S. (1997). *English Population History from Family Reconstitution 1580–1837.* Cambridge: Cambridge University Press.

Wu, H. Y. & Lin, Y. S. (1992). Life history variables of wild troops of Formosan macaques (*Macaca cyclopis*) in Kenting, Taiwan. *Primates*, **33**, 85–97.

Wynne-Edwards, K. E. & Lisk, R. D. (1984). Djungarian hamsters fail to conceive in the presence of multiple males. *Animal Behaviour*, **32**, 626–8.

Yamagiwa, J. (1986). Activity rhythms and the ranging of a solitary male mountain gorilla (*Gorilla gorilla beringei*). *Primates*, **27**, 273–82.

Yamamoto, M. E., Arruda, M. F., Sousa, M. B. C. & Alencar, A. I. (1996). Mating systems and reproductive strategies in *Callithrix jacchus* females. In *Abstracts from the XVI Conference of the American Society of Primatologists*, University of Wisconsin-Madison; Abstract no. 056.

Yamamura, N., Hasegawa, T. & Itô, Y. (1990). Why mothers do not resist infanticide: a cost–benefit genetic model. *Evolution*, **44**, 1346–57.

Yasui, Y. (1998). The 'genetic benefits' of female multiple mating reconsidered. *Trends in Ecology and Evolution*, **13**, 246–50.

Ylönen, H., Koskela, E. & Mappes, T. (1997). Infanticide in the bank vole (*Clethrionomys glareolus*): occurrence and the effect of familiarity on female infanticide. *Annales Zoologiae Fennici*, **34**, 259–66.

Yoerg, S. I. (1990). Infanticide in the Eurasian dipper. *Condor*, **92**, 775–6.

Yoshiba, K. (1968). Local and intertroop variability in ecology and social behavior of common Indian langurs. In *Primates: Studies in Adaptation and Variability,* ed. P. C. Jay, pp. 217–42. New York: Holt, Rinehart and Winston.

Zahavi, A. (1974). Communal nesting by the Arabian babbler. A case of individual selection. *Ibis*, **116**, 84–7.

Zar, J. H. (1984). *Biostatistical Analysis*. Englewood Cliffs, NJ: Prentice-Hall.

Zhao, Q. (1996). Male–infant–male interactions in Tibetan macaques. *Primates*, **37**, 135–43.

Ziegler, T. E., Epple, G., Snowdon, C. T., Porter, T. A., Belcher A. M. & Küderling, I. (1993). Detection of the chemical signals of ovulation in the cotton-top tamarin, *Saguinus oedipus*. *Animal Behaviour*, **45**, 313–22.

Zimmer, C. (1996). First, kill the babies. *Discover*, September, 73–8.

Zinner, D., Schwibbe, M. H. & Kaumans, W. (1994). Cycle synchrony and probability of conception in female hamadryas baboons, *Papio hamadryas*. *Behavioral Ecology and Sociobiology,* **35**, 175–83.

Zippelius, H.-M. (1992). *Die vermessene Theorie. Eine kritische Auseinandersetzung mit der Instinkttheorie von Konrad Lorenz und verhaltenskundlicher Forschungspraxis.* Braunschweig: Vieweg.

Zuckerman, S. (1932). *The Social Life of Monkeys and Apes*. London: Routledge and Kegan Paul. (Reprinted 1981.)

Zunino, G. E., Chalukian, S. C. & Rumiz, D. I. (1986). Infanticidio y desaparición de infantes asociados al reemplazo de machos en grupos de *Alouatta caraya*. In *A Primatologia No. Brasil – 2*, ed. M. Thiago de Mello, pp. 185–90. Campinas, SP: Sociedade Brasileira de Primatologia.

Species index

Subject index

abandonment
  of embryos, *see* abortion; Bruce effect
  of infants 246, 311, 447–65
abortion 218, 324, 353–4, 448–51, 457–64
adaptationism 13, 18, 23
adoption 209, 214–19, 246, 249, 276, 291
adoption avoidance 48
affiliative interactions (male–infant) 242,
  253, 269–92
age deception 231
agonistic buffering 279–82, 291–2
aggression 95, 154, 157, 199, 217, 224, 231
  by females 59, 429
  by males 45–6, 79, 95, 139, 147, 280, 281,
    282, 291, 393, 406–8; *see also* attacks
  maternal 249, 251, 266–67, 438, 445
alliances, *see* coalitions
allomaternal care 225, 424, 427, 442–3
altriciality 194, 431, 435, 437, 443
altruism (phenotypically altruistic) 13
attacks on infants 31–9, 43–51, 79, 82–3, 95,
  100–4, 106–9, 112, 113, 117, 127–39, 157,
  224, 272, 274, 281, 290
attractivity 363

behavioral freedom, *see* female control over
  mating
best male hypothesis 293–6, 309–12, 317–20
  genetic benefit 243, 319
  social/protection benefits 75, 95, 97, 148,
    243, 263
between-group encounters 91, 294, 298,
  300–5, 314, 317
birth
  rate 76, 78–9, 83–5, 87–90, 93, 97, 101, 106,
    109, 111, 119, 120, 310
  seasonality 96, 101
body mass (and infanticide) 229–35, 472–3

bridging 279
Bruce effect xiii, 178, 181, 186, 190, 192–4, 195,
  324, 330, 353

calls, estrous or mating 346, 350
Calvin 23
cannibalism 46–8, 71, 130–1, 143, 190, 427,
  429, 431–2, 439; *see also* infanticide,
  exploitation hypothesis
coalitions 78, 111–13, 148, 225–6, 241, 245,
  267, 283–6, 291, 293–4, 390–2, 439
coercion, sexual 3, 169–70, 214, 242, 261, 266,
  293, 375–81, 385–6, 392–3, 406–8, 410
coloniality 212, 249
communal breeding 199, 259
comparative analysis 63–70, 180, 229–36,
  286, 375–84, 388–412, 478–87, 492
competition 19, 199, 293–4
  between groups, *see* between-group
    encounters
  intra-sexual 77, 91, 94, 198, 200, 209, 211,
    213, 225, 295–6, 389–92, 395–6,
    400–1, 404–5
  reproductive, among males 77, 79, 89,
    93–4, 123, 150, 163–4, 167, 169–71, 200,
    230–1, 283
  reproductive, among females 77, 91, 148,
    200, 327, 423, 427, 434–45, 444–6
  scramble 75–7, 90–7; *see also* sperm
    competition
concentrated changes test 188, 286, 398, 401
conception rates 87–8, 91–2, 95, 358
conservation (and infanticide) 196, 493–4
convenience polyandry 328
cooperation with kin 13, 241, 266, 390–2
cooperative breeding 194, 199, 216, 225–35,
  426, 441–4
copulation, *see* mating